Lecture Notes in Artificial Intell

Subseries of Lecture Notes in Computer Scien

Edited by J. G. Carbonell and J. Siekmann

Lecture Notes in Computer Science

Edited by G. Goos, J. Hartmanis and J. van Leeuwen

Springer
Berlin
Heidelberg
New York
Barcelona
Budapest
Hong Kong
London
Milan
Paris
Santa Clara
Singapore
Tokyo

Didier Galmiche (Ed.)

Automated Reasoning with Analytic Tableaux and Related Methods

International Conference, TABLEAUX'97
Pont-à-Mousson, France, May 13-16, 1997
Proceedings

 Springer

Series Editors
Jaime G. Carbonell, Carnegie Mellon University, Pittsburgh, PA, USA
Jörg Siekmann, University of Saarland, Saarbrücken, Germany

Volume Editor

Didier Galmiche
CRIN-CNRS, Université Henri Poincaré - Nancy I
Campus Scientifique, BP 239
F-54506 Vandœuvre-les-Nancy Cedex, France
E-mail: galmiche@loria.fr

Cataloging-in-Publication Data applied for

Die Deutsche Bibliothek - CIP-Einheitsaufnahme

Automated reasoning with analytic tableaux and related methods
: international conference ; tableaux '97, Pont-á-Mousson, France,
May 13 - 16, 1997 ; proceedings / [International Conference
Tableaux]. Didier Galmiche (ed.). - Berlin ; Heidelberg ; New York ;
Barcelona ; Budapest ; Hong Kong ; London ; Milan ; Paris ; Santa
Clara ; Singapore ; Tokyo : Springer, 1997
   (Lecture notes in computer science ; Vol. 1227 : Lecture notes in
   artificial intelligence)
   ISBN 3-540-62920-3 kart.

CR Subject Classification (1991): F.4.1, I.2.3

ISBN 3-540-62920-3 Springer-Verlag Berlin Heidelberg New York

© Springer-Verlag Berlin Heidelberg 1997
Printed in Germany

Typesetting: Camera ready by author
SPIN 10549713   06/3142 – 5 4 3 2 1 0   Printed on acid-free paper

Foreword

The present volume contains the papers presented at the International Conference TABLEAUX'97 – Analytic Tableaux and Related Methods, held May 13 – 16, 1997 in Pont-à-Mousson (Abbaye des Prémontrés), France. This conference was a continuation of the international workshop on Theorem Proving with Analytic Tableaux and Related Methods held in Lautenbach near Karlsruhe (1992), Marseille (1993), Abingdon near Oxford (1994), St. Goar near Koblenz (1995), and Terrasini near Palermo (1996).

There were 49 papers submitted to TABLEAUX'97: 43 full papers, 5 system descriptions, and 1 position paper. Among these, 22 full papers and 2 system descriptions were selected for presentation at the conference and for inclusion in these proceedings together with two invited lectures. In addition, there were 3 tutorials and a presentation of 7 position papers. In fact TABLEAUX'97 attracted interest from many parts of the world with papers submitted from: Australia, Austria, Belgium, Canada, China, England, France, Germany, Italy, Japan, Lithuania, Netherlands, Palestine, Poland, Portugal, Scotland, Slovakia, Spain, USA, and Yugoslavia.

Given the increased interest in tableaux-based theorem proving and its applications, the conference intended to bring together researchers interested in all the aspects of mechanization of reasoning with tableaux and related methods (sequent calculi, connection method, model elimination,...) and working on theoretical foundations of methods, implementation techniques, system development, and applications. Interestingly enough, the spectrum of logics dealt with by the papers covers classical logic and several non-classical logics – including modal, intuitionistic, many-valued, and temporal logic.

Together with papers pertaining to the traditional domain of tableaux reasoning, the volume also contains papers dealing with applications to different domains of computer science and with other approaches of automated reasoning.

Finally, I would like to thank all the people who contributed to the success of the conference TABLEAUX'97, those responsible for local arrangements, the program committee for its efficient and rigorous scientific work, all the referees, the authors for their contributions and efforts, the invited speakers Ryuzo Hasegawa and Grigori Mints, the organizers of tutorials, and all the sponsors who made it possible to organize this conference in the *Abbaye des Prémontrés*.

Didier Galmiche
TABLEAUX'97 Program Chair

March 1997

Previous Tableaux Workshops

1992   Lautenbach, Germany   1993   Marseille, France
1994   Abingdon, England   1995   St. Goar, Germany
1996   Terrasini, Italy

Invited speakers

R. Hasegawa   Kyushu University, Japan
G. Mints   Stanford University, USA

Program Chair

D. Galmiche   CRIN-CNRS, UHP Nancy 1, France

Program Committee

C. Cellucci   Rome University, Italy
M. d'Agostino   Ferrara University, Italy
R. Dyckhoff   St Andrews University, U.K.
M. Fitting   CUNY, New York City, U.S.A.
U. Furbach   Koblenz University, Germany
D. Galmiche   CRIN-CNRS, UHP Nancy 1, France
R. Goré   Austr. National University, Australia
J. Goubault-Larrecq   GIE Dyade, France
R. Hähnle   Karlsruhe University, Germany
A. Leitsch   TU Wien, Austria
R. Letz   TU Munich, Germany
U. Moscato   Milan University, Italy
N. Murray   SUNY at Albany, U.S.A.
H. Ono   JAIST, Japan
J. Posegga   Deutsche Telekom AG, Germany
P. Schmitt   Karlsruhe University, Germany
C. Schwind   LIM-CNRS, Marseille, France
H. de Swart   Tilburg University, The Netherlands
L. Wallen   Oxford University, U.K.
G. Wrightson   Newcastle University, Australia

Referees

Wolfgang Ahrendt
Alessandro Avellone
Matthias Baaz
Peter Baumgartner
David Basin
Alan Bundy
George Becker
Bernhard Beckert
Chris Brink
Walter Carnielli
Carlo Cellucci
Marcello D'Agostino
Giorgio Delzanno
Jürgen Dix
Roy Dyckhoff
Norbert Eisinger
Christian Fermüller
Melvin Fitting
Ulrich Furbach
Didier Galmiche
Rajeev Goré
Guido Governatori
Jean Goubault-Larrecq
Andrew Haas

Reiner Hähnle
Ryuzo Hasegawa
Hugo Herbelin
Alain Heuerding
Joerg Hudelmaier
Dominique Larchey
Alexander Leitsch
Reinhold Letz
Christoph Lingenfelder
Ferrari Mauro
Dominique Méry
Pierangelo Miglioli
Ugo Moscato
Max Moser
Neil Murray
Ilkka Niemelä
Hans de Nivelle
Andreas Nonnengart
Nicola Olivetti
Hiroakira Ono
Mario Ornaghi
Joachim Posegga
Odile Papini
Jeremy Pitt

Jeremy Pitt
A. Puder
Frank Puppe
Greg Restall
Mark Reynolds
Maarten de Rijke
Vincent Risch
Holger Schlingloff
Manfred Schmidt-Schauß
Peter H. Schmitt
Camilla Schwind
Frieder Stolzenburg
Harrie de Swart
Mitio Takano
Simon Thompson
Christian Urban
Jacqueline Vauzeilles
Arild Waaler
Lincoln Wallen
Bernard Willems
Graham White
Graham Wrightson
Alberto Zanardo

Sponsors

CRIN-CNRS (Centre de Recherche en Informatique de Nancy)
INRIA-Lorraine
Université Henri Poincaré - Nancy I
Université Nancy II
INPL Nancy (Institut National Polytechnique de Lorraine)
SPI CNRS (Département des Sciences Pour l'Ingénieur du CNRS)
MENESR (Ministère de l'Education Nationale, de l'Enseignement Supérieur et de la Recherche)
FRANCE TELECOM - CNET
SUN Microsystems
AFIT (Association Française d'Informatique Théorique)
GDR PRC AMI
Conseil Régional de Lorraine
Communauté Urbaine du Grand-Nancy
Ville de Pont-à-Mousson

Position Papers

The regular conference program included the presentation of seven position papers. Informal proceedings containing these papers appeared as the internal scientific report CRIN 97-R-030 at CRIN-CNRS, Campus Scientifique, BP 239, 54506 Vandœuvre-les-Nancy Cedex, France.

Table of Contents

MGTP: A Model Generation Theorem Prover
– Its Advanced Features and Applications –

Ryuzo Hasegawa, Hiroshi Fujita and Miyuki Koshimura

Graduate School of Information Science and Electrical Engineering
Kyushu University
6-1, Kasuga-koen, Kasuga-shi, Fukuoka 816, Japan

Abstract This paper outlines a parallel model-generation based theorem-proving system MGTP that we have been developing, focusing on the recent developments including novel techniques for efficient proof-search and successful applications.

We have developed CMGTP (Constraint MGTP) to deal with constraint satisfaction problems. By attaining the constraint propagation with negative atoms, CMGTP makes it possible to reduce search spaces by orders of magnitude compared to the original MGTP.

To enhance the ability to prune search spaces, we have developed a new method called *non-Horn magic sets* (NHM) and incorporated its relevancy testing function into the folding-up (FU) method proposed by Letz. The NHM method suppresses useless model generation with clauses irrelevant to the goal. The FU method avoids generating duplicated sub-proofs after case-splitting. With these methods we can eliminate two major kinds of redundancies in model-generation based theorem provers. We have studied several applications in AI such as negation as failure, abductive reasoning and modal logic systems, through extensive use of MGTP. These studies share a basic common idea, that is, to use MGTP as a meta-programming system. We can build various reasoning systems on MGTP by writing the specific inference rules for each system in MGTP input clauses.

Furthermore, we are now working on other applications such as machine learning with MGTP and heuristic proof-search based on the genetic algorithm.

1 Introduction

Automated theorem proving is an important technology not only for reasoning about mathematical theorems but also for developing knowledge processing systems.

We started research on parallel theorem provers in 1989 in the FGCS project with the aim of integrating logic programming and theorem proving technologies. The immediate goal of this research was to develop a fast theorem-proving system on the PIM, by effectively utilizing KL1 languages[24] and logic programming techniques. We intended this system to be an "advanced general-purpose inference engine" that can facilitate the development of intelligent knowledge programming software in KL1.

We adopted the model generation method of SATCHMO[17] as the proof procedure. The main reason for this choice is that the method has a merit of not needing full unification; that is, it needs matching only.

MGTP exploits OR parallelism from non-Horn problems by independently exploring each branch of a proof tree caused by case splitting, whereas it exploits AND parallelism from Horn problems that do not cause case splitting. Almost linear speedup was achieved[9] for both Horn and non-Horn problems on a PIM/m 256 PE system. With MGTP, we succeeded in solving some open quasi-group problems in finite algebra. We also solved several hard *condensed detachment* problems[18] that could not be solved by OTTER with any strategy. Our success shows the effectiveness of a large-scale parallel theorem prover. Besides mathematical problems, MGTP is applicable to other AI-oriented applications such as constraint satisfaction problems, design and planning, and hardware/software verification.

Research on MGTP can be divided into three aspects: (1) parallel implementation, (2) extension of the MGTP features, and (3) application. The following sections describe the main results of the recent research activities, focusing on (2) and (3), that is the enhancement of the MGTP functions and its application.

Through research on proving quasi-group problems with MGTP, we found that MGTP lacks negative constraint propagation ability. Then, we developed CMGTP (Constraint MGTP) that can handle constraint propagations with negative atoms.

However, two major problems remain to be solved that are common to model-generation based provers; redundant inference caused by clauses that are irrelevant to the given goal, and duplication of the same subproof after case-splitting. To solve the former problem, we developed a method called *non-Horn magic sets* (NHM)[10]. For the latter problem, we implemented two sorts of factorization mechanisms: *static lemma generation* (SLG) and *dynamic lemma generation* (DLG). Then, we came up with a new method that combines the relevancy testing realized by NHM and SATCHMORE[16] with *folding-up* (FU) proposed by Letz[15]. We call this method *relevancy testing while folding-up* (RFU).

NHM is a natural extension of the magic sets developed in the deductive database field, and can be applied to non-Horn problems. The factorization mechanisms are to avoid repetitions of the same subproof after case-splitting. The RFU method, in particular, has a similar effect to NHM that prunes redundant inferences caused by irrelevant non-Horn clauses. With these advanced features being incorporated, MGTP now overcomes the major drawbacks of the model generation based provers.

Regarding applications, MGTP can be viewed as a meta-programming system. We can build various reasoning systems on MGTP by writing the inference rule used for each system as MGTP input clauses. Along this idea, we developed several techniques and reasoning systems necessary for AI applications. These include a method to incorporate negation as failure into MGTP[12], abductive reasoning systems[13], modal logic systems[14, 1], and machine learning.

INPUT   A set of clauses S

OUTPUT   The satisfiability status of S (*sat* or *unsat*)

PROCEDURE   Initialize a model candidate $M := \phi$, and a buffer for the set of consequents of violated clauses $D := \phi$. Find all violated clauses in S under M.

2-1) If a negative clause is found to be violated, then reject M. (M cannot be a candidate for a model of S)

2-2) For each non-negative clause that is found to be violated, add the consequent of the clause into D.

Look into D for model candidate extension.

3-1) If D is empty then return *sat*. (M is a model of S.)

3-2) Select a consequent Δ stored in D. $D := D \setminus \{\Delta\}$. Extend M to $M_j := M \cup \{B_j \sigma\}$ $(1 \leq j \leq m)$ where B_j is an atom in Δ. For each of the extended model candidate M_j, continue the procedure from (2). (case-splitting)

When all model candidates are rejected, then output *unsat* and terminate the procedure.

Fig 1. The basic MGTP procedure

2   Outline of MGTP

An MGTP input clause is represented as

$$A_1, \cdots, A_n \to B_1; \cdots; B_m \ (n, m \geq 0).$$

where A_i, B_j are atoms, "," denotes conjunction ";" denotes disjunction. the left-hand side of "$\to$" is called the antecedent of the clause, the right-hand side the consequent. If $n = 0$, the antecedent is *true* and the clause is called positive. If $m = 0$, the consequent is *false* and the clause is called negative. A clause is said to be violated under a set M of atoms if with some ground substitution σ the following condition holds:

$$(\forall i_{(1 \leq i \leq n)}.\ A_i \sigma \in M) \wedge (\forall j_{(1 \leq j \leq m)}.\ B_j \sigma \notin M).$$

The model generation method tries to construct models for a given set of clauses, starting from an empty set as a model candidate and extending it with consequent atoms in a violated clause. If there is a non-negative clause which is violated under the current model candidate M, then the method extends M with a consequent atom of the violated clause so as to satisfy it.

The MGTP proof procedure is sketched in Fig.1. In the following sections, the basic MGTP procedure will be modified as an enhancement is made on it.

$true \rightarrow dom(1), dom(2), dom(3), dom(4), dom(5).$
$dom(M), dom(N) \rightarrow$
   $p(M, N, 1); p(M, N, 2); p(M, N, 3); p(M, N, 4); p(M, N, 5).$
$p(Y, X, A), p(A, Y, B), p(B, Y, C), \{X \backslash = C\} \rightarrow false.$
$p(X, X, A), \{X \backslash = A\} \rightarrow false$
$p(X, Y1, A), p(X, Y2, A), \{Y1 \backslash = Y2\} \rightarrow false.$
$p(X1, Y, A), p(X2, Y, A), \{X1 \backslash = X2\} \rightarrow false.$
$p(X, 5, Y), \{Y < X - 1\} \rightarrow false.$

Fig 2. The MGTP input clauses for QG5.5

3 Advanced features of MGTP

3.1 Constraint handling

In this section, we present an extension of the MGTP system called CMGTP (Constraint MGTP), which can efficiently solve finite-domain constraint satisfaction problems such as quasi-group (QG) existence problems[3] in finite algebra.

In 1992, MGTP succeeded[8] in solving several open QG problems on a parallel inference machine PIM/m consisting of 256 processors. Later, other theorem provers or constraint solvers such as DDPP, FINDER and Eclipse solved other new open problems more efficiently than MGTP. Such research has revealed that the original MGTP lacks negative constraint propagation ability. This motivated us to develop an experimental system CP based on the CLP (constraint logic programming) scheme. Then we found that the constraint propagation mechanism used in CP can be realized by CMGTP[11, 22].

Quasi-group problems A Quasi-group is a pair $\langle Q, \circ \rangle$ where Q is a finite set, $\circ$ a binary operation on Q and for any $a, b, c \in Q$,

$$a \circ b = a \circ c \Rightarrow b = c \text{ and } a \circ c = b \circ c \Rightarrow a = b.$$

The multiplication table of this binary operation $\circ$ forms a Latin square. In a quasi-group, we can define the following inverse operations $\circ_{ijk}$ called *(ijk)-conjugate*:

$$x \circ_{123} y = z \Longleftrightarrow x \circ y = z$$
$$x \circ_{231} y = z \Longleftrightarrow y \circ z = x$$
$$x \circ_{312} y = z \Longleftrightarrow z \circ x = y$$

Multiplication tables of the inverse operations defined above also form Latin squares.

We tried to solve 7 categories of QG problems (called QG1, QG2,..., QG7), each of which is defined by adding some constraints to original quasi-group constraints. For example, QG5 constraint is defined as

$$\forall ab \in Q. \ ((ba)b)b = a.$$

CMGTP incorporates two mechanisms into MGTP: unit refutation and unit simplification. In CMGTP, negative atoms are used to represent negative constraints. If there exist P and $\neg P$ in M then $false$ is derived by unit refutation. If there exists a unit clause $\neg P_i(P_i) \in M$ and a disjunction $\Delta \in D$ which includes $P_i(\neg P_i)$, then $P_i(\neg P_i)$ is removed from Δ by unit simplification. The unit refutation and unit simplification mechanisms guarantee that for any atom $P \in M$, P and $\neg P$ are not both in the current M, and disjunctions in the current D have already been simplified by all unit clauses in M.

Fig.2 shows the original MGTP input clauses for QG5.5. These clauses are rewritten into CMGTP input clauses so that they enable negative information to be propagated along with negative atoms. For example, the third clause in Fig.2 will be rewritten as follows:

$$p(Y, X, A), p(A, Y, B) \rightarrow p(B, Y, X).$$
$$p(Y, X, A), \neg p(B, Y, X) \rightarrow \neg p(A, Y, B).$$
$$\neg p(B, Y, X), p(A, Y, B) \rightarrow \neg p(Y, X, A).$$

where the last two clauses are the contrapositives for the first clause, and they are to be used for negative information propagation.

3.2 Non-Horn magic sets

The basic MGTP procedure might choose a clause irrelevant to the goal to be solved and thus would lead to unnecessary model candidate extensions. In particular, it would cause an explosion of the number of model candidates when irrelevant non-Horn clauses are selected.

Concerning the above problem, SATCHMORE introduces a method called *relevancy testing* to avoid redundant model candidate extensions with irrelevant non-Horn clauses. The use of relevancy testing can restrict the selection of violated clauses for model generation to only those all of whose consequent literals are unifiable to some relevant literals. Here, a relevant literal is defined as a subgoal in a Prolog execution that is called in a failed search to prove $false$ or an antecedent literal of a non-Horn clause from the set of Horn clauses and the current model candidate. In relevancy testing, however, there are some overheads since the Prolog procedure to compute the set of relevant literals is performed whenever violated clauses are detected, thus the same subgoals may be evaluated repeatedly.

MGTP can avoid solving duplicated subgoals because it is based on bottom-up computation. However, it has the disadvantage of generating irrelevant atoms to prove the given goal. Thus, it is necessary to combine bottom-up with top-down so as to use goal information contained in negative clauses, and to avoid generating useless model candidates.

For this purpose, several methods such as *magic sets*[2], *Alexander template*[21], and *upside-down meta-interpretation* [4] have been proposed in the field of deductive databases. These methods transform the given Horn intensional databases to a more efficient one, and generate only ground atoms relevant to the given goal in extensional databases. However, these were restricted to Horn clauses.

We developed a new method called *non-Horn magic sets* (NHM) that extends the above mentioned methods to make them applicable to a general clause set including non-Horn clauses.

In the NHM method, each clause in a given clause set is transformed into two types of clauses. One is used to simulate backward reasoning, and the other is to control inferences in forward reasoning. The set of transformed clauses is proven by bottom-up theorem provers. We have developed two kinds of transformation methods: the *breadth-first NHM* and the *depth-first NHM*. The former simulates breadth-first backward reasoning, and the latter simulates depth-first backward reasoning.

Breadth-first NHM transformation Let S be a given set of clauses. We introduce a meta-logical predicate $goal/1$ which takes an atom in S as its argument. The literal $goal(A)$ means that an atom A is relevant to the goal and is necessary to be solved. A clause $A_1, \ldots, A_n \to B_1; \ldots; B_m$ in S is transformed into two clauses:

$$T_B^1 : goal(B_1), \ldots, goal(B_m) \to goal(A_1), \ldots, goal(A_n).$$
$$T_B^2 : goal(B_1), \ldots, goal(B_m), A_1, \ldots, A_n \to B_1; \ldots; B_m.$$

where T_B^1 is a clause for simulating backward reasoning, and T_B^2 is for forward reasoning. For a positive clause ($n = 0$), the first clause T_B^1 may be omitted. For a negative clause ($m = 0$), the conjunction of $goal(B_1)$, $\ldots$, $goal(B_m)$ becomes *true*.

An intuitive meaning of T_B^1 is that when it is necessary to solve the consequent $B_1, \ldots, B_m$ of the original clause, it is necessary to solve the antecedent $A_1, \ldots, A_n$ before doing that. The n antecedent literals are solved in parallel. T_B^2 means that a model candidate is extended with the consequent only when $A_1, \ldots, A_n$ are satisfied by the current model candidate and all the consequent atoms $B_1, \ldots, B_m$ are relevant to the given goal.

Depth-first NHM transformation A clause $A_1, \ldots, A_n \to B_1; \ldots; B_m$ in S is transformed into $n + 1$ clauses:

$$T_D^1 : goal(B_1), \ldots, goal(B_m) \to goal(A_1), cont_{k,1}(V_k).$$
$$T_D^2 : cont_{k,1}(V_k), A_1 \to goal(A_2), cont_{k,2}(V_k).$$
$$\vdots$$
$$T_D^n : cont_{k,(n-1)}(V_k), A_{n-1} \to goal(A_n), cont_{k,n}(V_k).$$
$$T_D^{n+1} : cont_{k,n}(V_k), A_n \to B_1; \ldots; B_m.$$

where k is the identifier of the original clause, V_k is the tuple of all variables appearing in the original clause. This transformation is a natural extension of magic sets for Horn clauses. Note that when $n = 0$, the original clause is transformed into a single clause

$$T_D^1 : goal(B_1), \ldots, goal(B_m) \rightarrow B_1; \ldots; B_m.$$

The $n + 1$ transformed clauses are interpreted as follows:

- When it is necessary to solve the consequent $B_1, \ldots, B_m$ of the original clause, we first attempt to solve the first atom A_1. At that time, "$k, 1$" is attached to the variable binding information V_k. The variable bindings obtained in the satisfiability checking of the antecedent are propagated to the next clause by the continuation literal $cont_{k,1}(V_k)$.
- If atom A_1 is solved under $cont_{k,1}(V_k)$, then we attempt to solve the second atom A_2. Then the augmented variable bindings obtained by solving A_1 are propagated to the next clause by the continuation literal. $cont_{k,2}(V_k)$.
- In a similar way, we attempt to solve every A_i. After all the antecedent atoms $A_1, \ldots, A_n$ are solved, the current model candidate is extended with the consequent $B_1; \ldots; B_m$.

Unlike the breadth-first NHM transformation, n antecedent atoms are being solved sequentially from A_1 to A_n. During this process, the variable binding information is propagated from A_1 to A_n in this order.

3.3 Factoring and lemmatization

There are two kinds of redundancies caused by case-splitting in MGTP. One is that the same subproof may be generated at several descendant nodes after a case-splitting occurs. Another is caused by useless model candidate extensions with irrelevant non-Horn clauses which can be avoided by the NHM method as explained in the previous section.

To remove the former kind of redundancy, we introduce into MGTP the methods similar to factorization and folding-up proposed by Letz[15]. The idea of factorization is to commit the subproof under one open branch to another if the same node appear in the both branches.

To implement the factorization operation, we have developed two approaches; one is a rather straightforward manner *static lemma generation* where the lemma is fixed in advance and attached to each disjunct in the consequent of the input clauses, the other is *dynamic lemma generation* where the lemma is not fixed in advance but determined on the fly at runtime.

On the other hand, we have developed a refined folding-up method which combines the lemmatization mechanism with the relevancy testing mechanism that has an effect similar to NHM. The new method not only prunes duplicated subproofs but also avoids irrelevant model extensions.

In the sequel, we assume that a node of a proof tree is labeled with a literal used for model candidate extension. Thus, a path from a leaf node to the root node (a branch) corresponds to a model candidate.

Static lemma generation A factorization mechanism can easily be realized with CMGTP by transforming the input clause $A \rightarrow B_1; B_2; \cdots; B_{m-1}; B_m$ into:

$$A \rightarrow B_1; (\neg B_1, B_2); \cdots; (\neg B_1, \cdots, \neg B_{m-1}, B_m). \quad \text{(left committed)}$$
$$A \rightarrow (B_1, \neg B_2, \cdots, \neg B_m); \cdots; (B_{m-1}, \neg B_m); B_m. \quad \text{(right committed)}$$

where $\neg B_i$ is a lemma. For example, consider $A \rightarrow B_1; (\neg B_1, B_2)$. If this clause is violated, the current branch (M) is split into the left branch (M_1) extended with B_1 and the right branch (M_2) extended with B_2. When an open branch extended under the node labeled with B_2 is further extended with the same literal B_1, its solution is committed to that of the node labeled with B_1 in M_1. Then, the branch is closed. This means that the proof of B_2 in the right branch M_2 depends on that of B_1 in the left branch M_1.

Dynamic lemma generation This is another implementation of the factorization mechanism where the dependency ordering of the disjuncts in a clause $A \rightarrow B_1; \cdots; B_m$ is not fixed in advance. Instead, we determine the order while performing inferences on the fly. When a case-splitting occurs, we create a process which manages the ordering among the consequent literals. When an open branch has just been extended with a node N_i labeled with a consequent literal B_i, we examine the branch starting from B_i to the root looking for a node N_a such that one of its siblings N_b has a label identical to B_i. If such N_a and N_b are found, we perform the following factorization operation.

1. If no ordering is defined between N_a and N_b, then define $N_b \prec N_a$ and factorize N_i with N_b.
2. Else if $N_b \prec N_a$, then factorize N_i with N_b.
3. Else if $N_a \prec N_b$, then continue to extend the node N_i, since factorization is not allowed for N_i.

Relevancy testing while folding-up The key to effectively implementing the folding-up procedure within MGTP is to consider carefully up to which node in the branch the lemma should be propagated.

To do this, we introduce three new parameters $Mark, ILem, OLem$ to the MGTP procedure. $Mark$ is the set of atoms which have actually been used to close branches under a node N. $ILems$ is the set of lemmas which may be used to reduce the number of disjuncts in the consequent. $OLems$ is the set of lemmas which may be propagated up to the ancestor nodes of N in the branch.

A unit lemmata is represented in the form $\Gamma \vdash p \rightarrow false$ where Γ is the set of atoms which must be satisfied in the context where the lemma is used.

Besides the basic function of folding-up, our method achieves an additional optimization whose effect is similar to those of the NHM method. This is possible by examining $Mark$ to see whether the clause used for a case-splitting is actually relevant to close the branches. If some B_i in the disjunction $B_1; \ldots; B_m$ that caused the case-splitting is irrelevant to close all the branches extended under

1. When a model candidate is rejected by a negative clause C, return *unsat* together with $Mark$ whose value is the set of atoms $\{A_1\sigma, \cdots, A_n\sigma\}$ in the instantiated antecedent of C, which is returned by the function $violated(Cs, M)$.
2. When a model candidate extension is initiated by a non-negative clause C, perform the three operations; simplification with lemma, new lemma generation, and removal of redundant subproofs which are irrelevant to the goal.

   For each model extending atom $B_i\sigma(i \leq i \leq m)$,

   2.1. If $B_i\sigma$ is refuted by a lemma in $ILem$, the model extension with $B_i\sigma$ is not executed. Instead, set $Mark_i$ the context Γ of the lemma.

   2.2. Otherwise, execute the model extension of M with $B_i\sigma$ and initiate the recursive call of the MGTP procedure on the extended model candidate.

   If Res_i is computed as *unsat*, then perform the following.

   a. If $B_i\sigma$ is relevant to the subproof under it, i.e. $B_i\sigma \in Mark_i$, create a lemma at that level, $\Gamma_i \vdash B_i\sigma \rightarrow false$, add it to $OLem_i$ which is obtained from the subproofs, then propagate the extended $OLem_i$ to the parent and the siblings of B_i. Note that the lemmas whose context includes $B_i\sigma$ should not be propagated.

   b. If $B_i\sigma$ is irrelevant to the subproof under it, i.e. $B_i\sigma \notin Mark_i$, return as the $Mark$ the $Mark_i$ that is given along with the subproof of $MGTP(Cs, M \cup \{B_i\sigma\}, ILem)$. The lemmas $OLem$ will become the union of the lemmas created from the already solved subproofs for $B_1\sigma, \cdots, B_i\sigma$.

   If 2.1 has been applied for every $B_i\sigma(1 \leq i \leq m)$ or 2.2 returns *unsat*, then return *unsat*.
3. Otherwise return *sat*.

Fig 3. The MGTP procedure with FUP

B_i, then it reveals that the whole disjunction $B_1; \ldots; B_m$ is irrelevant to the proof.

In this way, the method can remove the redundancy caused by irrelevant non-Horn clauses only partially, whereas the NHM method totally avoids selecting for model candidate extension both Horn and non-Horn clauses that are irrelevant to the goal.

Nonetheless, we can observe that the pseudo-NHM mechanism achieves significant performance.

The MGTP procedure augmented with the method is shown in Fig.3.

4 Applications

4.1 Bottom-up abduction

Abduction has recently been recognized as a very important form of reasoning for various AI problems. An *abductive framework* is a pair (Σ, Γ), where Σ is a set of formulas and Γ is a set of literals. Given a closed formula G, a set E of ground instances of Γ is an *explanation of G from (Σ, Γ)* if

1. $\Sigma \cup E \models G$, and
2. $\Sigma \cup E$ is consistent.

The computation of explanations of G from (Σ, Γ) can be seen as an extension of proof-finding by introducing a set of hypotheses from Γ that, if they could be proven by preserving the consistency of the augmented theories, would complete the proof of G.

Abduction by model generation Here, we introduce a method, which we call the Skip method, to implement abductive reasoning systems built on the MGTP. We consider the first-order abductive framework (Σ, Γ), where Σ is a set of *range-restricted Horn clauses* and Γ is a set of *atoms* (*abducibles*).

The simplest way to implement reasoning with hypotheses is as follows. For each hypothesis H in Γ, we supply a clause of the form:

$$\rightarrow H \mid \neg \mathsf{K} H, \tag{1}$$

where $\neg \mathsf{K} H$ means that "H is not assumed to be true in the model". Namely, each hypothesis is assumed either to hold or not to hold. To deal with $\neg \mathsf{K} H$, we need the axiom schema as an integrity constraint :

$$\neg \mathsf{K} H, H \rightarrow \qquad \text{for every hypothesis } H. \tag{2}$$

The above technique, however, may generate $2^{|\Gamma|}$ model candidates. To reduce them as much as possible, we can use a method that delays case-splitting for each hypothesis. That is, we do not supply any clause of the form (1) for any hypothesis of Γ, but, instead, introduce hypotheses when they are necessary. When abducibles $H_1, \ldots, H_n$ ($n \geq 0$) from Γ appear in the antecedent of a Horn clause in Σ as:

$$A_1 \wedge \ldots \wedge A_l \wedge \underbrace{H_1 \wedge \ldots \wedge H_n}_{\text{abducibles}} \rightarrow C,$$

we transform this clause into a non-Horn clause:

$$A_1, \ldots, A_l \rightarrow H_1, \ldots, H_n, C \mid \neg \mathsf{K} H_1 \mid \ldots \mid \neg \mathsf{K} H_n. \tag{3}$$

In this transformation, each hypothesis H_j in the antecedent is shifted to the right-hand side of the clause in the form of $\neg \mathsf{K} H_j$. Moreover, each H_j is *skipped* instead of being resolved, and is added to consequent C of the rule since C becomes true whenever all A_i's and H_j's are true.

αrule
$$f(F \vee G, W) \rightarrow f(F, W), f(G, W).$$
$$t(F \wedge G, W) \rightarrow t(F, W), t(G, W).$$
βrule
$$t(F \vee G, W) \rightarrow t(F, W); t(G, W).$$
$$f(F \wedge G, W) \rightarrow f(F, W); f(G, W)$$
νrule
$$t(\Box F, W), path(W, V) \rightarrow t(F, V).$$
πrule
$$f(\Box F, W) \rightarrow \{new\_world(V)\}, path(W, V), f(F, V).$$
close condition
$$t(F, W), f(F, W) \rightarrow false.$$

Fig 4. The MGTP input clauses for a modal logic

4.2 Modal logic

Modal logics have been gaining popularity in various domains of Computer Science. For such applications, modal logics require fast and efficient theorem provers. Recently, the translation proof method[20] which translates modal formulae into classical formulae has been proven to be useful because it can be applied to various modal systems. The translation approach has another merit in that it can employ many control strategies developed for theorem proving in classical logic. Unfortunately, the previously proposed methods do not address the issue of controlling inference to reduce the search space.

To take advantage of the above, we proposed a version of the translation method, called the *modal clause transformation method*. This method is based on the following meta-programming method.

The rewriting rules for the tableau method are written as schemata encoded in MGTP input clauses. Thus the modal tableau method[5] is simulated on MGTP. Fig.4 shows the MGTP input clauses representing the tableaux expansion rules and the close condition, where F, G are modal formulae; $t(F, W)/f(F, W)$ represents that F is true/false in the world W; $path(W, V)$ represents that V is accessible from W; $\{\cdots\}$ is a sequence of external predicates (evaluated in Prolog or KL1) and $new\_world(V)$ creates a new world V.

Due to a rather straightforward manner of simulation scheme as above, MGTP tends to generate too many branches by case-splitting. To improve this, we can apply a partial evaluation technique to the method, thereby suppressing the explosive generation of branches.

First, we represent modal formulae as sets of *modal clauses*. A modal clause is a disjunction of *modal atoms* and their negations, where a modal atom is a propositional atom or a modal formula with a modal operator $\Box$ followed by a modal clause. Any modal formula can be represented as a set of modal clauses.

Second, we translate the modal formulae so that the close condition testing is replaced with pattern matching in the antecedents of the translated clauses. Thus, we can suppress the generation of redundant branches that will immediately be closed.

In practical applications, formulae usually contain many sub-formulae irrelevant to the proof. To avoid the generation of irrelevant branches, we can apply the NHM method. Given an input modal formula, we analyze it and embed some control information specific to it into its translated formula. We call this translation method the *NHM modal clause transformation method.*

4.3 Machine learning

We tried to apply MGTP to machine learning[7] where MGTP is used to implement a parallel version of Progol.

Progol is an inductive logic programming system developed by Muggleton[19], which given examples and background knowledge constructs hypotheses on the basis of *inverse entailment* and general-to-specific search through the subsumption lattice. Progol has been used for many nontrivial applications with significant results.

Let B be a background knowledge, E a positive example, $\overline{\bot}$ a set of all ground literals which are the consequence of B and $\overline{E}$, then according to the inverse entailment,

$$B \wedge \overline{E} \models \overline{\bot} \models \overline{H}.$$

This implies $H \models \bot$, and $\bot$ is called the *most specific hypothesis* (MSH).

In the original Progol, $\overline{\bot}$ is constructed by repeatedly issuing many Prolog goals each of which is evaluated in a top-down manner one by one. In contrast, MGTP derives many consequent literals at the same time in a bottom-up manner. Therefore, significant amount of computation for duplicated (sub-)goals in the Prolog execution could be saved in the MGTP execution.

In addition, since MGTP already has very efficient parallel implementations on parallel inference machines, an efficient implementation of parallel-Progol will readily be realized as well.

We are also working on another MGTP-based ILP system, or Inductive-MGTP, in which we will not use MGTP just as an implementation tool but modify and extend it to do ILP within its own framework. The Inductive-MGTP would have some generalization mechanisms built in, and perform an *rlgg*-like operation on model elements being generated. The expected merits of such a system would include effective uses of non-Horn clauses in hypotheses, background knowledge and examples.

4.4 Use of genetic algorithm

The major causes of redundancy in the MGTP proof procedure are removed by the above described methods. Nevertheless, there are many difficult problems in theorem proving and AI applications which MGTP cannot solve easily since they have extremely large search spaces. The feature required to deal with such problems that MGTP lacks is a mechanism of performing heuristic search based on domain specific knowledge.

Recently it has been reported that genetic algorithm (GA) has succeeded in solving (though only approximately) problems in the field of optimization and scheduling which are typically NP-complete or NP-hard. The GA approach would be effective for use in heuristic search for proofs in a theorem proving process as well.

As a case study, we tried to solve the *condensed detachment* problems with MGTP employing a GA-like strategy for heuristic search. The condensed detachment rule is given as follows.

$$\begin{array}{ccc} father & mother & son/daughter \\ p(X), & p(i(X,Y)) \rightarrow & p(Y). \end{array}$$

Each atom $p(X)$ has a chromosome assigned to it. A chromosome is made up of several genes. Assume, for example, that the chromosome of $p(A)$ has $f_{p(A)}$ and $m_{p(A)}$, which are the identifiers of the father and the mother of $p(A)$, and that the chromosome of $p(B)$ has $f_{p(B)}$ and $m_{p(B)}$. Then we can obtain new sons/daughters by mating $f_{p(A)}$ with $m_{p(B)}$, and $f_{p(B)}$ with $m_{p(A)}$, or even $f_{p(A)}$ with $f_{p(B)}$, and $m_{p(A)}$ with $m_{p(B)}$. These correspond to the gene-crossing operations in GA. One can take into account some mutation operations as well.

The excellence of an atom $p(X)$ is evaluated on the basis of some comparative analysis of several properties of it. For instance, its size; smaller atom is superior than larger ones. The subsumption rates, both forward and backward, may also be counted as the preference measure of the atom; an atom that subsumes more atoms is preferred.

We can observe that even such a simple gene coding and selection scheme does work. For instance, the theorem LCL009-1[23] is proven after generating only 27 unit clauses by the MGTP system with GA whereas the system without GA generates 299 unit clauses.

We consider the GA approach as a promising method of heuristic search in particular to deal with a class of problems which have large number of alternative solutions.

5 Conclusion

To enhance the MGTP's ability of reducing search spaces, we developed the advanced features; constraint handling (CMGTP), non-Horn magic sets (NHM), factorization mechanisms such as SLG and DLG, and the refined folding-up mechanism (RFU). These are the key technologies for making MGTP a practical prover and applicable to various applications such as disjunctive databases, abductive reasoning and modal logic systems.

We have demonstrated that several logical systems such as abductive reasoning and modal logic systems can easily be realized on top of MGTP.

The final goal of the MGTP research is to integrate automated reasoning technology with logic programming technology. MGTP embodies an exhaustive searching property which KL1 lacks and has the potential capacity to give a new logic programming paradigm based on bottom-up computation.

We are now developing a new version of MGTP and programming environments such as a visualizer and debugger, with the KLIC system[6] running on a UNIX machine. This KLIC-based system will be provided as an IFS(ICOT Free Software) product.

Also we are aiming at further extending MGTP application areas. These include Inductive-MGTP for machine learning and GA-MGTP for dealing with heuristic search problems by using genetic algorithms.

References

1. J. Akahani, K. Inoue, and R. Hasegawa. Bottom-Up Modal Theorem Proving Based on Modal Clause Transformation. *J. IPS Japan*, 36(4):822–831, April 1995. (in Japanese).

2. F. Bancilhon, D. Maier, Y. Sagiv, and J.D. Ullman. Magic sets and other strange ways to implement logic programs. In *Proc. 5th ACM SIGMOD-SIGACT Symp. on Principles of Database Systems*, pages 1–15, 1986.

3. F. Bennett. Quasigroup Identities and Mendelsohn Designs. *Canadian Journal of Mathematics*, 41:341–368, 1989.

4. F. Bry. Query evaluation in recursive databases: bottom-up and top-down reconciled. *Data & Knowledge Engineering*, 5:289–312, 1990.

5. M. Fitting. First-Order Modal Tableaux. In *J. Automated Reasoning*, volume 4, pages 191–213, 1988.

6. T. Fujise, T. Chikayama, K. Rokusawa, and A. Nakase. KLIC: A Portable Implementation of KL1. In *Proc. Int. Symp. on Fifth Generation Computer Systems*, December 1994.

7. H. Fujita, N. Yagi, T. Ozaki, and K. Furukawa. A New Design and Implementation of PROGOL by Bottom-up Computation. In *Proc. of the 6th International Workshop on Inductive Logic Programming*, 1996.

8. M. Fujita, J. Slaney, and F. Bennett. Automatic Generation of Some Results in Finite Algebra. In *Proc. IJCAI-93*, 1993.

9. R. Hasegawa and M. Koshimura. An AND Parallelization Method for MGTP and Its Evaluation. In *Proc. First Int. Symp. on Parallel Symbolic Computation*, pages 194–203, Linz, 1994.

10. R. Hasegawa, Y. Ohta, and K. Inoue. Non-Horn Magic Sets and Their Relation to Relevancy Testing. Technical Report 834, ICOT, 1993. Dagstuhl Seminor on Deduction in Germany, 1993 Workshop on Finite Domain Theorem Proving, 1994.

11. R. Hasegawa and Y. Shirai. Constraint Propagation of CP and CMGTP: Experiments on Quasigroup Problems. In *Proc. Workshop 1C (Automated Reasoning in Algebra), CADE-12*, Nancy, France, 1994.

12. K. Inoue, M. Koshimura, and R. Hasegawa. Embedding Negation as Failure into a Model Generation Theorem Prover. In *Proc. 11th Int. Conf. on Automated Deduction*, pages 400–415. Springer-Verlag, 1992. LNAI 607.

13. K. Inoue, Y. Ohta, R. Hasegawa, and M. Nakashima. Bottom-Up Abduction by Model Generation. In *Proc. IJCAI-93*, 1993. ICOT TR-816.

14. M. Koshimura and R. Hasegawa. Modal Propositional Tableaux in a Model Generation Theorem Prover. In *Proc. Third Workshop on Theorem Proving with Analytic Tableaux and Related Methods*, pages 145–151, UK, 1994. also in ICOT TR-860.

15. R. Letz, K. Mayr, and C. Goller. Controlled integration of the cut rule into connection tableau calculi. *J. Automated Reasoning*, 13:297–337, 1994.

16. D. W. Loveland, D. W. Reed, and D. S. Wilson. SATCHMORE: SATCHMO with RElevancy. Technical report, Department of Computer Science, Duke University, Durham, North Carolina, 1993. CS-1993-06.

17. R. Manthey and F. Bry. SATCHMO: a theorem prover implemented in Prolog. In *Proc. 9th Int. Conf. on Automated Deduction*, Argonne, Illinois, 1988.

18. W. W. McCune and L. Wos. Experiments in Automated Deduction with Condensed Detachment. In *Proc. 11th Int. Conf. on Automated Deduction*, pages 209–223, Saratoga Springs, NY, 1992.

19. S. Muggleton. Inverse Entailment and Progol. *New Generation Computing*, 13:245–286, 1995.

20. H. J. Ohlbach. A resolution calculus for modal logics. In *Proc. 9th Int. Conf. on Automated Deduction*, pages 500–516, 1988.

21. J. Rohmer, R. Lescoeur, and J.M. Kerisit. The alexander method – a technique for the processing of recursive axioms in deductive databases. *New Generation Computing*, 4:273–285, 1986.

22. Y. Shirai and R. Hasegawa. Two Approaches for Finite-Domain Constraint Satisfaction Problem - CP and MGTP -. In L. Sterling, editor, *Proc. 12th Int. Conf. on Logic Programming*, pages 249–263. MIT Press, June 1995. Tokyo.

23. G. Sutcliffe, C. Suttner, and T. Yemenis. The tptp problem library. In *Proc. 12th Int. Conf. on Automated Deduction*, pages 252–266, 1994.

24. K. Ueda and T. Chikayama. Design of the Kernel Language for the Parallel Inference Machine. *Computer J.*, 33:494–555, December 1990.

Three Faces of Natural Deduction

G. Mints

Department of Philosophy, Stanford University,
Stanford, CA 94305, USA, mints@csli.stanford.edu

1 Introduction

Natural deduction was introduced by Gentzen [7] in 1934 as a formalization of rigorous human reasoning. It turned out to be especially suitable for extracting programs from proofs. This program interpretation (Curry-Howard isomorphism of natural deduction and lambda-calculus) gave rise to the slogan PROOFS=PROGRAMS. Close connections with formulations suitable for immediate proof search (Prawitz trasformation to sequent systems, [33] 1965, and Maslov transformation to resolution, [17] 1973) were discovered remarkably late.

This paper is a survey of several results in these directions. Some of the proofs are only sketched and will appear in other papers.

First we consider in the section 2 standard assignment of λ-terms to natural deductions in the $(\&, \rightarrow)$-fragment of the intuitionistic propositional logic (Curry-Howard isomorphism). We compare two familiar methods for extraction of programs from proofs in the intuitionistic sequent calculus. One method going back to Gentzen [7] is local: every step in the proof is transformed into a fixed construct of λ-calculus (added to what is already constructed). It introduces cuts. The second method (cf. [33, 26, 32]) is global: it makes a substitution for some variables and does not introduce cuts. It turns out (Theorem 1, cf. [28]) that these transformations differ only in a series of non-iterated β-conversions: the second one is obtained from the first by a complete development [4] of (redundant) redexes introduced to make the transformation global. Viewed as intertranslations between sequent system LJ and system NJ of natural deduction, local methods use the direct derivability of the rules of one system in the other, and so are linear in the number of inferences, while global methods use cut-free admissibility of the rules, and are exponential in the worst case.

Natural deduction is used both for exact formalization of reasoning and as a tool for extraction of programs from proofs and normalization of proofs (deductions) which exactly corresponds under Curry-Howard isomorphism to computation by corresponding programs. In both of these situations standard $\exists$-elimination rule presents well-known difficulties. The rule of existential instantiation $\exists x A[x] / A[a]$, which was introduced with the intension to bring formalization closer to human reasoning (the history is presented in [35, 33, 14]) allows to overcome some of these difficulties.

In the section 3 we describe a manageable formulation NJi of intuitionistic logic where restrictions on eigenvariables are made local and reduced to a minimum by a device from [31, sections 7,8], cf. also [24]. The only new syntactical objects are assumptions $< \Gamma \; '>_{\exists x F, a}$, which cannot be combined into more complicated formulae, and are eventually interpreted as $F[x/a]$. System $M\epsilon$ of [14] is probably the closest to NJi in the literature. Main differences with $M\epsilon$ are the use of ϵ-terms there and a different mechanism of restricting assumptions. Detailed comparison of $M\epsilon$ with other systems is presented in [14]. We present also a rule for disjunction which can play the same role for $\vee$ that $\exists i$ plays for $\exists$. Only intuitionistic predicate logic is considered here, but extension to classical or higher order case is straightforward. A translation * of NJi into NJ is defined and proved to preserve reductions (normalization steps) of proofs. This proves soundness and completeness of NJi and allows to derive strong normalization for NJi from strong normalization for NJ.

In the section 4 we describe an extension of Curry-Howard isomorphism CH [5, 11] to the !-free fragment $(\otimes, \&, -\circ, I)$ of intuitionistic propositional linear logic. Proofs are presented in [29]. The basis is provided by the results of two previous papers: [24] contains a proof of a normal form theorem for the second order Intuitionistic Predicate Linear Logic, and [22] treated λ-terms for (what is now called) $(\otimes, -\circ, I)$-fragment of intuitionistic propositional linear logic. Some of the related papers including [3, 23, 36, 39, 41] are discussed in [24]. We obtain a unique normal form for natural deduction in $(\otimes, \&, -\circ, I)$ fragment of the intuitionistic linear logic demonstrating a possibility to treat considerable fragment of linear logic in the framework of standard lambda calculus adding no new combinators. This possibility was realized in [22, 2] for the $(-\circ, \otimes I)$-fragment, but proofs in [22] are presented only for $(-\circ, I)$-fragment, and [2] is not easily accessible.

Section 5 illustrates connection between natural deduction and resolution for intuitionistic logic.

2 Translations of sequent calculus into natural deduction

In this section we consider $\&, \rightarrow$-fragment of intuitionistic logic.

Derivable objects of Gentzen's system LJ are sequents

$$A_1, \ldots, A_n \Rightarrow D \qquad (1)$$

In a term assignment assumptions $A_1, \ldots, A_n$ are assigned distinct variables $x_1, \ldots, x_n$ of types $A_1, \ldots, A_n$ and formula D receives a term u. A sequent (1) is transformed into a statement

$$x_1 : A_1, \ldots, x_n : A_n \Rightarrow u : D \quad \text{or} \quad \mathbf{x} : \Gamma \to u : D \qquad (2)$$

where $\mathbf{x} = x_1, \ldots, x_n$ and $\Gamma = A_1, \ldots, A_n$. Axiom $A \Rightarrow A$ is transformed into a statement $x : A \Rightarrow x : A$, and inference rules are transformed similarly.

2.1 Global term assignment T

Let us list the axioms and inference rules of the system LJ together with assignment of deductive terms to them (cf. [25, 32]).

System LJ and global assignment

Axioms: $\qquad\qquad\qquad x : A \Rightarrow x : A$

Inference rules

$$\Rightarrow \& \quad \frac{z : \Gamma \Rightarrow t : A \quad z : \Gamma \Rightarrow u : B}{z : \Gamma \Rightarrow p(t, u) : (A \& B)} \qquad\qquad \frac{x : A_i, z : \Gamma \Rightarrow t : D}{y : A_0 \& A_1, z : \Gamma \Rightarrow t[x/p_i y] : D}$$

$$\to \Rightarrow \quad \frac{z : \Gamma \Rightarrow u : A \quad x : B, z' : \Delta \Rightarrow t : D}{y : (A \to B), z, z' : \Gamma, \Delta \Rightarrow t[x/y(u)] \; : D} \qquad \Rightarrow \to \quad \frac{y : A, z : \Gamma \Rightarrow t : B}{z : \Gamma \Rightarrow \lambda yt \; : (A \to B)}$$

Structural rules:

$$Weakening \; \frac{z : \Gamma \Rightarrow t : D}{x : A, z : \Gamma \Rightarrow t : D} \qquad Contraction \; \frac{x : A, y : A, z : \Gamma \Rightarrow t : D}{x : A, z : \Gamma \Rightarrow t[y/x] : D}$$

$$Cut \; \frac{z : \Gamma \Rightarrow c : C \quad y : C, z' : \Delta \Rightarrow u : D}{z, z' : \Gamma, \Delta \Rightarrow u[y/c] : D}$$

All inference rules are by the definition invariant under permutation of formulas in the sequents. *Substitution* $t[x/u]$ of a term u for all free occurrences of a variable x in a term t (with renaming bound variables to avoid collision) is defined in a standard way.

For any derivation d in LJ the global method inductively assigns to d a deductive term $T(d)$ according to the rules above: if the last sequent of d is (2) then $T(d) = u$. Term $T(d)$ is defined up to a congruence, i.e. up to a choice of free variables $\mathbf{x}$ for assumptions and renaming bound variables.

2.2 Correspondence between λ-terms and natural deductions

Recall Curry-Howard isomorphism for NJ, i.e. assignment of terms to natural deductions.

Axioms: $\qquad\qquad\qquad x : A \Rightarrow x : A$

Inference rules

$$\&I \quad \frac{\mathbf{z} : \Gamma \Rightarrow t : A \quad \mathbf{z} : \Sigma \Rightarrow u : B}{\mathbf{z} : \Gamma, \Sigma \Rightarrow p(t, u) : (A \& B)} \qquad \&E \frac{\mathbf{z} : \Gamma \Rightarrow t : A_0 \& A_1}{\Gamma \Rightarrow p_i t] : A_i}$$

$$\rightarrow E \quad \frac{\mathbf{z} : \Gamma \Rightarrow u : A \quad \mathbf{z}' : \Delta \Rightarrow t : A \rightarrow B}{\mathbf{z}, \mathbf{z}' : \Gamma, \Delta \Rightarrow t(u) : D} \qquad \rightarrow I \quad \frac{y : A, \mathbf{z} : \Gamma \Rightarrow t : B}{\mathbf{z} : \Gamma \Rightarrow \lambda y t : (A \rightarrow B)}$$

and standard structural rules: weakening and contraction.

2.3 Local term assignment L

Direct or local translation **L** of *LJ*-derivations into terms differs from the global translation T described above only by the terms assigned to conclusions of antecedent logical rules and cut:

$$\& \Rightarrow : (\lambda x t)(p_i y) \quad \rightarrow \Rightarrow : (\lambda x t)(y(u)) \quad Cut : (\lambda y u)(c) \tag{3}$$

In other words the inductive definition of **L** is obtained from the definition of T if one replaces T by **L** and changes three clauses according to a familiar direct translation going back to Gentzen of antecedent rules into natural deduction, which we present here together with term assignment:

$$\frac{y : (A \& B) \Rightarrow y : (A \& B)}{y : (A \& B) \Rightarrow p_0 y : A} \qquad \frac{x : A, \mathbf{z} : \Gamma \Rightarrow t : D}{\mathbf{z} : \Gamma \Rightarrow \lambda x t : (A \rightarrow D)}$$
$$\frac{}{y : A \& B, \mathbf{z} : \Gamma \Rightarrow (\lambda x t)(p_0 y) : D}$$

$$\frac{\mathbf{z} : \Gamma \Rightarrow u : A \quad y : (A \rightarrow B) \Rightarrow y : (A \rightarrow B)}{y : (A \rightarrow B), \mathbf{z} : \Gamma \Rightarrow y(u) : B} \qquad \frac{x : B, \mathbf{z}' : \Delta \Rightarrow t : D}{\mathbf{z}' : \Delta \Rightarrow \lambda x t : B \rightarrow D}$$
$$\frac{}{y : (A \rightarrow B), \mathbf{z}, \mathbf{z}' : \Gamma, \Delta \Rightarrow (\lambda x t)(y(u)) : D}$$

$$\frac{\mathbf{z} : \Gamma \Rightarrow c : C \quad \frac{y : C, \mathbf{z}' : \Delta \Rightarrow u : D}{\mathbf{z}' : \Delta \Rightarrow \lambda y u : C \rightarrow D}}{\mathbf{z}, \mathbf{z}' : \Gamma, \Delta \Rightarrow (\lambda y u)(c) : D}$$

Recall [4, 5] that full development of lambda-terms is defined for terms of a new language Λ' which is obtained by adding new indexed lambda-symbols λ_i to the language of typed lambda calculus. Full development ϕ simply converts all redexes beginning with indexed λ-symbols:

$$\phi(x) = x, \quad \phi(\lambda x t) = \lambda x \phi(t), \quad \phi((\lambda_i x t)(u)) = t[x/u]$$

$$\phi(t(u)) = \phi(t)(\phi(u)) \; if \; t \neq \lambda_i x t'$$

Let us now define $\mathbf{L}'(d)$ as the result of indexing by 1 all λ-symbols introduced in (3).

The following statement (cf. [28, 29]) shows that local and global translations are cosely connected.

Theorem 1 *For every derivation d in LJ* $\quad T(d) = \phi(\mathbf{L}'(d))$.

Proof. Induction on d using defining equations for $T, \mathbf{L}', \phi$. $\square$.

3 Intuitionistic predicate logic NJi with existential instantiation

Formulas of first order predicate logic are constructed in a familiar way from atomic formulas (including propositional constant $\perp$) by the connectives $\&, \vee$, $\rightarrow, \forall, \exists$. Free and bound variables $FV(A), BV(A)$ of a formula A are defined as usual. There are no redundant quantifiers: $\exists x A, \forall x A$ are formulas only if $x \in FV(A)$.

Assumptions of the system NJi are formulas as well as expressions of the form

$$< \Gamma >_{\exists x A, a} \quad \text{and} \quad < \Gamma >_{A \vee B, C} \qquad (4)$$

where Γ is a multiset of assumptions, A, B are formulas, C is A or B and a is an individual variable. By the definition, $FV(< \Gamma >_{\exists x A, b}) = FV(A[x/b])$, $FV(< \Gamma >_{A \vee B, C}) = FV(C)$. Assumptions (4) are called $<>$-*formulas*. Note that they cannot be subformulas of (real) formulas. Expression $E[x/t]$ (or $E[t]$ if x is known) stands for the result of substituting t for all free occurrences of x into E. *Sequent* of NJi is any expression $\Gamma \Rightarrow A$ where Γ is a multiset of assumptions and A is a formula.

Inference rules

Standard axioms, contraction, introduction and elimination rules for $\&, \rightarrow, \forall$,

$$\exists I \; \frac{\Gamma \Rightarrow A[x/t]}{\Gamma \Rightarrow \exists x A} \qquad \exists i \; \frac{\Gamma \Rightarrow \exists x A}{< \Gamma >_{\exists x A, b} \Rightarrow A[x/b]} \qquad <> \exists \; \frac{< \Gamma >_{A, b}, \Delta \Rightarrow C}{\Gamma, \Delta \Rightarrow C}$$

with a proviso for the *eigenvariable* b: it is not free in the premise of $\exists i$, and in the conclusion of $<> \exists$.

$$\vee i \; \frac{\Gamma \Rightarrow A_i}{\Gamma \Rightarrow A_0 \vee A_1} \; i = 0, 1 \qquad\qquad \vee i \; \frac{\Gamma \rightarrow A_0 \vee A_1}{< \Gamma >_{A_0 \vee A_1, A_i} \Rightarrow A_i} \; i = 0, 1$$

$$<> \vee \; \frac{< \Gamma >_{A_0 \vee A_1, A_0}, \Delta \Rightarrow C \quad < \Gamma >_{A_0 \vee A_1, A_1}, \Sigma \Rightarrow C}{\Gamma, \Delta, \Sigma \Rightarrow C}$$

This concludes the description of the system NJi .

3.1 Soundness and completeness

To prove soundness we define translation of NJi into the standard natural deduction formulation NJ.

Lemma 1 *(a) Every deduction d in NJi of a sequent containing $< \Gamma >_{\exists x A, b}$ contains a subdeduction of $\Gamma \Rightarrow \exists x A$;*

(b) Every deduction d in NJi of a sequent containing $< \Gamma >_{A \vee B, C}$, contains a subdeduction of $\Gamma \Rightarrow A \vee B$

Proof. Easy induction on d: when d is traced from the bottom up, assumptions with $<>$ disappear at i-rules furnishing necessary subdeductions. □

Lemma 2 *If a sequent S in a deduction of a $<>$-free sequent contains assumptions $< \Gamma >_{F, a}$ and $< \Delta >_{G, a}$ with the same a, then these assumptions coincide, i.e. $\Gamma = \Delta$ (up to the order of elements) and $F = G$.*

Now consider the union of NJ and NJi . The system NJ+ is obtained by adding standard elimination rules for $\exists, \vee$ to NJi. Let us describe translations of NJ+ into NJ and NJi .

The rule $(\exists\ E)$ is eliminated in favor of $\exists i$ as follows:

$$\frac{\Gamma \Rightarrow \exists x A \quad \dfrac{A[x/b] \Rightarrow A[x/b]}{\quad \mid \quad} \atop A[x/b], \Delta \Rightarrow C}{\Gamma, \Delta \Rightarrow C} \qquad \frac{\dfrac{\Gamma \Rightarrow \exists x A}{< \Gamma >_{\exists x A, b} \Rightarrow A[x/b]} \atop \dfrac{\mid}{< \Gamma >_{\exists x A, b}, \Delta \Rightarrow C}}{\Gamma, \Delta \Rightarrow C}$$

The rule $\vee\ E$ is eliminated similarly. Instantiation and $<>$-rules are eliminated as follows in favor of the standard rules of NJ:

$$\frac{\dfrac{\Gamma \Rightarrow \exists x A}{< \Gamma >_{\exists x A, b} \Rightarrow A[x/b]} \atop \dfrac{\mid}{< \Gamma >_{\exists x A, b}, \Delta \Rightarrow C}}{\Gamma, \Delta \Rightarrow C} \qquad \frac{\Gamma \Rightarrow \exists x A \quad \dfrac{A[x/b] \Rightarrow A[x/b]}{\quad \mid \quad} \atop A[x/b], \Delta \Rightarrow C}{\Gamma, \Delta \Rightarrow C},$$

and similarly for $\vee$-rules, i.e. by the transformations inverse to elimination of standard elimination rules.

If $S = \Delta \Rightarrow G$ is a sequent of NJi, then $S*$ denotes the result of replacing the outermost occurrences (in Δ) of $< \Gamma >_{\exists x A, a}$ by $A[a]$ and of $< \Gamma >_{A \vee B, C}$ by C.

Theorem 2 *If S is derivable in NJ+ then $S*$ is derivable in NJ and S is derivable in NJi. In particular $NJ, NJi, NJ+$ are equivalent for $<>$-free sequents.*

Proof. Apply translations described above.

3.2 Normalization

Conversion (one-step reduction) relation d *conv* d' for deductions in NJi is defined in a natural way (cf. [30]). Reduction (by $n \geq 0$ conversions of subdeductions) is defined in a standard way .

Lemma 3 *Let d be a deduction in NJi. If d conv d' then d∗ reduces (in non-zero steps) to d'∗.*

Proof . Check all conversions.

Theorem 3 *NJi is strongly normalizable.*

Proof . Every sequence s of conversions of a deduction d in NJi induces by the previous Lemma a sequence of conversions s' of $d*$ with $length(s) \leq length(s')$. Since s' is finite, s is finite too.

4   Lambda terms and Natural Deductions in !-free propositional intuitionistic linear logic

4.1 $(\otimes, \&, \multimap)$-fragment

It is well-known that $\otimes$-elimination rule

$$\frac{\Gamma \Rightarrow A \otimes B \quad A, B, \Sigma \Rightarrow C}{\Gamma, \Sigma \Rightarrow C}$$

introduces into natural deduction the same kind of non-uniqueness as the $\otimes$-antecedent rule of the sequent calculus. This will be dealt with by a device going back to "central maps" of [13]. It introduces a kind of equivalence relation which corresponds to the commutativity of $\otimes$ and dispences with the permutations of adjacent $\otimes$-inferences. We use the ideas from [23] to obtain uniqueness of normal form for natural deductions in the intuitionistic constant-free multiplicative-additive linear logic without exponentials.

Let us describe term assignment and inference rules. We use variables typed by formulas A and denoted by x^A, y^A, z^A etc. We drop the type superscript when clear from the context.

Typed λ-terms with two kinds of pairing and projections are constructed in a standard way from typed variables by λ-abstraction, application, two kinds of pairing and corresponding projections. Types are indicated by superscripts.

Definition 1 $x^A \in \mathcal{T}$; if $t^A, u^B \in \mathcal{T}$, then $p(t, u)^{A \& B} \in \mathcal{T}$; if $t^{A_0 \& A_1} \in \mathcal{T}$, then $p_i t^{A_i} \in \mathcal{T}$, $i = 0, 1$; if $t^A, u^B \in \mathcal{T}$, then $\pi(t, u)^{A \otimes B} \in \mathcal{T}$; if $t^{A_0 \otimes A_1} \in \mathcal{T}$, then $\pi_i t^{A_i} \in \mathcal{T}$, $i = 0, 1$; if $t^{A \multimap B}, u^A \in \mathcal{T}$, then $t(u)^B \in \mathcal{T}$; if $t^B \in \mathcal{T}$, then $\lambda x^A . t^{A \multimap B} \in \mathcal{T}$.

The set $\mathcal{LT} \subset \mathcal{T}$ of *linear* $\{\otimes, \&, \multimap\}$-*terms* is defined in much more restrictive way. $FV(t)$, $BV(t)$ stand for the set of free and bound variables of the term t. Expression $u[t/v]$ is the result of substituting a term v for all occurrences of a subterm t in a term u. In most cases t will be a variable.

Definition 2 $x^A \in \mathcal{LT}$;

$\quad$ if $t^A, u^B \in \mathcal{LT}$, and $FV(t) = FV(u)$, then $p(t, u)^{A\&B} \in \mathcal{LT}$;

$\quad$ if $t^{A_0\&A_1} \in \mathcal{LT}$, then $p_i t^{A_i} \in \mathcal{LT}$, $i = 0, 1$;

$\quad$ if $t^A, u^B \in \mathcal{LT}$, and $FV(t) \cap FV(u) = \emptyset$ then $\pi(t, u)^{A\otimes B} \in \mathcal{LT}$;

$\quad$ if $t^{A\otimes B} \in \mathcal{LT}, u^G \in \mathcal{LT}$, $FV(t) \cap FV(u) = \emptyset$;

$\quad$ $x^A \neq y^B \in FV(u)$, and $FV(t) \cap BV(u) = \emptyset$ then $u[x^A/\pi_0 t, y^B/\pi_1 t]^G \in \mathcal{LT}$;

$\quad$ if $t^{A\multimap B}, u^A \in \mathcal{LT}$, and $FV(t) \cap FV(u) = \emptyset$, then $t(u)^B \in \mathcal{LT}$;

$\quad$ if $t^B \in \mathcal{LT}$ and $x^A \in FV(t)$, then $\lambda x^A.t^{A\multimap B} \in \mathcal{LT}$.

Since $\mathcal{T}, \mathcal{LT}$ are closed under type-preserving renaming of free and bound variables, we assume in the following that for every term t under consideration $FV(t) \cap BV(t) = \emptyset$. One can restate the definition of $\mathcal{LT}$ as a term assignment to natural deductions (cf. section 2.2). It works with *contexts* $z^A : A, \ldots, z^D : D$ or $z : \Gamma$ for short. We state only the rule $\otimes E$, where lists z, z' of variables consist of distinct variables and are disjoint; x, y are distinct and do not occur in z, z', t, but occur free in u.

$$\otimes E \quad \frac{z : \Gamma \Rightarrow t : A_0 \otimes A_1 \quad x : A_0, y : A_1, z' : \Sigma \Rightarrow u : C}{z, z' : \Gamma, \Sigma \Rightarrow u[x/\pi_0 t, y/\pi_1 t] : C}$$

Term assignment $T(d)$ to a natural deduction d is defined as before. Notation $\Gamma \Rightarrow t : G$ means that $t = T(d)$ for some natural deduction $d : \Gamma \Rightarrow G$. Notation $d \equiv d'$ (d is *equivalent* to d') means that $T(d) = T(d')$. Completeness and normalization theorem (but not strong normalization theorem), even for a much wider language of the second order intuitionistic linear logic, was established in [24]. These results imply that the second order system is a conservative extension of the first order fragment considered here.

As pointed out above, the clause in the definition of $\mathcal{LT}$ corresponding to $\otimes E$ is the main source of difficulties. Set

$$t \circ_{z^A z^B} u =_{def} u[z^A/\pi_0 t, z^B/\pi_1 t],$$

provided conditions for $\otimes E$ are satisfied.

Let us recall several definitions from [24]. A *segment* in a deduction is a sequence of the $\otimes E$ rules with one and the same conclusion:

$$\frac{d_1 : \Gamma_1 \Rightarrow t_1 : A_1 \otimes B_1 \quad d_0 : A_1, B_1, \Sigma_1 \Rightarrow u : G}{A_2, B_2, \Sigma_2 \Rightarrow t_1 \circ u : G}$$

$$\cdots \tag{5}$$

$$\frac{d_n : \Gamma_n \Rightarrow t_n : A_n \otimes B_n \quad A_n, B_n, \Sigma_n \Rightarrow t_{n-1} \circ \ldots \circ t_1 \circ u : G}{\Theta \Rightarrow t_n \circ t_{n-1} \circ \ldots \circ t_1 \circ u : G}$$

Maximal segment (cut) in a natural deduction is a segment beginning with a conclusion of an introduction rule and ending with the major premise of an elimination rule ($n = 0$ is possible). A deduction is *normal*, if it does not contain maximal segments . Hence we are dealing with β-conversions for linear logic. Recall the definition for ordinary typed λ-terms.

A term $t \in \mathcal{T}$ is *normal* if it does not contain β-*redexes*

$$\pi_i \pi(t_0, t_1), \; p_i p(t_0, t_1), \; (\lambda x t)(u) \tag{6}$$

A normalization step for a term $t \in \mathcal{T}$ is a *conversion* of a subterm which is a *redex* of the form (6) into the *reductum* t_i or $t[x/u]$:

$$\pi_i \pi(t_0, t_1) = t_i, \; p_i p(t_0, t_1) = t_i, \; (i = 0, 1), \; (\lambda x t)(u) = t[x/u] \tag{7}$$

Recall the following basic fact (cf. [8]).

Lemma 4 *Every* $t \in \mathcal{T}$ *is strongly normalizing (every sequence of conversions terminates) and the normal form is unique.*

Proof Let $t \in \mathcal{T}$ and t' be the result of identifying π and p, π_i and p_i. Every conversion of t is a conversion of t', and t' is strongly normalizable. $\quad\square$

Note. For $t \in \mathcal{LT}$ it seems more natural to define π-conversion corresponding more closely to conversion of natural deduction:

$$u[\pi_0 \pi(t_0, t_1), \pi_1 \pi(t_0, t_1)] \; conv \; u[t_0, t_1] \tag{8}$$

Fortunately it is possible to manage with the standard normalization steps mentioned above, and use the developed apparatus of simply typed λ-calculus in view of the following statement (cf. [22, 2]).

Lemma 5 *A deduction* d *is normal iff* $T(d)$ *is normal.*

Note. A normal term in $\mathcal{LT}$ does not contain subterms $t \circ_{z^A z^B} u \in \mathcal{LT}$ with $FV(t) = \emptyset$. Indeed, t should be of the form $\pi(t', t'')$ and hence $t \circ_{z^A z^B} u \in \mathcal{LT}$ contains a redex $\pi_0 t$.

All terms and deductions below are supposed to be normal unless explicitly stated otherwise.

Define for arbitrary terms $t, v \in \mathcal{T}$:

$$t | v \text{ iff } v = t \circ_{z^A z^B} u \text{ for some } z^A, z^B, \; u \in \mathcal{LT}$$

provided conditions for $\otimes E$ are satisfied;

$$v /_{z^A z^B} t =_{def} v[\pi_0 t / z^A, \pi_1 t / z^B], \text{ provided } z^A \neq z^B \notin FV(t, v)$$

t is a *factor* of v if $t | v$.

Lemma 7 below establishes that the following notion provides the necessary and sufficient condition of factorization.

Definition 3 *A term* $t \in \mathcal{LT}$ *is projected in a term* $v \in \mathcal{LT}$ *iff* t *occurs in* v *and every occurrence of* t *and of free variables of* t *in* v *comes from* $\pi_0 t$ *or* $\pi_1 t$.

In other words $\Gamma \Rightarrow t : A \otimes B$ for some Γ, A, B, and all occurrences of free variables of t in v (they exactly correspond to formulas in Γ) are concentrated inside $\pi_0 t, \pi_1 t$. Hence if t is projected in v then $v/_{z^A z^B} t$ for distinct $z^A, z^B \notin FV(v, t)$ contains neither t nor free variables of t: all occurrences of t are eliminated, and all occurrences of new subterms contain z^A or z^B.

Example. $x^{a \otimes b}$ is not projected in $p(\pi(\pi_1(x), \pi_0(x)), x) : (a \otimes b) \Rightarrow (b \otimes a) \& (a \otimes b)$ since the last occurrence of x is not in the scope of π_i.

Lemma 6 *If $v = t \circ_{z^A z^B} u$ and $FV(t) \neq \emptyset$ then*
  (a) $v/_{z^A z^B} t = u$,
  (b) t is projected in v and v contains both $\pi_0 t$ and $\pi_1 t$.

Lemma 7 *If $t, v \in \mathcal{LT}$ and v is normal, then t is projected in v iff $t|v$*

Previous results provide an algorithm for checking whether a normal term $\in \mathcal{T}$ is in $\mathcal{LT}$.

Notation

$$w = (t_1, \ldots, t_n) \circ_{z_{01}, z_{11}, \ldots, z_{0n}, z_{1n}} v \tag{9}$$

stands for

$$w = v[z_{01}/\pi_0 t_1, z_{11}/\pi_1 t_1], \ldots, [z_{0n}/\pi_0 t_n, z_{1n}/\pi_1 t_n]$$

provided $t_1, \ldots, t_n, v \in \mathcal{LT}$, lists $FV(t_i), FV(t_j), i \neq j$ are disjoint pairwise and with $FV(v)$, and $z_{01}, z_{11}, \ldots, z_{0n}, z_{1n}$ are pairwise distinct and belong to $FV(v) - FV(t_1, \ldots, t_n)$.

If (9) holds, we write

$$w =_{nf} (t_1, \ldots, t_n) \circ_{z_{01}, z_{11}, \ldots, z_{0n}, z_{1n}} v \tag{10}$$

if there are no further factors in v, i.e. there is no term t with $FV(t) \subset FV(w) - FV(t_1, \ldots, t_n)$ such that $t|v$.

A term is *prime* if it is not of the form $t \circ_{z, z'} v$ for any t, v.

Lemma 8 *If*

$$w =_{nf} (t_1, \ldots, t_n) \circ_{z_{01}, z_{11}, \ldots, z_{0n}, z_{1n}} v =_{nf} (s_1, \ldots, s_m) \circ_{y_{01}, y_{11}, \ldots, y_{0m}, y_{1m}} u$$

then $m = n$, $t_1, \ldots, t_n = s_1, \ldots, s_n$ up to the order of t_i, s_j, and $u = v$ up to renaming of variables z_{ik}.

Lemma 9 *Let $t_1, \ldots, t_n$ be complete list of maximal factors of a term w, and $v = w/t_1/ \ldots /t_n$. Then*

$$w =_{nf} (t_1, \ldots, t_n) \circ v \tag{11}$$

A deduction d is *prime* if $T(d)$ is prime.

Notation

$$d =_{nf} (d_1, \ldots, d_n) \circ_{A_{01}, A_{11}, \ldots, A_{0n}, A_{1n}} e \tag{12}$$

means that d is of the form:

$$\frac{d_1 : \Gamma_1 \Rightarrow A_1 \otimes B_1 \ldots d_n : \Gamma_n \Rightarrow A_n \otimes B_n \quad e : A_1, B_1, \ldots, A_n, B_n, \Delta \Rightarrow G}{d : \qquad\qquad\qquad\qquad\qquad\qquad \Gamma_1, \ldots, \Gamma_n, \Delta \Rightarrow G}$$

ending in a series of n inferences by the rule $\otimes E$, and $T(e)$ is not of the form $t \circ_{x,y} u$ with $FV(t) \subset \mathbf{z}^\Delta$. In this case $T(d) =_{nf} (T(d_1), \ldots, T(d_n)) \circ_{z_{01}, \ldots, z_{1n}} T(e)$.

A deduction d is *strongly normal* if it is normal and any subdeduction which is not prime has a form (12).

Definition 4 *d sym d' iff d, d' can be obtained from each other by moving $\otimes E$ up and down in an equivalent way.*

In fact d sym d' iff $T(d) = T(d')$ for normal deductions d, d'. First assign a deduction $\mathcal{D}(u)$ in strong normal form to any normal term $u \in \mathcal{LT}$ in such a way that

$$T(\mathcal{D}(u)) = u \tag{13}$$

Theorem 4 *(a) d sym $\mathcal{D}(T(d))$;*

(b) In particular, for every deduction d', $T(d) = T(d')$ implies d sym d'

Theorem 5 *Every deduction has an equivalent strongly normal form which is unique up to the order of $d_1, \ldots, d_n$ in (12).*

4.2 Adding Constant I to $(\otimes, \&, -\circ)$-fragment

The treatment of β-reduction in the presence of I is obtained by leaving out the most difficult points from the treatment of I in [22] which deals with η-conversion. Both sequent and natural deduction formulation for the language with the constant I are obtained by adding the axiom and an inference rule with corresponding expansion of the language $\mathcal{T}$ of λ-terms:

$$\Rightarrow \mathbf{I} : I \ (I - \text{axiom}) \qquad \frac{\mathbf{z} : \Gamma \Rightarrow u : G \quad \mathbf{z'} : \Sigma \Rightarrow t : I}{\mathbf{z}, \mathbf{z'} : \Gamma, \Sigma \Rightarrow \{u, t\} : G} \ (I - \text{rule}) \tag{14}$$

provided $\mathbf{z} \cap \mathbf{z'} = \emptyset$. I-rule is a structural rule, hence it is neither introduction, nor elimination rule.

Conversions of λ-terms are (7) plus

$$\{u, t\} \text{ conv } u \qquad \text{provided } FV(t) = \emptyset \tag{15}$$

Theorem 6 *(a) λ-terms with I are strongly normalizable;*
(b) The normal form is unique
(c) Every deduction has an equivalent strongly normal form which is unique up to the order of $d_1, \ldots, d_n$ in (12).

5 Natural Deduction and Resolution

Main tool of construction of resolution calculi for new logical system is Maslov Transformation [17, 21]. It allows to transform antecedent and succedent rules of Gentzen-type systems into resolution-style rules and easily establish completeness of such rules. In many cases natural deduction rules are even more closely connected with resolution. It is in this way that the author obtained first resolution formulation for non-classical logic in [18, 19] before he recognized potential of Maslov transformation. We consider here a resolution system **Rip** for the $(\&, \rightarrow)$-fragment of the intuitionistic propositional logic (cf. [19, 37]).

Clauses are formulas of the form

$$(p \rightarrow q) \rightarrow r, \quad p_1 \rightarrow (\ldots \rightarrow (p_n \rightarrow r) \ldots), \qquad n \geq 0 \qquad (16)$$

the latter abbreviated to $p_1, \ldots, p_n \rightarrow r$, where $p, q, r, p_1, \ldots, p_n$ are propositional variables. Γ, Δ, Σ stand for multisets of propositional variables. An expression like $(\Gamma, \Gamma', \ldots, A, A', \ldots)$ indicates the multiset obtained by taking the union of $\Gamma, \Gamma', \{A\}, \{A'\}, \ldots$ and contracting to multiplicity one.

Axioms are as always initial clauses and $p \rightarrow p$.

Inference rules are

$$\rightarrow \frac{(p \rightarrow q) \rightarrow r \quad \Delta, p^0 \rightarrow q}{\Delta \rightarrow r} \qquad Res \, \frac{\Delta \rightarrow p \quad p, \Delta' \rightarrow q}{(\Delta, \Delta') \rightarrow q}$$

where p^0 stands for p or empty list. Note there are no &-rules and that all object derivable by the rules are Horn clauses $\Delta \rightarrow p$. We are interested in deductions in **Rip** of statements $C_1, \ldots, C_n \vdash \rightarrow g$ where $C_1, \ldots, C_n$ are (initial) clauses, and g is a propositional variable (goal).

Theorem 7 *(a) Every formula F can be reduced by introduction of new propositional variables to a clause form $\&C_F \rightarrow g$ where C_F is a finite set of clauses and g is a variable:*

$F \rightarrow (\&C_F \rightarrow g)$ and $(\&C_F \rightarrow g)* \rightarrow F$ are intuitionistically derivable,

where star denotes the result of substituting new variables by the formulas for which they were introduced.

*(b) $\&C_F \rightarrow g$ is intuitionistically derivable iff $C_F \vdash \rightarrow g$ holds in **Rip**.*

Proof. (a) is well-known (cf. [21, 37]). (b). Since the rules of **Rip** are obviously sound, it is sufficient to remind that a normal natural deduction of $C_1, \ldots, C_n \Rightarrow g$ is already a resolution derivation up to a little patching.

References

[1] Abramsky S. Computational Interpretations of Linear Logic , Theoretical Computer Science 111, 1993, 3- 57

[2] Babaev A. Equality of Canonical Maps in Closed Categories. (Russian), Izvestija Azerb. Akad. Nauk, Ser. matem., 1980, N6

[3] N.Benton,G.Bierman,V. de Paiva, M. Hyland, Linear lambda-calculus and Categorical Models Revisited, Springer LNCS 702, 1993, 61-84

[4] H. Barendregt, The Lambda Calculus, Amsterdam, North-Holland, 1981

[5] H.Curry, R. Feys, Craig W. Combinatory Logic, Amsterdam, North-Holland, 1958

[6] G. Gentzen. Untersuchungen über das logische Schliessen. Mathematische Zeitschrift, 39, 1934, 176-210, 405-431

[7] G. Gentzen. Untersuchungen über das logische Schliessen. Mathematische Zeitschrift, 39, 1934, 176-210, 405-431

[8] J.Y. Girard, Y.Lafont, P. Taylor, Proofs and Types, Cambridge University Press, Cambridge 1988

[9] J.-Y. Girard, Linear Logic, Theoretical Computer Science, 1987, 50, 1-102

[10] H. Herbelin, A λ-calculus Structure Isomorphic to Gentzen-style Sequent Structure, CSL 94, 61-75

[11] W.Howard The Formulae-as-types Notion of Construction, in: To H. B. Curry: Essays on Combinatory Logic, Lambda Calculus and Formalism, Academic Press, 1980, 479-490

[12] Kleene S.C., Permutability of inferences in Gentzen's Calculi LK and LJ, Memoirs of the AMS, 1952, no 10, p.1-26

[13] G.M. Kelly , S. MacLane, Coherence in Closed Categories, Journal of Pure and Applied Algebra, 1971, 1, 97-140

[14] D.Leivant, Existential instantiation in a system of natural deduction, Mathematish Centrum-ZW 13-73, 1973

[15] P. Lincoln, J. Mitchell, Operational aspects of Linear Lambda Calculus, Proc. IEEE Symp. on Logic in Computer Science, 1992, 235-247

[16] F. Lamarche,Game Semantics for Full Propositional Linear Logic, Proc. IEEE Symp. on Logic in Computer Science, 1995, 464-473

[17] S. Maslov. The connection between tactics of inverse method and the resolution method. A.O.Slisenko (editor). Zapiski Nauchnykh Seminarov LOMI. v. 16, 1969. English translation in: J.Siekmann, G. Wrightson, eds. Automated Reasoning, Springer-Verlag, Berlin, 1983, v2., 264-27

[18] G.Mints. Resolution Calculi for Non-classical Logics (Russian), 9-th Soviet Symposium in Cybernetics, 1981, v.2, 34-36

[19] G.Mints. Resolution Calculi for Non-classical Logics (Russian). Semiotika i Informatika, 25, 1985, 120-135

[20] G.Mints. Gentzen-type Systems and Resolution Rule. Part I. Propositional Logic. Lecture Notes in Computer Sci. 417, 1990, 198-231

[21] G. Mints. Gentzen-type Systems and Resolution Rule. Part II. Logic Colloquium'90, 163-190

[22] G.Mints, Closed Categories and the Theory of Proofs, J.Soviet Mathematics, 1981, 15, 45-62, [27] 183-212 (Russian original 1977)

[23] G. Mints, Normal forms for sequent derivations, Kreiseliana, A. K. Peters, Wellesley, MA, 1996, 469-492

[24] G. Mints, Normal Deduction in the Intuitionistic Linear Logic, CSLI report CSLI-95-50, 1995, to appear in Arcive for Mathematical Logic

[25] G. Mints, On E-theorems, in: [27], 105-116 (Russian original 1974)

[26] G. Mints, Finite Investigations of Transfinite Derivations, J.Soviet Mathematics, 1978, 10, 548-596, [27], p. 17-72 (Russian original 1975)

[27] G. Mints, Selected Papers in Proof Theory, North-Holland-Bibliopolis, 1992

[28] G. Mints, Review of [32], Referativnyi Zhurnal Matematika, 1978, n. 10, 10A26 (Russian)

[29] G. Mints, Linear Lambda-terms and Natural Deduction, to appear in Studia Logica

[30] G. Mints, Existential instantiation and Strong Normalization, to appear

[31] Mints G., Lewis Systems and the System T. In: [27], 221-294 (Russian Original 1974)

[32] H. Pottinger, Normalization as a Homomorphic Image of Cutelimination, Annals of Pure and Applied Logic, 12, 1977, 323-357

[33] Prawitz D., Natural deduction, Almquist and Wiksell, 1965

[34] D.Prawitz, Ideas and Results in Proof Theory, in: Proc. 2-nd Scand.Logic Symp., North-Holland, 1972, 235-308

[35] Quine W.,V., On natural deduction, JSL,v.15,1950, 93-102

[36] S. Ronchi della Rocca, L. Roversi, Lambda Calculus and Intuitionistic Linear Logic (manuscript)

[37] H. Schwichtenberg,A. Troelstra, Basic Proof, Theory, Cambridge University Press, 1996

[38] A. Troelstra, D. van Dalen, Constructivism in Mathematics, An Introduction, v.2, 1988, North-Holland

[39] A.Troelstra, Natural Deductions for Intuitionistic Linear Logic , APAL 73, n1, 1995, 79-108

[40] A.Troelstra, Mathematical Investigation of Intuitionistic Arithmetic and Analysis, Springer Lecture Notes in Mathematics v. 344, 1973

[41] S. Valentini, The Judgement Calculus for Intuitionistic Linear Logic: Proof Theory and Semantics, Zeitschrift f. math. Log., 1992, 38, n1, 38-58

[42] J. Zucker, The Correspondence between Cut-elimination and Normalization, Annals of Pure and Applied Logic, 1974, 7, 1-156

Tableaux for Logic Programming with Strong Negation

Seiki Akama

Computational Logic Laboratory, Department of Information Systems,
Teikyo Heisei University, 2289 Uruido, Ichihara-shi,
Chiba, 290-01, Japan.
phone: +81-436-74-6134
e-mail: SJK15022@mgw.shijokyo.or.jp

Abstract. Logic programming with strong negation (LPS) was proposed by Pearce and Wagner (1991) to handle both explicit and implicit negative information in knowledge representation in AI. We describe tableau calculi for LPS and establish the completeness. The proposed tableau calculi can deal with a wider class of programs in LPS. We also discuss possible refinements of the tableau calculi to improve efficiency.

1. Introduction

Logic programming with strong negation (LPS) was proposed by Pearce and Wagner (1991) to handle both explicit and implicit negative information in knowledge representation in AI; also see Wagner (1991). Pearce and Wagner provided a theoretical basis for LPS within the framework of Nelson's (1949) *constructive logics with strong negation*. Recently, Nelson's logics found several applications in AI; see Akama (1987, 1989, 1995) and Pearce and Wagner (1990).

In this paper, we describe *tableau calculi* for LPS giving a proof-theoretic foundation and establish the completeness. The proposed tableau calculi can deal with a wider class of programs in LPS. Since Pearce and Wagner only considered the first-degree fragment of constructive logics, our formulation can be regarded as an extension of their work. We also discuss possible refinements of the tableau calculi to improve efficiency.

The organization of the paper is as follows. In section 2, we review Nelson's constructive logics with strong negation. Section 3 gives the backgrounds of Pearce and Wagner's LPS and its relations to constructive logics. In section 4, we introduce a tableau calculus TN and give a completeness proof. In section 5, we modify it for LPS to address the computational aspects. Finally, we conclude the paper in section 6 with the discussion on the future research topics.

2. Constructive Logic with Strong Negation

Nelson (1949) proposed an extension of positive intuitionistic logic with a new connective for *constructible falsity* or *strong negation* to overcome the non-constructive features of intuitionistic negation. Constructive logics with strong negation have been extensively studied by logicians for many years. Constructive predicate logic with strong negation N can be obtained by an axiomatization of positive intuitionistic logic with the following axioms for strong negation ($\sim$):

(N1) $A\& \sim A \to B$,

(N2) $\sim\sim A \leftrightarrow A$,

(N3) $\sim(A \to B) \leftrightarrow (A\& \sim B)$,

(N4) $\sim(A\&B) \leftrightarrow (\sim A\lor \sim B)$,

(N5) $\sim(A \lor B) \leftrightarrow (\sim A\& \sim B)$,

(N6) $\sim \forall x A(x) \leftrightarrow \exists x \sim A(x)$,

(N7) $\sim \exists x A(x) \leftrightarrow \forall x \sim A(x)$.

In N, we can define intuitionistic negation ($\neg$) as follows:

$\neg A \leftrightarrow (A \to\sim A)$.

Clearly, strong negation is stronger than intuitionistic negation, namely

$\sim A \to \neg A$,

but the converse does not hold. If we omit (N1) from N, we get N^- of Almukdad and Nelson (1984). Gentzen systems for N appear in Thomason (1969) for a natural dedcution system and Akama (1988a) for a sequent calculus.

A Kripke semantics for N was developed by Thomason (1969); also see Akama (1988b, 1990). A *strong negation model* is of the form $(W, \leq, w_0, val, D)$, where W is a set of worlds with the distinguished world w_0, $\leq$ is a reflexive and transitive relation on $W \times W$, val is a three-valued valuation assigning 1 (true), 0 (false), -1 (undefined) to the atomic formula $p(t)$ at $w \in W$ with a parameter $t \in D(w)$ satisfying:

if $val(w, p(t)) \neq -1$ and $w \leq v$ then $val(w, p(t)) = val(v, p(t))$.

and D is a domain function from W to a set of variables such that if $w \leq v$ then $D(w) \subseteq D(v)$. Then, we define the function $V(w, A)$ for any formula A:

$V(w, p(t)) = 1$ iff $val(w, p(t)) = 1$ for any atomic $p(t)$ with $t \in D(w)$,

$V(w, p(t)) = 0$ iff $val(w, p(t)) = 0$ for any atomic $p(t)$ with $t \in D(w)$,

$V(w, A\&B) = 1$ iff $V(w, A) = 1$ and $V(w, B) = 1$,

$V(w, A\&B) = 0$ iff $V(w, A) = 0$ or $V(w, B) = 0$,

$V(w, A \lor B) = 1$ iff $V(w, A) = 1$ or $V(w, B) = 1$,

$V(w, A \lor B) = 0$ iff $V(w, A) = 0$ and $V(w, B) = 0$,

$V(w, A \to B) = 1$ iff $\forall v \in W(w \leq v$ and $V(v, A) = 1$ implies

$V(v, B) = 1)$,

$V(w, A \to B) = 0$ iff $V(w, A) = 1$ and $V(w, B) = 0$,

$V(w, \sim A) = 1$ iff $V(w, A) = 0$,

$V(w, \sim A) = 0$ iff $V(w, A) = 1$,

$V(w, \forall x A(x)) = 1$ iff $\forall v \in W(w \leq v$ implies $V(v, A(t)) = 1$

for any $t \in D(v))$,

$V(w, \forall x A(x)) = 0$ iff $V(w, A(t)) = 0$ for some $t \in D(w)$,

$V(w, \exists x A(x)) = 1$ iff $V(w, A(t)) = 1$ for some $t \in D(w)$,

$V(w, \exists x A(x)) = 0$ iff $\forall v \in W(w \leq v$ implies $V(v, A(t)) = 0$

for any $t \in D(v))$.

Note here that $V(w, A) = -1$ if neither $V(w, A) = 1$ nor $V(w, A) = 0$ holds. We say that A is true iff $V(w_0, A) = 1$. A is *valid* iff it is true in all strong negation models. A strong negation model for N^- needs a further condition that $V(w, A) = V(w, \sim A) = 1$ for some w. A completeness proof for N may be found in Akama (1988b, 1990). Thomason (1969) proved that N with the axiom $\forall x(A(x) \vee B) \to (\forall x A(x) \vee B)$ (x is not free in B) has a Kripke semantics with constant domains.

Theorem 1
N (or N^-) is complete for all strong negation models.

3. Logic Programming with Strong Negation

Recently, there is a growing interest in extending logic programming with a new kind of negation in addition to *negation as failure* (NAF) of Clark (1978). For instance, Gelfond and Lifschitz (1990) proposed logic programs with "classical negation" to represent incomplete information; also see Kowalski and Sadri (1990). Alferes and Pereira (1996) claimed that logic programming with two kinds of negation can serve as a description language for common-sense reasoning. Independently of these approaches, Pearce and Wagner (1990, 1991) developed *logic programming with strong negation* (LPS) based on Nelson's constructive logic with strong negation; see Wagner (1991).

We now give a brief description of LPS. The language of LPS contains the logical symbols: $\wedge$ (conjunction), $\vee$ (disjunction), $\sim$ (strong negation) and 1 (truth). However, it does not have quantifiers. A *program* Π consists of *clauses* of the form $l \leftarrow F$ (where l is a literal and F is an arbitrary formula), which are considered as rules rather than formulas. A rule with premise 1 is called a *fact* abbreviated by l. A program Π containing a non-ground clause is viewed as a dynamic representation of the corresponding set of ground clauses formed by the current domain of individuals U denoted by $[\Pi]_U$, namely:
$$[\Pi]_U = \{l\sigma \leftarrow F\sigma \mid l \leftarrow F \in \Pi \text{ and } \sigma : Var(l, F) \to U\}$$
where σ ranges over all mappings from the set of variables of l and F into the Herbrand universe U. σ is called a *ground substitution* for $l \leftarrow F$ and $[\Pi]_U$ the *Herbrand expansion* of Π with respect to the Herbrand universe U_P of Π. $[\Pi]_{\mathcal{M}}$ abbreviates $[\Pi]_{U_{\mathcal{M}}}$, where $U_{\mathcal{M}}$ denotes the Herbrand universe of some interpretation $\mathcal{M}$. The *Herbrand base* of a program Π is denoted by At_P and its partial version by $Lit_P = At_p \cup \{\sim a \mid a \in At_P\}$.

A model for LPS uses a *partial Herbrand interpretation*. A partial Herbrand interpretation M is called a *model* of Π, written $\mathcal{M} \models \Pi$, if for all $l \leftarrow F \in [\Pi]_{\mathcal{M}}, \mathcal{M} \models F$ implies $\mathcal{M} \models l$. Let $Mod(\Pi) = \{M \mid \mathcal{M} \models \Pi\}$ and $M_P = \cap Mod(\Pi)$. Wagner (1991) showed that every program Π has a least model, i.e. $\mathcal{M}_P$. Say that F is a *logical consequence* of Π, denoted $\Pi \models F$, if every model of Π is also a model of F.

Theorem 2 (Wagner (1991))
$\Pi \models F$ iff $\mathcal{M}_P \models F$.

A proof theory for LPS can be given by a natural deduction system. We write $\Pi \vdash G$ to mean that a goal G is derived from the program Π. Instead of $\Pi \vdash F$ and $\Pi \vdash G$, we write $\Pi \vdash F, G$. Here are the rules:

$$\frac{\Pi \vdash F, G}{\Pi \vdash F \wedge G} \ (\wedge) \qquad \frac{\Pi \vdash \sim F, \sim G}{\Pi \vdash \sim (F \vee G)} \ (\sim \vee)$$

$$\frac{\Pi \vdash F}{\Pi \vdash F \vee G} \quad \frac{\Pi \vdash G}{\Pi \vdash F \vee G} \quad (\vee)$$

$$\frac{\Pi \vdash \sim F}{\Pi \vdash \sim (F \wedge G)} \quad \frac{\Pi \vdash \sim G}{\Pi \vdash \sim (F \wedge G)} \quad (\sim \wedge)$$

$$\frac{\Pi \vdash F}{\Pi \vdash \sim\sim F} \quad (\sim\sim)$$

(1) $\Pi \vdash 1$

Here, F and G are ground formulas. A non-ground formula is provable if some ground instance of it is, namely:

$$\frac{\Pi \vdash F(t) \text{ for some ground term } t}{\Pi \vdash F(x)} \quad (x)$$

To derive a ground literal, we need the following:

(l_1) $\Pi \vdash l$ iff $\exists (l \leftarrow F) \in [\Pi] : \Pi \vdash F$

Unfortunately, (l_1) works only for a well-founded program. For a complete search, we need to add the next:

$$\frac{\Pi \vdash F}{\Pi \vdash l} \quad (l_2)$$

provided that $l \leftarrow F \in [\Pi]$.

Pearce and Wagner (1991) established the connection of LPS and Nelson's constructive logics with strong negation as follows:

Theorem 3 (Pearce and Wagner (1991))
Let $P = \{F \rightarrow l \mid l \leftarrow F \in [\Pi]\}$ and $G \in L(\sim, \wedge, \vee)$. Then
$$\Pi \vdash G \text{ iff } P \vdash_{N^-} G.$$

In the theorem we can replace N^- by N if P is a consistent program. Wagner (1992) proposed *vivid logic* as a generalization of LPS with two kinds of negation for describing common-sense reasoning.

4. Tableau Calculi

Tableau calculi (or *semantic tableaux*) can be regarded as the variant of Gentzen systems; see Smullyan (1968). Tableau calculi are used as the proof method for both classical and non-classical logics (cf. Rautenberg (1979)). The main advantage of the use of tableau calculi is that we can deal directly with the full first-order logic. In addition, proofs in tableau calculi are easy to understand. It is thus natural to consider logic programming based on tableaux. However, the idea is not novel. In fact, Schöenfeld (1985) suggested tableau-based logic programming, but his exposition is far from complete. In this section, we first provide a tableau calculus TN for constructive logic with strong negation. Then, we extend it for LPS in the next section.

A tableau calculus for the propositional part of N can be found in Rautenberg (1979). We here give a first-order tableau in the different exposition. According to Smullyan (1968), we use the notion of *signed formula*. If X is a formula, then TX and FX are signed formulas. TX reads "X is provable" and FX reads "X is not provable", respectively. If S is a set of signed formulas and A is a signed formula, then we simply write $\{S, A\}$ for $S \cup \{A\}$. A tableau calculus TN consists of *axioms* and *reduction rules*. Let p be an atomic formula and A and B be formulas.

Axioms

(AX1) $\{Tp, Fp\}$
(AX2) $\{T \sim p, F \sim p\}$
(AX3) $\{Tp, T \sim p\}$

Reduction Rules

$$\frac{S, T(A\&B)}{S, TA, TB} \ (T\&) \qquad \frac{S, F(A\&B)}{S, FA; \ S, FB} \ (F\&)$$

$$\frac{S, T(A \vee B)}{S, TA; \ S, TB} \ (T\vee) \qquad \frac{S, F(A \vee B)}{S, FA, FB} \ (F\vee)$$

$$\frac{S, T(A \rightarrow B)}{S, FA; \ S, TB} \ (T \rightarrow) \qquad \frac{S, F(A \rightarrow B)}{S_T, TA, FB} \ (F \rightarrow)$$

$$\frac{S, T(\sim (A\&B))}{S, T(\sim A); \ S, T(\sim B)} \ (T \sim \&) \qquad \frac{S, F(\sim (A\&B))}{S, F(\sim A), F(\sim B)} \ (F \sim \&)$$

$$\frac{S, T(\sim (A \vee B))}{S, T(\sim A), T(\sim B)} \ (T \sim \vee) \qquad \frac{S, F(\sim (A \vee B))}{S, F(\sim A); \ S, F(\sim B)} \ (F \sim \vee)$$

$$\frac{S, T(\sim (A \rightarrow B))}{S, TA, T(\sim B)} \ (T \sim \rightarrow) \qquad \frac{S, F(\sim (A \rightarrow B))}{S, FA; \ S, F(\sim B)} \ (F \sim \rightarrow)$$

$$\frac{S, T(\sim\sim A)}{S, TA} \ (T \sim\sim) \qquad \frac{S, F(\sim\sim A)}{S, FA} \ (F \sim\sim)$$

$$\frac{S, T(\forall x A(x))}{S, T(A(t))} \ (T\forall) \qquad \frac{S, F(\forall x A(x))}{S_T, F(A(a))^*} \ (F\forall)$$

$$\frac{S, T(\exists x A(x))}{S, T(A(a))^*} \ (T\exists) \qquad \frac{S, F(\exists x A(x))}{S, F(A(t))} \ (F\exists)$$

$$\frac{S, T(\sim \forall x A(x))}{S, T(\sim A(a))^*} \ (T \sim \forall) \qquad \frac{S, F(\sim \forall x A(x))}{S, F(\sim A(t))} \ (F \sim \forall)$$

$$\frac{S, T(\sim \exists x A(x))}{S, T(\sim A(t))} \ (T \sim \exists) \qquad \frac{S, F(\sim \exists x A(x))}{S_T, F(\sim A(a))^*} \ (F \sim \exists)$$

Here, $*$ denotes the variable restriction that the parameter a introduced must not occur in any formula of S or in the formula $A(x)$. S_T stands for $\{TX \mid TX \in S\}$. A *proof* of a sentence X is a closed tableau for FX. A tableau is a tree constructed by the above reduction rules. A tableau is *closed* if each branch is closed. A branch is closed if it contains the axioms of the form (AX1), (AX2) and (AX3).

Now, we prove the completeness of the tableau calculus TN with respect to strong negation models. Although we know that N is complete, it would be interesting to show that TN is a complete proof proceduce. To our knowledge, the completeness of tableau calculi for N has not been established in the literature. Our strategy is similar to that for tableau for intuitionistic logic sketched in Fitting (1969), but the present treatment is more complicated.

Let $S = \{TX_1, ..., TX_n, \; FY_1, ..., FY_m\}$ be a set of signed formula, $(W, \leq, w_0, val, D)$ be a strong negation model, and $w \in W$. We say that w *refutes* S if:

$$V(w, X_i) = 1 \text{ if } TX_i \in S,$$
$$V(w, X_i) \neq 1 \text{ if } FX_i \in S.$$

A set S is *refutable* if something refutes it. If S is not refutable, it is *valid*.

Theorem 4 (Soundness of TN)

If A is provable, then A is valid.

(Proof): If A is of the form of axioms, it is easy to see that it is valid. For reduction rules, it suffices to check that they preserve validity. $\square$

We are now ready to prove the completeness of TN. We use the concept of Hintikka set to relate tableaux to the Henkin constructions. Before presenting the completeness proof, we recall that an *infinite* set of signed formulas is *consistent* if every finite subset is consistent. A finite set of signed formulas is consistent if no tableau for it is closed. Let Γ be a set of signed formulas and if $\Gamma \leq \Delta$ then $D(\Gamma) \subseteq D(\Delta)$ and $\Gamma_T \subseteq \Delta$. Then, we say that Γ is a *Hintikka set* if Γ is consistent, $\Gamma \in W$, and

$$T(A \& B) \in \Gamma \Rightarrow TA \in \Gamma \text{ and } TB \in \Gamma,$$
$$F(A \& B) \in \Gamma \Rightarrow FA \in \Gamma \text{ or } FB \in \Gamma,$$
$$T(A \vee B) \in \Gamma \Rightarrow TA \in \Gamma \text{ or } TB \in \Gamma,$$
$$F(A \vee B) \in \Gamma \Rightarrow FA \in \Gamma \text{ and } FB \in \Gamma,$$
$$T(A \rightarrow B) \in \Gamma \Rightarrow FA \in \Gamma \text{ or } TB \in \Gamma,$$
$$F(A \rightarrow B) \in \Gamma \Rightarrow TA \in \Delta \text{ and } FA \in \Delta \text{ for some } \Delta \text{ such}$$
$$\text{that } \Gamma \leq \Delta,$$
$$T(\sim (A \& B)) \in \Gamma \Rightarrow T(\sim A) \in \Gamma \text{ or } T(\sim B) \in \Gamma,$$
$$F(\sim (A \& B)) \in \Gamma \Rightarrow F(\sim A) \in \Gamma \text{ and } F(\sim B) \in \Gamma,$$
$$T(\sim (A \vee B)) \in \Gamma \Rightarrow T(\sim A) \in \Gamma \text{ and } T(\sim B) \in \Gamma,$$
$$F(\sim (A \vee B)) \in \Gamma \Rightarrow F(\sim A) \in \Gamma \text{ or } F(\sim B) \in \Gamma,$$
$$T(\sim (A \rightarrow B)) \in \Gamma \Rightarrow TA \in \Gamma \text{ and } T(\sim B) \in \Gamma,$$
$$F(\sim (A \rightarrow B)) \in \Gamma \Rightarrow FA \in \Gamma \text{ or } F(\sim B) \in \Gamma,$$
$$T(\sim\sim A) \in \Gamma \Rightarrow TA \in \Gamma,$$
$$F(\sim\sim A) \in \Gamma \Rightarrow FA \in \Gamma,$$
$$T(\forall x A(x)) \in \Gamma \Rightarrow T(A(a)) \in \Gamma \text{ for all } a \in D(\Gamma),$$
$$F(\forall x A(x)) \in \Gamma \Rightarrow F(A(a)) \in \Delta \text{ for some } \Delta$$
$$\text{and for some } a \in D(\Delta) \text{ such that } \Gamma \leq \Delta,$$
$$T(\exists x A(x)) \in \Gamma \Rightarrow T(A(a)) \in \Gamma \text{ for some } a \in D(\Gamma),$$
$$F(\exists x A(x)) \in \Gamma \Rightarrow F(A(a)) \in \Gamma \text{ for all } a \in D(\Gamma),$$
$$T(\sim \forall x A(x)) \in \Gamma \Rightarrow T(\sim A(a)) \in \Gamma \text{ for some } a \in D(\Gamma),$$
$$F(\sim \forall x A(x)) \in \Gamma \Rightarrow F(\sim A(a)) \in \Gamma \text{ for all } a \in D(\Gamma),$$
$$T(\sim \exists x A(x)) \in \Gamma \Rightarrow T(\sim A(a)) \in \Gamma \text{ for all } a \in D(\Gamma),$$
$$F(\sim \exists x A(x)) \in \Gamma \Rightarrow F(\sim A(a)) \in \Delta \text{ for some } \Delta \text{ and for}$$
$$\text{some } a \in D(\Delta) \text{ such that } \Gamma \leq \Delta.$$

If W is a Hintikka set, then we call $(W, \leq, w_0, val, D)$ a strong negation model for W if (i) $(W, \leq, w_0, val, D)$ is a strong negation model, (ii) for all $\Gamma \in W$, $TX \in \Gamma \Rightarrow V(\Gamma, X) = 1$ and $FX \in \Gamma \Rightarrow V(\Gamma, X) \neq 1$. It can be easily shown that there is a strong negation model for any Hintikka set.

Next, we define a *Hintikka element* with respect to a set of parameters $\mathbf{P}$ for a consistent set of signed formulas.

$$T(A \& B) \in \Gamma \Rightarrow TA \in \Gamma \text{ and } TB \in \Gamma,$$
$$F(A \vee B) \in \Gamma \Rightarrow FA \in \Gamma \text{ and } TB \in \Gamma,$$
$$T(\sim (A \rightarrow B)) \in \Gamma \Rightarrow TA \in \Gamma \text{ and } T(\sim B) \in \Gamma,$$
$$F(\sim (A \& B)) \in \Gamma \Rightarrow F(\sim A) \in \Gamma \text{ and } F(\sim B) \in \Gamma,$$
$$T(\sim (A \vee B)) \in \Gamma \Rightarrow T(\sim A) \in \Gamma \text{ and } T(\sim B) \in \Gamma,$$
$$T(A \vee B) \in \Gamma \Rightarrow TA \in \Gamma \text{ or } TB \in \Gamma,$$
$$F(A \& B) \in \Gamma \Rightarrow FA \in \Gamma \text{ or } FB \in \Gamma,$$
$$T(\sim (A \& B)) \in \Gamma \Rightarrow T(\sim A) \in \Gamma \text{ or } T(\sim B) \in \Gamma,$$
$$F(\sim (A \vee B)) \in \Gamma \Rightarrow F(\sim A) \in \Gamma \text{ or } F(\sim B) \in \Gamma,$$
$$T(A \rightarrow B) \in \Gamma \Rightarrow FA \in \Gamma \text{ or } TB \in \Gamma,$$
$$T(\forall x A(x)) \in \Gamma \Rightarrow T(A(a)) \in \Gamma \text{ for any } a \in \mathbf{P},$$
$$T(\sim \exists x A(x)) \in \Gamma \Rightarrow T(\sim A(a)) \in \Gamma \text{ for any } a \in \mathbf{P},$$
$$F(\exists x A(x)) \in \Gamma \Rightarrow F(A(a)) \in \Gamma \text{ for any } a \in \mathbf{P},$$
$$T(\exists x A(x)) \in \Gamma \Rightarrow T(A(a)) \in \Gamma \text{ for some } a \in \mathbf{P},$$
$$T(\sim \forall x A(x)) \in \Gamma \Rightarrow T(\sim A(a)) \in \Gamma \text{ for some } a \in \mathbf{P}.$$

Theorem 5

Let Γ be a countable and consistent set of signed formulas and $\mathbf{S}$ be the set of all parameters used in formulas in Γ. Let $a_1, a_2, ...$ be a countable list of parameters not in $\mathbf{S}$ and $\mathbf{P} = \mathbf{S} \cup \{a_1, a_2, ...\}$. Then, Γ can be extended to a Hintikka element with respect to $\mathbf{P}$.

(Proof): We can enumerate the countable set of all subformulas of formulas $X_1, X_2, ...$ in Γ using only parameters of $\mathbf{P}$. We define a sequence of sets of signed formulas. Let $\Gamma_0 = \Gamma$. Assume that Γ_n has been defined as a consistent extension of Γ_0 using only finitely many of $a_1, a_2, a_3, ...$. Let $\Delta_n' = \Gamma_n$. We then define $\Delta_n^2, ..., \Delta_n^{n+1}$ and let $\Gamma_{n+1} = \Delta_n^{n+1}$.

Assume here that we have defined Δ_n^k for some k ($1 \leq k \leq n$). For some formula X_k, one of TX_k or FX_k is in Δ_n^k for the consistency. If neither is, let $\Delta_n^{k+1} = \Delta_n^k$. If one is in Δ_n^k, we have several cases:

Case 1: X_k is $A \vee B$ and $TX_k \in \Delta_n^k$ in $(T\vee)$. Then, one of Δ_n^k, TA or Δ_n^k, TB is inconsistent. Let Δ_n^{k+1} be Δ_n^k, TA if consistent, and Δ_n^k, TB otherwise. Similarly, we can handle the cases: $(T \sim \&), (F \sim \vee), (F \sim \rightarrow), (T \rightarrow)$.

Case 2: X_k is $A \vee B$ and $FX_k \in \Delta_n^k$ in $(F\vee)$. Then, Δ_n^k, FA, FB is consistent. Let this be Δ_n^{k+1}. Similar reasoning can be applied to the following cases: $(T\&), (F \rightarrow), (F \sim \&), (T \sim \vee), (T \sim \rightarrow), (T \sim \sim), (F \sim \sim)$. $\square$

Theorem 6 (Completeness of TN)

If A is valid, then A is provable.

(Proof): Suppose A is not provable. Then, by theorem 5 we can obtain a Hin-

tikka set W such that $\Gamma \in W$ and $FA \in \Gamma$. So the completeness is established.
□

A tableau calculus TN^- for N^- can be obtained from TN by deleting the axiom (AX3). The completeness argument can be adapted without any difficulty.

5. Tableau Based Logic Programming with Strong Negation

Although we gain the completeness result for TN, we have said nothing about the features of TN in relation to logic programming. Indeed TN can deal with the full first-order language, but it needs a computation procedure to achieve an efficient proof of logic programs. Unfortunately, Pearce and Wagner did not focus on such features as unification. We here observe that the problem of finding suitable methods of computing answers to programs with strong negation is an important topic.

A *tableau based logic programming with strong negation* (TLPS) has the language of LPS enriched with $\rightarrow$ (implication), $\forall$ (universal quantifier) and $\exists$ (existential quantifier). A clause is of the form $l \leftarrow \varphi$. Observe here that φ may contain implications and we should distinguish $\rightarrow$ and $\leftarrow$. Our central idea in TLPS is to specify computation by a procedure call of a goal in the tableau in the sense of TN. For this purpose, we add the following rule to LPS:

(call) A tableau branch is closed if it contains the signed formula of a ground literal $F(l(t_1, ..., t_n))$ such that there is a signed formula of the clause $T(l(x_1, ..., x_n) \leftarrow \varphi(x_1, ..., x_n))$ and there is a closed tableau for the signed formula $F(\varphi(t_1, ..., t_n))$.

The rule (call) is in accordance with the basic mechanism of logic programming in which a goal can be seen as a query to the program. We call a tableau augmented with the rule (call) a LP-tableau. Then, a goal A succeeds in the program P if we can construct a closed LP-tableau for FA and A fails otherwise. This formulation dispenses with the essential changes in the computational mechanism (also semantics) of LPS in order to enhance the expressive power of the language.

To motivate TLPS as a logic programming system, we need a discussion of how rules that have eigenvariable restrictions would interact with rules without these restrictions. One way to resolve the problem is to employ the so-called *free variable tableau* due to Hähnle (1993). In the modified version, quantifier rules can be written as follows:

$$\frac{S, T(\forall x A(x))}{S, T(A(x))} \ (T\forall') \qquad \frac{S, F(\forall x A(x))}{S_T, F(A(f(x_1, ..., x_n)))^*} \ (F\forall')$$

$$\frac{S, T(\exists x A(x))}{S, T(A(f(x_1, ..., x_n)))^*} \ (T\exists') \qquad \frac{S, F(\exists x A(x))}{S, F(A(x))} \ (F\exists')$$

$$\frac{S, T(\sim \forall x A(x))}{S, T(\sim A(f(x_1, ..., x_n)))^*} \ (T \sim \forall') \qquad \frac{S, F(\sim \forall x A(x))}{S, F(\sim A(x))} \ (F \sim \forall')$$

$$\frac{S, T(\sim \exists x A(x))}{S, T(\sim A(x))} \ (T \sim \exists') \qquad \frac{S, F(\sim \exists x A(x))}{S_T, F(\sim A(f(x_1, ..., x_n)))^*} \ (F \sim \exists')$$

Here, in $(F\forall'), (T\exists'), (T \sim \forall'), (F \sim \exists'), x_1, ..., x_n$ are the free variables occurring on the current branch and f a new function symbol. A free variable tableau enables us to seek easily a suitable term via unification by delaying substitution.

Another improvement of TLPS is to avoid the duplicating formulas in tableau rules. Obviously, duplications of formulas increase the size of proofs and induce a non-deterministic search. Miglioli, Moscato and Ornaghi (1994) introduced a new sign F_C, together with T and F, into the tableau calculus for intuitionistic logic to overcome the issue of duplications of formulas. It is not difficult to extend TLPS with new rules.

Logic programming can be regarded as an interesting representation language. Gelfond and Lifschitz's (1990) *answer set semantics* provides a non-monotonic semantics for logic programs with two kinds of negation and clarifies the connections with existing non-monotonic formalisms like default and autoepistemic logics. There are some connections between constructive logics with strong negation and answer set semantics for extended logic programs due to Gelfond and Lifschitz (1990) and Kowalski and Sadri (1990). *Extended logic programs* have two kinds of negation, i.e. one for negation as failure "*not*" and the other for "classical" negation "$\sim$". Unfortunately, "classical" negation is not classical negation in classical logic, but strong negation in Nelson's constructive logics.

An extended logic program is viewed as a set of rules of the from:
$$L_0 \leftarrow L_1, ..., L_m, not \ L_{m+1}, ..., not \ L_n$$
where L_i is a literal (atom or strongly negated atom) and "*not*" is negation as failure. Let Π be an extended logic program. For any set of ground literals $S \subset Lit$ (where Lit is a set of literals), the program Π^S is the program without "*not*" obtainable from Π by deleting:

(i)   each rule containing $not \ L$ in its body with $L \in S$, and

(ii)   all formulas of the form $not \ L$ in the bodies of the remaining rules. Then, the transformed program does not contain "*not*". An *answer set* of Π is a minimal subset S of Lit such that:

(A1)   for each rule $L_0 \leftarrow L_1, ..., L_n$ of Π, if $L_1, ..., L_n \in S$ then $S_0 \in S$,

(A2)   If S contains a pair of complementary literals of the from

$A, \sim A$ (where A is an atom), then $S = Lit$.

The answer sets of Π can be interpreted as possible sets of beliefs that a rational agent may hold on the basis of the information expressed by the rules of Π. In Kowalski and Sadri (1990), the concept of extended answer set is defined as Gelfond and Lifschitz's definition without (A2). We say that an answer set S is Π is called *consistent* if it dose not contain a complementary pair of literals.

Unfortunately, answer sets are defined without explicit reference to any underlying logical system. But, if we identify a rule in extended programs with the corresponding first-order formula:

$$L_1 \& ... \& L_m \rightarrow L_0$$

then one can easily find that classical logic is not sound in answer sets. For instance, contraposition does not hold in the answer set semantics. Pearce (1993) established that constructive logics with strong negation can serve as the underlying logical basis of answer set semantics. In this regard, tableau calculi presented above can serve as an efficient proof method for the monotonic part of extended logic programming without "*not*".

Example

All birds can fly.

Tweety is a bird.

So, Tweety can fly.

The first two sentences can translate into the following extended program:

$fly(x) \leftarrow bird(x), not \sim fly(x)$

$\leftarrow bird(tweety)$

The conclusion is written as a goal:

$\leftarrow fly(tweety)$

After the Gelfond-Lifschitz transformation, we can interpret the inference as the following computation in $TLPS$:

(1) $T((fly(x) \leftarrow bird(x))\&bird(tweety))$

(2) $F(fly(tweety))$

(3) $T(bird(tweety))$ $(T\&)$

(4) $T(fly(x) \leftarrow bird(x))$

(5) $F(bird(tweety))$ $(call)$

(6) $false$ $((2), (5))$

Thus, the above goal succeeds. As you saw, this computation needs preprocessing as proposed by Gelfond and Lifschitz.

6. Conclusions

We have proposed a tableau calculus TN for Nelson's constructive logics with strong negation and expanded it with a procedure call as $TLPS$ to serve as a proof theory for logic programming with strong negation. We have also addressed computational issues related to unification and duplication of formulas. Our approach is a first step towards logic programming as the full constructive logics.

There are interesting avenues of further research. First, it would be worthwhile to focus attention on suitable fragments of the tableau calculus, suitable for the LP syntax, trying to make these efficient. Note here that N can be regarded as a monotonic base logic for systems of extended logic programming like that of Gelfond and Lifschitz as the above example indicates. Secondly, the tableau calculi can be used as a proof method for non-monotonic logics. In fact, there may be some connections of constructive logics and non-monotonic formalisms (cf. Pearce (1992, 1993)). It should be also pointed out that the Gelfond-Lifschitz approach can be based on extensions of N capable of representing non-monotonic reasoning (cf. Wagner (1992)). Third, it would be intriguing to extend LPS to higher-order constructive logics. One attractive way is to establish it by introducing Comprehension Scheme in the sense of naive set theory as described in Akama (1996).

References

Akama, S. (1987): Resolution in constructivism, *Logique et Analyse 120*, 385-399.

Akama, S. (1988a): On the proof method for constructive falsity, *Zeitschrift für mathematische Logik und Grundlagen der Mathematik 34*, 385-392.

Akama, S. (1988b): Constructive predicate logic with strong negation and model theory, *Notre Dame Journal of Formal Logic 29*, 18-27.

Akama, S. (1989): *Constructive Falsity: Foundations and Their Applications to Computer Science*, Ph.D. dissertation, Department of Administration Engineering, Keio University, Yokohama, Japan.

Akama, S. (1990): Subformula semantics for strong negation systems, *The Journal of Philosophical Logic 19*, 217-226.

Akama, S. (1995): Three-valued constructive logic and logic programs, *Proc. of the IEEE 25th International Symposium on Multiple-Valued Logic*, 276-281, Bloomington, USA, May.

Akama, S. (1996): Curry's paradox in contractionless constructive logic, *The Journal of Philosophical Logic 25*, 135-150.

Alferes, J. J. and Pereira, L. M. (1996): *Reasoning with Logic Programming*, LNAI 1111, Springer, Berlin.

Almukdad, A. and Nelson, D. (1984): Constructible falsity and inexact predicates, *The Journal of Symbolic Logic 49*, 8-37, 231-233.

Clark, K. L. (1978): Negation as failure, H. Gallaire and J. Minker (eds.), *Logic and Databases*, 293-322, Plenum Press, New York.

Fitting, M. (1969): *Intuitionistic Logic, Model Theory and Forcing*, North-Holland, Amsterdam.

Gelfond, M. and Lifschitz, V. (1990): Logic programs with classical negation, D. H. D. Warren and P. Szeredi (eds.), *Proc. of ICLP'90*, 579-597, MIT Press, Cambridge, Mass.

Hähnle, R. (1993): *Automated Deduction in Multiple-Valued Logics*, Oxford University Press, Oxford.

Kowalski, R. and Sadri, F. (1990): Logic programs with exceptions, D. H. D. Warren and P. Szeredi (eds.), *Proc. of ICLP'90*, 598-613, MIT Press, Cambridge, Mass.

Miglioli, P., Moscato, U. and Ornaghi, M. (1994): An improved refutation system for intuitionistic predicate logic, *Journal of Automated Reasoning 13*, 361-373.

Nelson, D. (1949): Constructible falsity, *The Journal of Symbolic Logic 14*, 16-26.

Pearce, D. (1992): Default logic and constructive logic, *Proc. of ECAI'92*, 309-313.

Pearce, D. (1993): Answer sets and constructive logic Part II: Extended logic programs and related nonmonotonic formalisms, L. M. Pereira and A. Nerode (eds.), *Logic Programming and Non-Monotonic Reasoning, Proc. of the 2nd International Workshop*, 457-475, MIT Press, Cambridge, Mass.

Pearce, D. and Wagner, G. (1990): Reasoning with negative information I: Strong negation in logic programs, *Acta Philosophica Fennica 49*, 430-453.

Pearce, D. and Wagner, G. (1991): Logic programming with strong negation, P. Schroeder-Heister (ed.), *Extensions in Logic Programming*, 311-326, Springer, Berlin.

Rautenberg, W. (1979): *Klassische and Nichtklassische Aussagenlogik*, Vieweg, Wiesbaden.

Schöenfeld, W. (1985): PROLOG extensions based on tableau calculus, *Proc. of IJCAI'85*, 730-732.

Smullyan, R. (1968): *First-Order Logic*, Springer, Berlin.

Thomason, R. H. (1969): A semantical study of constructible falsity, *Zeitschrift für mathematische Logik und Grundlagen der Mathematik 15*, 247-257.

Wagner, G. (1991): Logic programming with strong negation and inexact predicates, *Journal of Logic and Computation 1*, 835-859.

Wagner, G. (1992): *Vivid Logic: Knoweldge-Based Reasoning with Two Kinds of Negation*, Ph.D. dissertation, Freien Universität Berlin.

Generalized Tableau Systems for Intermediate Propositional Logics

ALESSANDRO AVELLONE   PIERANGELO MIGLIOLI
UGO MOSCATO   MARIO ORNAGHI

Dipartimento di Scienze dell'Informazione – Università degli Studi di Milano
via Comelico, 39/41, 20135 MILANO–ITALY
{avellone,miglioli,moscato,ornaghi}@dsi.unimi.it

Abstract. Given an intermediate propositional logic L (obtained by adding to intuitionistic logic INT a single axiom–scheme), a pseudo tableau system for L can be given starting from any intuitionistic tableau system and adding a rule which allows to insert in any line of a proof table suitable **T**–signed instances of the axiom-scheme. In this paper we study some sufficient conditions from which, given a well formed formula H, the search for these instances can be restricted to a suitable finite set of formulae related to H. We illustrate our techniques by means of some known logics, namely, the logic D of Dummett, the logics PR_k ($k \geq 1$) of Nagata, the logics FIN_m ($m \geq 1$), the logics G_n ($n \geq 1$) of Gabbay and de Jongh, and the logic KP of Kreisel and Putnam

Keywords: intermediate propositional logics, tableau systems, duplications, filtration techniques.

1 Introduction

In this paper we are interested in finding complete and correct tableau systems for intermediate propositional logics, i.e. logics including intuitionistic logic and included in classical logic. Problems of this kind, involving aspects exceeding the ordinary intuitionistic machinery, can be tackled by looking for appropriate tableau rules for single logics, as we did for the intermediate predicate logic of Kuroda [13, 14, 16] and for some propositional modal logics [15]. Here we want to outline a different perspective, i.e. a method to single out a wide class of propositional intermediate logics for which a reasonable tableau system can be given, even if the method will be illustrated only by a few examples. To do this, first of all we define the notion of *generalized tableau system*. One can observe that, given any propositional intermediate logic L whose specific axiomatization is provided by the axiom–scheme generated by a formula A, a tableau system for L can be defined starting from a tableau system for intuitionistic logic and adding a rule allowing to insert, in any line of a tableau, suitable T–signed instances of A. It is clear that such tableau systems are correct and complete for any L. But this characterization of tableau systems can be considered unsatisfactory, because the search for the instances of the axiom–scheme can be very unfeasible. So we want to give some constraints for such a search. In this sense, we present logics L for which, given a formula H, the search for the instances of the axiom–scheme of L (in order to construct a tableau for H) can be restricted to instances built up only using an appropriate finite set of formulae related to H. We recall that a problem of this kind has been analyzed also in [20], where a deep result is stated for a family of logics there called finite logics. Such logics, however, do not cover any of the logics treated in the present paper; moreover, the proofs given in [20] do not use semantical methods in an essential way, while, on the other hand, these methods are one of the main concerns of our paper (which intends also to illustrate a new application of the filtration techniques); finally, the results of [20] are not inserted in a context of tableau calculi.

We give different kinds of restrictions of the set of instances of the extraintuitionistic axiom–scheme, depending on the considered logic. The proofs of the completeness of such restrictions are provided by a suitable use of some known filtration techniques [2, 3, 4, 6, 7, 8, 12]. In this frame, a much larger family of logics could be treated along the same lines: the choice of the logics presented in the paper has only an illustrative purpose, and has been inspired by the fact that such logics are very well known in literature.

The paper is so organized.

In the next section we recall the basic notions, among which is our tableau system for propositional intuitionistic logic [13, 14, 16], and we explain how to use the latter (in a very nondeterministic way) in order to prove formulae of propositional intermediate logics; this provides what we call *pseudo tableaux*. Next we consider the problem of lowering the non determinism involved in the pseudo tableaux, and discuss the notion of *generalized tableau system*: this is intended as a pseudo tableau system enriched by a rule specifying how to restrict the set of the instances of the extraintuitionistic axiom–scheme to be inserted in the proofs, without affecting the completeness of the system; such a rule provides a finite set of instances depending on the formula to be proved and on the logic in hand. Indeed, we define four different kinds of restriction: the first restriction works for Dummett logic D [17] and Nagata logics PR_k ($k \geq 1$) [18]; the second restriction handles the logics FIN_k ($k \geq 1$) [2, 3]; the third restriction is appropriate for the logics G_n ($n \geq 1$) of Gabbay and de Jongh [8]; the fourth restriction is for the logic KP of Kreisel and Putnam [6, 10]. In Section 3 we introduce two filtration techniques and explain the results involving their use in connection with our generalized tableaux (which require additional care with respect to the usual applications of these techniques).

In Section 4 we apply the techniques of Section 3 and prove results that, in Section 5, immediately provide the completeness of the various generalized tableau systems defined in the paper. Finally, in Section 5, we state the completeness theorems for our generalized tableau systems and compare the various kinds of restrictions considered in the paper. For instance, it turns out that the restrictions introduced for D (which considerably lower the complexity of the proofs of the involved generalized tableau system as compared to the complexity of the proofs of the related pseudo tableau system) are too strong for KP: in order that the completeness of a system for KP be guaranteed, one has to allow the introduction of a larger class of instances of the axiom–scheme in the tableaux.

2 Basic definitions and generalized tableau calculi

The set of propositional *well formed formulae* (wff) is defined as usual, starting from an enumerable set of propositional variables and using the connectives ¬, ∧, ∨, →. We say that a wff A is *negated* iff $A = \neg B$ for some wff B.

A *substitution* will be any function σ associating a wff with every propositional variable. To denote the result of the application of the substitution σ to the wff A, we will write $\sigma(A)$ or, more simply, σA.

INT (respectively, CL) will denote both an arbitrary calculus for intuitionistic propositional logic (respectively, for classical propositional logic) and the set of intuitionistically valid wffs (respectively, the set of classically valid wffs).

As usual, an *intermediate propositional logic* [2, 3, 4, 12, 19] will be any set L of wffs satisfying the following conditions:

1. L is consistent;
2. $INT \subseteq L$;
3. L is closed under detachment;
4. L is closed under arbitrary substitutions.

In this paper the term "logic" will mean an intermediate propositional logic. As is well known, for any logic L we have $INT \subseteq L \subseteq CL$. Even if logics may be non–effective, we will be interested only in the ones which can be seen as sets of theorems of deductive systems with recursive sets of inference rules; this implies their recursive enumerability. In this line, if $\mathcal{A}$ is a set of axiom–schemes and L is a logic, the notation $L + \mathcal{A}$ will indicate both the deductive system closed under detachment and arbitrary substitutions obtained by adding to L the axiom–schemes of $\mathcal{A}$, and the set of theorems (which is a logic) of such a deductive system; of course, if L is recursively enumerable and $\mathcal{A}$ is a recursive set of axiom–schemes then $L + \mathcal{A}$ is recursively enumerable, while nothing can be said, in general, about the recursiveness of $L + \mathcal{A}$, even if L is recursive and $\mathcal{A}$ consists of a single axiom–scheme. If $\mathcal{A} = \{(A)\}$ consists of a single axiom–scheme (A) generated by a formula A, the notation $L + (A)$ will replace $L + \{(A)\}$.

If H is any wff, we will denote with $Sf(H)$ the set of subformulae of H and with $\mathcal{V}(H)$ the set of all propositional variables contained in H. Moreover, with $RSf(H)$ we will indicate the set $\{A \mid A$ is a propositional variable contained in H, or $A = K \rightarrow Q \in Sf(H)$, or $A = \neg K \in Sf(H)\}$.

As usual, if Γ is any set of wffs and L is a logic, we say that a wff A is L–*provable from* Γ, and we denote it by $\Gamma \vdash_L A$, iff there are $B_1, \ldots, B_n$ such that $\{B_1, \ldots, B_n\} \subseteq \Gamma$ and $B_1 \wedge \ldots \wedge B_n \rightarrow A \in L$. On the other hand, $\Gamma \not\vdash_L A$ will mean that $\Gamma \vdash_L A$ does not hold. Note that such a definition of L–provability applies also when no effective calculus is given (or can be given) to generate all the formulae of L. On the other hand, with every logic L treated in the paper, one can associate a calculus L–$INTNAT$ obtained by adding to intuitionistic natural calculus

INTNAT a zero–premises rule allowing to introduce, in any step of any proof, an arbitrary instance of the (single) axiom–scheme characterizing L (where, of course, such an instance is not taken as an undischarged assumption): one can easily see that $\Gamma \vdash_L A$ iff there is a proof in *L-INTNAT* whose undischarged assumptions belong to Γ.

We assume the reader to be familiar with the notion of *Kripke model* $\mathbf{K} = \langle P, \leq, \Vdash \rangle$, where $\mathbf{P} = \langle P, \leq \rangle$ is a *poset* or *frame* and $\Vdash$ is the *forcing relation*, defined between elements of P and atomic formulae, and extended in the usual way to arbitrary wffs; we say that $\mathbf{K}$ is *built on the poset* $\mathbf{P}$, or that $\mathbf{P}$ is the *underlying poset of* $\mathbf{K}$. If $\mathbf{P} = \langle P, \leq \rangle$ and $\alpha, \beta \in P$, then $\alpha < \beta$ will indicate the fact that $\alpha \leq \beta$ and $\alpha \neq \beta$. Moreover, we call *root of* $\mathbf{P}$ an element r (if it exists) such that, for every $\alpha \in P$, $r \leq \alpha$.

Given a Kripke frame $\mathbf{P} = \langle P, \leq \rangle$ and $\alpha, \beta \in P$, β is said to be an *immediate successor of* α in $\mathbf{P}$ iff $\alpha < \beta$ and, for all $\gamma \in P$ such that $\alpha \leq \gamma \leq \beta$, $\gamma = \alpha$ or $\gamma = \beta$. If $\mathbf{P} = \langle P, \leq \rangle$ and $\alpha \in P$, then α will be called *final in* $\mathbf{P}$ or, more simply, *final* iff, for every $\beta \in P$, $\alpha \leq \beta$ implies $\alpha = \beta$; moreover, for any $\gamma \in P$, $Fin(\gamma)$ will denote the set $\{\eta \mid \gamma \leq \eta$ and η is final in $\mathbf{P}\}$. Finally, if $\mathbf{P} = \langle P, \leq \rangle$, $\alpha \in P$ and $h \geq 1$, we say that α *has at least depth* h *in* $\mathbf{P}$ iff there exist $\beta_1, \ldots, \beta_h \in P$ such that $\beta_1 = \alpha$, β_h is a final element, and, for all $2 \leq i \leq h$, β_i is an immediate successor of β_{i-1}; we say that α *has at most depth* h *in* $\mathbf{P}$ iff it is not the case that α has at least depth $h + 1$ in $\mathbf{P}$; we say that α *has depth* h *in* $\mathbf{P}$ iff α has at least depth h in $\mathbf{P}$ and α has at most depth h in $\mathbf{P}$.

If $\mathcal{F}$ is a non empty class of posets, the set $\{\mathbf{K} = \langle P, \leq, \Vdash \rangle \mid \langle P, \leq \rangle \in \mathcal{F}\}$ will be denoted by $\mathcal{K}(\mathcal{F})$, and we will call it the *class of Kripke models built on* $\mathcal{F}$. Also, for every non empty class $\mathcal{F}$ of frames, $\mathcal{L}(\mathcal{F})$ will indicate the set of wffs $\{A \mid$ for every $\mathbf{K} = \langle P, \leq, \Vdash \rangle \in \mathcal{K}(\mathcal{F})$ and for every $\alpha \in P$, $\alpha \Vdash A\}$.

It is well known that, for every non empty class $\mathcal{F}$ of frames, $\mathcal{L}(\mathcal{F})$ is a logic [2, 3, 4, 12].

In the following we will consider only logics with a single extraintuitionistic axiom–scheme (this, of course, is equivalent to using a finite set of axiom–schemes). An axiom–scheme will be represented by a notation such as (A): here A is called a *generating formula* for (A), and (A) is the set of all the wffs of the form σA, where σ is any substitution. Any element of (A) will be called an *instance* of A.

The following fact is an immediate consequence of the definitions of intermediate logic and instance of an axiom–scheme:

Proposition 1 *Let* $L = INT + (A)$ *be any logic and let* H *be any wff; then* $H \in L$ *iff there is a finite set of instances* $A_1, \ldots, A_n$ *of* A *such that* $A_1 \wedge \ldots \wedge A_n \to H \in INT$.

The content of the above proposition can be immediately inserted in the context of our intuitionistic tableau system, giving rise to what we call the *pseudo tableau system* for the logic L. To do so, first of all we recall the machinery of the main tableau system $CINT$ of [13] (see also [16], where a variant of $CINT$ is considered).

The calculus $CINT$ works on *signed formula(e)* (s.f.) of the kind $\mathbf{T}A$, $\mathbf{F}A$, $\mathbf{F_c}A$, with A any wff. The meaning of the signs $\mathbf{T}$, $\mathbf{F}$, $\mathbf{F_c}$ in terms of their semantics with Kripke models $\mathbf{K} = \langle P, \leq, |\!\!\!-\rangle$ is the following:
let $S = \mathbf{T}A_1, \cdots, \mathbf{T}A_n, \mathbf{F}B_1, \ldots \mathbf{F}B_m, \mathbf{F_c}C_1, \ldots, \mathbf{F_c}C_k$ be a set of s.f.; we say that S is *realized* in $\alpha \in P$ iff:

a) for every h, $1 \leq h \leq n$, $\alpha |\!\!\!- A_h$;
b) for every i, $1 \leq i \leq m$, $\alpha |\!\!\!\not|\!\!- B_i$;
c) for every j, $1 \leq j \leq k$, $\alpha |\!\!\!- \neg C_j$.

By a *configuration* we mean a sequence $S_1/S_2/\ldots/S_j/\ldots$, where every S_j is a set of s.f..

The rules of $CINT$ are given in TABLE 1 and TABLE 2.

$$\frac{S, \mathbf{T}(A \wedge B)}{S, \mathbf{T}A, \mathbf{T}B}\mathbf{T}\wedge \qquad \frac{S, \mathbf{F}(A \wedge B)}{S, \mathbf{F}A/S, \mathbf{F}B}\mathbf{F}\wedge \qquad \frac{S, \mathbf{F_c}(A \wedge B)}{S_c, \mathbf{F_c}A/S_c, \mathbf{F_c}B}\mathbf{F_c}\wedge$$

$$\frac{S, \mathbf{T}(A \vee B)}{S, \mathbf{T}A/S, \mathbf{T}B}\mathbf{T}\vee \qquad \frac{S, \mathbf{F}(A \vee B)}{S, \mathbf{F}A, \mathbf{F}B}\mathbf{F}\vee \qquad \frac{S, \mathbf{F_c}(A \vee B)}{S, \mathbf{F_c}A, \mathbf{F_c}B}\mathbf{F_c}\vee$$

See TABLE 2 $\qquad \dfrac{S, \mathbf{F}(A \rightarrow B)}{S_c, \mathbf{T}A, \mathbf{F}B}\mathbf{F}\rightarrow \qquad \dfrac{S, \mathbf{F_c}(A \rightarrow B)}{S_c, \mathbf{T}A, \mathbf{F_c}B}\mathbf{F_c}\rightarrow$

$$\frac{S, \mathbf{T}\neg A}{S, \mathbf{F_c}A}\mathbf{T}\neg \qquad \frac{S, \mathbf{F}\neg A}{S_c, \mathbf{T}A}\mathbf{F}\neg \qquad \frac{S, \mathbf{F_c}\neg A}{S_c, \mathbf{T}A}\mathbf{F_c}\neg$$

TABLE 1: TABLEAU CALCULUS FOR INTUITIONISTIC PROPOSITIONAL LOGIC

$$\frac{S, \mathbf{T}(A \rightarrow B)}{S, \mathbf{F}A/S, \mathbf{T}B}\mathbf{T}\rightarrow_{AN} \quad \text{with } A \text{ atomic or negated.}$$

$$\frac{S, \mathbf{T}((A \wedge B) \rightarrow C)}{S, \mathbf{T}(A \rightarrow (B \rightarrow C))}\mathbf{T}\rightarrow\wedge \qquad \frac{S, \mathbf{T}((A \vee B) \rightarrow C)}{S, \mathbf{T}(A \rightarrow C), \mathbf{T}(B \rightarrow C)}\mathbf{T}\rightarrow\vee$$

$$\frac{S, \mathbf{T}((A \rightarrow B) \rightarrow C)}{S, \mathbf{F}(A \rightarrow B), \mathbf{T}(B \rightarrow C)/S, \mathbf{T}C}\mathbf{T}\rightarrow\rightarrow$$

TABLE 2: $T \rightarrow$ RULES

Now, some explanations are in order.

- Every rule of TABLE 1 and TABLE 2 is applied to a formula of a set S_i occurring in a configuration $S_1/\ldots/S_i/\ldots$; e.g., the notation $S, \mathbf{T}A \wedge B$ points out that the rule $\mathbf{T}\wedge$ is applied to the formula $\mathbf{T}A \wedge B$ of the set $S \cup \{\mathbf{T}A \wedge B\}$, where S is possibly empty.

- The rules $\mathbf{F} \to$, $\mathbf{F} \neg$, $\mathbf{F_c} \wedge$, $\mathbf{F_c} \to$, and $\mathbf{F_c} \neg$ narrow S to S_c, where S_c is obtained from S by deleting all the $\mathbf{F}$ s.f. (in other words, S_c is the certain part of S).
- We say that a rule gives rise to *duplications* if, in some application of the rule, the formula to which it has been applied must be rewritten after the application, to be used in further applications. In this sense, all the rules of the tableau calculus of [5] give rise to duplications. On the contrary, all the rules of TABLE 1 and TABLE 2 do not give rise to any duplication (for these rules we forbid duplications, even if the use of duplications is semantically sound). We recall that the problem of the elimination of duplications (there called *contractions*) has been raised also in the frame of natural and sequent calculi, where it has been independently solved by Dyckhoff [1] and Hudelmaier [9]. For a more comprehensive discussion on duplication in the frame of tableau calculi, see [13, 14] and especially [16].
- A *proof-table* in $CINT$ is a sequence of applications of the rules of TABLE 1 and TABLE 2 starting from some configuration; a proof-table is *closed* iff all the sets S_i of the final configuration are contradictory, i.e., for every i, there is some wff A such that S_i contains $\mathbf{T}A$ and $\mathbf{F}A$ or $\mathbf{T}A$ and $\mathbf{F_c}A$. A proof of a formula B in $CINT$ is a *closed proof-table* in $CINT$ starting with $\mathbf{F}B$.

Now, given a logic $L = INT + (A)$, the *pseudo tableau calculus PCL for L* is obtained by adding to $CINT$ the following rule:

$$\frac{S}{S, \mathbf{T}B_1, \ldots, \mathbf{T}B_n}(A)\text{–rule}$$

where $n \geq 1$ and $\mathbf{T}B_i$ $(1 \leq i \leq n)$ is any instance of A.

The notions of proof–table and of proof in PCL are defined, *mutatis mutandis*, as for $CINT$. We will call *pseudo tableau proof in PCL* any proof in PCL. From Proposition 1 and the results of [13] we immediately get the following result:

Theorem 1 *$H \in L$ iff there is a pseudo tableau proof in PCL of H.*

Theorem 1 is the desired translation of Proposition 1 in the frame of the pseudo tableau calculus PCL. Of course, the (A)–rule is quite non deterministic. Since we cannot foresee how many applications (and in which configurations) of such rule are needed to prove a formula H, the result stated by the theorem is rather unfeasible. To make reasonable the search for a proof of a formula H in the frame of the above calculus, we look for the possibility of fixing a well defined (and as narrow as possible) finite set I of instances of A with the following properties: I depends on the formula H to be proved; if H is provable, then there is a pseudo tableau proof in PCL of H where all the applications of the (A)–rule introduce only formulae of the form $\mathbf{T}B$ with $B \in I$; moreover, any such a $\mathbf{T}B$ is introduced only once in the pseudo tableau proof. This is what we will precisely do for a family of logics we will introduce in Section 4. Here we explain the kind of restrictions on the (A)–rule we will use for these logics.

Given a logic $L = INT + (A)$, we call *L–selection function* any function Σ_L so defined:

1. the domain of Σ_L is the set of all the wffs;
2. for every formula H, $\Sigma_L(H)$ is a finite subset of (A).

Let Σ_L be a L–selection function for the logic $L = INT + (A)$; then the $\Sigma_L(H)$–rule is the following restriction of the (A)–rule:

$$\frac{S}{S, \mathbf{TB}_1, \ldots, \mathbf{TB}_n} \Sigma_L(H)\text{–rule}$$

where $\Sigma_L(H) \supseteq \{B_1, ..., B_n\}$.

Given a L–selection function Σ_L, the $\Sigma_L(H)$–*generalized tableau calculus for* L, we call $\Sigma_L(H)$–GCL, is obtained by adding the $\Sigma_L(H)$–rule to $CINT$. The notions of configuration, proof–table, etc., for $\Sigma_L(H)$–GCL are defined in the obvious way, where any proof– table of the calculus must start with the configuration $\mathbf{F}H$. We call $\Sigma_L(H)$–*regular generalized tableau calculus for* L (denoted by $\Sigma_L(H)$–$RGCL$) the restriction of $\Sigma_L(H)$–GCL limiting the applications of the $\Sigma_L(H)$–rule according to the following prescription:

the $\Sigma_L(H)$–rule can be applied only once, at the beginning of a proof–table (starting with $\mathbf{F}H$), with S replaced by $\mathbf{F}H$ and $\{\mathbf{TB}_1, \ldots, \mathbf{TB}_n\}$ exactly coinciding with $\Sigma_L(H)$..

The main difference between $\Sigma_L(H)$–GCL and $\Sigma_L(H)$–$RGCL$ is that the latter eliminates any possibility of additional non determinism related to the extraintuitionistic parts of L: the only non determinism involved in a proof comes from the rules of $CINT$ (let us point out again that the rules of $CINT$ are duplication free). Notice, however, that in general not all the formulae of $\Sigma_L(H)$ need to be used to get a proof of H; thus the "more deterministic" calculus $\Sigma_L(H)$–$RGCL$ may give rise to extremely redundant and large proofs, which in $\Sigma_L(H)$–GCL can be avoided.

According to Theorem 1, given any logic $L = INT + (A)$ and any L–selection function Σ_L, the calculus $\Sigma_L(H)$–$RGCL$ (and the calculus $\Sigma_L(H)$–GCL) are sound, that is, if one can build a closed proof–table in the calculus starting from $\mathbf{F}H$ then $H \in L$. The problem is to single out (if any) L–selection functions Σ_L guaranteeing the completeness of $\Sigma_L(H)$–$RGCL$.

In the following points we define three kinds of L–selection functions which will give rise to complete $\Sigma_L(H)$–$RGCL$'s for a family of logics explained in Section 4.

Given any wff H and any logic $L = INT + \{(A)\}$, with $IstR_L(H)$ we will denote the set of all the formulae obtained by correctly instantiating the formula A generating the axiom–scheme (A) by the formulae of $RSf(H)$ (in other words, if A is a generating formula of (A) then $IstR_L(H)$ is the set of all the formulae obtained from A by means of substitutions replacing the variables of A with elements of $RSf(H)$). It is easy to see that $IstR_L(H)$ is a finite set.

Moreover, let $v = \{p_1, \ldots, p_n\}$ $(n \geq 1)$ be any finite non–empty set of propositional variables (in some prefixed order); then: by a v–*formula* we mean any wff containing only variables of v; by an *elementary* v–*formula* we mean any formula of the form $\bar{p}_1 \wedge \ldots \wedge \bar{p}_n$, where, for every $1 \leq i \leq n$, $\bar{p}_i$ is either p_i or $\neg p_i$; by an *elementary negated v–formula* we mean any formula of the form $\neg\neg H$, where H is an elementary v–formula; by a *representing negated v–formula* we mean either the formula $\neg\neg(p_1 \wedge \neg p_1)$ or any formula of the form $\neg\neg(H_1 \vee \ldots \vee H_j)$, with $j \geq 1$ and $H_1, \ldots, H_j$ distinct elementary v–formulae (where, to avoid repetitions of equivalent formulae, we assume that the elementary formulae are all distinct and taken according to some prefixed order). Of course, the set of the elementary negated v–formulae and the set of the representing negated v–formulae are finite; we denote the first set by $\mathcal{E}neg_v$ and the second set by $Rneg_v$. Since any negated formula belongs to INT iff it belongs to CL, it is not difficult to state the following fact [2, 3, 4]:

for every negated v–formula A there is $B \in Rneg_v$ such that $(A \rightarrow B) \wedge (B \rightarrow A) \in INT$.

In the following $\mathcal{E}neg(H)$ and $Rneg(H)$ will denote respectively $\mathcal{E}neg_{V(H)}$ and $Rneg_{V(H)}$.

With the notation $Ist_L^{\mathcal{E}neg}(H)$ we will indicate all the formulae obtained by correctly instantiating the formula A (generating the axiom–scheme (A)) by the formulae of $\mathcal{E}neg(H)$.

Finally, with $Ist_L^{\wedge\vee}(H)$ we will indicate all the wff obtained by correctly instantiating A by the formulae of $RSf_{\wedge\vee}(H)$, where $RSf_\wedge(H) = \{H_1 \wedge \ldots \wedge H_i \mid i \geq 1, H_1 \in RSf(H), \ldots, H_i \in RSf(H)$, and $H_1, \ldots, H_i$ are all distinct$\}$ and $RSf_{\wedge\vee}(H) = \{Z_1 \vee \ldots \vee Z_j \mid j \geq 1, Z_1 \in RSf_\wedge, \ldots, Z_j \in RSf_\wedge$, and $Z_1, \ldots, Z_j$ are all distinct$\}$.

Now, for any logic $L = INT + (A)$, we define the three following L–selection functions:

- Σ_L^1 is the function associating, with every wff H, the finite set of formulae $IstR_L(H)$;
- Σ_L^2 is the function associating, with every wff H, the finite set of formulae $Ist_L^{\mathcal{E}neg}(H)$;
- Σ_L^3 is the function associating, with every wff H, the finite set of formulae $Ist_L^{\wedge\vee}(H)$.

3 Filtration techniques

Let L be any logic; we introduce the following notions:

- A *L–saturated set* Γ is any consistent set of wffs closed under L–provability (i.e., $\Gamma \vdash_L A$ implies $A \in \Gamma$) and under the disjunction property (i.e., $A \vee B \in \Gamma$ implies $A \in \Gamma$ or $B \in \Gamma$).
- If Γ is a L–saturated set, by the *canonical model generated by* Γ, denoted by $\mathcal{C}_L(\Gamma)$, we mean the Kripke model $\mathbf{K} = \langle P, \leq, r, \Vdash \rangle$ satisfying the following properties:
  1. $P = \{\Gamma' \mid \Gamma \subseteq \Gamma'$ and Γ' is L–saturated$\}$;

2. for any two $\Gamma', \Gamma'' \in P$, $\Gamma' \leq \Gamma''$ iff $\Gamma' \subseteq \Gamma''$;

3. $r = \Gamma$;

4. for any $\Gamma' \in P$ and any propositional variable p, $\Gamma' \Vdash p$ iff $p \in \Gamma'$.

The two following propositions are well known [2, 3, 4, 12, 21]:

Proposition 2 *If L is a logic, A is any wff, and Δ is any set of wffs such that $\Delta \nvdash_L A$, then there exists a L-saturated set Γ such that $\Delta \subseteq \Gamma$ and $A \notin \Gamma$.*

Proposition 3 *If L is any logic, Γ is any L-saturated set, Γ' is any element of $C_L(\Gamma)$, and B is any wff, then $\Gamma' \Vdash B$ holds in $C_L(\Gamma)$ iff $B \in \Gamma'$.*

Following [2, 3, 4, 6, 7, 12, 17], given $\mathbf{K} = \langle P, \leq, \Vdash \rangle$ and $\alpha, \beta \in P$, we set $\alpha \subseteq_H \beta$ iff, for every $A \in RSf(H)$, if $\alpha \Vdash A$ then $\beta \Vdash A$. Moreover, $\alpha \equiv_H \beta$ iff $\alpha \subseteq_H \beta$ and $\beta \subseteq_H \alpha$. It is easy to see that $\equiv_H$ is an equivalence relation and that:

Proposition 4 *The set of equivalence classes of $\equiv_H$ on the set of elements of $\mathbf{K}$ is finite.*

As in [2, 3, 4, 6, 7, 12, 17], given $\mathbf{K} = \langle P, \leq, \Vdash \rangle$, we define the *quotient model* $\mathbf{K}/H$ as the Kripke model $\langle P', \leq', \Vdash' \rangle$ with the following properties:

1. P' is the set of equivalence classes generated by $\equiv_H$ on P;
2. if $[\alpha]$ and $[\beta]$ are two elements of P' (where $[\gamma]$ is the equivalence class of γ), then $[\alpha] \leq' [\beta]$ iff $\alpha \subseteq_H \beta$;
3. for every variable p such that $p \in RSf(H)$ and for every $[\alpha] \in P'$, $[\alpha] \Vdash' p$ in $\mathbf{K}/H$ iff $\alpha \Vdash p$ in $\mathbf{K}$;
4.. for every variable q such that $q \notin RSf(H)$ and for every $[\alpha] \in P$, $[\alpha] \Vdash' q$ in $\mathbf{K}/H$.

The main property of $\mathbf{K}/H$ is stated by the following proposition that can be proved by induction on the wff B as in [6, 7, 17]:

Proposition 5 *If $B \in Sf(H)$ then, for every element α of $\mathbf{K}$, $\alpha \Vdash B$ in $\mathbf{K}$ iff $[\alpha] \Vdash' B$ in $\mathbf{K}/H$.*

Let us denote with $ESf(H)$ the smallest set of formulae containing $RSf(H)$ and closed with respect to conjunctions and disjunctions; then $Sf(H) \subseteq ESf(H)$ and the following extension of Proposition 5 can be stated without any change in the proof:

if $B \in ESf(H)$ then, for every element α of $\mathbf{K}$, $\alpha \Vdash B$ in $\mathbf{K}$ iff $[\alpha] \Vdash' B$ in $\mathbf{K}/H$.

Let $L = INT + (A)$ be any logic and let us suppose that $L = \mathcal{L}(\mathcal{F})$ for some non empty class of Kripke frames $\mathcal{F}$. Then:

Theorem 2 *Let H be any wff such that $H \in L$. If, for every INT-saturated set Γ such that $IstR_L(H) \subseteq \Gamma$, $C_{INT}(\Gamma)/H$ is built on a frame of $\mathcal{F}$, then $IstR_L(H)\vdash_{\overline{INT}} H$.*

Proof: The proof is a consequence of Propositions 2 and 3. Let us suppose that there exists a wff $H \in L$ such that, for every INT-saturated set Γ containing $IstR_L(H)$, $C_{INT}(\Gamma)/H$ is built on a frame of $\mathcal{F}$, but $IstR_L(H)\nvdash_{\overline{INT}} H$. Then, by Proposition 2, there exists an INT-saturated set $\overline{\Gamma}$ such that $IstR_L(H) \subseteq \overline{\Gamma}$ and $H \notin \overline{\Gamma}$. Now, let us consider the Kripke model $C_{INT}(\overline{\Gamma})/H$; by hypothesis, it is built on a frame of $\mathcal{F}$; thus, $[\overline{\Gamma}]\|{-}H$ in $C_{INT}(\overline{\Gamma})/H$ and, by Proposition 5, since $H \in Sf(H)$, $\overline{\Gamma}\|{-}H$ in $C_{INT}(\overline{\Gamma})$. Therefore, by Proposition 3, $H \in \overline{\Gamma}$, absurd. $\qquad\square$

For any wff H, we set $H_{\mathcal{E}neg} = H \wedge \bigwedge \mathcal{E}neg(H)$. Since $H \in Sf(H) \subseteq Sf(H_{\mathcal{E}neg})$, we can state the following theorem essentially repeating the proof of Theorem 2:

Theorem 3 *Let H be any wff such that $H \in L$. If, for every INT-saturated set Γ such that $Ist_L^{\mathcal{E}neg}(H) \subseteq \Gamma$, $C_{INT}(\Gamma)/H_{\mathcal{E}neg}$ is built on a frame of $\mathcal{F}$, then $Ist_L^{\mathcal{E}neg}(H)\vdash_{\overline{INT}} H$.*

Given a logic $L = \mathcal{L}(\mathcal{F}) = INT + (A)$ for some non empty class of Kripke frames $\mathcal{F}$, we say that L has the *first property of the canonical quotient* ((f.p.c.q.) for short) iff, for every wff H and every INT-saturated set Γ such that $IstR_L(H) \subseteq \Gamma$, $C_{INT}(\Gamma)/H$ is built on a frame belonging to $\mathcal{F}$; on the other hand, we say that L has the *second property of the canonical quotient* ((s.p.c.q.) for short) iff, for every wff H and every INT-saturated set Γ such that $Ist_L^{\mathcal{E}neg}(H) \subseteq \Gamma$, $C_{INT}(\Gamma)/H_{\mathcal{E}neg}$ is built on a frame belonging to $\mathcal{F}$.

Now, we want to restate the results of Theorems 2 and 3 in terms of another, more sophisticated, technique of filtration. Thus, following [2, 4, 12], we present the filtration of the selective models.

Let $\mathbf{K} = \langle P, \leq, \|{-}\rangle$ have a root r, let $\alpha, \beta \in P$, and let H be a wff; we set $\alpha \leftarrow_H \beta$ iff the following conditions are satisfied:

1. $\alpha \equiv_H \beta$;
2. for every γ such that $\alpha \leq \gamma$ in $\mathbf{K}$ and $\alpha \not\equiv_H \gamma$, there exists $\delta \in P$ such that $\beta \leq \delta$ and $\gamma \equiv_H \delta$.

An $\alpha \in P$ is said to be $\leftarrow_H$-*terminal* iff, for every $\beta \in P$ such that $\alpha \leq \beta$ and $\alpha \equiv_H \beta$, we have $\alpha \leftarrow_H \beta$.

Let $\alpha \in P$ be $\leftarrow_H$-terminal. A $\beta \in P$ is called a $\leftarrow_H$-*immediate successor* of α in $\mathbf{K}$ iff:

1. $\alpha \not\equiv_H \beta$;
2. $\alpha \leq \beta$;
3. there is no γ in P such that $\alpha \leq \gamma \leq \beta$, $\alpha \not\equiv_H \gamma$ and $\beta \not\equiv_H \gamma$;
4. β is $\leftarrow_H$-terminal

Let $\alpha \in P$ be $\leftarrow_H$-terminal; a set $\{\alpha_1^s, \ldots, \alpha_k^s\}$ is called a *complete set of* $\leftarrow_H$-*immediate successors of* α *in* **K** iff the following conditions are satisfied:

1. $\alpha_1^s, \ldots, \alpha_k^s$ are $\leftarrow_H$-immediate successors of α in **K**;
2. for every i, j such that $1 \leq i, j \leq k$ and $i \neq j$, $\alpha_i^s \subseteq_H \alpha_j^s$ does not hold;
3. for every $\beta \in P$ such that β is a $\leftarrow_H$-immediate successor of α in **K**, there is $1 \leq i \leq k$ such that $\alpha_i^s \subseteq_H \beta$.

Finally, by $\mathbf{K}^{sel}/H$, called *a basic selective model of* **K** *with respect to* H, we will mean any Kripke model $\langle P', \leq', r', \Vdash' \rangle$ with the following properties:

1. the least element of the poset $\langle P', \leq' \rangle$ is a $\leftarrow_H$-terminal element r' of **K** belonging to the $\equiv_H$-equivalence class $[r]$ of the root r of **K**;
2. let α be any non final element of $\langle P', \leq', r' \rangle$; then the immediate successors $\alpha_1^s, \ldots, \alpha_k^s$ of α in $\langle P', \leq', r' \rangle$ are such that $\{\alpha_1^s, \ldots, \alpha_k^s\}$ is a complete set of $\leftarrow_H$-immediate successors of α in **K**;
3. let $\alpha \in P'$; then, if p is a variable of H, then $\alpha \Vdash' p$ iff $\alpha \Vdash p$ in **K**;
4. let $\alpha \in P'$; then, if p is a variable which does not occur in H, then $\alpha \Vdash' p$.

The model $\mathbf{K}^{sel}/H$ is not uniquely determined up to isomorphisms. However, for any wff H, at least a model $\mathbf{K}^{sel}/H$ exists, and all such models are finite; moreover, the set P' of the states of any $\mathbf{K}^{sel}/H$ is always included in the set P of the states of **K**. For an extensive treatment of the models $\mathbf{K}^{sel}/H$, the reader is referred to [2, 12]; as pointed out in [12], their properties are needed to handle the logics G_n $(n \geq 1)$ also considered in the present paper, where we will also use the models $\mathbf{K}^{sel}/H$ to treat the logics PR_k $(k \geq 1)$ (which cannot be handled by the quotient models), even if for such logics a simpler notion of filtration might have been introduced.

The following properties hold for the models $\mathbf{K}^{sel}/H$:

Proposition 6 *For every* α *and* β *of* $\mathbf{K}^{sel}/H$, *if* $\alpha \leq' \beta$ *then* $\alpha \leq \beta$.

Proposition 7 *For every* $A \in Sf(H)$ *and for every* α *of* $\mathbf{K}^{sel}/H$, $\alpha \Vdash' A$ *in* $\mathbf{K}^{sel}/H$ *iff* $\alpha \Vdash A$ *in* **K**.

In analogy with the case of Proposition 5, we can extend Proposition 7, without affecting its proof, into the following statement:

if $A \in ESf(H)$ then, for every α of $\mathbf{K}^{sel}/H$, $\alpha \Vdash' A$ in $\mathbf{K}^{sel}/H$ iff $\alpha \Vdash A$ in **K**.

Moreover, let $L = INT + (A)$ be any logic and let us suppose that $L = \mathcal{L}(\mathcal{F})$ for some non empty class of Kripke frames $\mathcal{F}$. Then the two following facts, whose proof is similar to the one of Theorem 2, hold:

Theorem 4 *Let* H *be any wff such that* $H \in L$. *If, for every* INT-*saturated set* Γ *such that* $IstR_L(H) \subseteq \Gamma$, $C_{INT}(\Gamma)^{sel}/H$ *is built on a frame of* $\mathcal{F}$, *then* $IstR_L(H) \vdash_{INT} H$.

Theorem 5 *Let H be any wff such that $H \in L$. If, for every INT-saturated set Γ such that $IstR_L^{\wedge\vee}(H) \subseteq \Gamma$, $C_{INT}(\Gamma)^{sel}/H$ is built on a frame of $\mathcal{F}$, then $IstR_L^{\wedge\vee}(H)\vdash_{INT} H$.*

Given a logic $L = \mathcal{L}(\mathcal{F}) = INT + (A)$ for some non empty class of Kripke frames $\mathcal{F}$, we say that L has the *first property of the canonical selective model* ((f.p.c.s.m.) for short) iff, for every wff H and every INT-saturated set Γ such that $IstR_L(H) \subseteq \Gamma$, $C_{INT}(\Gamma)^{sel}/H$ is built on a frame belonging to $\mathcal{F}$; on the other hand, we say that L has the *second property of the canonical selective model* ((s.p.c.s.m.) for short) iff, for every wff H and every INT-saturated set Γ such that $Ist_L^{\wedge\vee}(H) \subseteq \Gamma$, $C_{INT}(\Gamma)^{sel}/H$ is built on a frame belonging to $\mathcal{F}$.

4 The logics D, PR_k, FIN_k, G_n, and KP

In this section we study some intermediate logics which have the properties described in the previous sections.

The first logic we are dealing with is $D = INT + (A)$, where $A = (p \to q) \vee (q \to p)$ and p and q are propositional variables. In [17] it is shown that $D = \mathcal{L}(\mathcal{F}_D)$, where $\mathcal{F}_D$ is the class of rooted Kripke frames $\mathbf{P} = \langle P, \leq, r \rangle$ with the following properties:

for every $\alpha, \beta \in P$, $\alpha \leq \beta$ or $\beta \leq \alpha$.

Proposition 8 *D satisfies (f.p.c.q.).*

Proof: Let us suppose the contrary. Then, there exist an INT-saturated set Γ and a wff H such that $IstR_D(H) \subseteq \Gamma$, but the underlying frame of $C_{INT}(\Gamma)/H$ does not belong to $\mathcal{F}_D$.

Therefore, by definition of $\mathcal{F}_D$, there exist two elements $[\Gamma']$ and $[\Gamma'']$ of $C_{INT}(\Gamma)/H$ such that $[\Gamma'] \not\leq [\Gamma'']$ and $[\Gamma''] \not\leq [\Gamma']$. Thus, by definition of $C_{INT}(\Gamma)/H$, there exist two wffs A and B of $RSf(H)$ such that $[\Gamma']\|\!\!\vdash A$ and $[\Gamma']\|\!\!\not\vdash B$, and $[\Gamma'']\|\!\!\vdash B$ and $[\Gamma'']\|\!\!\not\vdash A$ in $C_{INT}(\Gamma)/H$; hence, by Proposition 5, $\Gamma'\|\!\!\vdash A$ and $\Gamma'\|\!\!\not\vdash B$, and $\Gamma''\|\!\!\vdash B$ and $\Gamma''\|\!\!\not\vdash A$ in $C_{INT}(\Gamma)$. It is easy to see that $\Gamma\|\!\!\not\vdash (A \to B)\vee(B \to A)$ in $C_{INT}(\Gamma)$; as a matter of fact, if $\Gamma\|\!\!\vdash (A \to B)\vee(B \to A)$ in $C_{INT}(\Gamma)$, then $\Gamma\|\!\!\vdash A \to B$ or $\Gamma\|\!\!\vdash B \to A$. Without loss of generality, let us suppose that $\Gamma\|\!\!\vdash A \to B$; hence, for all $\overline{\Gamma}$ in $C_{INT}(\Gamma)$, $\overline{\Gamma}\|\!\!\vdash B$ or $\overline{\Gamma}\|\!\!\vdash A$; but $\Gamma'\|\!\!\vdash A$ and $\Gamma'\|\!\!\not\vdash B$. Thus we conclude that $\Gamma\|\!\!\not\vdash (A \to B) \vee (B \to A)$; but $IstR_D(H) \subseteq \Gamma$, hence $(A \to B) \vee (B \to A) \in \Gamma$; therefore, by Proposition 3, $\Gamma\|\!\!\vdash (A \to B) \vee (B \to A)$, absurd. $\square$

Thus, we have proved:

Corollary 1 *For every wff H, $H \in D$ iff $IstR_D(H)\vdash_{INT} H$.*

The logics PR_k, known as Nagata logics [18, 19], are defined as: $PR_k = INT + (PR_k)$, where the axiom-scheme (PR_k) is defined inductively as follows:

- $(PR_1) = (p_1 \vee \neg p_1)$;
- $(PR_{i+1}) = (p_{i+1} \vee (p_{i+1} \to PR_i))$ $(i \geq 2)$;

for all i, p_i is a propositional variable. As is well known [2, 3, 4, 17, 18, 19] , $PR_k = \mathcal{L}(\mathcal{F}_{PR_k})$ for all $k \geq 1$, where:

$\mathcal{F}_{PR_k}$ is the class of Kripke frames $\langle P, \leq \rangle$ such that, for all $\alpha \in P$, α has at most depth k in $\langle P, \leq \rangle$.

Now we prove:

Proposition 9 *For all $k \geq 1$, PR_k satisfies (f.p.c.s.m.).*

Proof: Let us suppose the contrary; then there exist k, a wff H, and an INT-saturated set Γ such that $IstR_{PR_k}(H) \subseteq \Gamma$, but the underlying frame of $\mathcal{C}_{INT}(\Gamma)^{sel}/H$ does not belong to the class $\mathcal{F}_{PR_k}$. Therefore, by definition of $\mathcal{F}_{PR_k}$, there exist $k+1$ elements $\Gamma_1, \ldots, \Gamma_{k+1}$ of $\mathcal{C}_{INT}(\Gamma)^{sel}/H$ such that, for all $1 \leq i \leq k$, $\Gamma_{i+1} < \Gamma_i$; hence, by definition of $\mathcal{C}_{INT}(\Gamma)^{sel}/H$, there exist k wffs $A_1, \ldots A_k$ of $RSf(H)$ such that, for all $1 \leq i \leq k$, $\Gamma_{i+1}\|\!\!\not\vdash A_i$ and $\Gamma_i\|\!\!\vdash A_i$ in $\mathcal{C}_{INT}(\Gamma)^{sel}/H$. Thus, by Proposition 5, for all $1 \leq i \leq k$, $\Gamma_{i+1}\|\!\!\not\vdash A_i$ and $\Gamma_i\|\!\!\vdash A_i$ in $\mathcal{C}_{INT}(\Gamma)$, and, by Proposition 6, for all $1 \leq i \leq k$, $\Gamma_{i+1} \leq \Gamma_i$ in $\mathcal{C}_{INT}(\Gamma)$.

It is easy to see that $\Gamma_{k+1}\|\!\!\not\vdash \overline{PR}_k$ in $\mathcal{C}_{INT}(\Gamma)$, where $\overline{PR}_k$ is the instance of the generating formula of (PR_k) obtained by substituting $p_1, \ldots, p_k$ with $A_1, \ldots, A_k$ respectively. To this aim, we prove by induction that, for all $2 \leq i \leq (k+1)$, $\Gamma_i\|\!\!\not\vdash \overline{PR}_{i-1}$ in $\mathcal{C}_{INT}(\Gamma)$. Obviously, since $\Gamma_2 \leq \Gamma_1$, $\Gamma_1\|\!\!\vdash A_1$, and $\Gamma_2\|\!\!\not\vdash A_1$, we have $\Gamma_2\|\!\!\not\vdash A_1 \vee \neg A_1$ in $\mathcal{C}_{INT}(\Gamma)$. Now, let us suppose that, for all $2 \leq h \leq i$, $\Gamma_h\|\!\!\not\vdash \overline{PR}_{h-1}$ in $\mathcal{C}_{INT}(\Gamma)$. By hypothesis, $\Gamma_{i+1}\|\!\!\not\vdash A_i$ in $\mathcal{C}_{INT}(\Gamma)$; moreover, since, by hypothesis, $\Gamma_i\|\!\!\vdash A_i$ in $\mathcal{C}_{INT}(\Gamma)$ and $\Gamma_{i+1} \leq \Gamma_i$, and, by induction hypothesis, $\Gamma_i\|\!\!\not\vdash \overline{PR}_{i-1}$, we have: $\Gamma_{i+1}\|\!\!\not\vdash A_i \rightarrow \overline{PR}_{i-1}$. Therefore, $\Gamma_{i+1}\|\!\!\not\vdash \overline{PR}_i$, which concludes our induction. But $IstR_{PR_k}(H) \subseteq \Gamma$, hence $\overline{PR}_k \in \Gamma_{k+1}$. Therefore, by Proposition 3, $\Gamma_{k+1}\|\!\!\vdash \overline{PR}_k$ in $\mathcal{C}_{INT}(\Gamma)$, absurd. $\square$

Thus, for all $k \geq 1$, we have:

Corollary 2 *For every wff H, $H \in PR_k$ iff $IstR_{PR_k}(H)\Big|_{\overline{INT}} H$.*

The logics FIN_k [2, 3], for all $k \geq 1$, are defined as: $FIN_k = INT + (FIN_k)$, where the axiom-scheme (FIN_k) is defined as follows:

- $(FIN_1) = (\neg p_1 \vee \neg\neg p_1)$;
- $(FIN_2) = (\neg p_1 \vee (\neg p_1 \rightarrow \neg p_2) \vee (\neg p_1 \rightarrow \neg\neg p_2))$;
- $(FIN_i) = (\neg p_1 \vee (\neg p_1 \rightarrow \neg p_2) \vee (\neg p_1 \wedge \neg p_2 \rightarrow \neg p_3) \vee \ldots \vee (\neg p_1 \wedge \ldots \wedge \neg p_{n-1} \rightarrow \neg p_n) \vee (\neg p_1 \wedge \ldots \wedge \neg p_{n-1} \rightarrow \neg\neg p_n))$ $(i \geq 3)$;

$p_1, \ldots, p_n$ are propositional variables. As is well known [2, 3], $FIN_k = \mathcal{L}(\mathcal{F}_{FIN_k})$ for every $k \geq 1$, where:

$\mathcal{F}_{FIN_k}$ is the class of Kripke frames $\langle P, \leq \rangle$ such that, for all $\alpha \in P$, $| Fin(\alpha) | \leq k$.

Proposition 10 *For all $k \geq 1$, FIN_k satisfies (s.p.c.q.).*

Proof: Let us assume the contrary. Then there exist $k \geq 1$, a wff $H \in FIN_k$ and an INT–saturated set Γ such that $Ist_{FIN_k}^{\mathcal{E}neg}(H) \subseteq \Gamma$, but the underlying frame of $C_{INT}(\Gamma)/H_{\mathcal{E}neg}$ does not belong to $\mathcal{F}_{FIN_k}$.

Therefore, by definition of $\mathcal{F}_{FIN_k}$, $| Fin(\Gamma) |> k$. Let $[\Gamma_1], \ldots, [\Gamma_{k+1}]$ be $k+1$ distinct final elements of $C_{INT}(\Gamma)/H_{\mathcal{E}neg}$; hence, by definition of $C_{INT}(\Gamma)/H_{\mathcal{E}neg}$ and by the properties of its final states, for every $1 \leq i \leq k + 1$, there is $B_i \in \mathcal{E}neg(H)$ such that $[\Gamma_i]\|\!\!-B_i$ and, for every j with $1 \leq j \leq k + 1$ and $i \neq j$, $[\Gamma_j]\|\!\!\not\!-B_i$; thus, since the elements of $\mathcal{E}neg(H)$ and their negations are elements of $RSf(H_{\mathcal{E}neg})$, by Proposition 5, for all $1 \leq i,j \leq k + 1$ with $i \neq j$, $\Gamma_i\|\!\!-B_i$ and $\Gamma_j\|\!\!-\neg B_i$ in $C_{INT}(\Gamma)$. It is easy to see that $\Gamma\|\!\!\not\!-\overline{FIN}_k$ in $C_{INT}(\Gamma)$, where $\overline{FIN}_k$ is the instance of the generating formula of (FIN_k) obtained by substituting $p_1, \ldots, p_k$ with $B_1, \ldots, B_k$ respectively.

As a matter of fact, $\Gamma\|\!\!\not\!-\neg B_1$ since $\Gamma_1\|\!\!-B_1$; $\Gamma\|\!\!\not\!-\neg B_1 \to \neg B_2$ since $\Gamma_2\|\!\!-\neg B_1$ and $\Gamma_2\|\!\!\not\!-\neg B_2$; more generally, for all $2 \leq i \leq k$, $\Gamma\|\!\!\not\!-\neg B_1 \wedge \ldots \wedge \neg B_{i-1} \to \neg B_i$ since $\Gamma_i\|\!\!-\neg B_1 \wedge \ldots \wedge \neg B_{i-1}$ but $\Gamma_i\|\!\!\not\!-\neg B_i$; finally, $\Gamma\|\!\!\not\!-\neg B_1 \wedge \ldots \wedge \neg B_{k-1} \to \neg\neg B_k$ since $\Gamma_{k+1}\|\!\!-\neg B_1 \wedge \ldots \wedge \neg B_{k-1}$ but $\Gamma_{k+1}\|\!\!\not\!-\neg\neg B_k$. But $Ist_{FIN_k}^{\mathcal{E}neg}(H) \subseteq \Gamma$, hence $\overline{FIN}_k \in \Gamma$. Therefore, by Proposition 3, $\Gamma\|\!\!-\overline{FIN}_k$, absurd. $\qquad\square$

Corollary 3 *For every wff H, $H \in FIN_k$ iff $Ist_{FIN_k}^{\mathcal{E}neg}(H)\big|\!\!\frac{}{\text{INT}}\, H$.*

Remark 1: We recall that the elementary negated formulae substituted for the variables of the generating formula of the axiom–scheme (FIN_k) $(k \geq 1)$ (in order to provide the formulae of $Ist_{FIN_k}^{\mathcal{E}neg}(H)$) have the form $\neg\neg H$, with H an elementary formula. Hence, since all the variables of (FIN_k) occur negated and since any formula such that $\neg\neg\neg B$ is intuitionistically equivalent to $\neg B$, we can simplify the formulae of $IST_{FIN_k}^{\mathcal{E}neg}(H)$ by replacing in them any formula of the form $\neg\neg\neg H$ such that H is elementary with $\neg H$.

Now we deal with the logics G_n $(n \geq 1)$ of Gabbay and de Jongh [8]. These logics are defined as $G_n = INT + (G_n)$, where the generating formula of the axiom–scheme (G_n) is the following:

$$\bigwedge_{i=1}^{n+2} ((p_i \to \bigvee_{j=1,j\neq i}^{n+2} p_j) \to \bigvee_{j=1,j\neq i}^{n+2} p_j) \to \bigvee_{i=1}^{n+2} p_i$$

where, for all $1 \leq i \leq n + 2$, p_i is a propositional variable.

In [7, 8, 12] it is shown that, for all $n \geq 1$, $G_n = \mathcal{L}(\mathcal{F}_{G_n})$, where:

$\mathcal{F}_{G_n}$ is the class of finite Kripke frames $\langle P, \leq \rangle$ such that, for all $\alpha \in P$, there exist at most $n + 1$ immediate successors of α.

Proposition 11 *For all $k \geq 1$, G_k satisfies (s.p.c.s.m).*

Proof: Let us assume the contrary. Then there exist $k \geq 1$, a wff H and an INT–saturated set Γ such that $Ist_{G_k}^{\wedge\vee}(H) \subseteq \Gamma$, but the underlying poset of $C_{INT}(\Gamma)^{sel}/H$ does not belong to $\mathcal{F}_{G_k}$; therefore, there exist $\Gamma', \Gamma_1, \ldots, \Gamma_h$ such that $h \geq k+2$ and $\Gamma_1, \ldots, \Gamma_h$ are immediate successors of Γ'. Now, by definition of $C_{INT}(\Gamma)^{sel}/H$, for all $1 \leq i, j \leq h$ with $i \neq j$, we can find $A_{i,j} \in RSf(H)$ such that $\Gamma_i \|{-}A_{i,j}$ and $\Gamma_j \|{\not\vdash} A_{i,j}$; thus, let $A_i = \bigwedge_{j=1, j\neq i}^{h} A_{i,j}$ for all $1 \leq i \leq h$; it is easy to see that, for all $1 \leq i \leq h$, $A_i \in RSf_{\wedge}(H)$ and $\Gamma_i \|{-}A_i$ in $C_{INT}(\Gamma)^{sel}/H$; moreover, for $1 \leq i, j \leq h$ and $i \neq j$, $\Gamma_i \|{\not\vdash} A_j$ in $C_{INT}(\Gamma)^{sel}/H$. Let us set $B_i = A_i$ for $1 \leq i \leq k+1$, and let $B_{k+2} = A_{k+2} \vee \ldots \vee A_h$; then, for every $1 \leq j \leq k+2$, $B_j \in RSf_{\wedge\vee}(H)$. Finally, let $\overline{G}_k$ be the instance of the generating formula of (G_k) obtained by replacing, for all $1 \leq i \leq k+2$, p_i with B_i. With an easy refinement of the argument given in [12] (using in an essential way the properties of $C_{INT}(\Gamma)^{sel}/H$), it is possible to prove that $\Gamma' \|{\not\vdash} \overline{G}_k$ in $C_{INT}(\Gamma)$. Thus, since, by hypothesis, $\overline{G}_k \in \Gamma$, we get a contradiction. $\qquad \square$

Corollary 4 *For every wff H and any $n \geq 1$, $H \in G_n$ iff $Ist_{G_n}^{\wedge\vee}(H) \vdash_{INT} H$.*

Remark 2: An inspection of the proof of Proposition 11 show that $Ist_{G_n}^{\wedge\vee}(H)$ can be restricted to the set of formulae obtained by substituting the variables $p_1, \ldots, p_n, p_{n+1}$ of the generating formula of (G_n) with elements of $RSf_{\wedge}(H)$ and the last variable p_{n+2} with an element of $RSF_{\wedge\vee}$.

Now, we study the logic of Kreisel and Putnam [10], we denote by KP. This logic is defined as $KP = INT + (KP)$, where the generating formula of the axiom-scheme (KP) is the following:

$$(\neg p \to q \vee r) \to (\neg p \to q) \vee (\neg p \to r)$$

where p, q, and r are propositional variables.

This logic is well known to many people working in intermediate propositional logics [6, 7, 10, 12, 17]. Indeed, it was the first counterexample to Lukasiewicz's conjecture of 1952 (see [11]), asserting that intuitionistic logic is the greatest consistent propositional system with the disjunction property and closed under substitution and modus-ponens.

As stated in [12] (which simplifies what stated in [6, 7]), we have $KP = \mathcal{L}(\mathcal{F}_{KP})$, where:

> $\mathcal{F}_{KP}$ is the class of finite Kripke frames $\langle P, \leq \rangle$ with the following property: for all $\alpha, \beta, \gamma \in P$, if $\alpha \leq \beta$ and $\alpha \leq \gamma$, then there exists $\delta \in P$ such that $\alpha \leq \delta$, $\delta \leq \beta$, $\delta \leq \gamma$, and $Fin(\delta) = Fin(\beta) \cup Fin(\gamma)$.

To deal with KP, we need an appropriate selection function Σ_{KP}^4, requiring the following definitions:

- For every wff H, we set:
  1. $RSf_{\to}(H) = \{\neg\neg A \to B \mid \neg\neg A \in Rneg(H) \text{ and } B \in RSf(H_{\mathcal{E}neg})\}$;
  2. $RSf_{\to\vee}(H) = \{Z_1 \vee \ldots \vee Z_k \mid Z_1 \in RSf_{\to}(H), \ldots, Z_k \in RSf_{\to}(H), \text{ and } Z_1, \ldots, Z_k \text{ are all distinct}\}$.
- For every wff H, $Ist_{KP}^{\to\vee}(H)$ will be the set of all the instances of $(\neg p \to q \vee r) \to (\neg p \to q) \vee (\neg p \to r)$ having the form $(\neg\neg A \to B \vee C) \to (\neg\neg A \to B) \vee (\neg\neg A \to C)$, where $\neg\neg A \in Rneg(H)$ and $B, C \in RSf^{\to\vee}(H)$.

- For every wff H, the formula H^* will indicate the disjunction of all the elements of $RSf_{\to\vee}(H)$.
- For every wff H, we set $\Sigma^4_{KP}(H) = Ist^{\to\vee}_{KP}(H)$.
- We say that KP satisfies the *third property of the canonical quotient* ((t.p.c.q.) for short) iff, for every wff H and every INT–saturated set Γ such that $Ist^{\to\vee}_{KP}(H) \subseteq \Gamma$, $C_{INT}(\Gamma)/H^*$ is built on a frame belonging to $\mathcal{F}_{KP}$.

Since $RSf(H) \subseteq RSf(H^*)$ for every wff H, one has that $ESf(H) \subseteq ESf(H^*)$. Hence, if $B \in ESf(H)$, $\mathbf{K} = \langle P, \leq, \|\!\!-\rangle$ is any Kripke model and $\alpha \in P$, we have that $\alpha\|\!\!-B$ in $\mathbf{K}$ iff $[\alpha]\|\!\!-'B$ in $\mathbf{K}/H^*$; in particular, $\alpha\|\!\!-H$ in $\mathbf{K}$ iff $[\alpha]\|\!\!-'H$ in $\mathbf{K}/H^*$. Thus, according to the previous treatment, it suffices to prove that KP satisfies (t.p.c.q.) in order to be sure that, for every wff H, $H \in KP$ iff $Ist^{\to\vee}_{KP}(H)\vdash_{INT} H$.

Now, we have:

Proposition 12 KP *satisfies (t.p.c.q.).*

Proof (outline): Let H be any wff, let Γ be any INT–saturated set such that $IstR^{\to\vee}_{KP}(H) \subseteq \Gamma$, and let us consider $C_{INT}(\Gamma)/H^*$. Moreover, let $[\Gamma_1], [\Gamma_2], [\Gamma_3], [\Gamma^f_1], \ldots, [\Gamma^f_n]$ be elements of $C_{INT}(\Gamma)/H^*$ such that $[\Gamma_1] \leq [\Gamma_2]$, $[\Gamma_1] \leq [\Gamma_3]$, and $Fin([\Gamma_2]) \cup Fin([\Gamma_3]) = \{[\Gamma^f_1], \ldots, [\Gamma^f_n]\}$; then one can choose a formula $\neg\neg A \in Rneg(H)$ such that, for every final element $[\Gamma^f]$ of $C_{INT}(\Gamma)/H^*$, $[\Gamma^f]\|\!\!-\neg\neg A$ in $C_{INT}(\Gamma)/H^*$ iff $[\Gamma^f] \in \{[\Gamma^f_1], \ldots, [\Gamma^f_n]\}$.

Now, let us consider the set of formulae $\Gamma_1 \cup \{\neg\neg A\}$ (which is not, in general, a INT–saturated set); since $Ist^{\to\vee}_{KP}(H) \subseteq \Gamma_1$, by the particular forms of the formulae of $Ist^{\to\vee}_{KP}(H)$ it is possible to prove that there exists a INT–saturated set $\overline{\Gamma}$ satisfying the following properties: $\Gamma_1 \cup \{\neg\neg A\} \subseteq \overline{\Gamma}$; for every $B \in RSf(H^*)$, $B \in \overline{\Gamma}$ iff $\Gamma_1 \cup \{\neg\neg A\}\vdash_{INT} B$.

Starting from these properties of $\overline{\Gamma}$, one can prove that the equivalence class $[\overline{\Gamma}]$ is an element of $C_{INT}(\Gamma)/H^*$ with the following properties: $[\Gamma_1] \leq [\overline{\Gamma}]$, $[\overline{\Gamma}] \leq [\Gamma_2]$ and $[\overline{\Gamma}] \leq [\Gamma_3]$; $Fin([\overline{\Gamma}]) = Fin([\Gamma_2]) \cup Fin([\Gamma_3])$. Thus, our assertion holds $\qquad\square$

Corollary 5 *For every wff H, $H \in KP$ iff $Ist^{\to\vee}_{KP}(H)\vdash_{INT} H$.*

Remark 3: A more careful proof of Proposition 12 allows to replace the set of formulae $Ist^{\to\vee}_{KP}(H)$ with the smaller set $\underline{Ist}^{\to\vee}_{KP}(H)$ defined as follows:

1. $\overline{RSf_\to}(H) = \{\neg\neg A \to B \mid \neg\neg A \in Rneg(H) \text{ and } B \in RSf(H)\}$;
2. $\overline{RSf_{\to\vee}}(H) = \{Z_1 \vee \ldots \vee Z_k \mid Z_1 \in \overline{RSf_\to}(H), \ldots, Z_k \in \overline{RSf_\to}(H), \text{ and } Z_1, \ldots, Z_k \text{ are all distinct}\}$;
3. $\underline{Ist}^{\to\vee}_{KP}(H)$ is the set of all the instances of $(\neg p \to q \vee r) \to (\neg p \to q) \vee (\neg p \to r)$ having the form $(\neg\neg A \to B \vee C) \to (\neg\neg A \to B) \vee (\neg\neg A \to C)$, where $\neg\neg A \in Rneg(H)$ and $B, C \in \overline{RSf_{\to\vee}}(H)$.

5 Conclusions and Future Work

Now, let us consider the three classes of L–selection functions Σ_L^1, Σ_L^2, and Σ_L^3, introduced at the end of Section 2, and the selection function Σ_{KP}^4, introduced at the end of the previous section; let H be any wff, and let $\Sigma_L^1(H)$–$RGCL$, $\Sigma_L^2(H)$–$RGCL$, $\Sigma_L^3(H)$–$RGCL$, and $\Sigma_{KP}^4(H)$–$RGCL$ be the related regular generalized tableau calculi. Then, for L ranging in the set of logics defined in the previous section, combining the machinery developed in the two previous sections we get the following soundness and completeness results:

Theorem 6 *For every wff H, $H \in D$ iff there is a closed proof–table in the calculus $\Sigma_D^1(H)$–$RGCD$ starting from $\mathbf{F}H$.*

Theorem 7 *For every $k \geq 1$ and every wff H, $H \in PR_k$ iff there is a closed proof–table in the calculus $\Sigma_{PR_k}^1(H)$–$RGCPR_k$ starting from $\mathbf{F}H$.*

Theorem 8 *For every $k \geq 1$ and every wff H, $H \in FIN_k$ iff there is a closed proof–table in the calculus $\Sigma_{FIN_k}^2(H)$–$RGCFIN_k$ starting from $\mathbf{F}H$.*

Theorem 9 *For every $k \geq 1$ and every wff H, $H \in G_k$ iff there is a closed proof–table in the calculus $\Sigma_{G_k}^3(H)$–$RGCG_k$ starting from $\mathbf{F}H$.*

Theorem 10 *For every wff H, $H \in KP$ iff there is a closed proof–table in the calculus $\Sigma_{KP}^4(H)$–$RGCKP$ starting from $\mathbf{F}H$.*

As said in Section 1, other logics might have been chosen to illustrate our notion of generalized tableau calculus. For instance, other selection functions can be introduced for further interesting logics. In this context, we know a selection function working for Scott logic ST [3, 4, 6, 7, 12, 17], which is weaker than the above presented ones for the logics D, PR_k ($k \geq 1$), FIN_m ($m \geq 1$), and G_n ($n \geq 1$); we are looking for an improvement of such a selection function, which can be used also for other known logics; we are also looking for the possibility of singling out general rules allowing to associate selection functions with suitable families of logics on the unique basis of the syntactical form of the formulae generating the related axiom–schemes.

Of course, our results have, at the moment, mainly a theoretical interest, since the cardinalities of the sets of formulae in the ranges of our selection functions are still dramatically too great to be actually used in carrying out proofs; we only aim to give a contribution in the right direction in order to circumscribe realistic proof strategies, which surely need further ideas. In this perspective, even if our selection results can be applied as well to calculi different from tableau systems (for instance, they can be applied to the calculi of [1]), it seems to us that they can better suit tableau rules, which have, moreover, a "more semantical character".

Coming back to the previous results, we are investigating whether they are the best possible ones. For instance, we can prove that, differently from Σ_{KP}^4, the stronger selection function Σ_{KP}^1 (defined, *mutatis mutandis*, as the selection function provided for D) does not work for KP, i.e.:

Theorem 11 *There are formulae $H \in KP$ such that in the calculus $\Sigma^1_{KP}(H)-$ RGCKP no closed proof-table exists starting from $\mathbf{F}H$.*

We conclude the paper by proposing an interesting question.

First, we define the complexity c_H of a wff H as usual (for an atomic p we set $c_p = 0$; the complexity of $\neg A$ is c_A+1; the complexity of $A \wedge B$, $A \vee B$ and $A \to B$ is $max(c_A, c_B)+1$); then, given a logic $L = INT+(A(p_1, \ldots, p_n))$ with $p_1, \ldots, p_n$ all the variables of the generating formula A, a L–selection function Σ_L and a function f from the set of the natural numbers to the set of the natural numbers, we say that Σ_L is f–bounded iff, for every wff H and every $B \in \Sigma_L(H)$, there are wffs $H_1, \ldots, H_n$ such that $B = A(H_1, \ldots, H_n)$ and $max(c_{H_1}, \ldots, c_{H_n}) \leq f(c_H)$; finally, if $L = INT + (A)$ and Σ_L is a L–selection function, we say that Σ_L *is complete for L* iff, for every wff H, there is a closed proof-table in the calculus $\Sigma_L(H)$–RGCL starting from $\mathbf{F}H$.

Now, the question is:

For every recursive function f from the set of the natural numbers to the set of the natural numbers, is there some recursive logic $L = INT + (A)$ such that, for every f–bounded L–selection function Σ_L, Σ_L is not complete for L?

References

1. R. Dyckhoff. Contraction–free sequent calculi for intuitionistic logic. *Journal of Symbolic Logic*, 57(3):795–807, 1992.
2. M. Ferrari and P. Miglioli. Counting the maximal intermediate constructive logics. *Journal of Symbolic Logic*, 58(4):1365–1401, 1993.
3. M. Ferrari and P. Miglioli. A method to single out maximal intermediate propositional logics with the disjunction property I. *Annals of Pure and Applied Logic*, 76:1–46, 1995.
4. M. Ferrari and P. Miglioli. A method to single out maximal intermediate propositional logics with the disjunction property II. *Annals of Pure and Applied Logic*, 76:117–168, 1995.
5. M.C. Fitting. *Intuitionistic Logic, Model Theory and Forcing*. North–Holland, 1969.
6. D.M. Gabbay. The decidability of Kreisel–Putnam system. *Journal of Symbolic Logic*, 35:431–437, 1970.
7. D.M. Gabbay. *Semantical Investigations in Heyting's Intuitionistic Logic*. Reidel, Dordrecht, 1981.
8. D.M. Gabbay and D.H.J. de Jongh. A sequence of decidable finitely axiomatizable intermediate logics with the disjunction property. *Journal of Symbolic Logic*, 39:67–78, 1974.
9. J. Hudelmaier. An $o(n \, log \, n)$–space decision procedure for intuitionistic propositional logic. *Journal of Logic and Computation*, 3(1):63–75, 1993.
10. G. Kreisel and H. Putnam. Eine Unableitbarkeitsbeweismethode für den Intuitionistischen Aussagenkalkül. *Archiv für Mathematische Logik und Grundlagenforschung*, 3:74–78, 1957.

11. J. Lukasiewicz. On the intuitionistic theory of deduction. *Indagationes Mathematicae*, 14:69–75, 1952.
12. P. Miglioli. An infinite class of maximal intermediate propositional logics with the disjunction property. *Archive for Mathematical Logic*, 31(6):415–432, 1992.
13. P. Miglioli, U. Moscato, and M. Ornaghi. How to avoid duplications in a refutation system for intuitionistic logic and Kuroda logic. In K. Broda, M. D'Agostino, R. Goré, R. Johnson, and S. Reeves, editors, *Proceedings of 3rd Workshop on Theorem Proving with Analytic Tableaux and Related Methods*. Abingdon, U.K., May 4–6, 1994. Imperial College of Science, Technology and Medicine TR-94/5, 1994, pp. 169-187.
14. P. Miglioli, U. Moscato, and M. Ornaghi. An improved refutation system for intuitionistic predicate logic. *Journal of Automated Reasoning*, 12:361–373, 1994.
15. P. Miglioli, U. Moscato, and M. Ornaghi. Refutation systems for propositional modal logics. In P. Baumgartner, R. Hähnle, and J. Posegga, editors, *Theorem Proving with Analytic Tableaux and Related Methods: 4th International Workshop, Schloss Rheinfels, St. Goar, Germany*, volume 918 of *LNAI*, pages 95–105. Springer–Verlag, 1995.
16. P. Miglioli, U. Moscato, and M. Ornaghi. Avoiding duplications in tableau systems for intuitionistic and Kuroda logics. *L.J. of the IGPL*, 5(1):145–167, 1997.
17. P. Minari. *Indagini semantiche sulle logiche intermedie proposizionali*. Bibliopolis, 1989.
18. S. Nagata. A series of successive modifications of Peirce's rule. *Proceedings of the Japan Academy, Mathematical Sciences*, 42:859–861, 1966.
19. H. Ono. Kripke models and intermediate logics. *Publications of the Research Institute for Mathematical Sciences, Kyoto University*, 6:461–476, 1970.
20. K. Sasaki. The simple substitution property of the intermediate propositional logics on finite slices. *Studia Logica*, 52:41–62, 1993.
21. C.A. Smorynski. Applications of Kripke models. In A.S. Troelstra, editor, *Metamathematical Investigation of Intuitionistic Arithmetic and Analysis*, volume 344 of *Lecture Notes in Mathematics*. Springer–Verlag, 1973.

Lean Induction Principles for Tableaux*

Matthias Baaz[1] and Uwe Egly[2] and Christian G. Fermüller[3]

[1] Institut für Algebra und Diskrete Mathematik E118.2
Technische Universität Wien
Wiedner Hauptstraße 8–10, A–1040 Wien, Austria
e-mail: baaz@logic.tuwien.ac.at
[2] Institut für Informationssysteme E184.3
Technische Universität Wien
Treitlstraße 3, A–1040 Wien, Austria
e-mail: uwe@kr.tuwien.ac.at
[3] Institut für Computersprachen E185.2
Technische Universität Wien
Resselgasse 3/1, A–1040 Wien, Austria
e-mail: chrisf@logic.tuwien.ac.at

Abstract. In this paper, we deal with various induction principles incorporated in an underlying tableau calculus with equality. The induction formulae are restricted to literals. Induction is formalized as modified closure conditions which are triggered by applications of the δ-rule. Examples dealing with (weak forms of) arithmetic and strings illustrate the simplicity and usability of our induction handling. We prove the correctness of the closure conditions and discuss possibilities to strengthen the induction principles.

1 Introduction

The use of induction principles in various forms certainly is an important and prominent topic in Automated Deduction, as witnessed, e.g., by [15, 6, 1] and quite recently in [17]. The complexity of the problem — both, in terms of proof search and formulation of appropriate deduction mechanisms — is well known. Our aim is to demonstrate that various forms of induction can be incorporated into classical free variable tableaux in an elegant way. We were partly inspired by the mechanism described in [3] where the principle for Noetherian induction is implemented as a combination of function introduction and a special cut rule. Surprisingly, the tableau calculus allows for a considerably simpler incorporation of induction principles. Since we restrict our attention here to elegant and *computationally simple* induction principles, we have called these principles *lean*.

The paper is organized as follows. Section 2 provides the basic notations concerning free variable tableaux. In Section 3 we present various induction

---

* The first author was supported by the FWF under grant P11934-MAT. The authors would like to thank Hans Tompits and the referees for their useful comments on an earlier version of this paper.

schemata for which Section 4 introduces corresponding tableau closure rules. Moreover, these rules are illustrated by detailed examples. The correctness theorems and remarks on completeness can be found in Section 5. We conclude with methodological observations in Section 6.

2 The Underlying Tableau Calculus

We assume familiarity with classical semantic tableaux and, in particular, with the "free variable" version that has been described, e.g., in Fitting's book [12]. For sake of convenience we remind the reader that — using Smullyan's elegant *uniform notation* — the propositional rules take the form:

$$\frac{\alpha}{\begin{array}{c}\alpha_1\\\alpha_2\end{array}} \quad \text{and} \quad \frac{\beta}{\beta_1 | \beta_2} \quad \text{and} \quad \frac{\neg\neg X}{X}$$

where the Greek letters stand for the following types of formulae:

α	α_1	α_2
$X \wedge Y$	X	Y
$\neg(X \vee Y)$	$\neg X$	$\neg Y$
$\neg(X \rightarrow Y)$	X	$\neg Y$

and

β	β_1	β_2
$\neg(X \wedge Y)$	$\neg X$	$\neg Y$
$X \vee Y$	X	Y
$X \rightarrow Y$	$\neg X$	Y

Rules for arbitrary truth-functional connectives can be determined analogously. For the analysis of quantifiers we distinguish the following types of formulae:

γ	$\gamma_0(t)$
$(\forall x)\phi(x)$	$\phi(t)$
$\neg(\exists x)\phi(x)$	$\neg\phi(t)$

and

δ	$\delta_0(t)$
$\neg(\forall x)\phi(x)$	$\neg\phi(t)$
$(\exists x)\phi(x)$	$\phi(t)$

Using these notations, the quantifier rules can be stated as:

$$\frac{\gamma}{\gamma_0(y))}[\gamma\text{-rule}]$$

where y is a new free variable.

$$\frac{\delta}{\delta_0(f(\mathbf{x}))}[\delta\text{-rule}]$$

where $\mathbf{x}$ is the vector of the free variables occurring in δ and f is a function symbol uniquely associated with the set of formulae that are identical to δ up to renaming of variables.

Remark. In fact, the above rule is not Fitting's original δ-rule, but the improved version called δ^{++}-rule in [4]. We emphasize that the choice of the type of δ-rule is independent from the induction mechanisms introduced below. For an analysis of the effects of different types of δ-rules on proof complexity see [2].

Theories with induction usually contain the equality predicate. This is obvious for successor induction; but even if order induction is used, a sensible handling of equality is pertinent in all realistic scenarios. In [12] a "free variable tableau replacement rule" is described, a version of which is called "tableau paramodulation" in [8]. Our results are largely independent from the choice of a mechanism for equality handling. However, to achieve reasonably efficient proof search for non-trivial examples it is advisable to employ one of the more sophisticated equality reasoning methods for tableaux as introduced, e.g., in [9] and [7] and nicely summarized in [8]. In our central examples in Section 4, we shall make use of the following *paramodulation rule*:

$$\frac{\begin{array}{c} A[s'] \\ s = t \end{array}}{A[t]}$$

where the mgu of s and s' is applied
to the whole tableau immediately after
adding $A[t]$ to the current branch.

For completeness we also have to require that every branch containing a formula of the form $s \neq t$ can be closed if there exists an mgu σ such that $s\sigma = t\sigma$. The substitution σ is applied to the whole tableau immediately after closing the current branch.

Whatever versions of the δ- and equality-rules are chosen, we shall concentrate on additional rules for the "underlying tableau calculus" that correspond to the application of induction principles.

3 Induction Principles

We emphasize the usefulness of *different* forms of induction.
The first-order schema

$$[A(0) \wedge \forall x(A(x) \rightarrow A(s(x)))] \rightarrow \forall y A(y) \tag{SI}$$

is best known as an axiom schema of Peano Arithmetic. For our purposes another form of successor induction is more suitable:

$$\exists x A(x) \rightarrow [\exists z \, (A(z) \wedge \forall y \, (A(y) \rightarrow z \neq s(y)))]. \tag{S2}$$

The fact that S2 corresponds to SI in arithmetic is most readily seen if one take the contraposition of (S2) and using case distinction $z = s(y)$ and $z \neq s(y)$. The resulting schema is (SI) except for the negation sign in front of the schema variable A. In the context of Hilbert's ε-calculus (see [14]), the induction schema is simply:

$$A(y) \rightarrow e \neq s(y) \qquad \text{where } e = \varepsilon_z(A(z)). \tag{ε-IND}$$

In Peano Arithmetic, *successor induction* is equivalent to the principle of *complete induction*:

$$\forall xy\, ([(y < x \rightarrow A(y)) \rightarrow A(x)] \rightarrow \forall z A(z)) \qquad \text{(CI)}$$

which in turn can be reformulated to take the form of the *least number principle*:

$$\exists x A(x) \rightarrow \exists z\, [A(z) \wedge \forall y\, (y < z \rightarrow \neg A(y))]. \qquad \text{(O1)}$$

We want to point out that not only (fragments of) arithmetic but also many other theories ask for a proper handling of various forms of induction. In particular CI and O1 make sense for all Noetherian relations "$<$"; but also different interpretations of the function symbol s (and the constant 0) lead to various interesting theories. We refer to Example 3 in Section 4 for a problem which requires induction with respect to a non-total ordering. There we shall use an induction rule that corresponds to the following schema O2, which holds in many theories for which O1 is not valid:

$$\exists x A(x) \rightarrow \exists z\, [A(z) \wedge \forall y\, (A(y) \rightarrow (z \neq y \vee \neg(y < z)))] \qquad \text{(O2)}$$

O2 encodes the fact that, for every property A, there is a witness for which "$<$" is irreflexive. O2 is very weak in the sense that it is implied by O1 but not vice versa. Still other useful induction schemata arise by combining the successor function and the order relation. Consider, for example,

$$\exists x A(x) \rightarrow \exists z\, [A(z) \wedge \forall y\, (A(y) \rightarrow z < s(y))] \qquad \text{(SO)}$$

which is valid for the natural numbers and a few other interesting structures.

4 Induction Closure Rules for Tableaux

How can the induction principles of Section 3 be incorporated into free variable tableau calculi? In principle, one could explicitly augment the prefix of each tableau with all relevant instances of the appropriate induction schema. Obviously, this naive strategy will in general blow up the search space in a forbidding manner. As a more elegant way of dealing with induction, we suggest to add conditions for the closure of branches. These conditions are triggered by applications of the δ-rule.

Consider order induction in the form of O1. Informally, we may understand the schema to express the validity of the following principle in the context of tableaux:

Suppose a term f is introduced by a δ-rule to replace $\exists x A(x)$ by $A(f)$. By the induction principle, we may assume that f is the *smallest* witness for $A(\cdot)$. Now, assume that on some branch containing $A(f)$ we also find $A(t)$ as well as $t < f$ for some term t. This expresses that $A(\cdot)$ also holds for some element smaller than f, contradicting the minimality assumption for f. Therefore this branch gets closed.

To remain in the realm of feasible tableau proof search procedures we require $A(x)$ to be a literal, i.e. a negated or unnegated atomic formula. Moreover, we incorporate the unification principle into this new closure rule.

Definition 1 (O1-rule). Let $A(f(\mathbf{x}))$ be the δ_0-formula introduced by an application of a δ-rule, where A is a literal. Whenever a branch contains the formulae $A(f(\mathbf{x}))$, $A(t)$ and $t' < r'$, and there is a substitution σ such that $t\sigma = t'\sigma$ as well as $f(\mathbf{x})\sigma = r'\sigma$, the branch can be closed by applying the most general substitution σ with this property to the whole tableau.

The restriction to atomic induction formulae leads to a calculus that may be weaker for some theories than, e.g., Gentzen's **LK** with all instances of the induction schema O1 as additional axioms. Moreover, as is well known, cuts cannot be eliminated in presence of induction. However, our main goal is to demonstrate that even the simple rule of Definition 2 added to a standard tableau calculus like the one in [12] suffices to solve many problems that require induction. To this aim we provide detailed examples. A discussion how the resulting calculus can be strengthened can be found in Section 5.

Example 1. Consider a weak fragment of arithmetic, characterized by the following three axioms for the successor function "s" and the order predicate "$<$".

$$\forall x \, \neg(x < 0) \tag{1}$$

$$\forall xy \, (s(x) < s(y) \rightarrow x < y) \tag{2}$$

$$\forall x \exists y \, [x \neq 0 \rightarrow (x = s(y) \land y < x)] \tag{3}$$

Although it is easy to refute $s^{(n)}(0) < s^{(n)}(0)$ for all values of n, one cannot prove $\forall x \, \neg(x < x)$ without induction in this theory as the following model $\mathcal{N}$ of (1), (2), and (3) demonstrates.

As the domain of $\mathcal{N}$ we take the set $\mathbb{N}$ of natural numbers augmented by an additional element c. The function s is interpreted as successor function for all $n \in \mathbb{N}$; $val_{\mathcal{N}}(s)(c) = c$. The constant symbol 0 is interpreted as the number 0. Finally, $val_{\mathcal{N}}(<)$ is the extension of the predicate "smaller than" on $\mathbb{N}$ and $val_{\mathcal{N}}(n < c) = \mathbf{true}$ for $n = c$ or $n \in \mathbb{N}$ but $val_{\mathcal{N}}(c < n) = \mathbf{false}$ for all $n \neq c$.

Obviously, $\forall x \, \neg(x < x)$ is false in $\mathcal{N}$. In contrast, the O1-rule leads to a fairly short refutation of $\exists x \, x < x$ in our theory of successor and "smaller than" as demonstrated in Figure 1.

Consider successor induction in the form of S2. In the same way as the order induction schema O1 translates into the tableau closure rule of Definition 1, schema S2 can be transformed into the following rule.

Definition 2 (S2-rule). Let $A(f(\mathbf{x}))$ be the δ_0-formula introduced by an application of the δ-rule, where A is a literal. Whenever a branch contains the formulae $A(f(\mathbf{x}))$, $A(t)$ and $r' = s(t')$, and there is a substitution σ such that $t\sigma = t'\sigma$ and $f(\mathbf{x})\sigma = r'\sigma$, the branch can be closed by applying the most general substitution σ with this property to the whole tableau.

$$1 : \forall x \, \neg(x < 0)$$
$$2 : \forall xy \, (s(x) < s(y) \rightarrow x < y)$$
$$3 : \forall x \exists y \, [x \neq 0 \rightarrow (x = s(y) \wedge y < x)]$$
$$C : \exists x \, x < x$$
$$4[\delta(C)] : \quad c < c \qquad \Leftarrow \text{triggers induction rule}$$
$$5[\gamma(3)] : \exists y \, [x_1 \neq 0 \rightarrow (x_1 = s(y) \wedge y < x_1)]$$
$$6[\delta(5)] : \quad x_1 \neq 0 \rightarrow (x_1 = s(p(x_1)) \wedge p(x_1) < x_1))$$

$7[\beta_1(6)] : \ \neg(x_1 \neq 0)$	$8[\beta_2(6)] : \ x_1 = s(p(x_1)) \wedge p(x_1) < x_1$
$9[\neg\neg(7)] : \ x_1 = 0$	$8'[\text{sub}(\sigma)] : \ c = s(p(c)) \wedge p(c) < c$
$10[P(9{\to}4)] : \ c < 0$	$13[\alpha_1(8')] : \ c = s(p(c))$
$\sigma = \{x_1 \backslash c\}$	$14[\alpha_2(8')] : \ p(c) < c$
$11[\gamma(1)] : \ \neg(x_2 < 0)$	$15[P(13{\to}4)] : \ c < s(p(c))$
$X[10/11] : \ \{x_2 \backslash c\}$	$16[P(13{\to}15)] : \ s(p(c)) < s(p(c))$
	$17[\gamma(2)] : \ s(x_3) < s(y_3) \rightarrow x_3 < y_3$

$$18[\beta_1(17)] : \ \neg(s(x_3) < s(y_3)) \qquad 19[\beta_2(17)] : \ x_3 < y_3$$
$$X[18/16] : \rho = \{x_3\backslash p(c), y_3\backslash p(c)\} \quad 19'[\text{sub}(\rho)] : \ p(c) < p(c)$$
$$\text{closure by O1} : [4/14/19']$$

Fig. 1. Tableau proof for Example 1 (ordering induction O1).

To assist the reader in understanding the correctness of this rule (with respect to theories containing S2 as axiom schema), we state once more a corresponding informal justification. (Formal correctness proofs are outlined in Section 5.)

Let f be the (Skolem-)term introduced by the δ-rule to analyze $\exists x A(x)$. By the induction principle, we may assume that f is the *smallest* witness for $A(\cdot)$. Now, suppose that on some branch containing $A(f)$ we also find $A(t)$ as well as $f = s(t)$ for some term t. This expresses that $A(\cdot)$ also holds for the predecessor of f, which contradicts the assumption that f is minimal with respect to $A(\cdot)$. Therefore this branch gets closed.

Example 2. In order to illustrate Definition 2, consider Robinson's arithmetic **Q**, i.e., Peano Arithmetic without induction. In fact, we do not need all seven of Robinson's axioms but only the successor axiom

$$\forall x[x \neq 0 \rightarrow \exists y(x = s(y))] \tag{1}$$

and the usual inductive definition for addition:

$$\forall x \, x + 0 = x \tag{2}$$

$$\forall xy \, x + s(y) = s(x + y) \tag{3}$$

It is well known that one cannot prove $\forall x\ 0+x = x$ in $\mathbf{Q}$, although $0+t = t$ is derivable for all ground terms t. Figure 2 shows that the negation — $\exists x\ 0+x \neq x$ — of this theorem gets refutable if we apply our tableau closure rule for successor induction.

$1:\ \forall x\ [x \neq 0 \rightarrow \exists y(x = s(y))]$

$2:\ \forall x\ x + 0 = x$

$3:\ \forall xy\ x + s(y) = s(x + y)$

$C:\ \exists x\ 0 + x \neq x$

$E:\ \forall xy\ [x = y \rightarrow s(x) = s(y)]$

$4[\delta(C)]:\ 0 + c \neq c$ $\Leftarrow$ triggers induction rule

$5[\gamma(2)]:\ x_1 + 0 = x_1$

$6[\gamma(1)]:\ x_2 \neq 0 \rightarrow \exists y\ x_2 = s(y)$

$7[\beta_1(6)]:\ x_2 = 0$

$8[P(7{\rightarrow}5)]:\ x_2 = 0 + 0$

     $\{x_1\backslash 0\}$

$9[P(7{\rightarrow}8)]:\ x_2 = 0 + x_2$

    $\mathsf{X}[4/9]:\ \sigma = \{x_2\backslash c\}$

$10[\beta_2(6)]:\ \exists y\ x_2 = s(y)$

$10'[\mathrm{sub}(\sigma)]:\ \exists y\ c = s(y)$

$11[\delta(10')]:\ c = s(d)$

$12[\gamma(E)]:\ x_3 = y_3 \rightarrow s(x_3) = s(y_3)$

$13[\beta_1(12)]:\ x_3 \neq y_3$

closure by S2 : $[4/11/13]$

$\rho = \{x_3\backslash 0 + d, y_3\backslash d\}$

$14[\beta_2(12)]\ s(x_3) = s(y_3)$

$14'[\mathrm{sub}(\rho)]:\ s(0 + d) = s(d)$

$15[\gamma(3)]:\ s(x_4 + y_4) = x_4 + s(y_4)$

$16[P(15{\rightarrow}14')]:\ 0 + s(d) = s(d)$

$17[P(11{\rightarrow}16)]:\ 0 + c = c$

    $\mathsf{X}[4/17]$

Fig. 2. Tableau proof for Example 2 (successor induction).

Observe that we have to make use of the equality axiom E to obtain a closed tableau. This is only the case because the δ_0-formula that triggers the induction closure rule is a negated equality atom. Since the proper instance of the induction schema S2 is not present as a formula in the tableau (but only as a condition against which other literals are matched), the equality atom (corresponding to the negation of $A(t)$) does not occur explicitly on the branch. Thus it cannot be used for paramodulation. This is the reason why we had to state axiom E explicitly. .

In a similar manner as for O1 and S2, one can also translate all other induction schemata of the appropriate (existential) form into tableau closure rules. The important point here is that these principles are not at all equivalent in general. We illustrate this with a more complex example referring to the induction schema O2.

The tableau closure rule obtained from O2 is as follows.

Definition 3 (O2-rule). Let $A(f(\mathbf{x}))$ be the δ_0-formula introduced by an application of a δ-rule, where A is a literal. Whenever a branch contains the formulae $A(f(\mathbf{x}))$, $A(t)$, $t' = r'$ and $t'' < r''$, and there is a substitution σ such that $t\sigma = t'\sigma = t''\sigma$ as well as $f(\mathbf{x})\sigma = r'\sigma - r''\sigma$, the branch can be closed by applying the most general substitution σ with this property to the whole tableau.

The informal justification of this induction principle is completely analogous to that of the O1-rule.

Example 3. To illustrate the O2-induction closure rule we choose an example for which the order along which we have to induce is only partial in general. To emphasize that our rules are not only useful in the context of arithmetic, we investigate axioms that specify properties of binary strings.[4] We code binary strings over the alphabet $\{0, 1\}$ as terms consisting of the unary function symbols $0(\cdot)$ and $1(\cdot)$, and the constant symbol ϵ, which denotes the empty string. Thus, e.g., the string 100 is represented as $1(0(0(\epsilon)))$, or rather 100ϵ, as we prefer to omit the parentheses.

The axiom

$$\forall x[x \neq \epsilon \to \exists y(x = 0y \lor x = 1y)] \tag{1}$$

states that all non-empty strings start either with letter 0 or 1. Obviously, we also want to have

$$\forall x(x \neq 0x \land x \neq 1x) \tag{2}$$

The intended interpretation of "$s < t$" is: "s is a proper prefix of t". Therefore we state

$$\forall xy[(0x < 0y \lor 1x < 1y) \to x < y] \tag{3}$$

$$\forall xy[x < y \to (0x < 0y \land 1x < 1y)] \tag{4}$$

$$\forall x[\neg(0x < x) \land \neg(1x < x)] \tag{5}$$

as well as

$$\forall x \neg(x < \epsilon). \tag{6}$$

Although one can easily prove $\neg(t < t)$ for all concrete strings, i.e. ground terms t, induction is needed to show $\forall x \neg(x < x)$. This is demonstrated by the following (non-standard) model $\mathcal{M}$ for $(1) \land \ldots \land (6)$. The domain of $\mathcal{M}$ is $\{e, a, b, c, d\}$; $val_{\mathcal{M}}(\epsilon) = e$ and $val_{\mathcal{M}}(1)$ and $val_{\mathcal{M}}(0)$ are given by the following table:

---

[4] Actually, we do not need all axioms in the induction proof.

s	$val_{\mathcal{M}}(1)(s)$	$val_{\mathcal{M}}(0)(s)$
e	a	a
a	b	b
b	a	a
c	d	d
d	c	c

Moreover, $val_{\mathcal{M}}(c < c) = val_{\mathcal{M}}(d < d) = \textbf{true}$; and $val_{\mathcal{M}}(s < t) = \textbf{false}$ for all other pairs of terms s, t. It can easily be checked that $\mathcal{M}$ satisfies all of the above axioms, but, by definition, the statement $\forall x \neg (x < x)$ is false in $\mathcal{M}$. Observe that O2 does not imply O1. Different induction principles are — not only proof theoretically, but even semantically — non-equivalent in general.

Using the induction closure rule of Definition 3, we can refute the negation $\exists x \; x < x$ of $\forall x \neg (x < x)$ in our theory of binary strings as shown in the following Figure 3 and Figure 4.

$$
\begin{aligned}
&1: \; \forall x \; (x \neq \epsilon \rightarrow \exists y \; (x = 0y \lor x = 1y)) \\
&2: \; \forall x \; (x \neq 0x \land x \neq 1x) \\
&3: \; \forall xy \; ((0x < 0y \lor 1x < 1y) \rightarrow x < y) \\
&4: \; \forall xy \; (x < y \rightarrow (0x < 0y \land 1x < 1y)) \\
&5: \; \forall x \; (\neg(0x < x) \land \neg(1x < x)) \\
&6: \; \forall x \; \neg(x < \epsilon) \\
&C: \; \exists x \; x < x \\
&8[\delta(C)]: c < c \qquad\qquad\qquad \Leftarrow \text{triggers induction rule} \\
&9[\gamma(1)]: x_1 \neq \epsilon \rightarrow \exists y \; (x_1 = 0y \lor x_1 = 1y)
\end{aligned}
$$

$10[\beta_1(9)]: \neg(x_1 \neq \epsilon)$ $\qquad\qquad$ $11[\beta_2(9)]: \exists y \; (x_1 = 0y \lor x_1 = 1y)$
$12[\neg\neg(10)]: x_1 = \epsilon$ $\qquad\qquad\qquad$ see Figure 4
$13[P(12{\rightarrow}8)]: c < \epsilon$
$\quad \sigma = \{x_1 \backslash c\}$
$14[\gamma(6)]: \neg(x_6 < \epsilon)$
$\quad \mathsf{X}[13/14] : \{x_6 \backslash c\}$

Fig. 3. Tableau proof for Example 3 – Part 1

The partial substitution obtained for the tableau in Figure 3 is $\sigma = \{x_1 \backslash c, x_6 \backslash c\}$. The tableau construction is continued on the open (right) branch of Figure 3. After instantiating with σ, we get the expansion of the tableau as shown in Figure 4. The final substitution obtained for the whole tableau is σ. The open (right) path can be expanded as the left part with two differences, namely (i) $0d$ has to be replaced by $1d$ and (ii) the formula numbers in the closure have to be adapted accordingly.

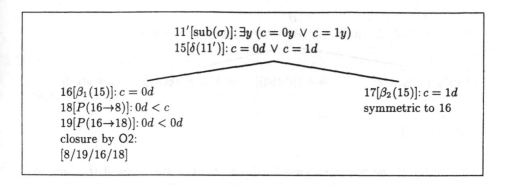

Fig. 4. Tableau proof for Example 3 – Part 2

Like in Example 2, we may have to augment our axioms by equality axioms. There, we have to add equality axioms if the δ_0-formula that triggers the induction closure condition is a negated equality atom.

As an easy but very useful exercise the ambitious reader should formalize the tableau closure rule corresponding to schema SO and find a problem for which this SO-rule is essential to get a proof.

5 Correctness and Completeness

We already have stated informal justifications for our induction closure rules in Section 4. We provide a more formal and informative correctness proof for the S2-rule. The other principles can be shown to be correct in an analogous way.

Theorem 4. *The S2-rule is correct. More exactly, any closed tableau constructed using the S2-rule (Definition 2) and standard propositional, quantifier and equational rules can be transformed into a closed tableau without S2-rule applications but with all adequate instances of schema S2 in the prefix.*

Proof. We show that, under the assumption of Definition 2, any branch B closable by the S2-rule is also closed by a subtableau where the induction axiom S2 is used. Assume that

$$l_1 : A(f(\mathbf{x})), \quad l_2 : A(t), \quad \text{and} \quad l_3 : r' = s(t')$$

occur on B which is closed by the S2-rule. Recall that there is an mgu σ which simultaneously makes t and t' as well as r' and $f(\mathbf{x})$ identical. Instead of using the S2-rule, B is extended by a closed tableau constructed step-wise as follows.

$$B$$
$$S2: \ \exists x A(x) \ \rightarrow \ \exists z[A(z) \wedge \forall y(A(y) \rightarrow z \neq s(y))]$$

$m[\beta_1(S2)]: \ \neg \exists x A(x)$

$m+2[\gamma(m)]: \ \neg A(u)$

$\mathsf{X}[l_1/m+2]:$

$\mu = \{u \backslash f(\mathbf{x})\}$

$m+1[\beta_2(S2)]: \ \exists z[A(z) \wedge \forall y(A(y) \rightarrow z \neq s(y))]$

$m+3[\delta(m+1)]: \ A(k(\mathbf{v})) \wedge \forall y(A(y) \rightarrow k(\mathbf{v}) \neq s(y))]$

$m+4[\alpha_1(m+3)]: \ A(k(\mathbf{v}))$

The arity of the function symbol k is identical to the arity of the function symbol f. Now we derive $A(g)$ and $h' = s(g')$ using $A(k(\mathbf{v}))$ instead of $A(f(\mathbf{x}))$. The term g corresponds to the term t and $h' = s(g')$ corresponds to $r' = s(t')$. Obviously, $g\tau = g'\tau$ and $h'\tau = k(\mathbf{v})\tau$. Then we proceed as follows.

$$\vdots$$

$$m+4[\alpha_1(m+3)]: \ A(k(\mathbf{v}))$$

$$\vdots$$

$$l_2'[\ldots]: \ A(g)$$

$$\vdots$$

$$l_3'[\ldots]: \ h' \stackrel{.}{=} s(g')$$

$$\vdots$$

$$n[\alpha_2(m+3)]: \ \forall y(A(y) \rightarrow k(\mathbf{v}) \neq s(y))$$

$$n+1[\gamma(n)]: \ A(y_1) \rightarrow k(\mathbf{v}) \neq s(y_1)$$

$n+2[\beta_1(n+1)]: \ \neg A(y_1)$

$\mathsf{X}[l_2'/n+2]:$

$\lambda = \{y_1 \backslash g\}$

$n+3[\beta_2(n+1)]: \ k(\mathbf{v}) \neq s(y_1)$

Hence, we have an inequality $k(\mathbf{v}) \neq s(g)$ if we instantiate y_1 in $n+3$. Since $g'\tau = g\tau$, we obtain $k(\mathbf{v})\tau \neq h'\tau$. The path gets closed by reference to the reflexivity axiom because $h'\tau$ is identical to $k(\mathbf{v})\tau$. $\qquad \square$

Analogously to Theorem 4, one can show the correctness of the other two rules defined in Section 4.

Theorem 5. *The O1-rule and the O2-rule are correct.*

Let us continue with a discussion of the strength of the introduced tableau calculus with lean induction principles.

Remarks on completeness (strength)[5] **:** It is well known that theories containing induction principles generally do not admit cut-free proof systems.

---

[5] Because of the essential incompleteness of sufficiently strong theories it should be clear that the issue here is the relative strength of different proof systems and not model theoretical completeness.

However, observe that the explicit use of equality axioms like $S = \forall xy\,((x = y \wedge A(x)) \rightarrow A(y))$ (together with the axiom $\forall x\, x = x$) is a substitute for the "bivalence principle" for tableau, which in turn is nothing else but "cut". More exactly, the use of S allows to extend any branch B of a tableau by the following tableau figure:

$$\vdots$$

$$E_1 : \forall x\, x = x$$
$$E_2 : \forall xy\,((x = y \wedge A(x)) \rightarrow A(y))$$
$$l[\gamma(E_2)] : (x = y \wedge A(x)) \rightarrow A(y)$$

$$l_1[\beta_1(l)] : \neg(x = y \wedge A(x)) \qquad\qquad m[\beta_2(l)] : A(y)$$
$$m'[\mathrm{sub}(\sigma)] : A(x)$$

$$l_2[\beta_1(l_1)] : x \neq y \qquad\qquad l_3[\beta_2(l_1)]\,\neg A(x)$$
$$l_2 + 1[\gamma(E_1)] : x_1 = x_1$$
$$\mathsf{X}[l_2 + 1/l_2]$$
$$\sigma = \{x_1 \backslash x, y \backslash x\}$$

In a similar way one can generate arbitrary instantiations of the cut formula $A(x)$ by use of equality axioms.

A basic fact of mathematical logic is the existence of the arithmetical hierarchy, which expresses that the restriction to induction formulae of limited logical complexity leads to strictly weaker calculi [13]. Since our rules only represent induction on literals, it may seem that the resulting calculi are very weak. This observation should be qualified as follows.

(1) We point out that for sufficiently strong theories — like arithmetic with a function symbol "$\dot{-}$" for $\lambda xy[\max\{x - y, 0\}]$ — the restriction to literals as induction formulae is inessential as far as only the propositional structure is concerned: We can bring all formulae $Q_1...Q_n\mathbf{x}\, A(\mathbf{x})$, where A is quantifier free, into the form

$$Q_1...Q_n\mathbf{x} \bigvee_i \bigwedge_j r_i^j = t_i^j \wedge \bigwedge_k \hat{r}_i^k \neq \hat{t}_i^k$$

which is equivalent to

$$Q_1...Q_n\mathbf{x} \bigvee_i \bigwedge_j r_i^j = t_i^j \wedge \bigwedge_k \exists y\,(r_i^k + s(y) = t_i^k \vee r_i^k = t_i^k + s(y))$$

and, consequently,

$$Q_1...Q_{n'}\mathbf{x'} \bigvee_l \bigwedge_m (r_l^m \dot{-} t_l^m) + (t_l^m \dot{-} r_l^m) = 0$$

which reduces to the literal

$$\prod_i \sum_j (r_i^j \dot{-} t_i^j) + (t_i^j \dot{-} r_i^j) = 0.$$

(2) Independently from the theory itself, one can recover the full strength of analytic induction and cut within analytic tableau (with equality axioms) and our (atomic) induction closure rules by expanding all formulas to definitional form by definitional (or structure-preserving) translations [10, 11, 16]. Of course, this is only possibly by the fact that the expansion to definitional forms extends the original language.

Naturally, we do not suggest to proceed according to remarks (1) and (2) in realistic scenarios of Automated Deduction. We rather want to emphasize (especially with a hint to our examples in Section 4) that even without such "tricks", our simple induction rules turn out to be surprisingly strong.

6 Methodological Postscript

The work documented by the above results was in part motivated by the following question: In which situations and why should one opt for a tableau-based calculus instead of a saturation-based one like resolution or superposition? We feel that this and related important *methodological issues* receive too few or too shallow consideration at least in the case of classical logic.

Concerning induction one should keep in mind that Herbrand's theorem does not hold in sufficiently strong theories (like Peano Arithmetic)[6]. Therefore, no analytic proof systems exist, as long as we do not consider extensions of the language. This even holds if induction is restricted to literals only, like in our examples above. Nevertheless, we have been able to provide an essentially *analytic* presentation of such theories in tableau format. This is only possible since the δ-rules of tableau procedures represent a form of Skolemization that generates the necessary *extension of the language* in form of Skolem terms *during proof search*. As a general principle, this flexibility of the Skolemization mechanism in tableaux represents a sometimes decisive advantage over calculi like resolution and superposition, where Skolemization is understood as a preprocessing step.

References

1. R. Aubin. Mechanizing Structural Induction. *Theoretical Computer Science*, 9:329–362, 1980.
2. M. Baaz and C.G. Fermüller. Non-elementary speed-ups between Different Versions of Tableaux. In R. Hähnle P. Baumgartner and J. Posegga, editors, *Theorem Proving with Analytic Tableaux and Related Methods (4th International Workshop TABLEAUX'95)*, pages 217–230. Springer, 1995.
3. M. Baaz and A. Leitsch. Methods of Functional Extension. In *Collegium Logicum: Annals of the Kurt Gödel Society*. Springer, 1994.
4. B. Beckert, R. Hähnle, and P.H. Schmitt. The Even More Liberalized δ-Rule in Free Variable Semantic Tableaux. In A. Leitsch G. Gottlob and D. Mundici, editors, *Proceedings of the Kurt Gödel Colloquium*. Springer, 1993.

---

[6] Observe that the existence of an Herbrand expansion for Peano Arithmetic would imply that all ϵ_0-recursive functions were composable by case distinction from primitive recursive ones.

5. W. Bibel. *Automated Theorem Proving.* Vieweg, Braunschweig, second edition, 1987.
6. R. S. Boyer and J. S. Moore. *A Computational Logic.* Academic Press, 1979.
7. A. Degtyarev and A. Voronkov. Equality elimination for the tableau method. In *Proceedings of* DISCO'96, 1996.
8. A. Degtyarev and A. Voronkov. Equality Reasoning in Sequent-Based Calculi. Technical Report No. 127, CS Department, Uppsala University, 1996.
9. A. Degtyarev and A. Voronkov. What you always wanted to know about rigid *E*-unification. In *Proceedings of* JELIA'96, 1996.
10. E. Eder. An Implementation of a Theorem Prover Based on the Connection Method. In W. Bibel and B. Petkoff, editors, *AIMSA 84, Artificial Intelligence - Methodology, Systems, Applications, Varna, Bulgaria.* North-Holland Publishing Company, 1984.
11. E. Eder. *Relative Complexities of First Order Calculi.* Vieweg, Braunschweig, 1992.
12. M. Fitting. *First-Order Logic and Automated Theorem Proving.* Springer, second edition, 1996.
13. P. Hájek and P. Pudlák. *Metamathematics of First-Order Arithmetic.* Springer Verlag, 1993.
14. D. Hilbert and P. Bernays. *Grundlagen der Mathematik II.* Springer, 1939.
15. D. R. Musser. On Proving Inductive Properties of Abstract Data Types. In *Proc. Principles of Programming Languages*, pages 154–162, 1980.
16. D. A. Plaisted and S. Greenbaum. A Structure-Preserving Clause Form Translation. *J. Symbolic Computation*, 2:293–304, 1986.
17. H. Zhang, editor. *J. Automated Reasoning: Special Issue on Inductive Theorem Proving*, volume 16, nos 1-2, 1996.

Tableaux for Diagnosis Applications

Peter Baumgartner[1], Peter Fröhlich[2], Ulrich Furbach[1], Wolfgang Nejdl[2]

[1] Universität Koblenz, Institut für Informatik
E-mail: {peter,uli}@informatik.uni-koblenz.de
[2] Universität Hannover
E-mail: {froehlich,nejdl}@kbs.uni-hannover.de

Abstract. In [NF96] a very efficient system for solving diagnosis tasks has been described, which is based on belief revision procedures and uses first order logic system descriptions. In this paper we demonstrate how such a system can be rigorously formalized from the viewpoint of deduction by using the calculus of hyper tableaux [BFN96]. The benefits of this approach are twofold: first, it gives us a clear logical description of the diagnosis task to be solved; second, as our experiments show, the approach is feasible in practice and thus serves as an example of a successful application of deduction techniques to real-world applications.

1 Introduction

In this paper we will demonstrate that model generation theorem proving is very well suited for solving consistency-based diagnosis tasks. According to Reiter ([Rei87]) a simulation model of the technical device under consideration is constructed and is used to predict its normal behavior. By comparing this prediction with the actual behavior it is possible to derive a diagnosis.

This work was motivated by the study of the diagnosis system DRUM-2 [FN96,NF96]. This system is in essence a model generation procedure which takes into account the particularities of logical descriptions of a diagnosis task. In this paper we demonstrate how such a system can be rigorously formalized from the viewpoint of tableau based theorem proving by using the calculus of hyper tableaux [BFN96]. The resulting system is quite efficient. We know of no other general purpose theorem prover which has been used to solve large diagnosis problems.

Notation. We assume that the reader is familiar with the basic concepts of propositional logic. *Clauses*, i.e. multisets of literals, are usually written as the disjunction $A_1 \vee \cdots \vee A_m \vee \neg B_1 \vee \cdots \vee \neg B_n$ or as an implication $A_1, \ldots, A_m \leftarrow B_1, \ldots, B_n$ $(m \geq 0, n \geq 0)$. With $\overline{L}$ we denote the complement of a literal L. Two literals L and K are *complementary* if $\overline{L} = K$.

2 Model–Based Diagnosis

Heuristic rule–based expert systems were the first approach to automated diagnosis. The knowledge bases of such systems could not be easily modified or verified to be correct and complete. These difficulties have been overcome by the introduction of model–based diagnosis [Rei87], where a simulation model of the device under consideration is used to predict its normal behavior, given the observed input parameters. Diagnoses are computed by comparison of predicted vs. actual behavior (Fig. 1).

This approach uses an extendible logical description of the device, called the system description (SD), formalized by a set of first–order formulas. The system description consists of a set of axioms characterizing the behavior of system components of certain

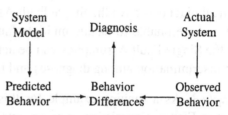

Fig. 1. Model–Artifact Difference

types. The topology is modeled separately by a set of facts. We will now define the diagnostic concept mathematically. The diagnostic problem is described by system description SD, a set $COMP$ of components and a set OBS of observations (logical facts). With each component we associate a behavioral mode: $Mode(c, Ok)$ means that component c is behaving correctly, while $Mode(c, Ab)$ (abbreviated by $Ab(c)$) denotes that c is faulty.

A *Diagnosis* D is a set of faulty components, such that the observed behavior is consistent with the assumption, that exactly the components in D are behaving abnormally.

Definition 1 (Reiter 87). A *Diagnosis* of $(SD, COMP, OBS)$ is a set $\Delta \subseteq COMP$, such that $SD \cup OBS \cup \{Ab(c) \mid c \in \Delta\} \cup \{\neg Ab(c) \mid c \in COMP - \Delta\}$ is consistent. Δ is called a *Minimal Diagnosis*, iff it is the minimal set (wrt. $\subseteq$) with this property.

Minimal Diagnoses are a natural concept, because we do not want to assume that a component is faulty, unless this is necessary to explain the observed behavior. See [CT91] for an overview of alternative diagnostic definitions. To compute diagnoses of $(SD, COMP, OBS)$ it is sufficient to compute models of $SD \cup OBS$:

Corollary 2. *Let M be a model of $SD \cup OBS$. Then $\{c \in COMP \mid M(Ab(c)) = true\}$ is a diagnosis of $(SD, COMP, OBS)$.*

The set of all minimal diagnoses can be large for complex technical devices. Therefore, stronger criteria than minimality are often used to further discriminate among the minimal diagnoses. These criteria are usually based on the probability or cardinality of diagnoses. In the remainder of this paper we will use restrictions on the cardinality of diagnoses. We say that a given diagnosis Δ *satisfies the n-Faults assumption* iff $|\Delta| \leq n$. A widely used example of the n–Fault–Assumption is the 1–Fault–Assumption or *Single Fault Assumption*. Many specialized systems for technical diagnosis have the Single Fault Assumption implicit and are unable to handle multiple faults. In model–based diagnosis systems the Single Fault Assumption can be activated explicitly in order to provide more discrimination among diagnoses and to speed up the diagnosis process.

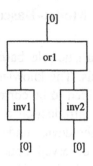

Fig. 2. A Circuit

Example 3. As a running example for this paper consider the simple digital circuit in Figure 2 consisting of an or–gate (*or1*) and two inverters (*inv1* and *inv2*). Its function can be described by the following propositional formulas[1].

$$OR1 : \neg(ab(or1)) \rightarrow high(or1, o) \leftrightarrow (high(or1, i1) \vee high(or1, i2))$$
$$INV1 : \neg(ab(inv1)) \rightarrow high(inv1, o) \leftrightarrow \neg(high(inv1, i))$$
$$INV2 : \neg(ab(inv2)) \rightarrow high(inv2, o) \leftrightarrow \neg(high(inv2, i))$$
$$CONN1 : high(inv1, o) \leftrightarrow high(or1, i1)$$
$$CONN2 : high(inv2, o) \leftrightarrow high(or1, i2)$$

Thus we have $SD = \{OR1, INV1, INV2, CONN1, CONN2\}$ as the system description. We observe that both inputs of the circuit have low voltage and the output also has low voltage, i.e. the clause set of OBS is given by $\{\neg(high(inv1, i)), \neg(high(inv2, i)), \neg(high(or1, o))\}$. $(SD, COMP, OBS)$ admits exactly two minimal diagnosis: the single fault diagnosis $\{ab(or1)\}$ and the two fault diagnosis $\{ab(inv1), ab(inv2)\}$.

The expected behavior of the circuit given that both inputs are low would be high voltage at the outputs of both inverters and consequently high voltage at the output of the or–gate. This model of the correctly functioning device

$$I_0 = \{high(inv1, o), high(inv2, o), high(or1, i1), high(or1, i2), high(or1, o\}$$

can be computed very efficiently even for large devices by domain–specific tools, e.g. circuit simulators.

---

[1] These formulas can be obtained by instantiating a first order description of the gate functions with the structural information. The second parameters – iX and o – mean input and output, respectively

3 Hyper Tableaux Calculus

In [BFN96] we introduced a variant of clausal normal form tableaux called "hyper tableaux". Hyper tableaux keep many desirable features of analytic tableaux (structure of proofs, reading off models in special cases) while taking advantage of central ideas from (positive) hyper resolution. We refer the reader to [BFN96] for a detailed discussion. In order to make the present paper self-contained we will recall a simplified ground version of the calculus. Then we will show how it can be used to generate models for a diagnosis task.

From now on S always denotes a finite ground clause set, and Σ denotes its signature, i.e. the set of all predicate symbols occurring in it. We consider finite ordered trees T where the nodes, except the root node, are labeled with literals. In the following we will represent a branch b in T by the sequence $b = L_1,\ldots,L_n$ ($n \geq 0$) of its literal labels, where L_1 labels an immediate successor of the root node, and L_n labels the leaf of b. The branch b is called *regular* iff $L_i \neq L_j$ for $1 \leq i,j \leq n$ and $i \neq j$, otherwise it is called *irregular*. The tree T is *regular* iff every of its branches is regular, otherwise it is *irregular*. The set of *branch literals* of b is $\text{lit}(b) = \{L_1,\ldots L_n\}$. For brevity, we will write expressions like $A \in b$ instead of $A \in \text{lit}(b)$. In order to memorize the fact that a branch contains a contradiction, we allow to label a branch as either *open* or *closed*. A tableau is *closed* if each of its branches is closed, otherwise it is *open*.

A *selection function* is a total function f which maps an open tableau to one of its open branches. If $f(T) = b$ we also say that b *is selected in T by f*. Fortunately, there is no restriction on which selection function to use. For instance, one can use a selection function which always selects the "leftmost" branch.

Definition 4 (Hyper tableau). *Hyper tableaux for S are inductively defined as follows:*

Initialization step: The empty tree, consisting of the root node only, is a hyper tableau for S. Its single branch is marked as "open".

Hyper extension step: If (1) T is an open hyper tableau for S with open branch b, and (2) $C = A_1,\ldots,A_m \leftarrow B_1,\ldots,B_n$ is a clause from S ($m \geq 0$, $n \geq 0$), called *extending clause* in this context, and (3) $\{B_1,\ldots,B_n\} \subseteq b$ (equivalently, we say that C is *applicable to* b) then the tree T' is a hyper tableau for S, where T' is obtained from T by *extension of b by C*: replace b in T by the *new* branches $(b,A_1)\ldots,(b,A_m),(b,\neg B_1)\ldots,(b,\neg B_n)$ and then mark every inconsistent new branch[2] as "closed", and the other new branches as "open".

---

[2] A branch is *inconsistent* iff among its labels are two complementary literals.

We say that a branch b is *finished* iff it is either closed, or else whenever C is applicable to b, then extension of b by C yields some irregular new branch.

The applicability condition of an extension expresses that *all* body literals have to be satisfied by the branch to be extended (like in hyper *resolution*). From now on we consider only regular hyper tableaux. This restriction guarantees that for finite (ground) clause sets no branch can be extended infinitely often. This property is essential to guarantee that our proof procedure terminates (cf. Section 6).

Definition 5 (Branch Semantics). As usual, we represent an interpretation I for given domain Σ as the set $\{A \in \Sigma \mid I(A) = true, A \text{ atom}\}$. *Minimality* of interpretations is defined via set-inclusion.

Given a tableau with consistent branch b. The branch b is mapped to the interpretation $[\![b]\!]_\Sigma := \mathrm{lit}(b)^+$, where $\mathrm{lit}(b)^+ = \{A \in \mathrm{lit}(b) \mid A \text{ is a positive literal }\}$. Usually, we write $[\![b]\!]$ instead of $[\![b]\!]_\Sigma$ and let Σ be given by the context.

Example 6 (Hyper Tableau). The following is the clause set generated from our running Example 3.

ORE1:
$$ab(or1), high(or1, i1), high(or1, i2) \leftarrow high(or1, o)$$
$$ab(or1), high(or1, o) \leftarrow high(or1, i1)$$
$$ab(or1), high(or1, o) \leftarrow high(or1, i2)$$

INV1:
$$ab(inv1) \leftarrow high(inv1, o), high(inv1, i)$$
$$ab(inv1), high(inv1, i), high(inv1, o) \leftarrow$$

INV2:
$$ab(inv2) \leftarrow high(inv2, o), high(inv2, i)$$
$$ab(inv2), high(inv2, i), high(inv2, o) \leftarrow$$

CONN1:
$$high(or1, i1) \leftarrow high(inv1, o)$$
$$high(inv1, o) \leftarrow high(or1, i1)$$

CONN2:
$$high(or1, i2) \leftarrow high(inv2, o)$$
$$high(inv2, o) \leftarrow high(or1, i2)$$

LOW_INV1_I: $\leftarrow high(inv1, i)$ LOW_INV2_I: $\leftarrow high(inv2, i)$

OBSERVATION: $\leftarrow high(or1, o)$

Figure 3 contains a hyper tableau. Each open branch corresponds to a partial model. The highlighted model can be understood as an attempt to construct a model for the whole clause set, without assuming unnecessary ab-predicates. Only for making the clauses from *OR1* true is it necessary to include $ab(or1)$ into the model. $high(or1, o)$ cannot be assumed, as this contradicts the observation $\leftarrow high(or1, o)$.

In order to make the generation of models efficient enough for the large benchmark circuits used in this paper, additional knowledge has to be used to guide this model generation. This is done by starting from a model of the correct behavior of the device SD and revising the model only where necessary. This idea of a "semantically guided " model generation has been introduced first in the DRUM-2 system [FN96,NF96].

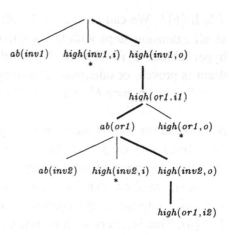

Fig. 3. Hyper derivation from the clause set in Example 6.

For every tableau for a clause set S with finished open branch b the central property is that $[\![b]\!]$ is a model for S.

Conversely, if S is unsatisfiable then there is no tableaux containing a finished open branch. In other words, *refutational completeness* holds. This is an instance of a more general result in [BFN96]. For computing diagnoses (i.e. models), however, we need also a model completeness result:

Theorem 7 (Model Completeness of Hyper Tableaux.). *Let T be a hyper tableau for S such that every open branch is finished. Then, for every minimal model I of S there is an open branch b in T such that $I = [\![b]\!]$.*

Proof. If no minimal model for S exists then the theorem holds vacuously. Otherwise let I be a minimal model for S. In a first step we show that there is an open branch b such that $I \subseteq [\![b]\!]$. It trivially holds that

$$S \cup I \cup \neg \overline{I} \text{ is satisfiable,} \tag{1}$$

where $\overline{I} := \Sigma \setminus I$, and $\neg M := \{\neg A \mid A \in M\}$. It is not too hard to see that 1 is equivalent to $S \cup \neg \overline{I} \models I$. (the minimality of I is essential here). This holds if and only if

$$S \cup \neg \overline{I} \cup \{\bigvee_{A \in I} \neg A\} \text{ is unsatisfiable.} \tag{2}$$

Hence, by refutational completeness of Hyper tableaux there is a closed hyper tableau T' for this clause set. Further, by 1, the subset $S \cup \neg \overline{I}$ is satisfiable. Hence, for the construction of T' the clause $\bigvee_{A \in I} \neg A$ must be at used once for an extension step, say at branch b. But, by definition of hyper extension step this is possible only if the complementary literals are on the branch b, i.e.

$I \subseteq \mathrm{lit}(b)^+$. We can omit from T' all extension steps with $\bigvee_{A \in I} \neg A$, as well as all extension steps with the negative unit clauses $\neg \overline{I}$. The result is an open hyper tableau T for S alone. Now, either the branch b is finished, and the theorem is proven, or otherwise T can repeatedly be extended so that at least one open finished branch b'' with $\mathrm{lit}(b) \subseteq \mathrm{lit}(b'')$ comes up. Reason: otherwise every such extension b'' of b would be closed, meaning that we could find a closed hyper tableau for $S \cup \neg \overline{I}$ alone, which by soundness of hyper tableau contradicts the satisfiability of $S \cup \neg \overline{I}$. Thus, b'' is the desired branch with $I \subseteq [\![b'']\!]$. This concludes the proof of the first step.

Next, we show that for some branch b in T with $I \subseteq [\![b]\!]$ we have even $I = [\![b]\!]$. Suppose, to the contrary, for every branch b with $I \subseteq [\![b]\!]$ we have $I \subset [\![b]\!]$. That is, every such branch b contains a literal A with $A \notin I$. But then trivially $A \in \overline{I}$. Hence b can be closed with $\neg A \in \neg \overline{I}$. If this is done for every such branch b, we can find a closed hyper tableau for $S \cup \neg \overline{I}$ alone, which, by soundness of hyper tableaux contradicts the satisfiability of $S \cup \neg \overline{I}$. Hence, $I = [\![b]\!]$ for some branch b in T.

This theorem enables us to compute in particular minimal diagnosis by simply collecting all ab-literals along $[\![b]\!]$, because every minimal diagnosis must be contained in some minimal model of S.

4 The DRUM-2 System

Since Reiter's seminal paper [Rei87], several generic systems for model–based diagnosis have been developed using logical inference, assumption based truth maintenance and conflicts as their underlying principles (see [dKW87] and many others). Efficiency problems due to administration overhead inherent to this approach have only recently been solved [RdKS93]. DRUM-2 ([FN96,NF96]) has emerged from a different line of research, where models serve as a data structure for the reasoning process. DRUM-2 has adapted and extended implementation ideas from model-based belief revision systems ([CW94]).

The basic idea of DRUM-2 is to start with a model of the correct behavior of the device under consideration, i.e. with an interpretation I_0, such that $I_0 \models SD \cup \{\neg Ab(c) | c \in COMP\}$. Then the system description SD is augmented by an observation of abnormal behavior OBS, such that the assumption that all components are working correctly is no longer valid. Thus, I_0 is no model of $SD \cup OBS$, but DRUM-2 uses I_0 to guide the search for models of $SD \cup OBS$. The models are computed iteratively by inverting truth values of literals in I_0 which contradict formulas in $SD \cup OBS$. We will now describe DRUM-2's algorithm intuitively using a simple circuit as an example. A formalization of DRUM-2 using belief revision vocabulary can be found in [FN96].

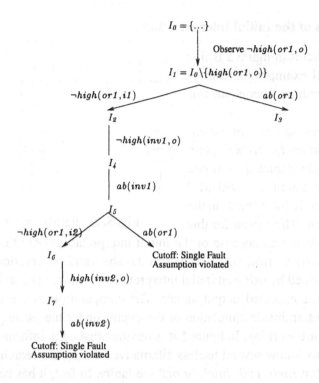

$I_0 = \{\ldots\}$

Observe $\neg high(or1, o)$

$I_1 = I_0 \setminus \{high(or1, o)\}$

$\neg high(or1, i1)$ $ab(or1)$

I_2 I_3

$\neg high(inv1, o)$

I_4

$ab(inv1)$

I_5

$\neg high(or1, i2)$ $ab(or1)$

I_6 Cutoff: Single Fault
Assumption violated

$high(inv2, o)$

I_7

$ab(inv2)$

Cutoff: Single Fault
Assumption violated

Fig. 4. "Repairing" an interpretation in DRUM-2.

We will now show how to compute diagnoses under the single fault assumption. n–Fault–Assumptions are a concept of the DRUM-2 diagnosis engine; they are therefore not represented as part of the theory. Figure 4 shows the steps performed by DRUM-2 during the search for consistent models of $SD \cup OBS$ from our example. In the first step the output observation $\neg high(or1, o)$ is incorporated into I_0 by deleting $high(or1, o)$ from the interpretation. The new interpretation I_1 contradicts the formula $OR1$. There are two possible repair steps, which can remove the violation of this formula: removing $high(or1, i1)$ from the interpretation (leading to I_2) or adding $ab(or1)$ (leading to I_3). Since $I_3 \models SD \cup OBS$ we have found a diagnosis $\{or1\}$. In the left branch of the tree the search for diagnoses continues. However, since both inverters would have to be abnormal to explain the low voltage at the output of the inverter no other single fault diagnosis is found.

The changes to the model performed by DRUM-2 are focused by the initial interpretation I_0. Using this simple mechanism DRUM-2 is currently one of the fastest generic systems for model–based diagnosis as recently reported in [NF96].

4.1 Effects of the Initial Interpretation

The small circuit in figure 2 is fine as a minimal example for clarifying the algorithms throughout this paper. However, it is too small to show the focusing effect of the initial interpretation I_0. In the slightly larger example depicted in figure 5 the computation of DRUM-2 would be exactly the same as in the smaller circuit. The reason for this

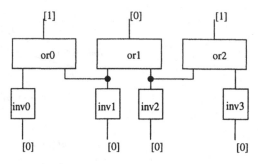

Fig. 5. A slightly larger circuit

efficiency gain is that because of the initial interpretation only those gates appear in the revision tree, which influence the abnormal observation. Note, that the benefit gained by using an initial interpretation is more than saving the computation of the expected output values. The computation of the initial model is just one deterministic simulation of the circuit under the assumption that all components are working. In figure 5 it is obvious that an uninformed procedure would have to follow several useless alternatives during the search for models, i.e. assume that $inv0$, $or0$, $inv2$, or $or2$ are faulty. In fact, it has been shown in [NG94] that the use of an initial model leads to a constant diagnosis time for a circuit consisting of n sequentially connected full adders, whereas the diagnosis time of uninformed algorithms is quadratic in n.

5 Formalizing the Diagnosis Task with Semantic Hyper Tableaux

Recall from Section 4 that DRUM-2 uses an initial interpretation I_0 to focus on certain clauses from the clause set to be candidates for extending or generating new models. In this section we discuss how to incorporate initial interpretations into the hyper tableaux calculus.

Our first technique by *cuts* should be understood as the semantics of the approach; an efficient implementation by a *compilation technique* is presented afterwards.

5.1 Initial Interpretations via Cuts

The use of an initial interpretation in DRUM-2 can be approximated in the hyper tableau calculus by the introduction of an additional inference rule, the *atomic cut rule*.

Definition 8. The inference rule **Atomic cut (with atom A)** is given by: if

> T is an open hyper tableau for S, $f(T) = b$ (i.e. b is selected in T by f), where b is an open branch,

then the literal tree T' is a hyper tableau for S, where T' is obtained from T by extension of b by $A \vee \neg A$ (cf. Def. 4).

Note that in regular hyper tableau it cannot occur that a cut with atom A is applied, if either A or $\neg A$ is contained on the branch. As a consequence it is impossible to use the "same cut" twice on a branch.

We approximate initial interpretations by applying atomic cuts at the beginning:

Definition 9. An *initial tableau* for an interpretation I_0 is given by a regular tableau which is constructed by a applying atomic cuts with atoms from I_0 as long as possible.

The branches of an initial tableau for an interpretation I_0 consist obviously of all interpretations with atoms from I_0. In our running example the initial interpretation is

$$I_o = \{high(inv1, o), high(inv2, o),$$
$$high(or1, i1), high(or1, i2),$$
$$high(or1, o)\} \ .$$

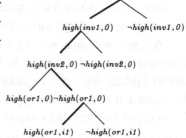

Fig. 6. Initial tableau.

A part of the initial tableau for this I_0 is depicted in Figure 6. Note that the highlighted branch corresponds to the highlighted part in Figure 3. If this branch is extended in successive hyper extension steps, the diagnosis $ab(or1)$, which was contained in the model from Figure 3 can be derived as well.

Note that the cuts introduce negative literals into a branch. The Definition 5 of branch semantics applies to the calculus with cut as well: the interpretation associated to a branch assigns *true* to an atom if it occurs positive on the branch, and a negated atom is interpreted as *false*, just as all atoms which are not on the branch.

In the following we take an initial tableau as the initialization step of the hyper tableaux construction. Since this initial tableau represents semantics, we call the result *semantic hyper tableaux*.

Definition 10 (Semantic Hyper Tableau – SHT). A *semantic hyper tableau* for I_0 and S is a hyper tableau which is generated according to Definition 4, except that the empty tableau in the initialization step is replaced by an initial tableau for I_0.

It is easy to derive an open tableau starting from the initial tableau for I_0 in Figure 6, such that it contains the model from Figure 3.

The model completeness follows directly from the corresponding Theorem 7:

Proposition 11 (Model Completeness of SHT). *Let T be a semantic hyper tableau for I_0 and S, such that every open branch is finished. Then, for every minimal model I of S there is an open branch b in T such that $I = [\![b]\!]$.*

5.2 Initial Interpretations via Renaming

The just defined "semantical" account for initial interpretations via cut is unsuited for realistic examples. This is, because all the $2^n - 1$ possible deviations from the initial interpretation will have to be investigated as well[3]. Hence, in this section we introduce a compilation technique which implements the deviation from the initial interpretation only by need.

Assume we have an initial interpretation $I_0 = \{a\}$ and a clause set which contains $b \leftarrow$ and $c \leftarrow a \wedge b$. By the only applicable atomic cut we get the initial tableau with two branches, namely $\{a\}$ and $\{\neg a\}$. The first branch can be extended twice by an hyper extension step, yielding $\{a, b, c\}$. The second branch can be extended towards $\{\neg a, b\}$. No more extension step is applicable to this tableau. Let T_{cut} be this tableau.

Let us now transform the clause set with respect to I_0, such that every atom from I_0 occurring in a clause is shifted to the other side of the $\leftarrow$ symbol and complemented. In our example we get the clause $c \vee \neg a \leftarrow b$; the fact $b \leftarrow$ remains, because b is not in I_0. Using $b \leftarrow$ we construct a tableau consisting of the single branch $\{b\}$, which can be extended in a successive hyper step by using the renamed clause. We get a tableau consisting of two branches $\{b, c\}$ and $\{b, \neg a\}$. Let T be that tableau. Now, let us interpret a branch in T as usual, except that we set an atom from I_0 to *true* if its negation is not contained in the branch. Under this interpretation the branch $\{b, c\}$ in T corresponds to the usual interpretation of $\{a, b, c\}$ in T_{cut}. Likewise, the second branch $\{b, \neg a\}$ in T_{cut} corresponds to the second model in T.

Note that by this renaming we get tableaux where atoms from I_0 occur only *negatively* on open branches; such cases just mean deviations from I_0. Each such deviation represents a change of the current interpretation along the branch such that an input clause is rendered as *true* which was *false* before the deviation. That is, in contrast to the cut approach, which enumerates all possible

---

[3] On the other side, the semantic tableau approach permits to avoid that the same model be computed several times.

deviations, in the renaming approach the deviations are now brought into the tableau by need.

The following definition introduces the just described idea formally. Since we want to avoid unnecessary changes to the hyper calculus, a new predicate name neg_A instead of $\neg A$ will be used.

Definition 12 (I-transformation). Let $C = L_1 \vee \cdots \vee L_n$ be a clause and I be a set of atoms. The I-*transformation of* C is the clause obtained from C by replacing every positive literal A with $A \in I$ by $\neg neg\_A$, and by replacing every negative literal $\neg A$ with $A \in I$ by neg_A. The I-*transformation of* S is defined as the I-transformation of every clause in S.

It is easy to see that every I-transformation preserves the models of a clause set, in the sense that every model for the non-transformed clause set constitutes a model for the transformed clause set by setting neg_A to *true* iff A is *false*, for every $A \in I$.

As explained informally above, the branch semantics of tableaux derived from a renamed, i.e. I_0-transformed clause set, is changed to assign *true* to every atom from I_0, unless its negation is on the branch. This is a formal definition:

$$[\![b]\!]^I = (I \setminus \{A \mid neg\_A \in \mathrm{lit}(b)\}) \cup (\mathrm{lit}(b) \setminus \{neg\_A \mid neg\_A \in \mathrm{lit}(b)\})$$

The connection of semantic hyper tableaux to hyper tableaux and renaming is given by the next theorem.

Theorem 13. *Let T be a semantic hyper tableau for S and I_0 where every open branch is finished; let T^{I_0} be a hyper tableau for the I_0-transformation of S where every open branch is finished. Then, for every open branch b^{I_0} in T^{I_0} there is an open branch b in T such $[\![b^{I_0}]\!]^{I_0} = [\![b]\!]$. The converse does not hold.*

The theorem tells us that with the renamed clause set we compute some deviation of the initial interpretation. The value of the theorem comes from the fact that the converse does not hold in general. That is, not every possible deviation is examined by naive enumeration of all combinations.

In order to see that the converse does not hold, take e.g. $S = \{a \leftarrow\}$ and $I_0 = \{b\}$. There is only one semantic hyper tableau of the stated form, namely the one with the two branches $\{b, a\}$ and $\{\neg b, a\}$. On the other side, the I_0-transformation leaves S untouched, and thus the sole hyper tableau for S consists of the single branch $\{a\}$ with semantics $[\![\{a\}]\!]^{I_0} = \{a, b\}$. However, the semantics of the branch $\{\neg b, a\}$ in the former tableau is different.

6 Implementation and Experiments

We have implemented a proof procedure for the hyper tableaux calculus of [BFN96], modified it slightly for our diagnosis task, and applied it to some benchmark examples from the diagnosis literature.

The Basic Proof Procedure. A basic proof procedure for the plain hyper tableaux calculus for the propositional case is simple, and coincides with e.g. SATCHMO [MB88]. Initially let T be a tableau consisting of the root node only. Let T be the tableau constructed so far. *Main loop:* if T is closed, stop with "unsatisfiable". Otherwise select an open branch b from T (branch selection) which is not labeled as "finished" and select a clause $H \leftarrow B$ (extension clause selection) from the input clause set such that $B \subseteq b$ (applicability) and $H \cap b = \{\}$ (regularity check). If no such clause exists, b is labeled as "finished" and $[\![b]\!]$ is a model for the input clause set. If every open branch is labeled as "finished" then stop, otherwise enter the main loop again.

In the diagnosis task it is often demanded to compute every diagnosis. Hence the proof procedure does not stop after the first open branch is found, but only marks it as "finished" and enters the main loop again. Consequently, the "branch selection function" is not of real significance because every unfinished open branch will be selected eventually. However, the "extension clause selection" *is* an issue. A standard heuristic for tableau procedures is to preferably select clauses which avoid branching. For our diagnosis experiments, however, a clause selection function, which prefers clauses with some body literal being equal to the leaf of the branch to be extended, is superior for the benchmark circuits.

Adaption for the Diagnosis Task. Recall that our diagnosis task requires to bias the proof search with hyper tableau in two ways: incorporation of the initial interpretation, and implementing the n–Fault-Assumption (cf. Section 2). While the former is treated by renaming predicates in the input clause set (Section 5), the latter is dealt with by the following new inference rule: "any branch containing $n + 1$ (due to regularity necessarily pairwise different) ab-literals is closed immediately". Notice that this inference rule has the same effect as if the $\binom{|COMP|}{n+1}$ clauses

$$\leftarrow ab(C_1), \ldots, ab(C_{n+1}) \qquad \text{for } C_i \in COMP, \, C_i \neq C_j,$$
$$\text{where } 1 \leq i, j \leq n+1 \text{ and } i \neq j.$$

specifying the n-Fault-Assumption would be added to the input clause set. Since even for the smallest example (c499) and the 1-Fault-Assumption the clause set would blow up from 1600 to 60000 clauses, the inference rule solution is mandatory.

No more changes to the basic proof procedure are required.

Implementation. Our prover is called *NIHIL* (*New Implementation of Hyper In Lisp*) and is a re-implementation in SCHEME of a former Prolog implementation. Because the SCHEME code is compiled to C, the basic performance is quite good. Obviously, since NIHIL is a first-order prover and the data structures are arranged for this case, there is a significant overhead when dealing with propositional formulas. We are confident that this could be improved considerably by using standard techniques for propositional logic provers (Gallier/Downing algorithm). Some operations would have constant instead of linear complexity then. In fact, this pay-off is demonstrated in the DRUM-2 system.

Experiments. For our experiments we ran parts of the ISCAS-85 benchmarks [Isc85] from the diagnosis literature. This benchmark suite includes combinatorial circuits from 160 to 3512 components. Table 7 describes the characteristics of the circuits we tested. NIHIL was set up as described above. The abovementioned optimizations *factorization* and *level cut* were tried, but had no influence on these examples.

The results are summarized in Table 7 on the right. *Time* denotes proof time proper in seconds, and thus excludes time for reading in and setup (which is less than about 10 seconds in any case).

Name	# Gates	# Inputs	# Outputs	Time	# Clauses	# Steps
C499	202	41	32	2	1685	2050
C880	383	60	26	1	2776	158
C1355	546	41	32	40	3839	21669
C2670	1193	233	140	4	8260	425
C3540	1669	50	22	1403	10658	253543
C5315	2307	178	123	13	16122	3024

Fig. 7. ISCAS'85 Circuits and NIHIL results.

The times are taken on a SparcStation 20. *# Clauses* is the number of input clauses; *# Steps* denotes the number of hyper extension steps to obtain the final tableau. We emphasize that the results refer to the clause sets with renamed predicates according to Section 5. Without renaming, and thus taking advantage of the initial interpretation, only c499 was solvable (in 174 seconds); all other examples could not be solved within 2 hours, whatever flag settings/heuristic we tried!

7 Conclusions

In this paper we analyzed the relationship between logic-based diagnostic reasoning and tableaux based theorem proving. We showed how to implement diagnostic reasoning efficiently using a hyper tableaux based theorem prover. We identified the use of an initial model as the main optimization technique in the diagnostic reasoning engine DRUM-2 and showed how to apply this technique within a hyper tableaux calculus. The resulting theorem prover NIHIL very efficiently diagnoses large benchmark circuits from the diagnosis literature. There are some open theoretical questions, e.g. we have to prove formally that by renaming we again get a model complete calculus and it would be interesting to characterize the computed models more exactly. Further work will also include a closer examination of multiple faults (n-Fault-Assumption) and further efficiency improvements.

References

[BFN96] P. Baumgartner, U. Furbach, and I. Niemelä. Hyper Tableaux. In *JELIA 96*. European Workshop on Logic in AI, Springer, LNCS, 1996.

[CT91] Luca Console and Pietro Torasso. A spectrum of logical definitions of model-based diagnosis. *Computational Intelligence*, 7(3):133–141, 1991.

[CW94] T. S-C. Chou and M. Winslett. A model–based belief revision system. *Journal of Automated Reasoning*, 12:157–208, 1994.

[dKW87] Johan de Kleer and Brian C. Williams. Diagnosing multiple faults. *Artificial Intelligence*, 32:97–130, 1987.

[FN96] Peter Fröhlich and Wolfgang Nejdl. A model–based reasoning approach to circumscription. In *Proceedings of the 12th European Conference on Artificial Intelligence*, 1996.

[Isc85] The ISCAS-85 Benchmarks. http://www.cbl.ncsu.edu/ www/CBL_Docs/iscas85.html, 1985.

[MB88] R. Manthey and F. Bry. SATCHMO: a theorem prover implemented in Prolog. In *Proc. 9th CADE*. Argonnee, Illinois, Springer LNCS, 1988.

[NF96] Wolfgang Nejdl and Peter Fröhlich. Minimal model semantics for diagnosis – techniques and first benchmarks. In *Seventh International Workshop on Principles of Diagnosis*, Val Morin, Canada, October 1996.

[NG94] Wolfgang Nejdl and Brigitte Giefer. DRUM:Reasoning without conflicts and justifications. In *5th International Workshop on Principles of Diagnosis (DX-94)*, pages 226–233, October 1994.

[RdKS93] Olivier Raiman, Johan de Kleer, and Vijay Saraswat. Critical reasoning. In *Proceedings of the International Joint Conference on Artificial Intelligence*, pages 18–23, Chambery, August 1993.

[Rei87] Raymond Reiter. A theory of diagnosis from first principles. *Artificial Intelligence*, 32:57–95, 1987.

[Rob65] J. A. Robinson. Automated deduction with hyper-resolution. *Internat. J. Comput. Math.*, 1:227–234, 1965.

Free Variable Tableaux for Propositional Modal Logics

Bernhard Beckert*

Imperial College
Department of Computing
180 Queen's Gate, London SW7, England
beckert@ira.uka.de
http://i12www.ira.uka.de/~beckert

Rajeev Goré

Automated Reasoning Project
Australian National University
Canberra, ACT, 0200, Australia
rpg@arp.anu.edu.au
http://arp.anu.edu.au/

Abstract. We present a sound, complete, modular and *lean* labelled tableau calculus for many propositional modal logics where the labels contain "free" and "universal" *variables*. Our "lean" Prolog implementation is not only surprisingly short, but compares favourably with other considerably more complex implementations for modal deduction.

1 Introduction

Free variable semantic tableaux are a well-established technique for first-order theorem proving—both theoretically and practically. Free variable quantifier rules [19, 7] are crucial for efficiency since free variables act as a meta-linguistic device for tracking the eigenvariables used during proof search.

Traditional tableau-based theorem provers developed during the last decade for first-order logic have been complex and highly sophisticated, typified by systems like Setheo [16] and $_3T^AP$ [2]. On the other hand, free variable tableaux, and their extensions like universal variable tableaux, have been used successfully for *lean* Prolog implementations, as typified by leanT^AP [3]. A "lean" implementation is an extremely compact (and efficient) program that exploits Prolog's built-in clause indexing scheme and backtracking mechanisms instead of relying on elaborate heuristics. Such compact lean provers are much easier to understand than their more complex stablemates, and hence easier to adapt to special needs.

Simultaneously, Kanger's meta-linguistic indices for non-classical logics [15] have been generalised by Gabbay into Labelled Deductive Systems [10]. And Massacci [17] and Russo [20] have recently shown the utility of using *ground* labels for obtaining *modular* modal tableaux and natural deduction systems (respectively); see [11] for an introduction to labelled modal tableaux.

By allowing labels to contain free (and universal) variables, we obtain *efficient* and *modular* tableaux systems for all the 15 basic *propositional* modal logics. Furthermore, our leanT^AP style implementation compares favourably with existing fast implementations of modal tableau systems like LWB [13].

---

* On leave from University of Karlsruhe, Institute for Logic, Complexity and Deduction Systems, D-76128 Karlsruhe, Germany.

Our object language uses *labelled formulae* like $\sigma : A$, where σ is a label and A is a formula, with intuitive reading "the possible world σ satisfies the formula A"; see [6, 18, 11] for details. Thus, $1 : \Box p$ says that the possible world 1 satisfies the formula $\Box p$. Our box-rule then reduces the formula $1 : \Box p$ to the labelled formula $1.(x) : p$ which contains the *universal* variable x in its label and has an intuitive reading "the possible world $1.(x)$ satisfies the formula p". Since different instantiations of x give different labels, the labelled formula $1.(x) : p$ effectively says that "all successors of the possible world 1 satisfy p", thereby capturing the usual Kripke semantics for $\Box p$ (almost) exactly. But the possible world 1 may have *no* successors; so we enclose the variable in parentheses and read $\sigma : A$ as "for all instantiations of the variables in σ, if the world corresponding to that instantiation of σ exists then the world satisfies the formula A".

Similar approaches using labels containing variables have been explored by Governatori [12] and D'Agostino et al. [5]. But D'Agostino et al. relate the labels to modal algebras, instead of to first-order logic as we do. And whereas Governatori uses string unification over labels to detect complementary formulae, we use Prolog's matching, since string unification cannot be implemented in a lean way. Our variables are of a simpler kind: they capture all *immediate* children of a possible world (in a rooted tree model), but do not capture *all R-successors*; see [17, 11]. As a consequence, we can make extensive use of Prolog features like unification and backtracking in our implementation. Note, however, that a non-lean extension of our calculi using string unification is perfectly feasible.

The following techniques, in particular, are crucial:

Free variables: Applying the traditional ground box-rule requires guessing the correct eigenvariables. Using (free) variables in labels as "wildcards" that get instantiated "on demand" during branch closure allows more intelligent choices of these eigenvariables. To preserve soundness for worlds with no R-successors, variable positions in labels must be conditional.

Universal variables: Under certain conditions, a variable x introduced by a formula like $\Box A$ is "universal" in that an instantiation of x on one branch need not affect the value of x on other branches, thereby localising the effects of a variable instantiation to one branch. The technique entails creating and instantiating local duplicates of labelled formulae instead of the originals.

Finite diamond-rule: Applying the diamond-rule to $\Diamond A$ usually creates a new label. By using (a Gödelisation of) the formula A itself as the label instead, we guarantee that only a finite number of different labels (of a certain length) are used in the proof. In particular, different (identically labelled) occurrences of $\Diamond A$ generate the same unique label.

The paper is structured as follows: In Sections 2 and 3 we introduce the syntax and semantics of labelled modal tableaux. In Section 4 we introduce our calculus, formalise its soundness and completeness (full proofs can be found in [1]), and present an example. In Section 5 we describe our implementation and present experimental results; and in Section 6 we present our conclusions and discuss future work.

2 Syntax

The formulae of modal logics are built-in the usual way; see [11]. To reduce the number of tableau rules and the number of case distinctions in proofs, we restrict all considerations to implication-free formulae in negated normal form (NNF); thus negation signs appear in front of primitive propositions only. Using NNF formulae is no real restriction since *every* formula can be transformed into an equivalent NNF formula in linear time.

Labels are built from natural numbers and variables, with variables intended to capture the similarities between the ∀ quantifier of first-order logic and the □ modality of propositional modal logic. However, whereas first-order logic forbids an empty domain, the □ modality tolerates possible worlds with no successors.[2] To capture this (new) behaviour, variable positions in labels are made "conditional" on the existence of an appropriate successor by enclosing these conditional positions in parentheses.

Definition 1. Let Vars be a set of variables, and let $\mathbf{N}$ be the set of natural numbers. Let x, y, z range over arbitrary members of Vars, let n and m range over arbitrary members of $\mathbf{N}$, and let l range over arbitrary members of Vars $\cup \mathbf{N}$. Then, the string 1 is a **label**; and if σ is a label, then so are $\sigma.m$ and $\sigma.(l)$. The **length** of a label σ is the number of dots it contains plus one, and is denoted by $|\sigma|$. The constituents of a label σ are called **positions** in σ and terms like "the 1st position" or "the n-th position" are defined in the obvious way. A position is **conditional** if it is of the form (l), and a label is conditional if it contains a conditional position. By ipr(σ) we mean the set of all non-empty **initial prefixes** of a label σ, excluding σ itself. A label is **ground** if it consists of (possibly conditional) members of $\mathbf{N}$ only. Let $\mathcal{L}$ be the set of all ground labels.

When dealing with ground labels, we often do not differentiate between the labels $\sigma.n$ and $\sigma.(n)$, and we use $\sigma.[n]$ to denote that the label may be of either form. Note also that $\sigma.x$ (parentheses around x omitted) is not a label: the parentheses mark the positions that contain variables, or that used to contain variables before a substitution was applied.

Definition 2. A set Γ of labels is **strongly generated** if: (a) there is some (root) label $\rho \in \Gamma$ such that $\rho \in$ ipr(σ) for all $\sigma \in \Gamma \setminus \{\rho\}$; and (b) $\sigma \in \Gamma$ implies $\tau \in \Gamma$ for all $\tau \in$ ipr(σ).

Since we deal with mono-modal logics with semantics in terms of rooted frames (see Section 3), we always assume that our labels form a strongly generated set with root $\rho = 1$. In any case, our definition of labels guarantees that all our labels begin with 1, and it is easy to see that the labels that appear in any of our tableaux are strongly generated.

---

[2] To that extent, modal logics are similar to free logic, i.e., first-order logic where the domains of models may be empty [4].

Definition 3. A **labelled tableau formula** (or just tableau formula) is a structure of the form $X : \Delta : \sigma : A$, where X is a subset of Vars $\cup$ N, Δ is a set of labels, σ is a label, and A is a formula in NNF. If the set Δ is empty, we use $X : \sigma : A$ as an abbreviation for $X : \emptyset : \sigma : A$. A tableau formula $X : \Delta : \sigma : A$ is **ground**, if σ and all labels in Δ are ground. If $\mathcal{F}$ is a set of labelled tableau formulae, then lab($\mathcal{F}$) is the set $\{\sigma \mid X : \Delta : \sigma : A \in \mathcal{F}\}$.

The intuitions behind the different parts of our "tableau formulae" are as follows: The fourth part A is just a traditional modal formula. The third part σ is a label, possibly containing variables introduced by the reduction of $\Box$ modalities. If the label σ is ground, then it corresponds to a particular path in the intended rooted tree model; for example, the ground label 1.1.1 typically represents the leftmost child of the leftmost child of the root 1. If σ contains variables, then it represents all the different paths (successors) that can be obtained by different instantiations of the variables, thereby capturing the semantics of the $\Box$ modalities that introduced them. Our rule for splitting disjunctions allows us to retain these variables in the labels of the two disjuncts, but because $\Box$ does not distribute over $\vee$, such variables then lose their "universal" force, meaning that these "free" variables can be instantiated only *once* in a tableau proof. We use the first component X to record the variables in the tableau formula ϕ that are "universal", meaning that ϕ can be used multiply in the same proof with different instantiations for these variables. The free variables in ϕ (that do not appear in X) can be used with only one instantiation since they have been pushed through the scope of an $\vee$ connective. The second part Δ, which can be empty, has a significance only if our calculus is applied to one of the four logics **KB**, **K5**, **KB4**, and **K45** (that are non-serial, but are symmetric or euclidean, see Section 3). It is empty for the other logics. The intuition of Δ is that the formula A has to be true in the possible world called σ only if the labels in Δ name legitimate worlds in the model under consideration. This feature has to be used, if (a) rule applications may shorten labels, which is the case if the logic is symmetric or euclidean, and (b) the logic is non-serial and, thus, the existence of worlds is not guaranteed. The set Δ can contain both universal and free variables, and some of them may appear in σ.

Definition 4. Given a tableau formula $\phi = X : \Delta : \sigma : A$, Univ($\phi$) $= X$ is the set of **universal variables** of ϕ, while Free(ϕ) $= \{x$ appears in σ or $\Delta \mid x \notin X\}$ is the set of **free variables** of ϕ. These notions are extended in the obvious way to obtain the sets Free($\mathcal{T}$) and Univ($\mathcal{T}$) of free and universal variables of a given tableau $\mathcal{T}$ (see Def. 5).

Definition 5. A **tableau** is a (finite) binary tree whose nodes are tableau formulae. A **branch** in a tableau $\mathcal{T}$ is a maximal path in $\mathcal{T}$.[3] A branch may be marked as being **closed**. If it is not marked as being closed, it is **open**. A tableau branch is **ground** if every formula on it is ground, and a tableau is ground if all its branches are ground.

---

[3] Where no confusion can arise, we identify a tableau branch with the set of tableau formulae it contains.

Since we deal with propositional modal logics, notions from first-order logic like variables and substitutions are needed only for handling meta-linguistic notions like the accessibility relation between worlds. Specifically, whereas substitutions in first-order logic assign terms to variables, here they assign numbers or other variables (denoting possibles worlds) to variables.

Definition 6. A **substitution** is a (partial) function μ : Vars $\rightarrow$ N $\cup$ Vars. Substitutions are extended to labels and formulae in the obvious way. A substitution is **grounding** if its range is the (whole) set Vars and its range is N; that is, if it maps all variables in Vars to natural numbers. A substitution is a **variable renaming** if its range is Vars, and it replaces distinct variables by other distinct variables only. The **restriction** of a substitution μ to a set X of variables is denoted by $\mu_{|X}$.

Definition 7. Given a tableau $\mathcal{T}$ containing a tableau formula $X : \Delta : \sigma : A$, a tableau formula $X' : \Delta' : \sigma' : A$ is a $\mathcal{T}$-**renaming** of $X : \Delta : \sigma : A$ if there is a variable renaming μ such that $X' : \Delta' : \sigma' : A = (X : \Delta : \sigma : A)\mu$, and every variable introduced by μ is new to the tableau $\mathcal{T}$.

3 Semantics

To save space, we assume familiarity with Kripke semantics for propositional normal modal logics; see [11] for details of any undefined terms. A world $w \in W$ is **idealisable** if it has a successor in W. To illustrate the modularity of our method we concentrate on the five basic axioms known as (T) $\Box A \rightarrow A$, (D) $\Box A \rightarrow \Diamond A$, (4) $\Box A \rightarrow \Box \Box A$, (5) $\Diamond A \rightarrow \Box \Diamond A$, (B) $A \rightarrow \Box \Diamond A$, and the 15 extensions of the basic propositional normal modal logic **K** obtained as shown in the first two columns of Table 1. The following properties of the reachability relation R characterise these axioms, (T): reflexivity, (D): seriality, (4): transitivity, (5): euclideanness, and (B): symmetry; see [11] for details. We therefore obtain:

Definition 8. For any logic **L** from Table 1, $\langle W, R \rangle$ is an **L**-frame if each axiom of **L** is valid in $\langle W, R \rangle$. A model $\langle W, R, V \rangle$ is an **L**-model if $\langle W, R \rangle$ is an **L**-frame.

It is also well-known that finer characterisation results for these logics can be given in terms of *finite rooted tree* frames; see [11] for details. These results are built into the following definition of **L**-accessibility which views a set of strongly generated ground labels as a tree with root ρ where $\sigma.[n]$ is an immediate child of σ (hence the name "strongly generated").

Definition 9. Given a logic **L** and a set Γ of strongly generated ground labels with root $\rho = 1$, a label $\tau \in \Gamma$ is **L-accessible** from a label $\sigma \in \Gamma$, written as $\sigma \rhd \tau$, if the conditions set out in Table 1 are satisfied. A label $\sigma \in \Gamma$ is an **L-deadend** if no $\tau \in \Gamma$ is **L**-accessible from σ.

The following lemma (see [11] for a proof) shows that the **L**-accessibility relation $\rhd$ on labels has the properties like reflexivity, transitivity, etc. that are appropriate for **L**-frames.

Lemma 10. *If Γ is a strongly generated set of ground labels with root $\rho = 1$, then $\langle \Gamma, \triangleright \rangle$ is an L-frame.*

Logic	Axioms	τ is L-accessible from σ	Logic	Axioms	τ is L-accessible from σ
K	(K)	$\tau = \sigma.[n]$	**KT**	(KT)	$\tau = \sigma.[n]$ or $\tau = \sigma$
KB	(KB)	$\tau = \sigma.[n]$ or $\sigma = \tau.[m]$	**K4**	(K4)	$\tau = \sigma.\theta$
K5	(K5)	$\tau = \sigma.[n]$, or $\lvert\sigma\rvert \geq 2$ and $\lvert\tau\rvert \geq 2$	**K45**	(K45)	$\tau = \sigma.\theta$, or $\lvert\sigma\rvert \geq 2$ and $\lvert\tau\rvert \geq 2$
KD	(KD)	**K**-condition, or σ is a **K**-deadend and $\sigma = \tau$	**KDB**	(KDB)	**KB**-condition, or $\lvert\Gamma\rvert = 1$ and $\sigma = \tau = 1$
KD4	(KD4)	**K4**-condition, or σ is a **K**-deadend and $\sigma = \tau$	**KD5**	(DK5)	**K5**-condition, or $\lvert\Gamma\rvert = 1$ and $\sigma = \tau = 1$
KD45	(KD45)	**K45**-condition, or $\lvert\Gamma\rvert = 1$ and $\sigma = \tau = 1$	**KB4**	(KB4)	$\lvert\Gamma\rvert \geq 2$
B	(KTB)	$\tau = \sigma$, or $\tau = \sigma.[n]$, or $\sigma = \tau.[m]$	**S4**	(KT4)	$\tau = \sigma.\theta$ or $\tau = \sigma$
S5	(KT5)	all σ, τ			

Table 1. Basic logics, axiomatic characterisations, and L-accessibility $\triangleright$.

Traditionally, the notion of satisfaction relates a world in a model with a formula or a set of formulae. For formulae annotated with ground labels, this notion must be extended by a further "interpretation function" mapping ground labels to worlds; see [7, 11]. If labels contain free variables and, in particular, universal variables, then this notion must also cover all possible instantiations of the universal variables, thus catering for many different "interpretation functions". We extend the notion of satisfiability so it is naturally preserved by our tableau expansion rules, and so that a "closed tableau" is not satisfiable.

We proceed incrementally by defining satisfiability for: ground labels; ground tableau formulae; non-ground tableau formulae; and finally for whole tableaux. But first we enrich models by the "interpretation function" that maps labels to worlds. Note that such interpretations give a meaning to *all* ground labels, not just to those that appear in a particular tableau.

Definition 11. An L-interpretation is a pair $\langle \mathbf{M}, \mathbf{I} \rangle$, where $\mathbf{M} = \langle W, R, V \rangle$, is an L-model and $\mathbf{I} : \mathcal{L} \to W \cup \{\bot\}$ is a function interpreting ground labels such that: (a) $\mathbf{I}(1) \in W$; (b) $\mathbf{I}(\sigma.(n)) = \mathbf{I}(\sigma.n)$ for all $\sigma.n$ and $\sigma.(n)$ in $\mathcal{L}$; (c) for all $\sigma \in \mathcal{L}$, if $\mathbf{I}(\tau) = \bot$ for some $\tau \in \mathrm{ipr}(\sigma)$ then $\mathbf{I}(\sigma) = \bot$; (d) if $\sigma \triangleright \tau$, $\mathbf{I}(\sigma) \in W$, $\mathbf{I}(\tau) \in W$, and $\mathbf{I}(\sigma)$ is idealisable, then $\mathbf{I}(\sigma) \, R \, \mathbf{I}(\tau)$.

Definition 12. An L-interpretation $\langle \mathbf{M}, \mathbf{I} \rangle$, where $\mathbf{M} = \langle W, R, V \rangle$, **satisfies** a ground label σ, if for all labels $\tau.n \in \mathrm{ipr}(\sigma) \cup \{\sigma\}$ (that end in an unconditional label position): $\mathbf{I}(\tau) \in W$ implies $\mathbf{I}(\tau.n) \in W$. The L-interpretation $\langle \mathbf{M}, \mathbf{I} \rangle$ **satisfies** a ground tableau formula $X : \Delta : \sigma : A$, if (a) $\mathbf{I}(\sigma) = \bot$, or $\mathbf{I}(\tau) = \bot$ for some $\tau \in \Delta$, or $\mathbf{I}(\sigma) \models A$; and (b) if $\mathbf{I}(\tau) \in W$ for all $\tau \in \Delta$, then $\langle \mathbf{M}, \mathbf{I} \rangle$ satisfies σ.

Thus, a tableau formula is satisfied by default if its label σ is undefined (that is, if $\mathbf{I}(\sigma) = \bot$) or if one of the labels in Δ is undefined. But because we deal only with strongly generated sets of labels with root 1, Definitions 11 and 12 force $\mathbf{I}$ to "define" as many members of $\mathrm{ipr}(\sigma)$ as is possible. However, for a conditional ground label of the form $\tau.(n)$, where n is parenthesised, it is perfectly acceptable to have $\mathbf{I}(\tau.(n)) = \bot$ even if $\mathbf{I}(\tau) \in W$.

Example 1. If $\langle \mathbf{M}, \mathbf{I} \rangle$ satisfies $\sigma = 1.1.1$, then $\mathbf{I}(1)$, $\mathbf{I}(1.1)$, and $\mathbf{I}(1.1.1)$ must be defined. If $\sigma = 1.(1).1$, then $\mathbf{I}(1.(1))$ need not be defined; but if it is, then $\mathbf{I}(1.(1).1)$ must be defined.

The domain of every interpretation function $\mathbf{I}$ is the set of all *ground* labels $\mathcal{L}$, but our tableaux contain labels with variables. We therefore introduce a definition of satisfiability for non-ground tableau formulae capturing our intuitions that a label $\sigma.(x)$ stands for *all* possible successors of the label σ, and taking into account the special nature of universal variables.

Definition 13. Given an L-interpretation $\langle \mathbf{M}, \mathbf{I} \rangle$ and a grounding substitution μ, a (non-ground) tableau formula $\phi = X : \Delta : \sigma : A$ is **satisfied** by $\langle \mathbf{M}, \mathbf{I}, \mu \rangle$, written as $\langle \mathbf{M}, \mathbf{I}, \mu \rangle \models \phi$, if, for all grounding substitutions λ, the ground formula $\phi \lambda_{|X} \mu$ is satisfied by $\langle \mathbf{M}, \mathbf{I} \rangle$ (Def. 12). A set $\mathcal{F}$ of tableau formulae is satisfied by $\langle \mathbf{M}, \mathbf{I}, \mu \rangle$, if every member of $\mathcal{F}$ is simultaneously satisfied by $\langle \mathbf{M}, \mathbf{I}, \mu \rangle$.

In the above definition, a ground formula $\phi \lambda_{|X} \mu$ is constructed from ϕ in two steps such that the definition of satisfiability for ground formulae can be applied. To cater for the differences between the free variables and universal variables, we use two substitutions: a fixed substitution μ and an arbitrary substitution λ. The first step, applying $\lambda_{|X}$ to ϕ, instantiates the universal variables $x \in X$. The second step, applying μ to $\phi \lambda_{|X}$, instantiates the free variables. Therefore, the instantiation of universal variables $x \in X$ is given by the arbitrary substitution λ, and the instantiation of free variables $x \notin X$ is given by the fixed substitution μ.

Note, that in the following definition of satisfiable tableaux, there has to be a single satisfying L-interpretation for *all* grounding substitutions μ.

Definition 14. A tableau $\mathcal{T}$ is L-satisfiable if there is an L-interpretation $\langle \mathbf{M}, \mathbf{I} \rangle$ such that for *every* grounding substitution μ there is some *open* branch $\mathcal{B}$ in $\mathcal{T}$ with $\langle \mathbf{M}, \mathbf{I}, \mu \rangle \models \mathcal{B}$.

4 The Calculus

We now present an overview of our calculus, highlighting its main principles.

Our calculus is a refutation method. That is, to prove that a formula A is a theorem of logic $\mathbf{L}$, we first convert its negation $\neg A$ into NNF obtaining a formula B, and then test if B is L-unsatisfiable. To do so, we start with the initial tableau whose single node is $\emptyset : \emptyset : 1 : B$ and repeatedly apply the tableau expansion rules, the substitution rule, and the closure rule until a closed tableau has been constructed. Since our rules preserve L-satisfiability of tableaux, a closed tableau indicates that B is indeed L-unsatisfiable, and hence that its negation A is L-valid. Since L-frames characterise the logic L we then know that A is a theorem of logic L. Constructing a tableau for $\emptyset : \emptyset : 1 : B$ can be seen as a search for an L-model for B. Each branch is a partial definition of a possible L-model, and different substitutions give different L-models. Our tableau rules extend *one* particular branch using *one* particular formula, thus differing *crucially* from the systematic methods in [6, 11] where a rule extends *all* branches that pass through one particular formula.

Free variables are used in the labels so that when the box-rule is applied in a world, the actual *ground* label of the successor world does not have to be guessed. Instead, free variables can be instantiatedly before a branch is closed to make that closure possible. Note, however, that one single instantiation of the free variables has to be found that allows us to close all branches of a tableau *simultaneously*, and that instantiating a free variable (in the wrong way) to close one branch, can make it impossible to close other branches.

Because a world may have no successor, variable positions in labels have to be conditional to preserve soundness for non-serial logics.

Every variable is introduced into a label by the reduction of a box-formula like $\Box A$. Such a variable x in a tableau formula ϕ on branch $\mathcal{B}$ is "universal" if a renaming $\phi' = \phi\{x := x'\}$ of ϕ could be added to $\mathcal{B}$ without generating additional branches. That is, the modified tableau would be no more difficult to close than the original. An easy way to generate the renaming is to repeat the rule applications that lead to the generation of ϕ, starting from the box-rule application that created x. Once the renaming ϕ' is present on $\mathcal{B}$, the variable x never has to be instantiated to close $\mathcal{B}$ because ϕ' could be used instead of ϕ, thus instantiating x' instead of x. However, if x occurs on two separate branches in the tableau, then x is not universal because repeating these rule applications would generate at least one additional branch. Since the only rule that causes branching is the disjunctive rule, the two separate occurrences of x must have been created by a disjunctive rule application. Therefore, an application of the disjunctive rule to a formula ψ causes the universal variables of ψ to become free variables. Thus, all free variables are a result of a disjunction within the scope of a $\Box$, corresponding to the fact that $\Box$ does not distribute over $\vee$.

When the disjunctive rule "frees" universal variables, additional copies of the box-formula that generated them are needed. However, these additional copies are not generated by the box-rule, but by the disjunctive rule itself.

Our diamond-rule does not introduce a *new* label $\sigma.n$, when it is applied to $X : \Delta : \sigma : \Diamond A$. Instead, each formula $\Diamond A$ is assigned its own unique label $\lceil A \rceil$ which is a Gödelisation of A itself. This rule is easier to implement than the traditional one; and it guarantees that the number of different labels (of a certain length) in a proof is finite, thus restricting the search space.

The box-rule for symmetric and euclidean logics can shorten labels. For example, the tableau formula $X' : \Delta' : 1 : A$ is derived from $X : \Delta : 1.(1) : \Box A$ if the logic is symmetric. The semantics for serial logics guarantee that all labels define worlds, but in non-serial logics, the label 1 may be defined even though 1.(1) is undefined. To ensure that the formula $X' : \Delta' : 1 : A$ or one of its descendants is used to close a branch only if the label 1.(1) *is* defined, the label 1.(1) is made part of Δ' (see Section 4.2). Such problems do not occur when rule applications always lengthen labels since τ has to be defined if $\tau.l$ is defined.

All expansion rules are sound and *invertible* (some denominator of each rule is **L**-satisfiable *iff* the numerator is **L**-satisfiable). Thus, unlike traditional modal tableau methods where the order of (their non-invertible) rule applications is crucial [6, 11], the order of rule application is *immaterial*.

The differences in the calculi for different logics **L** is mainly in the box-rule, with different denominators for different logics. In addition, a simpler version of the closure rule can be used if the logic is serial.

4.1 Tableau Expansion Rules

There are four expansion rules, one for each type of complex (non-literal) formula. If we wanted to avoid NNF we would have four formula classes (α, β, ν, π) a la Smullyan [6], and an extra rule for double negation. Since we assume that all our formulae are in NNF, we need just one representative for each of the four classes.

As usual, in each rule, the formula above the horizontal line is its *numerator* (the premiss) and the formula(e) below the horizontal line, possibly separated by vertical bars, are its *denominators* (the conclusions). All expansion rules (including the box-rule) are "destructive"; that is, once the (appropriate) rule has been applied to a formula occurrence to expand a branch, that formula occurrence is not used again to expand that branch. Note that we permit multiple occurrences of the same formula on the same branch.

Definition 15. Given a tableau $\mathcal{T}$, a new tableau $\mathcal{T}'$ may be constructed from $\mathcal{T}$ by applying one of the **L-expansion rules** from Table 2 as follows: If the numerator of a rule occurs on a branch $\mathcal{B}$ in $\mathcal{T}$, then the branch $\mathcal{B}$ is extended by the addition of the denominators of that rule. For the disjunctive rule the branch splits and the formulae in the right and left denominator, respectively, are added to the two resulting sub-branches instead.

The box-rule(s) shown in Table 2 require explanation. The form of the rule is determined by the index **L** in the accompanying table. But some of the denominators have side conditions that determine when they are applicable. For example, the constraint $\sigma_6 = 1.l_6$ means that (5) is part of the denominator only when the numerator of the box-rule is of the form $X : \Delta : 1.l_6 : \Box A$. Similarly, the constraints $\sigma_3 = \tau_3.l_3$ and $\sigma_5 = \tau_5.l_5$ for the (4^r) and (B) denominators mean these rules can be used only for a numerator of the form $X : \Delta : \sigma : \Box A$ where $|\sigma| \geq 2$, thereby guaranteeing that the *strictly shorter* labels τ_3 and τ_5 that appear in the respective denominators are properly defined. Note that the (4^d) denominator is the restriction of the (4) denominator to the case where $|\sigma| \geq 2$. The table indicates that the rules for a logic **L** and its serial version **LD** are identical because

$$\frac{X : \Delta : \sigma : A \wedge B}{\begin{array}{c} X : \Delta : \sigma : A \\ X' : \Delta' : \sigma' : B \end{array}}$$

Conjunctive rule. $X' : \Delta' : \sigma' : B$ is a $\mathcal{T}$-renaming of $X : \Delta : \sigma : B$.

$$\frac{X : \Delta : \sigma : A \vee B}{\begin{array}{c|c} \emptyset : \Delta_1 : \sigma_1 : A & \emptyset : \Delta_1 : \sigma_1 : B \\ X_2 : \Delta_2 : \sigma_2 : A \vee B & X_3 : \Delta_3 : \sigma_3 : A \vee B \end{array}}$$

Disjunctive rule. $\psi_i = X_i : \Delta_i : \sigma_i : A \vee B$ are $\mathcal{T}$-renamings of $\psi = X : \Delta : \sigma : A \vee B$ for $1 \leq i \leq 3$ (where the X_i are pairwise disjoint). If $X = \emptyset$ then ψ_2, ψ_3 are omitted.

$$\frac{X : \Delta : \sigma : \Diamond A}{X : \Delta : \sigma.\lceil A \rceil : A}$$

Diamond-rule. $\lceil \cdot \rceil$ is an arbitrary but fixed bijection from the set of formulae to $\mathbf{N}$.

$$\frac{X : \Delta : \sigma : \Box A}{\begin{array}{ll} X \cup \{x\} : \Delta : \sigma.(x) : A & \text{(K)} \\ X_1 \cup \{x_1\} : \Delta_1 : \sigma_1.(x_1) : \Box A & \text{(4)} \\ X_2 \cup \{x_2\} : \Delta_2 : \sigma_2.(x_2) : \Box A & (4^d) \\ X_3 : \Delta_3 \cup \{\sigma_3\} : \tau_3 : \Box A & (4^r) \\ X_4 : \Delta_4 : \sigma_4 : A & \text{(T)} \\ X_5 : \Delta_5 \cup \{\sigma_5\} : \tau_5 : A & \text{(B)} \\ X_6 : \Delta_6 \cup \{\sigma_6\} : 1 : \Box\Box A & \text{(5)} \end{array}}$$

Box-rule. For $1 \leq i \leq 6$, $X_i : \Delta_i : \sigma_i : \Box A$ are $\mathcal{T}$-renamings of $X : \Delta : \sigma : \Box A$. The variables $x, x_1, x_2 \in$ Vars are new to $\mathcal{T}$. The sets $X \cup \{x\}$, $X_1 \cup \{x_1\}$, $X_2 \cup \{x_2\}$, X_3, X_4, X_5, and X_6 are pairwise disjoint. In addition, $\sigma_3 = \tau_3.l_3$, $\sigma_5 = \tau_5.l_5$, $\sigma_6 = 1.l_6$, and $|\sigma_2| \geq 2$. The form of the denominator depends on the logic $\mathbf{L}$, and is determined by including every denominator corresponding to the entry for $\mathbf{L}$ in the table below.

Logics	Box-rule denominator	Logics	Box-rule denominator
K, D	(K)	**K45, K45D**	(K), (4), (4^r)
T	(K), (T)	**K4B,**	(K), (B), (4), (4^r)
KB, KDB	(K), (B)	**B**	(K), (T), (B)
K4, KD4	(K), (4)	**S4**	(K), (T), (4)
K5, KD5	(K), (4^d), (4^r), (5)	**S5**	(K), (T), (4), (4^r)

Table 2. Tableau expansion rules.

these logics are distinguished by the form of our closure rule; see Definition 18. Various other ways to define the calculi for serial logics exist; see [11].

4.2 The Substitution Rule and the Closure Rule

By definition, the substitution rule allows us to apply *any* substitution at *any* time to a tableau. In practice, however, it makes sense to apply only "useful" substitutions; that is, those most general substitutions which allow to close a branch of the tableau.

Definition 16. Substitution rule: Given a tableau $\mathcal{T}$, a new tableau $\mathcal{T}' = \mathcal{T}\sigma$ may be constructed from $\mathcal{T}$ by applying a substitution σ to $\mathcal{T}$ that instantiates free variables in $\mathcal{T}$ with other free variables or natural numbers.

In tableaux for modal logics without free variables as well as in free-variable tableaux for first-order logic, a tableau branch is closed if it contains complementary literals since this immediately implies the existence of an inconsistency. Here, however, this is not always the case because the labels of the complementary literals may be conditional. For example, the (apparently contradictory) pair $\emptyset : 1.(1) : p$ and $\emptyset : 1.(1) : \neg p$ is not necessarily inconsistent since the world $\mathbf{I}(1.(1))$ may not exist in the chosen model. Before declaring this pair to be inconsistent, we therefore have to ensure that $\mathbf{I}(1.(1)) \neq \perp$ for all L-interpretations satisfying the tableau branch $\mathcal{B}$ that is to be closed. Fortunately, this knowledge can be deduced from other formulae on $\mathcal{B}$. Thus in our example, a formula like $\psi = X : 1.1 : A$ on $\mathcal{B}$ would "justify" the use of the literal pair $\emptyset : 1.(1) : p$ and $\emptyset : 1.(1) : \neg p$ for closing the branch $\mathcal{B}$ since any L-interpretation $\langle \mathbf{M}, \mathbf{I} \rangle$ satisfying $\mathcal{B}$ has to satisfy ψ, and, thus, $\mathbf{I}(1.(1)) = \mathbf{I}(1.1) \neq \perp$ has to be a world in the chosen model $\mathbf{M}$. The crucial point is that the label 1.1 of ψ is *unconditional* exactly in the *conditional* positions of $\emptyset : 1.(1) : p$ and $\emptyset : 1.(1) : \neg p$. These observations are now extended to the general case of arbitrary *ground* labels.

Definition 17. A ground label σ with j-th position $[n_j]$ $(1 \leq j \leq |\sigma|)$ is **justified** on a branch $\mathcal{B}$ if there is some set $\mathcal{F} \subseteq \mathcal{B}$ of tableau formulae such that for every j: (a) some label in $\mathrm{lab}(\mathcal{F})$ has (an unconditional but otherwise identical) j-th position n_j; and (b) for all $\tau \in \mathrm{lab}(\mathcal{F})$: if $|\tau| \geq j$ then the j-th position in τ is n_j or (n_j).

Definition 18. Given a tableau $\mathcal{T}$ and a substitution $\rho : \mathrm{Univ}(\mathcal{T}) \rightarrow \mathbf{N}$ that instantiates universal variables in $\mathcal{T}$ with natural numbers, the **L-closure rule** allows to construct a new tableau $\mathcal{T}'$ from $\mathcal{T}$ by marking $\mathcal{B}$ in $\mathcal{T}$ as closed provided that: (a) the branch $\mathcal{B}\rho$ of $\mathcal{T}\rho$ contains a pair $X : \Delta : \sigma : p$ and $X' : \Delta' : \sigma : \neg p$ of complementary literals; and (b1) the logic $\mathbf{L}$ is serial, or (b2) all labels in $\{\sigma\} \cup \Delta \cup \Delta'$ are ground and justified on $\mathcal{B}\rho$.

Note that the substitution ρ that instantiates universal variables is not actually applied to the tableau when the branch is closed; it only has to exist.

By definition, only complementary *literals* close tableau branches, but in theory, pairs of complementary *complex formulae* could be used as well.

4.3 Tableau Proofs

We now have all the ingredients we need to define the notion of a tableau proof.

Definition 19. A sequence $\mathcal{T}^0, \ldots, \mathcal{T}^r$ of tableaux is an **L-proof** for the **L**-unsatisfiability of a formula A if: (a) $\mathcal{T}^0$ consists of the single node $\emptyset : \emptyset : 1 : A$; (b) for $1 \leq m \leq r$, the tableau $\mathcal{T}^m$ is constructed from $\mathcal{T}^{m-1}$ by applying an L-expansion rule (Def. 15), the substitution rule (Def. 16), or the L-closure rule (Def. 18); and (c) all branches in $\mathcal{T}^r$ are marked as closed.

Theorems 20 and 22 state soundness and completeness for our calculus with respect to the Kripke semantics for logic $\mathbf{L}$; the proofs can be found in [1].

Theorem 20 (Soundness). *Let A be a formula in NNF. If there is an L-proof $\mathcal{T}^0, \ldots, \mathcal{T}^r$ for the L-unsatisfiability of A (Def. 19), then A is L-unsatisfiable.*

We prove completeness for the non-deterministic and unrestricted version of the calculus, and also for all tableau procedures based on this calculus that deterministically choose the next formula for expansion (in a *fair* way) and that only apply most general closing substitutions.

Definition 21. Given an open tableau $\mathcal{T}$, a *tableau procedure* Ψ deterministically chooses an open branch $\mathcal{B}$ in $\mathcal{T}$ and a non-literal tableau formula ψ on $\mathcal{B}$ for expansion.

The tableau procedure Ψ is fair if, in the (possibly infinite) tableau that is constructed using Ψ (where no substitution is applied and no branch is closed), every formula has been used for expansion of every branch on which it occurs.

Theorem 22 (Completeness). *Let Ψ be a fair tableau procedure, and let A be an L-unsatisfiable formula in NNF. Then there is a (finite) tableau proof $\mathcal{T}^0, \ldots, \mathcal{T}^r$ for the L-unsatisfiability of A, where $\mathcal{T}^i$ is constructed from $\mathcal{T}^{i-1}$ ($1 \leq i \leq r$) by (a) applying the appropriate L-expansion rule to the branch $\mathcal{B}$ and the formula ψ on $\mathcal{B}$ chosen by Ψ from $\mathcal{T}^{i-1}$; or (b) applying a most general substitution such that the L-closure rule can be applied to a previously open branch in $\mathcal{T}^{i-1}$.*

Example 2. We prove that $A = \Box(p \to q) \to (\Box p \to (\Box q \wedge \Box p))$ is a K-theorem. To do this, we first transform the negation of A into NNF; the result is $B = \text{NNF}(\neg A) = \Box(\neg p \vee q) \wedge \Box p \wedge (\Diamond \neg q \vee \Diamond \neg p)$. The (fully expanded) tableau $\mathcal{T}$, that is part of the proof for the K-unsatisfiability of B is shown in Figure 1. The nodes of the tableau are numbered; a pair $[i; j]$ is attached to the i-th node, the number j denotes that node i has been created by applying an expansion rule to the formula in node j. Note, that by applying the disjunctive rule to 6, the nodes 11 to 14 are added; 13 and 14 are renamings of 6. The variable y_1 is no longer universal in 11 and 12.

When the substitution $\sigma = \{y_1/\lceil \neg q \rceil\}$ is applied to $\mathcal{T}$, the branches of the resulting tableau $\mathcal{T}\sigma$ can be closed as follows, thereby completing the tableau proof: The left branch $\mathcal{B}_1$ of $\mathcal{T}\sigma$ can be closed by the universal variable substitution $\rho_1 = \{x/\lceil \neg q \rceil\}$ because $\mathcal{B}_1\rho_1$ then contains the complementary pair $\{\lceil \neg q \rceil\} : 1.(\lceil \neg q \rceil) : p$ and $\emptyset : 1.(\lceil \neg q \rceil) : \neg p$ in nodes 7 and 11, respectively. The label $1.(\lceil \neg q \rceil)$ of these literals is justified on $\mathcal{B}_1\rho_1$ by label $1.\lceil \neg q \rceil$ of formula 10. In this case, the complementary literals contain conditional labels which are only justified by a third formula on the branch, so checking for justification is indispensable. The middle branch $\mathcal{B}_2$ of $\mathcal{T}\sigma$ can be closed using the same universal variable substitution $\rho_2 = \rho_1 = \{x/\lceil \neg q \rceil\}$ as for the left branch. The branch $\mathcal{B}_2\rho_2$ then contains the complementary literals $\{\lceil \neg q \rceil\} : 1.(\lceil \neg q \rceil) : q$ and $\emptyset : 1.(\lceil \neg q \rceil) : \neg q$ in nodes 10 and 12. The label is again justified by formula 10, which in this case is one of the complementary literals. Note that the middle branch in $\mathcal{T}$ can be closed only by the substitution $\sigma = \{y_1/\lceil \neg q \rceil\}$, other choices will not suffice. The right branch $\mathcal{B}_3$ of $\mathcal{T}\sigma$ can be closed using the universal variable substitution $\rho_3 = \{x/\lceil \neg p \rceil\}$ as $\mathcal{B}_3\rho_3$ then contains the pair $\{\lceil \neg p \rceil\} : 1.(\lceil \neg p \rceil) : p$ and $\{\lceil \neg p \rceil\} : 1.\lceil \neg p \rceil : \neg p$ of complementary literals in nodes 7 resp. 15. The label $1.(\lceil \neg p \rceil)$ of node 7 is justified on $\mathcal{B}_3$ by formula 15.

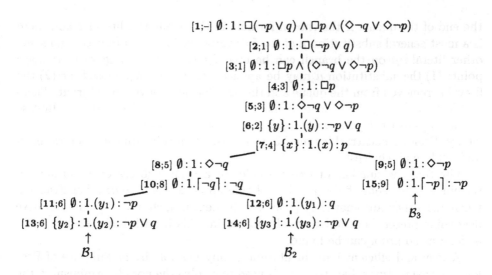

Fig. 1. The tableau $\mathcal{T}$ from Example 2.

The universal variable substitution $\rho_1 = \rho_2 = \{x/\lceil\neg q\rceil\}$ that closes $\mathcal{B}_1$ and $\mathcal{B}_2$ is incompatible with the substitution $\rho_3 = \{x/\lceil\neg p\rceil\}$ that closes $\mathcal{B}_3$. Therefore, if the variable x were not universal in formula 7, the tableau could not be closed; a second instance of formula 7 would have to be added (which in this example would not do much harm).

5 leanK: An Implementation

We have implemented our calculus as a "lean" theorem prover written in Prolog (the source code is available at `http://i12www.ira.uka.de/modlean` on the *World Wide Web*). The basic version leanK, for the logic K, consists of just eleven Prolog clauses and 45 lines of code. The version for the logic **KD**, which allows unjustified labels, is even shorter: it consists of only 6 clauses and 27 lines of code.

The leanK program employs the following fair tableau procedure: Given a tableau $\mathcal{T}$, the branch that is expanded next is the left-most open branch, with the formulae on any particular branch organised as a queue. The first formula in the chosen branch/queue is removed from the queue and is used to update the tableau as follows: If the chosen formula is not a literal then some (one) rule is applicable to it, and the formulae created by that rule application are added to the queue as follows: if the (traditional part of the) created formula is strictly less complex than the numerator, this new formula is added to the front of the queue, otherwise it is added to the end of the queue. In particular, this means that renamings of formulae added by the disjunctive rule, and the formula labelled (4) and (4^r) in the denominator of the box-rule, are added to the end of the queue. If the chosen formula in the queue is a literal ϕ and there is a most general substitution μ of the free variables in ϕ such that $\phi\mu$ and some other literal $\psi\mu$ on the branch can be used for closure, then there is a choice point: (1) the substitution μ may be applied and the branch closed, or (2) the literal is removed from the queue and the next formula moves to the front. There is a further choicepoint if there is more than one closing substitution μ. In case no closing substitution μ exists, option (2) is used deterministically: If there is a choice, Prolog's backtracking mechanism is used to resolve this non-determinism and explore all choices.

Limiting the number of free variables in a branch forces every branch to terminate after some finite number of rule applications. Prolog's backtracking mechanism then automatically processes the next branch in the queue. Iterative deepening preserves completeness by increasing this branch limit, step by step, as long as no proof can be found.

A lean and efficient implementation is only possible by making use of Prolog's special features: backtracking is used to resolve the non-determinism in the tableau procedure; built-in unification is used for finding most general closing substitutions and for the justification test; and first-argument clause indexing is employed to quickly determine the appropriate tableau rule for the next formula.

To avoid generating useless renamings of disjunctive formulae, the version of leanK used to obtain statistics uses the following restriction: when the disjunctive rule is applied to a formula $\psi = X : \Delta : \sigma : A \vee B$, the renaming ψ_2 (resp. ψ_3) created by the disjunctive rule is "protected" from further applications of the disjuntive rule until one of the variables in X_1 has been instantiated. That is, a renaming is useful only when one of the variables in X_1 has been used to close a branch using (a descendant of) $\emptyset : \Delta_1 : \sigma_1 : A$ or $\emptyset : \Delta_1 : \sigma_1 : B$.

The following table shows statistics for a set of 72 K-theorems kindly provided by A. Heuerding. Of these, leanK could prove 61 in the allotted time of 15sec, with 52 in less than 10msec (not shown in the table). The program was terminated if no proof had been found after 15sec. The table shows the number of branches that were closed, the maximal number of free variables in a branch, and the proof time (running under SICStus Prolog on a SUN Ultra 1 workstation).

No.	24	44	46	50	52	55	56	67	72
Branches	22251	90	137	43	56	1011	68	26565	154
Var.-Limit	10	5	5	4	4	11	4	11	6
Time [msec]	4400	50	80	20	30	1000	30	9520	90

The examples that took several seconds to prove show an advantage of lean implementations: the very high inference rate of about 2500 closed branches per second. The complexity of these formulae is non-trivial; one of the more complex ones, No. 55, is:

$$((\Box(\Box(p \to \Box p) \to p) \to p) \wedge (\Box(\Box((\Box(p \to \Box p) \to p) \wedge (\Box(\Box(p \to \Box p) \to p) \to$$
$$\Box\Box(\Box(p \to \Box p) \to p)) \to \Box((\Box(p \to \Box p) \to p) \wedge (\Box(\Box(p \to \Box p) \to p) \to \Box\Box(\Box(p \to \Box p) \to p)))) \to$$
$$(\Box(p \to \Box p) \to p) \wedge (\Box(\Box(p \to \Box p) \to p) \to \Box\Box(\Box(p \to \Box p) \to p))) \to (\Box(p \to \Box p) \to p) \wedge$$
$$(\Box(\Box(p \to \Box p) \to p) \to \Box\Box(\Box(p \to \Box p) \to p))) \to \Box(\Box(p \to \Box p) \to p) \to p \vee \Box p)$$

6 Conclusion and Future Work

Our initial results, presented in the last section, are very encouraging. We believe that labels with variables deliver the following advantages:

- The use of variables generates a smaller search space since a label can now stand in for all its ground instances. This is in stark contrast to the modular systems of [17], where only ground labels are used.
- The use of a Gödelisation function in the diamond-rule leads to a smaller number of labels than in other labelled tableau methods since two different occurrences of the formula $X : \Delta : \sigma : \Diamond A$ lead to the same formula $X : \Delta : \sigma. \lceil A \rceil : A$. We therefore do not need to delete duplicate occurrences of a formula as is done in some tableau implementations for modal logics. This is particularly important since the world $\sigma. \lceil A \rceil$ may be the root of a large sub-model and duplicating it is likely to be extremely inefficient.
- The use of universal variables can exponentially shorten the length of proofs (see [1]), with only a minor increase in the implementation overheads.
- Our "lean" implementation is perfect for applications where the deductive engine must be transparent and easily modifiable.

Our method is really a very clever translation of propositional modal logics into first-order logic, and most of the complications arise because some worlds may have no successors. The new notion of conditional labels allows us to keep track of these complications, and thus handle the non-serial logics that frustrate other "general frameworks" [9, 14]. Nevertheless, our method can also handle *second order* "provability" logics like **G** and **Grz**; see [11]. Furthermore, specialised versions of these tableau systems can match the theoretical lower bounds for particular logics like **K45**, **G** and **Grz** if we give up modularity; see [11, 17]. We intend to extend our initial implementation of leanK along these lines.

The 15 basic modal logics are known to be decidable and techniques from [6, 11, 17, 13] can be used to extend our method into a decision procedure. However, it is not clear that this is possible in a *lean* way. The extra implementation restriction mentioned in the previous section is of vital importance here since it is essentially a demand driven *contraction rule* on box-formulae since box-formulae get copied only as the required free variables get instantiated. And controlling contraction is often the key to decidability.

Fitting [8] has recently shown how to view the original leanT^AP program for classical propositional logic as an unusual sequent calculus `dirseq`. He has also shown how to extend `dirseq` to handle the modal logics **K**, **KT**, **K4**, and **S4**. As with traditional modal tableaux, however, `dirseq` does not handle the symmetric logics like **S5** and **B**. We are currently extending our work to give a modular free variable version of `dirseq` that does handle these logics.

It is also possible to extend our method to deal with the notions of global and local logical consequence [6].

An alternative variable label approach [12] uses different unification algorithms to find complementary literals for branch closure. However, the interactions between modalities, variable labels, and unification algorithms is by no means easy to disentangle. Extending our method to utilise special unification algorithms is perfectly possible, now that correctness and completeness have been worked out for the interactions between modalities and variable labels alone.

References

1. B. Beckert and R. Goré. Free variable tableaux for propositional modal logics. Interner Bericht 41/96, Universität Karlsruhe, Fakultät für Informatik, 1996.
2. B. Beckert, R. Hähnle, P. Oel, and M. Sulzmann. The tableau-based theorem prover $_3T^AP$, version 4.0. In *Proc. CADE-13*, LNCS 1104. Springer, 1996.
3. B. Beckert and J. Posegga. leanT^AP: Lean tableau-based deduction. *Journal of Automated Reasoning*, 15(3):339–358, 1995.
4. E. Bencivenga. Free logic. In D. Gabbay and F. Günthner, editors, *Handbook of Philosophical Logic*, volume 3. Kluwer, Dordrecht, 1986.
5. M. D'Agostino, D. Gabbay, and A. Russo. Grafting modalities onto substructural implication systems. *Studia. Logica*, 1996. To appear.
6. M. Fitting. *Proof Methods for Modal and Intuitionistic Logics*, volume 169 of *Synthese Library*. D. Reidel, Dordrecht, Holland, 1983.
7. M. Fitting. *First-Order Logic and Automated Theorem Proving*. Springer, second edition, 1996.
8. M. Fitting. Leantap revisited. Draft Manuscript, Jan. 1996.
9. A. Frisch and R. Scherl. A general framework for modal deduction. In J. Allen, R. Fikes, and E. Sandewall, editors, *Proc. 2nd Conference on Principles of Knowledge Representation and Reasoning*. Morgan-Kaufmann, 1991.
10. D. Gabbay. *Labelled Deductive Systems*. Oxford University Press, 1996.
11. R. Goré. Tableau methods for modal and temporal logics. In M. D'Agostino, D. Gabbay, R. Hähnle, and J. Posegga, editors, *Handbook of Tableau Methods*, chapter 7. Kluwer, Dordrecht, 1997. To appear.
12. G. Governatori. A reduplication and loop checking free proof system for S4. In *Short Papers: TABLEAUX'96*, number 154-96 in RI-DSI, Via Comelico 39, 20135 Milan, Italy, 1996. Department of Computer Science, University of Milan.
13. A. Heuerding, M. Seyfried, and H. Zimmermann. Efficient loop-check for backward proof search in some non-classical logics. In P. Miglioli, U. Moscato, D. Mundici, and M. Ornaghi, editors, *Proc. TABLEAUX'96*, LNCS 1071. Springer, 1996.
14. P. Jackson and H. Reichgelt. A general proof method for first-order modal logic. In *9th Int. Joint Conference on Artificial Intelligence*, pages 942–944, 1987.
15. S. Kanger. *Provability in Logic*. Stockholm Studies in Philosophy, University of Stockholm. Almqvist and Wiksell, Sweden, 1957.
16. R. Letz, J. Schumann, S. Bayerl, and W. Bibel. SETHEO: A high-performance theorem prover. *Journal of Automated Reasoning*, 8(2):183–212, 1992.
17. F. Massacci. Strongly analytic tableaux for normal modal logics. In A. Bundy, editor, *Proc., CADE-12*, LNCS 814. Springer, 1994.
18. G. Mints. *A Short Introduction to Modal Logic*. CSLI, Stanford, 1992.
19. S. Reeves. Semantic tableaux as a framework for automated theorem-proving. In C. Mellish and J. Hallam, editors, *Advances in Artificial Intelligence*. Wiley, 1987.
20. A. Russo. Generalising propositional modal logic using labelled deductive systems. In F. Baader and K. Schulz, editors, *Proceedings FroCoS*. Kluwer, 1996.

A Sequent Calculus for Skeptical Default Logic

P. A. Bonatti and N. Olivetti

Dip. di Informatica - Università di Torino
Corso Svizzera 185, I-10149 Torino, Italy
E-mail: {bonatti,olivetti}@di.unito.it

Abstract. In this paper, we contribute to the proof-theory of Reiter's Default Logic by introducing a sequent calculus for skeptical reasoning. The main features of this calculus are simplicity and regularity, and the fact that proofs can be surprisingly concise and, in many cases, involve only a small part of the default theory.

1 Introduction

Non-monotonic logics play a fundamental role in knowledge representation and commonsense reasoning, as well as in the theory of programming languages.[1] The semantic and algorithmic aspects of non-monotonic reasoning have been extensively investigated (e.g. see [22, 26, 13, 17, 18, 9, 29, 33, 25] and [30, 27, 3, 4, 7, 35, 1, 2, 31, 36]). On the other hand, the proof-theoretic aspects are not yet completely understood.

The fundamental papers by Gabbay [14], Makinson [24] and Kraus, Lehmann and Magidor [19], focus their attention on general properties of non-monotonic inference, rather than on specific formalisms. In particular, they do not axiomatize any form of non-monotonic assumption making. A similar consideration holds for the papers by Bochman [5] and Nait Abdallah [28]. The only complete axiomatizations are Levesque's Hilbert-style system for skeptical reasoning in a generalized autoepistemic logic [20],[2] Olivetti's sequent calculus for minimal entailment [32], and Bonatti's sequent calculi for credulous reasoning in default logic and normal autoepistemic logic [6]. A novel feature of [6] is the use of an *axiomatic rejection method* (cf. [21, 37, 38, 10, 11, 40, 41, 39, 8]) for checking the consistency of the defaults' justifications, and for deriving negative autoepistemic literals (disbeliefs).

In the present paper, we proceed along the line of research initiated in [6]. We introduce a sequent calculus for skeptical default reasoning, with two main goals in mind. First, we are aiming at a terse, abstract characterization of default inference, without committing to any specific proof strategy. The resulting calculus combines the constructive nature of algorithmic approaches with the declarative nature of axiomatic systems. Among the possible applications of such a theoretical tool, we mention the following.

---

[1] We mention the semantics of negation in logic programming, inheritance in object oriented languages, and the structured operational semantics of process algebras.

[2] From Levesque's system, Jiang [16] derived a resolution principle for clausal autoepistemic theories.

- It constitutes an intermediate step toward a general framework, where different proof strategies and heuristics can be explored and compared.
- It is a useful tool for investigating the power and the efficiency of nonmonotonic reasoning (e.g., it can be used to prove non-elementary speed-up results [12]).
- It facilitates the understanding of nonmonotonic logics, and constitutes a promising didactic tool.

Secondly, we want our calculus to yield concise proofs, where the defaults which are irrelevant to the conclusion play no essential role, by analogy with the following examples.

Example 1. Consider an arbitrary default theory T containing (among other components) the default

$$\delta = \frac{:A}{\neg A}.$$

We claim that T skeptically entails $\neg A$. To see this, note that δ cannot be applied, because its consequent denies the premise. Thus, in order to block δ, each extension of T must contain $\neg A$, which means that $\neg A$ must be skeptically derivable. Note that the other components of T (propositional formulae and defaults different from δ) play no role in the above argument.

Example 2. Consider an arbitrary default theory T containing (among other components) the following sentence and default rule:

$$\frac{:B}{\neg A}, \quad \neg A \vee \neg B \to C.$$

Any such T entails skeptically C. Indeed, if the default is applicable, then its conclusion and the implication $\neg A \vee \neg B \to C$ suffice to prove C; alternatively, if the default is not applicable, then the negation of its justification (i.e. $\neg B$) must be derivable (we do not care how), and through $\neg A \vee \neg B \to C$ we may conclude C, also in this case. None of T's sentences and defaults, except the above ones, play any role in this argument.

Note that ignoring part of the given theory is not a trivial task in nonmonotonic reasoning. For instance, it seems impossible to achieve a similar behavior in credulous default logic. Difficulties are strictly related to the following property.

Proposition 1. *For all sentences C and default theories T, such that T credulously entails C, there exists a default δ such that $T \cup \{\delta\}$ does not entail credulously C.*

In other words, the answer to a credulous reasoning problem cannot be a function of a strict subset of the given theory T. All the current approaches to skeptical reasoning are based on credulous reasoning, in the sense that they enumerate all the extensions of the given theory—with the exception of [27], where autoepistemic theories are translated into classical propositional theories (usually much larger than the given theory) which can be queried through classical theorem proving. The enumeration-based approaches are inefficient in two respects; first, the number of extensions can be exponential in the size of the default theory; secondly, by the above proposition, all the rules of the given theory need to be considered. The calculus introduced in this paper does not have the above limitations, and by this very fact it is fundamentally different from all the previous enumeration-based approaches.

The paper is organized as follows. In the next section we recall some basic definitions and properties concerning default logic and the rejection method introduced in [8]. In Section 3 we introduce the skeptical default sequent calculus and demonstrate its main properties. The paper is concluded by a brief comparison between the skeptical calculus and the credulous calculus of [6].

2 Preliminaries

2.1 Propositional Default Logic

Here, only some basic notions are recalled; for more details see [34, 22]. Let $\mathcal{L}$ be a standard propositional language. A *default* is an inference rule of the form:

$$\frac{\alpha : \beta_1, \cdots, \beta_n}{\gamma} ,$$

sometimes denoted by $\alpha : \beta_1, \cdots, \beta_n / \gamma$, where $\alpha, \beta_1, \cdots, \beta_n, \gamma \in \mathcal{L}$. Roughly speaking, the intuitive meaning of the above default is: if α can be derived and each β_i is consistent with the rest of the theory, then conclude γ. For all defaults δ having the above structure, the precondition α, which is called *prerequisite*, will be denoted by $p(\delta)$; the set of sentences $\{\beta_1, \cdots, \beta_n\}$, called *justification*, will be denoted by $j(\delta)$; and γ, which is the *conclusion* of the default, will be denoted by $c(\delta)$. We will employ $\neg j(\delta)$ as an abbreviation for $\{\neg\beta_1, \cdots, \neg\beta_n\}$. A *default theory* is a pair $\langle W, D \rangle$, where W is a set of sentences and D is a set of defaults. We shall often identify a default theory $\langle W, D \rangle$ with the set $W \cup D$. A theory[3] E is a *default extension* (or simply an extension) of $\langle W, D \rangle$ if and only if $E = \bigcup_{i<\omega} E_i$, where

$$E_0 = W,$$
$$E_{i+1} = \text{Th}(E_i) \cup \{c(\delta) \mid \delta \in D, E_i \vdash p(\delta), E \cap \neg j(\delta) = \emptyset\} .$$

A default theory Γ *skeptically entails* a sentence ψ if ψ belongs to all the extensions of Γ. Γ *credulously entails* a sentence ψ if ψ belongs to at least one extension of Γ.[4]

Example 3. (Nixon's Diamond) The statements

---

[3] By theory, we mean a set of sentences, closed under classical entailment.
[4] Sometimes "credulous" and "skeptical" are replaced by "brave" and "cautious", in the literature.

Nixon is a quaker (Q);
Nixon is a republican (R);
If Nixon is a republican then, if possible, assume that Nixon is not a pacifist $(\neg P)$;
If Nixon is a quaker then, if possible, assume that Nixon is a pacifist (P);

can be represented by the default theory $\langle W, D \rangle$ where $W = \{Q, R\}$ and $D = \{(R : \neg P/\neg P), (Q : P/P)\}$. The reader may easily verify that this theory has two extensions: $E' = \text{Th}(\{Q, R, \neg P\})$ and $E'' = \text{Th}(\{Q, R, P\})$. Intuitively, in E', the first default is applied; its consequent blocks the second default. Symmetrically, in E'', the second default is applied, and blocks the first one. Clearly, Q and R are skeptically entailed by $\langle W, D \rangle$, while P and $\neg P$ are entailed only credulously.

Definition 2. A default δ is **active** in a set of sentences E if and only if $E \vdash p(\delta)$ and $E \cap \neg j(\delta) = \emptyset$.

Intuitively, a default is active in E if its preconditions are satisfied in the context defined by E.[5]

Example 4. In the above example, the default $R : \neg P/\neg P$ is active in E', while $Q : P/P$ is active in E''.

Lemma 3. [6] *Assume that δ is not active in a theory E. Then E is an extension of $\langle W, D \rangle$ iff E is an extension of $\langle W, D \cup \{\delta\} \rangle$.*

2.2 Anti-Sequent Calculus

An *anti-sequent* is a pair of sets of sentences $\langle \Gamma, \Sigma \rangle$, denoted by $\Gamma \nvdash \Sigma$. As usual, Γ, α is an abbreviation for $\Gamma \cup \{\alpha\}$. The intended meaning of $\Gamma \nvdash \Sigma$ is: there exists a model of Γ where all the sentences of Σ are false. If M is such a model, then we say that the anti-sequent is *true* and that M is an *anti-model* for it.

An anti-sequent $\Gamma \nvdash \Sigma$ is an axiom of the anti-sequent calculus if, and only if, Γ and Σ are disjoint sets of propositional variables. The rules of the calculus are listed in Fig. 1.

Note that most rules coincide with their classical counterparts. The only difference is that the classical rules with two premises are split into pairs of rules with one premise (e.g. $\nvdash \bullet \wedge$ and $\nvdash \wedge \bullet$). Intuitively, this means that what is exhaustive search in classical sequent calculus, becomes nondeterminism in anti-sequent calculus. The above proof-system is sound and complete.

---

[5] The notion of active default is essentially similar to the notion of *generating default* which, however, is defined for extensions only. When E is an arbitrary set of sentences, the set of active defaults does not necessarily generate E; hence the change of terminology.

$$\frac{\Gamma \not\vdash \Sigma, \alpha}{\Gamma, \neg\alpha \not\vdash \Sigma} \; (\neg \not\vdash) \qquad\qquad \frac{\Gamma, \alpha \not\vdash \Sigma}{\Gamma \not\vdash \Sigma, \neg\alpha} \; (\not\vdash \neg)$$

$$\frac{\Gamma, \alpha, \beta \not\vdash \Sigma}{\Gamma, \alpha \wedge \beta \not\vdash \Sigma} \; (\wedge \not\vdash) \qquad \begin{array}{c} \dfrac{\Gamma \not\vdash \Sigma, \alpha}{\Gamma \not\vdash \Sigma, \alpha \wedge \beta} \; (\not\vdash \bullet\wedge) \\[2ex] \dfrac{\Gamma \not\vdash \Sigma, \beta}{\Gamma \not\vdash \Sigma, \alpha \wedge \beta} \; (\not\vdash \wedge\bullet) \end{array}$$

$$\frac{\Gamma, \alpha \not\vdash \Sigma}{\Gamma, \alpha \vee \beta \not\vdash \Sigma} \; (\bullet\vee \not\vdash)$$

$$\frac{\Gamma, \beta \not\vdash \Sigma}{\Gamma, \alpha \vee \beta \not\vdash \Sigma} \; (\vee\bullet \not\vdash) \qquad \frac{\Gamma \not\vdash \Sigma, \alpha, \beta}{\Gamma \not\vdash \Sigma, \alpha \vee \beta} \; (\not\vdash \vee)$$

$$\frac{\Gamma \not\vdash \Sigma, \alpha}{\Gamma, \alpha \to \beta \not\vdash \Sigma} \; (\bullet \to\not\vdash)$$

$$\frac{\Gamma, \beta \not\vdash \Sigma}{\Gamma, \alpha \to \beta \not\vdash \Sigma} \; (\to \bullet \not\vdash) \qquad \frac{\Gamma, \alpha \not\vdash \Sigma, \beta}{\Gamma \not\vdash \Sigma, \alpha \to \beta} \; (\not\vdash \to)$$

Fig. 1. Rules of anti-sequent calculus

Theorem 4 [8]. *An anti-sequent $\Gamma \not\vdash \Sigma$ is true if and only if it is provable.*

The anti-sequent calculus preserves many properties of the standard sequent calculus. For example, the above rules are perfectly symmetric, and satisfy the subformula property; for each connective there exist rules for introducing it in the left-hand side and in the right-hand side of anti-sequents; moreover, as we already pointed out, the new rules are strikingly similar to their classical counterparts. The rule below is a counterpart of the classical Cut rule, and generalizes Lukasiewicz's detachment rule.

$$\frac{\Gamma \not\vdash \Sigma \qquad \Gamma, \alpha \vdash \Sigma}{\Gamma \not\vdash \Sigma, \alpha} \; \text{Cut 2} \, .$$

This rule is manifestly sound. By Theorem 4, we have that the calculus without Cut 2 is complete, so we get the admissibility of Cut 2 for free. In the anti-sequent calculus, proofs correspond to counterexamples. Note that a derivation in the anti-sequent calculus is linear, and hence contains exactly one axiom, that constitutes a partial anti-model for the conclusion of the proof. In some other rejection methods there is no clear correspondence between proofs and counterexamples (see [8] for further details).

3 The Skeptical Calculus for Default Logic

3.1 Residual Rules

As an intermediate step towards a skeptical calculus for default logic, in this section we develop a calculus for propositional logic extended with ordinary monotonic inference rules of the form α/β. We call such a rule *residue* because, for our purpose, it is what is left of a default rule after deleting all justifications. Let $\mathcal{L}$ denote the propositional language, we define $\mathcal{L}^{res}$, the language of residues, as follows:
$$\mathcal{L}^{res} = \mathcal{L} \cup \{\alpha/\beta \mid \alpha, \beta \in \mathcal{L}\}.$$
Given a subset S of $\mathcal{L}^{res}$, we are interested in the deductive closure of S under classical provability *and* residual rules.

Definition 5. Let $S \subseteq \mathcal{L}^{res}$; the deductive closure of S, denoted by $\text{Th}^{res}(S)$ is the least set $S' \subseteq \mathcal{L}$ which satisfies the following conditions:

a) $S \cap \mathcal{L} \subseteq S'$;
b) if $S' \vdash \alpha$, then $\alpha \in S'$;
c) if $\alpha \in S'$, and $\alpha/\gamma \in S$, then $\gamma \in S'$.

We now show that given S, $\text{Th}^{res}(S)$ exists and can be generated inductively.

Proposition 6. *Given S, let*
$$S_0 = S \cap \mathcal{L};$$
$$S_{i+1} = \text{Th}(S_i) \cup \{\beta \mid \alpha/\beta \in S \wedge S_i \vdash \alpha\}.$$
Then $\text{Th}^{res}(S) = \bigcup_i S_i$.

In Fig. 2 we give a sequent and anti-sequent calculus for residual rules, that is for Th^{res}. Sequents and antisequents have respectively the form $\Gamma \vdash \Delta$ and $\Gamma \nvdash \Delta$, where Γ is a finite subset of $\mathcal{L}^{res}$ and Δ is a finite subset of $\mathcal{L}$.

In the next lemma we state some easy properties of the closure operator Th^{res} which are needed in the next theorem.

Proposition 7. *Let $S \subseteq \mathcal{L}^{res}$, then*

1. $\text{Th}(S) \subseteq \text{Th}^{res}(S)$;
2. *if $S \subseteq S'$, then $\text{Th}^{res}(S) \subseteq \text{Th}^{res}(S')$;*
3. $\text{Th}^{res}(S \cup \{\alpha/\beta\}) \subseteq \text{Th}^{res}(S \cup \{\beta\})$;
4. *if $\alpha \in \text{Th}^{res}(S)$, then $\text{Th}^{res}(S \cup \{\alpha/\beta\}) = \text{Th}^{res}(S \cup \{\beta\})$;*
5. *if $\alpha \notin \text{Th}^{res}(S)$, then $\text{Th}^{res}(S \cup \{\alpha/\beta\}) = \text{Th}^{res}(S)$.*

Theorem 8. *The standard sequent calculus and the anti-sequent calculus extended with (Re1)-(Re4) are complete w.r.t. residual rules. That is*

(i) *$\Gamma \vdash \Delta$ is derivable iff $\bigvee \Delta \in \text{Th}^{res}(\Gamma)$;*
(ii) *$\Gamma \nvdash \Delta$ is derivable iff $\bigvee \Delta \notin \text{Th}^{res}(\Gamma)$.*

$$\textbf{(Re1)} \ \frac{\Gamma \vdash \Delta}{\Gamma, \alpha/\gamma \vdash \Delta} \qquad\qquad \textbf{(Re2)} \ \frac{\Gamma \vdash \alpha \quad \Gamma, \gamma \vdash \Delta}{\Gamma, \alpha/\gamma \vdash \Delta}$$

$$\textbf{(Re3)} \ \frac{\Gamma \not\vdash \Delta \quad \Gamma \not\vdash \alpha}{\Gamma, \alpha/\gamma \not\vdash \Delta} \qquad\qquad \textbf{(Re4)} \ \frac{\Gamma, \gamma \not\vdash \Delta}{\Gamma, \alpha/\gamma \not\vdash \Delta}$$

Plus the rules of classical sequent calculus and anti-sequent calculus, restricted to $\mathcal{L}$.

Fig. 2. Classical calculi extended with residues

3.2 The Skeptical Sequent Calculus

In order to simplify the presentation, we first introduce a basic version of the skeptical calculus; a generalized version (which leads to more efficient deductions) will be introduced in the next section.

The skeptical sequent calculus exploits *constraints* of the form $\mathbf{M}\alpha$ or $\mathbf{L}\alpha$, where $\alpha \in \mathcal{L}$. Intuitively, $\mathbf{M}$ and $\mathbf{L}$ are analogous to a possibility modality and to a necessity modality, respectively. We say that a set of sentences E *satisfies* a constraint $\mathbf{M}\alpha$ if $E \not\vdash \neg\alpha$; we say that E satisfies $\mathbf{L}\alpha$ if $E \vdash \alpha$.

A *skeptical default sequent* is a 3-tuple $\langle \Sigma, \Gamma, \Delta \rangle$, denoted by $\Sigma; \Gamma \vdash \Delta$, where Σ is a set of constraints, Γ is a propositional default theory, and Δ is a set of propositional sentences. The intended meaning of the above sequent is: $\bigvee \Delta$ belongs to all the extensions of Γ that satisfy the constraints Σ. When this is the case, we say that the sequent is *true*.

The *skeptical sequent calculus* for default logic comprises the axioms and rules illustrated in Fig. 3. Intuitively, (Sk1) explores the alternative cases where the justifications $\beta_1 \ldots \beta_n$ are/are not consistent; in the first case (first premise) the default is equivalent to the residue α/γ (and is replaced with the latter); the other premisses correspond to all the possible ways of contradicting the justifications; clearly, in these cases, the default cannot be applied and can therefore be removed. When the set of constraints cannot be possibly satisfied, the skeptical sequent is vacuously true; the rules (Sk2) and (Sk3) detect this condition. Finally, (Sk4) captures the property that default logic extends classical logic.

Theorem 9. *The skeptical calculus is sound and complete, that is, a skeptical sequent is derivable if, and only if, it is true.*

Example 5. Consider the default theory $\Gamma = \{ : B/\neg A, : A/\neg B, \neg A \lor \neg B \to C \}$. This theory skeptically entails C. Fig. 4 illustrates a proof of C (the classical part of the derivations is omitted).

$$\text{(Sk1)} \quad \frac{M\beta_1,\ldots,M\beta_n,\Sigma\,;\,\Gamma,\,\alpha/\gamma \vdash \Delta \qquad L\neg\beta_1,\Sigma\,;\,\Gamma \vdash \Delta \quad \ldots \quad L\neg\beta_n,\Sigma\,;\,\Gamma \vdash \Delta}{\Sigma\,;\,\Gamma,\,\dfrac{\alpha:\beta_1\ldots\beta_n}{\gamma} \vdash \Delta}$$

$$\text{(Sk2)} \quad \frac{\Gamma \vdash \neg\alpha}{M\alpha,\Sigma\,;\,\Gamma \vdash \Delta}\ (\Gamma \subseteq \mathcal{L}^{res}) \qquad\qquad \text{(Sk3)} \quad \frac{\Gamma \not\vdash \alpha}{L\alpha,\Sigma\,;\,\Gamma \vdash \Delta}\ (\Gamma \subseteq \mathcal{L}^{res})$$

$$\text{(Sk4)} \quad \frac{\Gamma \vdash \Delta}{\Sigma\,;\,\Gamma \vdash \Delta}\ (\Gamma \subseteq \mathcal{L}^{res})$$

Plus the rules for residues, restricted to $\mathcal{L}^{res}$.

Fig. 3. Skeptical sequent calculus

4 Enhanced Calculus

According to the skeptical calculus introduced in the previous section, the rules for residues (i.e. the monotonic part of the calculus) cannot be applied until all *proper* defaults—that is, defaults with nonempty justification—have been eliminated. Intuitively, we are forced to verify, for each possible subset of the defaults, whether it generates an extension or not. This causes proof trees to be exponentially large in the size of the default theory; more precisely, each proof tree has at least 2^n nodes, where n is the number of defaults occurring in the root. However, in general, it is not necessary to consider every default, in order to derive a skeptical conclusion (cf. examples 1 and 2). In this section, we show that a sound generalization of the skeptical rules can be used to reduce dramatically the proof size.

The generalized rules are illustrated in Fig. 5 (they are meant to replace (Sk2)-(Sk4)). The basic idea behind (Sk2') and (Sk4') is that each extension of Γ that satisfies Σ, contains both the propositional sentences of Γ and all the sentences α such that $L\alpha \in \Sigma$; moreover, these sentences are closed under classical entailment and the residues occurring in Γ. Therefore, the sentences in this closure can be used to prove the conclusion of a skeptical sequent, as in (Sk4'), or to prove that no extension of Γ can possibly satisfy a constraint $M\alpha$, as in (Sk2'). In order to understand (Sk3'), note that any extension of Γ is the closure of the propositional sentences of Γ, plus the consequents of some of its defaults. The set $\Gamma^{res+cons}$ is an upper approximation of these sentences; any sentence which does not follow from $\Gamma^{res+cons}$ cannot belong to any extension of Γ; this observation is used in (Sk3') to conclude that no extension of Γ satisfies

$$\vdots \; \Pi_1 \qquad\qquad\qquad\qquad \vdots \; \Pi_2$$

$$\frac{MB;\ \neg A, \dfrac{:A}{\neg B},\ \neg A \vee \neg B \to C \vdash C \quad L\neg B;\ \dfrac{:A}{\neg B},\ \neg A \vee \neg B \to C \vdash C}{;\ \dfrac{:B}{\neg A},\ \dfrac{:A}{\neg B},\ \neg A \vee \neg B \to C \vdash C} \; \text{(Sk1)}$$

where Π_1 is

$$\frac{\neg A, \neg B, \neg A \vee \neg B \to C \vdash \neg A}{MA, MB;\ \neg A, \neg B, \neg A \vee \neg B \to C \vdash C} \; \text{(Sk2)} \qquad \frac{\genfrac{}{}{0pt}{}{\vdots}{\neg A, \neg A \vee \neg B \to C \vdash C}}{L\neg A, MB;\ \neg A, \neg A \vee \neg B \to C \vdash C} \; \begin{matrix}\text{(Sk4)}\\[2pt]\text{(Sk1)}\end{matrix}$$

$$\underline{} \; MB;\ \neg A, \dfrac{:A}{\neg B},\ \neg A \vee \neg B \to C \vdash C$$

and Π_2 is

$$\frac{\dfrac{\neg B, \neg A \vee \neg B \to C \vdash C}{MA, L\neg B;\ \neg B, \neg A \vee \neg B \to C \vdash C}\ \text{(Sk4)} \qquad \dfrac{\dfrac{\dfrac{A, C \nvdash}{C \nvdash \neg A}\ (\nvdash\neg)}{\neg A \vee \neg B \to C \nvdash \neg A}\ (\to\bullet\nvdash)}{L\neg A, L\neg B;\ \neg A \vee \neg B \to C \vdash C}\ \begin{matrix}\text{(Sk3)}\\[2pt]\text{(Sk1)}\end{matrix}}{L\neg B;\ \dfrac{:A}{\neg B},\ \neg A \vee \neg B \to C \vdash C}$$

Fig. 4. An example

$L\alpha$, and hence the conclusion of (Sk3') is vacuously true.[6]

Theorem 10. *The rules* (Sk2')-(Sk4') *are sound.*

Remark. The rules (Sk2)-(Sk4) are special cases of (Sk2')-(Sk4'), therefore, the enhanced calculus is complete, by Theorem 9.

The next examples show the effectiveness of the generalized rules in reducing the length of the proofs.

---

[6] In the extended version of the paper we introduce also a modification of (Sk1), called (Sk1'), where each premise $L\neg\beta_i, \Sigma; \Gamma \vdash \Delta$ is replaced with $M\beta_1 \ldots M\beta_{i-1}, L\neg\beta_i, \Sigma; \Gamma \vdash \Delta$ (the order of the β_is is irrelevant, it can be any permutation of the justifications of the default). This modification affects neither soundness nor completeness. The advantage of (Sk1') is that *a justification β may occur in many defaults without affecting the size of the proofs* (up to a constant). Intuitively, after some assumption on β_i has been done, (that is, either $M\beta_i \in \Sigma$ or $L\neg\beta_i \in \Sigma$), the corresponding constraint can be immediately used to simplify (possibly eliminate) the selected default.

$$(\text{Sk2}') \; \frac{\Sigma', \Gamma' \vdash \neg\alpha}{\mathbf{M}\alpha, \Sigma; \Gamma \vdash \Delta} \qquad (\text{Sk3}') \; \frac{\Gamma^{res+cons} \not\vdash \alpha}{\mathbf{L}\alpha, \Sigma; \Gamma \vdash \Delta} \qquad (\text{Sk4}') \; \frac{\Sigma', \Gamma' \vdash \Delta}{\Sigma; \Gamma \vdash \Delta}$$

where $\Sigma' \subseteq \{\alpha \mid \mathbf{L}\alpha \in \Sigma\}$, $\Gamma' \subseteq \Gamma^{res} \overset{\text{def}}{=} \Gamma \cap \mathcal{L}^{res}$, and $\Gamma^{res+cons} = \Gamma^{res} \cup \{c(\delta) \mid \delta \in \Gamma, \, j(\delta) \neq \emptyset\}$.

Fig. 5. Enhanced rules

$$\frac{\vdots}{\begin{array}{c} \neg A, \neg A \vee \neg B \to C \vdash C \\ \hline \mathbf{M}B; \neg A, \dfrac{:A}{\neg B}, \neg A \vee \neg B \to C \vdash C \end{array}} (\text{Sk4}') \qquad \frac{\vdots}{\begin{array}{c} \neg B, \neg A \vee \neg B \to C \vdash C \\ \hline \mathbf{L}\neg B; \dfrac{:A}{\neg B}, \neg A \vee \neg B \to C \vdash C \end{array}} (\text{Sk4}')$$
$$\frac{}{; \dfrac{:B}{\neg A}, \dfrac{:A}{\neg B}, \neg A \vee \neg B \to C \vdash C} (\text{Sk1})$$

Fig. 6. An example

Example 6. Consider the default theory $\Gamma = \{ : B/\neg A, \; : A/\neg B, \; \neg A \vee \neg B \to C \}$ of Ex. 5. With the enhanced calculus, the proof of C can be greatly simplified, as shown by Fig. 6 (cf. Fig. 4). Note that $: A/\neg B$ plays no essential role in the proof, and that it might be replaced by any default theory Γ, as in Fig. 7. The latter proof, is actually a formalization of the argument presented in Ex. 2; the leftmost branch shows that if $: A/\neg B$ is applied, then C can be derived from its conclusion; the other branch shows that if $: A/\neg B$ cannot be applied, then $\neg B$ must be derivable, which suffices to obtain C. An important feature of this proof is that it has constant length for all Γ, no matter how many defaults are in Γ.

$$\frac{\vdots}{\begin{array}{c} \neg A, \neg A \vee \neg B \to C \vdash C \\ \hline \mathbf{M}B; \neg A, \Gamma, \neg A \vee \neg B \to C \vdash C \end{array}} (\text{Sk4}') \qquad \frac{\vdots}{\begin{array}{c} \neg B, \neg A \vee \neg B \to C \vdash C \\ \hline \mathbf{L}\neg B; \Gamma, \neg A \vee \neg B \to C \vdash C \end{array}} (\text{Sk4}')$$
$$\frac{}{; \dfrac{:B}{\neg A}, \Gamma, \neg A \vee \neg B \to C \vdash C} (\text{Sk1})$$

Fig. 7. A generic proof

Example 7. Consider an arbitrary default theory T containing the default $\delta = (: A/\neg A)$, as in Ex. 1. The informal proof of $\neg A$ presented there, can be formalized as shown in Fig. 8(a). The leftmost branch shows that the default cannot be applied; the rightmost branch derives $\neg A$ from the assumption that the default is blocked ($\neg A$, in the left-hand side of the upper right sequent, is obtained from the constraint $\mathbf{L}\neg A$ immediately below). Note that all the defaults of Γ are indeed ignored.

If we further assume that A does not occur in $T \setminus \{\delta\}$, then δ cannot be blocked and T has no extensions. Consequently, any contradiction $\perp$ is skeptically derivable. Fig. 8(b) contains the schema of the formal proof (the completeness of the anti-sequent calculus guarantees that the right branch of the proof can be successfully completed).

$$
\frac{\dfrac{\neg A \vdash \neg A}{\mathbf{M}A; \Gamma, \neg A \vdash \neg A}\text{(Sk2')} \quad \dfrac{\dfrac{\neg A \vdash \neg A}{\mathbf{L}\neg A; \Gamma \vdash \neg A}\text{(Sk4')}}{}\text{(Sk1)}}{; \Gamma, \dfrac{:A}{\cdot A} \vdash \neg A}
$$

$$
\frac{\dfrac{\neg A \vdash \neg A}{\mathbf{M}A; \Gamma, \neg A \vdash \perp}\text{(Sk2')} \quad \dfrac{\Gamma^{res+cons} \not\vdash \neg A}{\dfrac{}{\mathbf{L}\neg A; \Gamma \vdash \perp}}\text{(Sk3')}}{; \Gamma, \dfrac{:A}{\neg A} \vdash \perp}\text{(Sk1)}
$$

(a)   (b)

Fig. 8. An example

Two interesting rules, which together capture the notion of *cumulativity* (cf. [14, 24, 19]) are shown in Fig. 9. It is well known that (Cut) is sound for skeptical default inference, while (WM) is not [24]. Although (Cut) is not needed for the completeness of the skeptical calculus, it can be useful for capturing certain proof strategies. These aspects will be tackled in an extended version of the paper.

$$
\frac{\Sigma; \Gamma \vdash \alpha \quad \Sigma; \Gamma, \alpha \vdash \Delta}{\Sigma; \Gamma \vdash \Delta}\text{(Cut)} \qquad \frac{\Sigma; \Gamma \vdash \alpha \quad \Sigma; \Gamma \vdash \Delta}{\Sigma; \Gamma, \alpha \vdash \Delta}\text{(WM)}
$$

Fig. 9. Cumulativity rules (Cut and Weak Monotony)

5   A comparison with the credulous calculus

In this section we reformulate the credulous calculus proposed in [6] in order to make an informal comparison with the skeptical one presented in this paper.

The calculus presented in [6] makes use of sequents with a different structure and does not use the constraints. In Fig. 10 we give the rules of the credulous calculus, rephrased to match the structure of the skeptical sequents.

$$\textbf{(Cr1)}\quad \frac{\Gamma \vdash \Delta}{; \Gamma \vdash \Delta}\ (\Gamma \subseteq \mathcal{L})$$

$$\textbf{(Cr2)}\quad \frac{\Gamma \vdash \alpha \quad \Sigma; \Gamma \vdash \Delta}{\mathbf{L}\alpha, \Sigma; \Gamma \vdash \Delta}\ (\Gamma \subseteq \mathcal{L}) \qquad \textbf{(Cr3)}\quad \frac{\Gamma \nvdash \neg\alpha \quad \Sigma; \Gamma \vdash \Delta}{\mathbf{M}\alpha, \Sigma; \Gamma \vdash \Delta}\ (\Gamma \subseteq \mathcal{L})$$

$$\textbf{(Cr4)}\quad \frac{\mathbf{M}\neg\alpha, \Sigma; \Gamma \vdash \Delta}{\Sigma; \Gamma, (\alpha : \beta_1 \ldots \beta_n/\gamma) \vdash \Delta} \qquad \textbf{(Cr5)}\quad \frac{\mathbf{L}\neg\beta_i, \Sigma; \Gamma \vdash \Delta}{\Sigma; \Gamma, (\alpha : \beta_1 \ldots \beta_n/\gamma) \vdash \Delta}$$

$$\textbf{(Cr6)}\quad \frac{\Gamma \cap \mathcal{L} \vdash \alpha \quad \mathbf{M}\beta_1, \ldots \mathbf{M}\beta_n, \Sigma; \Gamma, \gamma \vdash \Delta}{\Sigma; \Gamma, (\alpha : \beta_1 \ldots \beta_n/\gamma) \vdash \Delta}$$

Plus the rules for classical sequents and anti-sequents restricted to $\mathcal{L}$.

Fig. 10. Credulous sequent calculus

The most prominent difference with respect to the skeptical calculus is that residual rules are not needed. A similar approach might be taken in the skeptical framework, at the price of a certain loss of elegance. The major reason for adopting residues, however, is flexibility, as explained in Ex. 8 below.

To improve the understanding of the relations between the two calculi, we note that the constraints in the skeptical case are simply assumptions, whereas in the credulous case they must be satisfied. This explains a certain duality between the rules of the credulous calculus and the rules of the skeptical one. Rule $(Cr1)$ is dual of $(Sk4)$, rules $(Cr2)$ and $(Cr3)$ are duals of $(Sk2)$ and $(Sk3)$. Rules $(Cr4)$, $(Cr5)$, $(Cr6)$ corresponds to $(Sk1)$. A default can be introduced if it is unapplicable (either by rule $(Cr4)$: its prerequisite cannot be proved, or by $(Cr5)$: one of its justification is inconsistent), or it is applicable (rule $(Cr6)$). While, in the skeptical case, we must prove the conclusion of the sequent in both cases, in the credulous case, we choose one of the alternatives, and we keep it.

Example 8. $\Gamma = \{ : A/\neg B,\ \neg B : C/D,\ : B/\neg A \}$. We have that Γ credulously entails D, but not skeptically. Here below is a derivation of $;\ \Gamma \vdash D$, in the credulous calculus:

$$
\cfrac{
  \cfrac{
    \cfrac{
      \neg B \vdash \neg B \qquad
      \cfrac{
        \cfrac{
          \neg B, D \not\vdash \neg A \qquad
          \cfrac{
            \neg B, D \not\vdash \neg C \;\; ; \neg B, D \vdash D \quad
            \cfrac{\neg B, D \vdash D}{} Cr1
          }{MC; \neg B, D \vdash D} Cr3
        }{MA, MC; \neg B, D \vdash D} Cr3
      }{MA, MC; \neg B, D \vdash D} Cr2
    }{L\neg B, MA, MC; \neg B, D, : B/\neg A \vdash D} Cr5
  }{
    \neg B \vdash \neg B \qquad
    \cfrac{MA, MC; \neg B, D, \cfrac{:B}{\neg A} \vdash D}{} Cr6
  }
}{
  \vdash True \qquad
  MA; \neg B, \cfrac{\neg B : C}{D}, \cfrac{:B}{\neg A} \vdash D
} Cr6
$$
$$
\cfrac{}{; \cfrac{:A}{\neg B}, \cfrac{\neg B : C}{D}, \cfrac{:B}{\neg A} \vdash D} Cr6
$$

Note that in $(Cr6)$, the prerequisite must be proved by the propositional part of the theory. In this way, the provability of the prerequisite is checked immediately, and residual rules are not needed. This restriction, however, forces an ordering in the introduction (or elimination, if we inspect a derivation backwards) of defaults. For example, in the shown derivation (inspecting it backwards) we cannot "eliminate" $\neg B : C/D$ before $: A/\neg B$, since we need the consequent of the latter default to prove the prerequisite of the former. This lack of flexibility may prevent some interesting proof-strategies from being represented in the above credulous calculus.

Since this calculus is just a restatement of the one presented in [6], we have soundness and completeness wrt. credulous reasoning.

Theorem 11. $\Sigma; \Gamma \vdash \Delta$ *is derivable in the credulous calculus iff there is an extension E of Γ, which satisfies all constraints in Σ, and such that $\bigvee \Delta \in E$.*

We conclude by noting that the above calculus is not the only possible formulation of credulous default inference; it is possible to give an alternative formulation which makes use of residual rules and is *perfectly symmetric*, or dual, to the skeptical calculus. A presentation of this latter formulation and a discussion of the relationships between the dual calculi will be deferred to a full paper.

6 Further Work

We are currently adapting the skeptical calculus to other non-monotonic formalisms, such as Autoepistemic Logic, Circumscription and cumulative variants of Default Logic. Moreover, we are exploring proof-strategies which proceed in a goal-directed fashion as long as possible (the cut rule of Fig. 9 plays an important role, in this respect). A strictly related topic is the development of refinements of the calculus, suitable for Logic Programming languages.

An interesting direction for further research concerns first-order default logic. Tiomkin [39] introduced an anti-sequent calculus, complete w.r.t. finite models, that can be used in place of the anti-sequent calculus adopted here. One interesting problem is the identification of a class of default theories for which the resulting proof-system is complete. On the other hand, one may explore variants of Default Logic based on semi-decidable notions of consistency, stronger than the classical one.

References

1. G. Amati, L. Carlucci Aiello, D. Gabbay, F. Pirri. A proof theoretical approach to default reasoning I: tableaux for default logic. *Journal of Logic and Computation*, 6(2):205-231, 1996.
2. F. Baader, B. Hollunder. Embedding defaults into terminological knowledge representation formalisms. *Journal of Automated Reasoning*, 14(1):149-180, 1995.
3. C. Bell, A. Nerode, R. Ng and V.S. Subrahmanian. Implementing deductive databases by linear programming. In *Proc. of ACM-PODS*, 1992.
4. C. Bell, A. Nerode, R. Ng and V.S. Subrahmanian. Implementing stable semantics by linear programming. In [33].
5. A. Bochman. On the relation between default and modal consequence relations. In *Proc. of KR'94*, 63-74, Morgan Kaufmann, 1994.
6. P.A.Bonatti. Sequent calculi for default and autoepistemic logics. In *Proc. of TABLEAUX'96*, LNAI 1071, pp. 127-142, Springer-Verlag, Berlin, 1996.
7. P.A. Bonatti. Autoepistemic logic programming. *Journal of Automated Reasoning*, 13:35-67, 1994.
8. P.A. Bonatti. A Gentzen system for non-theorems. Technical Report CD-TR 93/52, Christian Doppler Labor für Expertensysteme, Technische Universität, Wien, September 1993.
9. G. Brewka. Cumulative default logic: in defense of nonmonotonic inference rules. *Artificial Intelligence* 50:183-205, 1991.
10. X. Caicedo. A formal system for the non-theorems of the propositional calculus. *Notre Dame Journal of Formal Logic*, 19:147-151, (1978).
11. R. Dutkiewicz. The method of axiomatic rejection for the intuitionistic propositional calculus. *Studia Logica*, 48:449-459, (1989).
12. U. Egly, H. Tompits. Non-elementary speed-ups in default logic. Submitted.
13. D. Gabbay et al. (eds). *Handbook of Logic in Artificial Intelligence and Logic Programming*, Vol.III, Clarendon Press, Oxford, 1994.
14. D. Gabbay. Theoretical foundations for non-monotonic reasoning in expert systems. In K.R. Apt (ed.) *Logics and Models of Concurrent Systems*. Springer-Verlag, Berlin, 1985.
15. M.L. Ginsberg. A circumscriptive theorem prover. *Artificial Intelligence*, 39(2):209-230, (1989).
16. Y.J. Jiang. A first step towards autoepistemic logic programming. *Computers and Artificial Intelligence*, 10(5):419-441, (1992).
17. K. Konolige. On the Relationship between Default and Autoepistemic Logic. *Artificial Intelligence*, 35:343-382, 1988. + Errata, same journal, 41:115, 1989/90.
18. K. Konolige. On the Relation Between Autoepistemic Logic and Circumscription. In *Proceedings IJCAI-89*, 1989.

19. S. Kraus, D. Lehmann and M. Magidor. Nonmonotonic reasoning, preferential models and cumulative logics. *Artificial Intelligence*, 44(1):167-207, (1990).
20. H.J. Levesque. All I know: a study in autoepistemic logic. *Artificial Intelligence*, 42:263-309, (1990).
21. J. Lukasiewicz. *Aristotle's syllogistic from the standpoint of modern formal logic*. Clarendon Press, Oxford, 1951.
22. W. Lukaszlewicz. *Non-Monotonic Reasoning*. Ellis Horwood Limited, Chichester, England, 1990.
23. J. McCarthy. Circumscription: a form of non-monotonic reasoning. *Artificial Intelligence*, 13:27-39, (1980).
24. D. Makinson. General theory of cumulative inference. In M. Reinfrank, J. De Kleer, M.L. Ginsberg and E. Sandewall (eds.) *Non-monotonic Reasoning*, LNAI 346, Springer-Verlag, Berlin, 1989, 1-18.
25. W. Marek, A. Nerode, M. Truscyński (eds). *Logic Programming and Non-monotonic Reasoning: Proc. of the Third Int. Conference*. LNAI 928, Springer-Verlag, Berlin, 1995.
26. W. Marek, M. Truszczyński. *Nonmonotonic Logics – Context-Dependent Reasoning*. Springer, 1993.
27. W. Marek, M. Truszczyński. Computing intersections of autoepistemic expansions. In [29].
28. M.A. Nait Abdallah. An extended framework for default reasoning. In *Proc. of FCT'89*, LNCS 380, 339-348, Springer-Verlag, 1989.
29. A. Nerode, W. Marek, V.S. Subrahmanian (eds.). *Logic Programming and Non-monotonic Reasoning: Proc. of the First Int. Workshop*, MIT Press, Cambridge, Massachusetts, 1991.
30. I. Niemela. Decision procedures for autoepistemic logic. *Proc. CADE-88*, LNCS 310, Springer-Verlag, 1988.
31. I. Niemela. Toward efficient default reasoning. Proc. IJCAI'95, 312-318, Morgan Kaufmann, 1995.
32. N. Olivetti. Tableaux and sequent calculus for minimal entailment. *Journal of Automated Reasoning*, 9:99-139, (1992).
33. L. M. Pereira, A. Nerode (eds.). *Logic Programming and Non-monotonic Reasoning: Proc. of the Second Int. Workshop*, MIT Press, Cambridge, Massachusetts, 1993.
34. R. Reiter. A logic for default reasoning. *Artificial Intelligence*, 13:81-132, (1980).
35. V. Risch, C.B. Schwind. Tableau-based characterization and theorem proving for default logic. *Journal of Automated Reasoning*, 13:223-242, 1994.
36. T. Schaub. A new methodology for query answering in default logics via structure-oriented theorem proving. *Journal of Automated Reasoning*, 15(1):95-165, 1995.
37. D. Scott. Completeness proofs for the intuitionistic sentential calculus. *Summaries of Talks Presented at the Summer Institute for Symbolic Logic (Itaha, Cornell University, July 1957)*, Princeton: Institute for Defense Analyses, Communications Research Division, 1957, 231-242.
38. J. Słupecki, G. Bryll, U. Wybraniec-Skardowska. Theory of rejected propositions. *Studia Logica*, 29:75-115, (1971).
39. M. Tiomkin. Proving unprovability. In *Proc. of LICS'88*, 1988.
40. A. Varzi. Complementary sentential logics. *Bulletin of the Section of Logic*, 19:112-116, (1990).
41. A. Varzi. Complementary logics for classical propositional languages. *Kriterion. Zeitschrift für Philosophie*, 4:20-24 (1992).

A Fast Saturation Strategy for Set-Theoretic Tableaux

Domenico Cantone*

*Università di Catania, Dipartimento di Matematica,
Viale A. Doria, 6, I-95125 Catania, Italy
e-mail: cantone@dipmat.unict.it*

Abstract. In this paper we present a fast tableau saturation strategy which can be used as an optimized decision procedure for some fragments of set theory.

Such a strategy is based on the use of a model checking technique which guides the saturation process. As a result, it turns out that the saturation process converges much faster than previous decision algorithms either to a closed tableau or to a model satisfying the input formula.

For the sake of simplicity, the strategy is illustrated for the extension **MLSS** of **MLS** with the finite enumeration operator $\{\bullet, \bullet, \cdots, \bullet\}$.

1 Introduction

In the last few years, the decision problem for various classes of set-theoretic formulae has been studied very actively as part of a project aimed at the design and implementation of a set-theoretically based proof verifier with an *inferential core* comprising, among others, decision procedures for sublanguages of set theory. Several theoretical results originated from such research have been collected in [CFO89]. Some of these procedures have already been implemented using *ad hoc* techniques within the system ETNA, a set-theoretically based verification system under development at the University of Catania and New York University (cfr. [CF95]).

Recently, in [CRC95, CRC97, CF95] we started an investigation aimed at the discovery of tableau calculi and effective saturation strategies for solving the satisfiability problem for fragments of set theory and related areas.

The advantages of implementing decision procedures by means of complete and effective saturation strategies for tableau calculi over *ad hoc* methods are manyfold:

- tableaux naturally maintain information about proof attempts; such information can then be used either to reconstruct *proofs* or, in case of formulae which are not theorems, to construct *counter-examples* (this can be particularly useful in didactic applications, such as in the use of computers for teaching logic or set theory courses);

---

* This research was partially supported by C.N.R. of Italy, Research Project No. 95.00411.CT12 (S.E.T.A.), and by MURST 40% project *Calcolo algebrico e simbolico*.

– tableau calculi can easily be extended by new rules, thus allowing, in favourable cases, smooth generalizations to more expressive decidable fragments;
– implementations of saturation strategies for tableau calculi can be equipped with heuristics; for instance a user could easily deactivate some of the rules, impose restrictions on their applicability, etc.

Effective saturation strategies for tableau calculi have already been found for the extensions of the basic unquantified Multi-Level Syllogistic theory (in short **MLS**) respectively with a predicate expressing finiteness and with a predicate expressing the transitive closure of membership (cf. [CRC95, CRC97]).[2] A saturation strategy for a tableau calculus to solve the satisfiability problem for a quantified fragment of set theory is contained in [CF95].

Roughly speaking, in all cases, the basic idea has been to consider all possible membership and equality relationships among *old* variables, initially present in a suitably *normalized* formula, and *new* variables, introduced to distinguish pairs of variables which are related to each other by inequalities. All membership and equality relationships are used in turn to deduce additional relations by means of the remaining literals present in the formula, and the process continues until no further membership or equality can be deduced (subject to certain restrictions which prevent the introduction of infinitely many new variables), i.e. until the tableau has been saturated. If all possibilities lead to a contradiction, then one can prove that the input formula is *injectively* unsatisfiable, namely that it does not have any set model which maps distinct variables into distinct sets. Since the ordinary satisfiability problem can be reduced to the injective satisfiability problem, one has also a satisfiability test and, dually, a test for validity in all set models for the given collection of formulae.

The major drawbacks with the above approach are the following:
– the normalization process usually introduces new variables;
– the reduction from ordinary satisfiability to injective satisfiability is very expensive;
– the height of a saturated tableau, though in the case of the theories briefly mentioned above can be kept polynomial in the size of the input formula, can be quite large, even for simple formulae.

In this paper we present a fast tableau saturation strategy which allows to address directly the ordinary satisfiability problem for some decidable fragments of set theory and which requires a much less expensive normalization preprocessing.

Such a strategy is based on the use of a model checking technique which guides the saturation process. As a result, it turns out that the saturation process converges much faster either to a closed tableau or to a set model satisfying the root formula.

The basic idea is somewhat similar to that used in [CZ92].

For the sake of simplicity, the strategy is illustrated for the extension **MLSS**

---

[2] We recall that **MLS** is the unquantified theory involving the constructs $\in$ (membership), $=$ (equality), $\subseteq$ (set inclusion), $\cup$ (binary union), $\cap$ (binary intersection), and $\setminus$ (set difference). Its decision problem has been first solved in [FOS80].

of **MLS** with the finite enumeration operator $\{\bullet, \bullet, \cdots, \bullet\}$, though it can be adapted to other cases as well. Moreover, again for simplicity, we shall assume that formulae are normalized, though this is not strictly necessary, as will be pointed out in an example.

The paper is organized as follows. Section 2 introduces some basic definitions and tools which are needed in the rest of the paper. The tableau calculus for **MLSS** is presented in Section 3 together with a complete and effective saturation strategy, which is illustrated by way of a simple example. A section with some final remarks and directions for future research concludes the paper.

2 Preliminaries

We introduce some basic concepts such as the von Neumann standard hierarchy of sets. Also the concept of realization of a given graph is introduced, as well as some of its properties.

2.1 Hierarchies of sets, assignments, set models, and the decision problem

In the present paper we are mainly interested in the satisfiability problem for set theoretic formulae in the von Neumann standard cumulative hierarchy $\mathcal{V}$ of sets defined by:

$$\mathcal{V}_0 = \emptyset$$
$$\mathcal{V}_{\alpha+1} = \mathcal{P}(\mathcal{V}_\alpha), \quad \text{for each ordinal } \alpha,$$
$$\mathcal{V}_\lambda = \bigcup_{\mu < \lambda} \mathcal{V}_\mu, \quad \text{for each limit ordinal } \lambda,$$
$$\mathcal{V} = \bigcup_{\alpha \in On} \mathcal{V}_\alpha,$$

where $\mathcal{P}(S)$ is the powerset of S and On is the class of all ordinals (cf. [Jec78]).

It can easily be seen that there can be no membership cycle in $\mathcal{V}$, namely sets in $\mathcal{V}$ are well-founded w.r.t. membership.

An ASSIGNMENT over a collection of variables V is any function from V into $\mathcal{V}$. A set theoretic formula φ is said to be SATISFIED by an assignment M over its variables if the formula resulting from φ by substituting in it sets Mx in place of free occurrences of x and by interpreting set theoretic operators and predicates according to their standard meaning is true. An assignment which satisfies a given formula φ is said to be a SET MODEL for φ. A formula φ is said to be SATISFIABLE if it has a set model.

The DECISION PROBLEM or SATISFIABILITY PROBLEM for a collection $\mathcal{C}$ of set-theoretic formulae is the problem of establishing for any given formula in $\mathcal{C}$ whether it has a set model or not.

2.2 Realizations

We shall prove the completeness of the tableau calculus and of the related saturation strategy, to be given in the next section, by showing that any saturated open branch carries enough information, in terms of positive membership relations, to guide the construction of a set model for all formulae present in the branch. The technique used for the actual model construction relies on *realizations* of a given (membership) graph.

The notion of realization, which is defined below, will therefore play an important role in what follows.

Let $G = (N, \widehat{\in})$ be a directed *acyclic* graph, where $\widehat{\in}$ is a binary relation over N, and let (V, T) be a partition of N. Also, let $\{u_t : t \in T\}$ be a family of sets.

Definition 1. The REALIZATION of $G = (N, \widehat{\in})$ relative to $\{u_t : t \in T\}$ and to a partition (V, T) of N is the assignment R over N recursively defined by:

$$
\begin{aligned}
Rx &= \{Rz : z \in V \cup T \text{ and } z\widehat{\in}x\}, \quad \text{for } x \text{ in } V; \\
Rt &= \{u_t\}, \quad\quad\quad\quad\quad\quad\quad\quad \text{for } t \text{ in } T.
\end{aligned}
\tag{1}
$$

Observe that R is well defined by (1) above, since the graph G is acyclic.

Next we introduce a notion of height, for all x in $V \cup T$, by putting

$$
height(x) = \begin{cases} 0 & \text{if } x \in T, \text{ or} \\ & \text{if } y \,\widehat{\notin}\, x, \text{ for } y \in V \cup T \\ \max\{height(y) + 1 : y \in V \cup T \wedge y\widehat{\in}x\} & \text{otherwise.} \end{cases}
$$

Also we put, for all $x \in N$,

$$
G(x) = \{y \in N : y\widehat{\in}x\}.
$$

The following elementary lemma states the main properties of realizations.

Lemma 2. *Let $G = (V \cup T, \widehat{\in})$ be a directed acyclic graph, with $V \cap T = \emptyset$. Also, let $\{u_t : t \in T\}$ and R be respectively a family of sets and the realization of G relative to $\{u_t : t \in T\}$ and (V, T). Let $\in^R$ be the binary relation defined by $x \in^R y$ iff $Rx \in Ry$ and put $G^R = (V \cup T, \in^R)$. Also, let $\sim^R$ be the equivalence relation defined on $V \cup T$ by $x \sim^R y$ iff $Rx = Ry$. Assume that*

(a) $u_t \neq u_{t'}$, for all distinct t, t' in T;

(b) $u_t \neq Rv$, for all $t \in T$ and $v \in V \cup T$.

Then

(i) $|\{x \in V \cup T : x \sim^R t\}| = 1$, for all t in T;

(ii) $Rx = Ry \cup Rz$ iff $G^R(x) = G^R(y) \cup G^R(z)$, for x, y, z in V;

(iii) $Rx = Ry \cap Rz$ iff $G^R(x) = G^R(y) \cap G^R(z)$, for x, y, z in V;

(iv) $Rx = Ry \setminus Rz$ iff $G^R(x) = G^R(y) \setminus G^R(z)$, for x, y, z in V;

(v) $Rx = \{Ry_1, \ldots, Ry_k\}$ iff $\{y_1, \ldots, y_k\} \subseteq G^R(x) \subseteq \bigcup_{i=1}^{k}\{y \in V : y \sim^R y_i\}$, for x in V and $y_1, \ldots, y_k$ in $V \cup T$, with $k \geq 0$;

(vi) if $x \sim^R y$, then $height(x) = height(y)$, for x, y in $V \cup T$;

(vii) if $Rx \in Ry$, then $height(x) < height(y)$, for x, y in $V \cup T$;

(viii) there can be no membership cycle in $\{Rv : v \in V \cup T\}$. □

Remark 3. Notice that in the above lemma, conditions (a) and (b) can always be satisfied by letting the u_ts be pairwise distinct sets of cardinality no less than $|V \cup T|$, since $|Rx| < |V \cup T|$. □

3 A decidable tableau calculus for MLSS

In this section we describe a simple, yet in practice effective, tableau saturation strategy which is guided by interleaved model checking steps. Such a strategy allows to prune considerably the tableau by eliminating most of the irrelevant branches.

The strategy to be presented will be exemplified for the fragment **MLSS** (Multi-Level Syllogistic with Singleton), but it could easily be adapted to other cases as well.

We recall that the language of **MLSS** contains:
- a denumerable infinity of individual variables $x, y, z, \ldots$;
- the predicate symbols $\in, =, \subseteq$;
- the operators $\cap, \cup, \setminus, \{\bullet, \bullet, \cdots, \bullet\}$;
- the constant $\emptyset$ (empty set);
- parentheses (to form compound terms);
- the logical connectives $\neg, \wedge, \vee, \rightarrow, \leftrightarrow$ (to form compound formulae).

Notice that explicit quantification is not allowed in **MLSS**-formulae.

Though not strictly necessary, it is convenient to limit our analysis to **MLSS**-formulae of a special form, namely to conjunctions of normalized literals each of which has one of the following types

$$x \in y\,,\, x \notin y\,,\, x = y\,,\, x \neq y\,,\, z = x \cup y\,,\, z = x \cap y\,,\, z = x \setminus y\,,\, x = \{y_1, \ldots, y_k\}\,, \quad (2)$$

where $x, y, z, y_1, \ldots, y_k$ are individual variables. We shall call such formulae *normalized* **MLSS***-conjunctions*.

To see that the satisfiability problem for **MLSS** is equivalent to the satisfiability problem for normalized **MLSS**-conjunctions, we consider the following procedure $Normalize(\varphi)$ (cf. Table 1).

Let $\langle \varphi_1, \ldots, \varphi_k \rangle$ be the k-tuple returned by the call $Normalize(\varphi)$. Then it is an easy matter to show that
- each formula φ_i is a normalized conjunction of **MLSS**;
- φ is satisfiable if and only if at least one among $\varphi_1, \ldots, \varphi_k$ is satisfiable.

Therefore we have

Lemma 4. *The satisfiability problem for* **MLSS***-formulae is equivalent to the satisfiability problem for normalized* **MLSS***-conjunctions.* □

Remark 5. We resort to normalized conjunctions just for the sake of simplicity. The saturation process to be described below would work with just minor modifications even for unrestricted **MLSS**-formulae. This fact will be further argued on in an example. □

The tableau rules to be presented in this paper are listed in Table 2.

A similar collection of rules was introduced in [CF95]. Besides rules (1)–(10), such collection comprised also the following rules:

$$\frac{}{x = y \mid x \neq y} \,(11') \qquad \frac{x \neq y}{\begin{array}{c|c} w \in x & w \notin x \\ w \notin y & w \in y \end{array}} \,(12') \qquad \frac{}{w \in y' \mid w \notin y'} \,(13')$$

```
Procedure Normalize(φ);
Comment: φ is an MLSS-formula.
     Φ := true;
     ψ := φ;
     - let q₀ be a new variable not occurring in φ;
     Φ := Φ ∧ q₀ = q₀ \ q₀;
     - substitute in ψ each occurrence of the constant ∅ with the variable q₀;
     while ψ contains terms of the form x ∪ y, x ∩ y, x \ y, {y₁, ..., yₖ},
               with x, y, y₁, ..., yₖ individual variables do
               - let t be any such term and let xₜ be a newly introduced variable;
               - substitute in ψ each occurrence of the term t with the variable xₜ;
               Φ := Φ ∧ (xₜ = t);
     end while;
     - let ψ₁ ∨ ... ∨ ψₖ be a disjunctive normal form of ψ;

     Comment: at this point each ψᵢ is a propositional combination of atoms of type
          x ∈ y, x = y, x ⊆ y;

     for i := 1 to k do
          for each conjunct of type ¬(x ⊆ y) in ψᵢ do
               - let z_{xy} be a newly introduced variable;
               - substitute in ψᵢ each occurrence of ¬(x ⊆ y) with (z_{xy} ∈ x ∧ z_{xy} ∉ y);
          end for;
          for each conjunct of type x ⊆ y in ψᵢ do
               - substitute in ψᵢ each occurrence of x ⊆ y with y = x ∪ y;
          end for;
     end for;
     return (ψ₁ ∧ Φ, ..., ψₖ ∧ Φ);
end procedure.
```

Table 1. Procedure *Normalize*

Thus our new set of rules results from the old one by merging rules $(11')$ and $(12')$ into the new rule (12), and by restricting the applicability of rule $(13')$ through the addition of premises $x = y \setminus y'$ and $w \in y$ in rule (11) (see Table 2).

Definition 6. Let S be a finite collection of normalized **MLSS**-literals. An INITIAL **MLSS**-TABLEAU for S is a one-branch tree whose nodes are labeled by the literals in S.

An **MLSS**-TABLEAU for S is a tableau labeled with **MLSS**-literals which can be constructed from the initial tableau for S by a finite number of applications of the rules (1)–(12) of Table 2.

For simplification purposes, we assume that no literal can occur more than once on any given branch, namely we assume that a rule which would add a literal already present in a branch has no effect. □

Definition 7. Let $\mathcal{T}$ be an **MLSS**-tableau for a given finite collection S of normalized **MLSS**-literals.

A branch ϑ of $\mathcal{T}$ is said to be

- STRICT, if no rule has been applied more than once on ϑ to the same literal occurrences;
- SATURATED WITH RESPECT TO A GIVEN RULE, if the rule has been applied at least once to each instance of its premises on ϑ;

$$\frac{\begin{array}{c} z = y \cup y' \\ x \in y \end{array}}{x \in z} \quad (1)$$

$$\frac{\begin{array}{c} z = y \cup y' \\ x \in y' \end{array}}{x \in z} \quad (2)$$

$$\frac{\begin{array}{c} z = y \cup y' \\ x \in z \end{array}}{x \in y \,|\, x \in y'} \quad (3)$$

$$\frac{\begin{array}{c} z = y \cap y' \\ x \in z \end{array}}{\begin{array}{c} x \in y \\ x \in y' \end{array}} \quad (4)$$

$$\frac{\begin{array}{c} z = y \cap y' \\ x \in y \\ x \in y' \end{array}}{x \in z} \quad (5)$$

$$\frac{\begin{array}{c} x = y \setminus y' \\ w \in x \end{array}}{\begin{array}{c} w \in y \\ w \notin y' \end{array}} \quad (6)$$

$$\frac{\begin{array}{c} x = y \setminus y' \\ w \in y \\ w \notin y' \end{array}}{w \in x} \quad (7)$$

$$\frac{y = \{x_1, \ldots, x_k\}}{\begin{array}{c} x_1 \in y \\ \vdots \\ x_k \in y \end{array}} \quad (8)$$

$$\frac{\begin{array}{c} y = \{x_1, \ldots, x_k\} \\ z \in y \end{array}}{z = x_1 \,|\, \cdots \,|\, z = x_k} \quad (9)$$

$$\frac{\begin{array}{c} x = y \\ \varphi \end{array}}{\begin{array}{c} \varphi_x^y \\ \varphi_y^x \end{array}} \quad (10)^a$$

$$\frac{\begin{array}{c} x = y \setminus y' \\ w \in y \end{array}}{w \in y' \;|\; w \notin y'} \quad (11)$$

$$\frac{}{x = y \left|\begin{array}{c} w \in x \\ w \notin y \end{array}\right| \begin{array}{c} w \notin x \\ w \in y \end{array}} \quad (12)^b$$

---

[a] We denote by φ_x^y the formula resulting from φ by substituting in it each occurrence of y with x.

[b] w must be a new variable not occurring on the branch to which the rule is applied.

Table 2. Tableaux rules for **MLSS**

- SATURATED WITH RESPECT TO A GIVEN COLLECTION OF RULES, if ϑ is saturated with respect to each rule of the collection;
- CLOSED, if either ϑ contains a set of literals of the form $x \in x_1 \in \cdots \in x_n \in x$, for some variables $x, x_1, \ldots, x_n$ with $n \geq 0$, or it contains a pair of complementary literals $X, \neg X$;
- SATISFIABLE, if there exists a set model for the literals occurring on ϑ; any such model will be called a SET MODEL FOR ϑ.

A tableau $\mathcal{T}$ is said to be

- ANNOTATED, if some information may be stored on branches and/or nodes;
- STRICT, or SATURATED WITH RESPECT TO A SET OF RULES, or CLOSED, if such are all its branches;
- SATISFIABLE, if at least one of its branches is satisfiable. □

Remark 8. (**A**) Notice that a closed branch, and therefore a closed tableau, is unsatisfiable.

(**B**) Soundness of rules (1)–(12) is evident; that is, if a tableau $\mathcal{T}$ is satisfiable, then any tableau obtained from $\mathcal{T}$ by any number of applications of rules (1)–(12) is satisfiable as well. For the sake of brevity we comment upon the soundness of rules (1) and (12) only, since the other can be treated similarly. Concerning rule (1), let us assume that M is a set model satisfying a certain branch ϑ containing the literals $z = y \cup y'$ and $x \in y$. Then $Mz = My \cup My'$ and $Mx \in My$, so that $Mx \in Mz$. Thus the branch $\vartheta' = \vartheta; x \in z$, which can be obtained from ϑ by an application of rule (1), is satisfiable as well. Concerning rule (12), let ϑ be a tableau branch and let us assume as before that M satisfies ϑ. Let x, y be any two variables, and let w be a new variable not occurring on ϑ. Let us put:

$$\vartheta_1 = \vartheta; x = y;$$
$$\vartheta_2 = \vartheta; w \in x; w \notin y;$$
$$\vartheta_3 = \vartheta; w \notin x; w \in x.$$

If $Mx = My$, then obviously M satisfies ϑ_1. On the other hand, if $Mx \neq My$, then by extensionality there exists $s \in (Mx \setminus My) \cup (My \setminus Mx)$. Let us assume first that $s \in Mx \setminus My$ and define the following assignment

$$M'z = \begin{cases} s & \text{if } z \text{ coincides with } w \\ Mz & \text{otherwise} \end{cases}$$

Then it is an easy matter to check that M' satisfies the branch ϑ_2. Similarly, if $s \in My \setminus Mx$, then one can easily build a model M'' satisfying φ_3. Hence, if ϑ is satisfiable, then at least one among ϑ_1, ϑ_2, and ϑ_3 is satisfiable, i.e. rule (12) is sound.

As an immediate consequence of the soundness of rules (1)-(12), any tableau for a satisfiable collection of **MLSS**-literals must be satisfiable. $\square$

In Table 3 we shall give a decision procedure for normalized conjunctions of **MLSS** in the form of a saturation strategy for the given tableau calculus. This consists of calls to an auxiliary procedure *Saturate* interleaved with attempts to construct a set model for a branch which is still not closed. Candidate set models are constructed by means of the realization of the graph of the positive membership literals present in the branch. If the realization so constructed fails to satisfy the branch, it will not satisfy some of its literals (in fact just literals of type $x \notin y$, or $x = y \setminus z$, or $x \neq y$). This negative information is used to single out a pair of variables which though modeled by the same set, they would need to be distinct. An application of rule (12) to such a pair of variables will force them to be modeled differently.

The interleaving of deduction and checking steps has the overall effect to speed up the saturation process.

Next, in Table 4 we give the procedure $Saturate(\mathcal{T}, V)$, which systematically saturates the non-closed branches of the input tableau $\mathcal{T}$ with respect to rules (1)–(11) only of Table 2, under strictness hypothesis. It is to be noted that further efficiency of the saturation process could be achieved by maintaining suitable global data structures.

Remark 9. In procedures *MLSS_Tableau_Test(S)* and *Saturate(T, V)*, we shall make use of the following notation. Let ϑ be a saturated open branch of the tableau T for a finite collection S of normalized **MLSS**-literals, constructed during an execution of procedure *MLSS_Tableau_Test(S)*. By V_S, T, $\sim_S$, T', V', $\in_\vartheta$, G_ϑ, R_ϑ we denote the following objects:

V_S: is the collection of variables occurring in S;

T: is the collection of variables occurring on ϑ other than V_S;

$\sim_S$: is the equivalence relation induced on $V_S \cup T$ by equality literals $x = y$ in ϑ;

T': is the set $\{t \in T : t \not\sim_S x$, for all $x \in V_S\}$;

V': is the set $(V_S \cup T) \setminus T'$;

$\widehat{\in}_\vartheta$: is the binary relation on $V' \cup T'$ defined by

$$x \widehat{\in}_\vartheta y \quad \text{iff} \quad \text{the literal } x \in y \text{ is in } \vartheta, \text{ for } x, y \in V';$$

G_ϑ: is the oriented graph $(V' \cup T', \widehat{\in}_\vartheta)$ (this is also referred to as the *membership graph* relative to ϑ;

R_ϑ: is a realization of G_ϑ relative to the partition (V', T') and to pairwise distinct sets u_t, for $t \in T'$, each having cardinality no less than $|V' \cup T'|$. ☐

Remark 10. Each branch ϑ in the tableau T, constructed by procedures *MLSS_Tableau_Test(S)* and *Saturate(T, V)*, is annotated with the attributes *Diversified*$_\vartheta$ and *To_be_diversified*$_\vartheta$. Roughly speaking, the set *Diversified*$_\vartheta$ collects all pairs (x_1, x_2) of variables in V_S for which there exists another variable $u \in V_S \cup T$ such that $u \in x_i$ and $u \notin x_{3-i}$ are in ϑ, for some $i \in \{1, 2\}$. Notice that if all variables of a lesser height than that of x_i are "diversified", then it can be shown that for any realization R_ϑ of G_ϑ relative to sufficiently large and pairwise disjoint sets u_t, for $t \in T'$, one has $R_\vartheta x_1 \neq R_\vartheta x_2$.

On the other hand, the set *To_be_diversified*$_\vartheta$ collects all pairs (x_1, x_2) of variables in V_S such that the branch ϑ contains the literals $x_i \in y$ and $x_{3-i} \notin y$, for some $y \in V_S$. Notice, therefore, that if a model M satisfies ϑ, then $M x_1 \neq M x_2$, for each $(x_1, x_2) \in V_S \times V_S$.

In view of what has been just said, it follows that the set Δ_ϑ defined in line M12 of procedure *MLSS_Tableau_Test* consists of all those pairs (x_1, x_2) of variables which are mapped into the same set by the realization R_ϑ, though they would need to be diversified. Therefore line M13 just asserts that if at any moment the branch ϑ is not closed and R_ϑ does not satisfy ϑ, then there must exist a pair of variables (x_1, x_2) which need to be diversified but which do not yet belong to the set *Diversified*$_\vartheta$. ☐

Example. As already observed, for the sake of simplicity we have presented a collection of tableaux rules, namely rules (1)-(12) of Table 2, which can be applied only to normalized conjunctions of **MLSS**. In practice, however, one can add to rules (1)-(12) also other rules to handle formulae of **MLSS** which are not necessarily normalized. What is needed are rules able to deal with propositional connectives and with compound set terms as well.

Procedure $MLSS\_Tableau\_Test(S)$;
    Comment: S is a finite collection of normalized **MLSS**-*literals.*

M1   let V_S be the collection of variables occurring in S;
M2   let T be a one branch tableau induced by S;
M3   **for** $\vartheta \in T$ **do**
M4 $Diversified_\vartheta := \emptyset$;
M5   **end for**;
M6 $T := Saturate(T, V_S)$;
M7   **while** there exists a non-closed branch ϑ in T **do**
M8   let T, $\sim_\vartheta$, T', $\widehat{\in}_\vartheta$, G_ϑ, R_ϑ be defined as in Remark 9;
M9   **if** R_ϑ satisfies ϑ **then**
M10   **return** "input formula is satisfied by R_ϑ";
M11   **else**
M12   put $\Delta_\vartheta = \{(w,w') \in V_S \times V_S : R_\vartheta\, w = R_\vartheta\, w'\} \cap To\_be\_diversified_\vartheta$;

M13   **Assert:** the set $\Delta_\vartheta \setminus Diversified_\vartheta$ is non-empty;

M14   pick a pair $(x,y) \in \Delta_\vartheta \setminus Diversified_\vartheta$;
M15   apply rule (12) to the pair of variables x, y, namely
M16   - let u be the next unused variable;
M17   - split the branch ϑ in three branches ϑ_1, ϑ_2, and ϑ_3, where
M18 $\vartheta_1 = \vartheta; u \in x, u \notin y$;
M19 $\vartheta_2 = \vartheta; u \notin x, u \in y$;
M20 $\vartheta_3 = \vartheta; x = y$;
M21   - and put:
M22 $To\_be\_diversified_{\vartheta_1} := To\_be\_diversified_{\vartheta_2} := To\_be\_diversified_{\vartheta_3}$
                        $:= To\_be\_diversified_\vartheta$;
M23 $Diversified_{\vartheta_1} := Diversified_{\vartheta_2}$
                        $:= Diversified_\vartheta \cup \{(x',y'), (y',x') \in V_S \times V_S : x' \sim_\vartheta x$ and $y' \sim_\vartheta y\}$;
M24 $Diversified_{\vartheta_3} := Diversified_\vartheta$;
M25   - *Comment: Notice that after saturation w.r.t. rule (10), all branches*
                from ϑ_3 will be closed.
M26   assign to T the resulting tableau;
M27 $T := Saturate(T, V_S)$;
M28   **end if**;
M29   **end while**;
M30   **return** "input formula is unsatisfiable, as proved by the closed tableau T";

end procedure.

Table 3. Procedure $MLSS\_Tableau\_Test$

In the present example, besides standard tableau rules for propositional connectives (see for instance [Fit90]), we shall also allow ourselves to use the following rules, among others:

$$\frac{x \notin t_1 \setminus t_2}{x \in t_1 \mid x \notin t_1} \ (13) \qquad \frac{x \in t_1 \setminus t_2}{\begin{array}{c} x \in t_1 \\ x \notin t_2 \end{array}} \ (14) \qquad \frac{}{t_1 = t_2 \left| \begin{array}{c|c} w \in t_1 & w \notin t_1 \\ w \notin t_2 & w \in t_2 \end{array} \right.} \ (15)$$

where x is a variable, t_1, t_2 are set terms, w is a new variable not occurring in the branch to which rule (15) is applied.

The formula that we intend to test for validity in set theory is

$$\Phi \ : \ (x = \{y\} \wedge x = y \cup z) \rightarrow (y = y \setminus y \wedge x = z).$$

Table 5 contains a closed tableau for $\neg\Phi$ which can be constructed by means of procedure $MLSS\_Tableau\_Test$ (therefore Φ is valid in set theory).

We indicate with φ_{ij} the j-th formula of row i in Table 5. Closed branches

---

Procedure $Saturate(\mathcal{T}, V)$;

Comment: $\mathcal{T}$ *is an annotated* **MLSS-tableau, with attributes** $To\_be\_diversified_\vartheta$ *and* $Diversified_\vartheta$, *for each of its branches* ϑ. V *is a subset of the variables occurring in* $\mathcal{T}$.

- strictly saturate the non-closed branches of $\mathcal{T}$ with respect to rules (1)–(11) of Table 2;
- let $\mathcal{T}$ be the resulting annotated tableau, where it is supposed that during saturation the attributes are inherited by the new branches;

for each branch ϑ of $\mathcal{T}$ **do**

$\quad To\_be\_diversified_\vartheta :=$

$\qquad \{(x_1, x_2) \in V \times V : x_i \in z$ and $x_{3-i} \notin z$ are in ϑ, for some z in V

$\qquad$ and for some $i \in \{1, 2\}$, or $x_1 \neq x_2$ is in $\vartheta\}$;

$\quad Diversified_\vartheta := \{(x, y) \in V \times V : (x', y') \in Diversified_\vartheta$, for some $x' \sim_\vartheta x, y' \sim_\vartheta y\}$;

end for;

return $\mathcal{T}$;

end procedure.

Table 4. Procedure *Saturate*

1.	$\neg((x = \{y\} \wedge x = y \cup z) \rightarrow (y = y \setminus y \wedge x = z))$
2.	$x = \{y\} \wedge x = y \cup z$
3.	$\neg(y = y \setminus y \wedge x = z)$
4.	$x = \{y\}$
5.	$x = y \cup z$
6.	$y \in x$

	7.	$y \neq y \setminus y$			$x \neq z$	
8.	$y \in y$		$y \in z$	$y \in y$		$y \in z$
9.	$\bot$	$w \in y$	$w \notin y$	$\bot$	$w \in x$	$w \notin x$
10.		$w \notin y \setminus y$	$w \in y \setminus y$		$w \notin z$	$w \in z$
11.		$w \in y$ $\mid$ $w \notin y$	$w \in y$		$w = y$	$w \in x$
12.		$w \in x$ $\mid$ $\bot$	$w \notin y$		$y \notin z$	$\bot$
13.		$w = y$	$\bot$		$\bot$	
14.		$y \in y$				
		$\bot$				

Table 5. Tableau proof of $(x = \{y\} \wedge x = y \cup z) \rightarrow (y = y \setminus y \wedge x = z)$

are marked with the symbol $\bot$. In order to save space, branches corresponding to the first alternative of rules (12) and (15) have not been indicated, since they are closed.

The deductions can be justified as follows.

- $\varphi_{31}, \varphi_{41}, \varphi_{51}, \varphi_{71}$, and φ_{72} are obtained by propositional rules.
- φ_{61} is obtained from φ_{41} by means of rule (8).
- φ_{81} and φ_{82} (resp. φ_{83} and φ_{84}) are obtained from φ_{51} and φ_{61} by applying rule (1). Notice that branches ending with φ_{81} and φ_{83} are closed because they contain a membership cycle.
- The partial tableau obtained up to the application of the above rules cannot further be extended by means of rules (1)-(14) only. In order to choose the right pair of variables (resp. terms) to which to apply rule (12) (resp. (15)),

we construct realizations of the membership graphs relative to the branches which are not closed; for instance, for the branch ϑ ending with φ_{82} we have:
$V_S = V' = \{x, y, z\}, \quad T = T', \quad \widehat{\mathcal{E}}_\vartheta = \{(y, x), (y, z)\},$
$R_\vartheta = \{(y, \emptyset), (x, \{\emptyset\}), (z, \{\emptyset\})\}.$
Plainly, R_ϑ does not satisfy φ_{71}, so that we can apply rule (15) on ϑ to the terms y and $y \setminus y$, obtaining $\varphi_{91}, \varphi_{10,1}, \varphi_{92}$, and $\psi_{10,2}$.
- The remaining deduction steps can easily be justified. $\qquad\qquad\square$

3.1 Correctness proof

To prove the correctness of procedure $MLSS\_Tableau\_Test$, we shall first make the assumption that the assert-statement M13 evaluates always to true whenever it is encountered. We shall briefly refer to such an assumption as the "correctness of M13".

Then under the correctness assumption of M13, we shall show partial correctness of procedures $MLSS\_Tableau\_Test$ and $Saturate$, as well as their termination, namely we shall prove their total correctness. We shall later discharge our temporary assumption by proving the correctness of all executions of the assert-statement M13.

Thus, let us assume for the time being that during the execution of procedure $MLSS\_Tableau\_Test$ the assert-statement M13 evaluates to true whenever it is encountered.

Partial correctness Under the assumption of the correctness of M13, it is immediate to check that procedure $MLSS\_Tableau\_Test$ can only return its control either when it finds a realization R_ϑ satisfying one of the tableau branches, or when all branches are closed. In the first case the realization R_ϑ must in particular satisfy the input collection S of **MLSS**-literals; in the latter case, because of the soundness of rules (1)-(12) (cf. Remark 8(B)), it follows immediately that the input set S must be unsatisfiable (cf. Remark 8(A)).

Proof of termination We begin by showing that procedure $Saturate$ always terminates.

Lemma 11. *Each call to procedure Saturate terminates.*

Proof. It is enough to observe that none of the rules (1)-(11) which are applied by procedure $Saturate$ introduces new variables. Therefore the number of possible literals which can occur on any branch during saturation is finite. Since no literal can have multiple occurrences on any branch, it follows that any fair saturation strategy under strictness hypothesis must terminate. $\qquad\qquad\square$

In view of the preceding lemma, in order to prove termination also for $MLSS\_Tableau\_Test$, we only need to show that its while-loop can only be executed finitely many times.

Let $\mathcal{T}^*$ be the tableau limit constructed by $MLSS\_Tableau\_Test$. We begin by observing that $\mathcal{T}^*$ must be finite. Indeed, if $\mathcal{T}^*$ were infinite, then by König Lemma it would have an infinite branch ϑ^*. This is possible only if branch ϑ^* is processed infinitely many times by statements M14-M24 with the consequence that the set $Diversified_\vartheta$. would be properly incremented infinitely many times. But this would lead to a contradiction, since plainly $Diversified_\vartheta \subseteq V_S \times V_S$ and moreover for each branch ϑ the value of the attribute $Diversified_\vartheta$ is monotonic increasing, during execution of $MLSS\_Tableau\_Test$.

Having proved that $\mathcal{T}^*$ is finite, in order to show termination it is now enough to observe that under the assumption of correctness of M13, at least one new node is added to the tableau at each iteration of the while-loop of $MLSS\_Tableau\_Test$.

In conclusion we have proved that under the correctness assumption of M13 we have the total correctness of procedure $MLSS\_Tableau\_Test$. In the following subsection we shall discharge such an assumption.

Correctness proof of the assert-statement M13

We begin by stating some elementary facts which will be needed later on.

Lemma 12. *At any step during the construction of a tableau by means of procedures $MLSS\_Tableau\_Test$ and $Saturate$, the following facts hold:*

(a) *if a literal $x \in y$ is added to a branch ϑ, then $y \sim_\vartheta y'$, for some y' in V_S.*

(b) *if $R_\vartheta\, x = R_\vartheta\, y$, with x and y distinct variables, then there exist x', y' in V_S such that $x \sim_\vartheta x'$ and $y \sim_\vartheta y'$.*

Proof. (a) can easily be proved by induction on the number of applications of rules (1)–(12).

Concerning (b), observe that if for instance $x \not\sim_\vartheta z$, for all $z \in V_S$, then $R_\vartheta\, x$ would contain among its elements a distinctive set u_x. Such set then can be a member only of the set $R_\vartheta\, x$. $\qquad\qquad\qquad\Box$

Let ϑ be a non-closed branch of the tableau $\mathcal{T}$ returned by some call to procedure $Saturate$ during the execution of $MLSS\_Tableau\_Test(S)$. Let V_S, T, $\sim_\vartheta$, T', V', G_ϑ, R_ϑ be defined as in Remark 9, respectively. Let also $To\_be\_diversified_\vartheta$, $Diversified_\vartheta$ be the attributes of ϑ computed by procedures $MLSS\_Tableau\_Test$ and $Saturate$.

Furthermore, for $(u_1, u_2), (w_1, w_2) \in \Delta_\vartheta$, let us put $(u_1, u_2) \prec_\vartheta (w_1, w_2)$ if there exist $i, j \in \{1, 2\}$ such that the literals $u_i \in w_j$ and $u_{3-i} \notin w_j$ are in ϑ, where we recall that:

$$\Delta_\vartheta = \{(w, w') \in V_S \times V_S : R_\vartheta\, w = R_\vartheta\, w'\} \cap To\_be\_diversified_\vartheta\,,$$

$$To\_be\_diversified_\vartheta = \{(x_1, x_2) \in V_S \times V_S : x_i \in z \text{ and } x_{3-i} \notin z \text{ are in } \vartheta,$$
$$\text{for some } z \text{ in } V_S \text{ and for some } i \in \{1, 2\}\,,$$
$$\text{or } x_1 \neq x_2 \text{ is in } \vartheta\}\,.$$

(cf. line M12 of procedure $MLSS\_Tableau\_Test$).

We have then the following elementary properties.

Lemma 13. *Relation $\prec_\vartheta$ is acyclic.*

Proof. Notice that if $(u_1, u_1') \prec_\vartheta (u_2, u_2')$, then $R_\vartheta\, u_1 \in R_\vartheta\, u_2$. Therefore any $\prec_\vartheta$-cycle would induce a membership cycle among the sets $R_\vartheta\, u_i$, for $i = 1, \ldots, k$, which would contradict the well-foundedness of sets $R_\vartheta\, v$ (cf. Lemma 2(viii)). $\qquad\Box$

Lemma 14. *If $(w, w') \in \Delta_\vartheta \cap Diversified_\vartheta$, then there exists $(u, u') \in \Delta_\vartheta$ such that $(u, u') \prec_\vartheta (w, w')$.*

Proof. Let $(w, w') \in \Delta_\vartheta \cap \text{Diversified}_\vartheta$. Then the following properties hold:

(a) $R_\vartheta\, w = R_\vartheta\, w'$;

(b) for some variable u either

    (b_1) the literals $u \in w$ and $u \notin w'$ are in ϑ, or

    (b_2) the literals $u \notin w$ and $u \in w'$ are in ϑ, or

    (b_3) the literal $w \neq w'$ is in ϑ.

Assume that we are in case (b_1). Then, since $R_\vartheta\, u \in R_\vartheta\, w'$, whereas $u \notin w'$ is in ϑ, there must be a variable u' such that $R_\vartheta\, u = R_\vartheta\, u'$ and the literal $u' \in w'$ is in ϑ. Notice that by Lemma 12(b) we can assume w.l.o.g. that both u and u' are in V_S. Thus we have that $(u, u') \in \Delta_\vartheta$ and $(u, u') \prec_\vartheta (w, w')$.

Case (b_2) is similar.

Finally, assume that we are in case (b_3). As $(w, w') \in \text{Diversified}_\vartheta$, then either the literals $u \in w$ and $u \notin w'$ are in ϑ, or the literals $u \notin w$ and $u \in w'$ are in ϑ, for some $u \in V_S$. In the former case, since $R_\vartheta\, u \in R_\vartheta\, w'$, it follows that there must exist $u' \in V_S$ such that $R_\vartheta\, u' = R_\vartheta\, u$, with $u' \in w'$ in ϑ. Thus, again, $(u, u') \in \Delta_\vartheta$ and $(u, u') \prec_\vartheta (w, w')$. The latter case is similar. $\square$

The preceding lemmas allow us to also prove the following result.

Lemma 15. *If $\Delta_\vartheta \neq \emptyset$, then $\Delta_\vartheta \setminus \text{Diversified}_\vartheta \neq \emptyset$.*

Proof. Assume that $\Delta_\vartheta \neq \emptyset$. If by contradiction $\Delta_\vartheta \setminus \text{Diversified}_\vartheta = \emptyset$, then $\emptyset \neq \Delta_\vartheta \subseteq \text{Diversified}_\vartheta$. Thus, by the preceding lemma there would exist an infinite descending $\prec_\vartheta$-chain in $\Delta_\vartheta \cap \text{Diversified}_\vartheta$. But $\Delta_\vartheta \cap \text{Diversified}_\vartheta$ is clearly finite, therefore there would exist a $\prec_\vartheta$-cycle, which is forbidden by Lemma 13. Hence we must have $\Delta_\vartheta \setminus \text{Diversified}_\vartheta \neq \emptyset$. $\square$

Before proving the correctness of the assert-statement M13, we show that due to saturation w.r.t. rules (1)–(11), most of the literals in ϑ are automatically satisfied by R_ϑ.

Lemma 16. *Under the above hypothesis that ϑ is saturated w.r.t. rules (1)–(11), the realization R_ϑ satisfies all literals occurring in ϑ of the following types:*

$$x = y \cup z,\ x = y \cap z,\ x = \{y_1, \ldots, y_k\}$$
$$x = y,\quad x \in y.$$

Moreover, for each literal $x = y \setminus z$ in ϑ we have $R_\vartheta\, y \setminus R_\vartheta\, z \subseteq R_\vartheta\, x$.

Proof. **Case:** $x = y \cup z$. Let $x = y \cup z$ be in ϑ. W.l.o.g., we can assume that $x, y, z \in V_S$ (otherwise ϑ would contain a literal $x' = y' \cup z'$, with $x', y', z' \in V_S$ and such that $x \sim_\vartheta x'$, $y \sim_\vartheta y'$, and $z \sim_\vartheta z'$; then $x' = y' \cup z'$ could play the role of $x = y \cup z$ in the following discussion). By saturation of ϑ w.r.t. rules (1)–(3), it follows that $G_\vartheta(x) = G_\vartheta(y) \cup G_\vartheta(z)$. Therefore by Lemma 2(ii) we have $R_\vartheta\, x = R_\vartheta\, y \cup R_\vartheta\, z$.

Case: $x = y \cap z$. This case is analogous to the preceding one.

Case: $y = \{x_1, \ldots, x_k\}$. Let $y = \{x_1, \ldots, x_k\}$ be in ϑ. Again, w.l.o.g. we can assume that $y, x_1, \ldots, x_k$ are in V_S. By saturation w.r.t. rules (8) and (9), it turns out easily that $G_\vartheta(y) = \bigcup_{i=1}^{k}\{x_i' : x_i' \sim_\vartheta x_i\}$. Hence $R_\vartheta\, y = \{R_\vartheta\, x_1, \ldots, R_\vartheta\, x_k\}$.

Case: $x = y$. Let $x = y$ be in ϑ. Then in particular $x \sim_\vartheta y$, so that $R_\vartheta\, x = R_\vartheta\, y$.

Case: $x \in y$. Let $x \in y$ be in ϑ. Then plainly $R_\vartheta\, x \in R_\vartheta\, y$.

So far we have proved that R_ϑ satisfies all literals in ϑ of the above types.

Case: $x = y \setminus z$. Assume now that the literal $x = y \setminus z$ is in ϑ. Once more, we can assume w.l.o.g. that $x, y, z \in V_S$. Let $s \in R_\vartheta\, y \setminus R_\vartheta\, z$. Thus $s = R_\vartheta\, w$, for some variable w for which the literal $w \in y$ is in ϑ. But then, by saturation w.r.t. rule (11), it would follow that either $w \in z$ is in ϑ, or that $w \notin z$ is in ϑ. The first case is not possible, because we would have $s = R_\vartheta\, w \in R_\vartheta\, z$, a contradiction. Thus the latter case must hold, namely the literal $w \notin z$ must be in ϑ. But then, by saturation w.r.t. rule (7), we have that the literal $w \in x$ must also be in ϑ, so that $s = R_\vartheta\, w \in R_\vartheta\, x$, proving that $R_\vartheta\, y \setminus R_\vartheta\, z \subseteq R_\vartheta\, x$.

$\square$

We can now prove our main lemma.

Lemma 17. *The assert-statement M13 evaluates always to true whenever it is encountered, namely $\Delta_\vartheta \setminus Diversified_\vartheta \neq \emptyset$.*

Proof. Assume by way of contradiction that the assert-statement M13 evaluates to false during a certain execution. Therefore we have that at a certain step our partial tableau $\mathcal{T}$ must have a non-closed branch ϑ such that the realization R_ϑ (of G_ϑ relative to some pairwise distinct sets of sufficiently large cardinality) does not satisfy ϑ, namely it does not satisfy a literal ℓ occurring on ϑ, and such that

$$\Delta_\vartheta \setminus Diversified_\vartheta = \emptyset. \tag{3}$$

Notice that ϑ is saturated w.r.t. rules (1)–(11). Hence by the preceding lemma, the literal ℓ can only be of type $x \notin y$ or $x = y \setminus z$, or $x \neq y$, so that we have the following cases.

Case: $x \notin y$. If ℓ is the literal $x \notin y$, then our assumption that R_ϑ does not satisfy ℓ implies that $R_\vartheta\, x \in R_\vartheta\, y$. Thus either $x \in y$ is in ϑ, or for some variable x' distinct from x and such that $R_\vartheta\, x' = R_\vartheta\, x$ the literal $x' \in y$ is in ϑ. The first case cannot occur, since by hypothesis the branch ϑ is not closed. Thus the latter case must hold. By Lemma 12 we can assume w.l.o.g. that x, x', y are in V_S. Hence we have that $(x, x') \in \Delta_\vartheta$, so that by Lemma 15 we can deduce $\Delta_\vartheta \setminus Diversified_\vartheta \neq \emptyset$, contradicting (3).

Case: $x = y \setminus z$. Assume that R_ϑ does not satisfy a literal of type $x = y \setminus z$, where we can assume w.l.o.g. that $x, y, z \in V_S$. Thus, by the preceding lemma, $R_\vartheta\, x \nsubseteq R_\vartheta\, y \setminus R_\vartheta\, z$. Hence there must exist $s \in R_\vartheta\, x$ such that $s \notin R_\vartheta\, y \setminus R_\vartheta\, z$. We have that $s = R_\vartheta\, w$, for some variable w for which the literal $w \in x$ is in ϑ. But then, by saturation w.r.t. rule (6), it would follow that $w \in y$ and $w \notin z$ are in ϑ, so that $s = R_\vartheta\, w \in R_\vartheta\, y$. Thus we must also have $s \in R_\vartheta\, z$, which yields the existence of a variable w' distinct from w such that the literal $w' \in z$ is in ϑ and such that $R_\vartheta\, w = R_\vartheta\, w'$. Thus, $(w, w') \in \Delta_\vartheta$, and again by Lemma 15 we can deduce $\Delta_\vartheta \setminus Diversified_\vartheta \neq \emptyset$, contradicting (3).

Case: $x \neq y$. Assume that R_ϑ does not satisfy a literal of type $x \neq y$. W.l.o.g. we can assume that $x, y \in V_S$. Therefore, since $R_\vartheta\, x = R_\vartheta\, y$, it follows at once that $(x, y) \in \Delta_\vartheta$.

Since in any case we derive a contradiction, it follows that in fact the assert-statement M13 must evaluate to true whenever it is encountered. □

4 Conclusions

We have described a tableau saturation strategy for the fragment **MLSS** of set theory. Such a strategy is based on the use of a model checking step which either discovers a set model for the input formula at some early stage or suggests a correct choice of the pairs of variables which need to be distinguished.

We intend to implement the proposed strategy and to compare it with a blind search strategy.

Also, we intend to investigate the applicability of the saturation strategy presented in other contexts besides set theory.

Acknowledgements

The author wishes to thank S. Battiato, B. Beckert, E. Omodeo, A. Policriti, and three anonymous referees for pointing out some imprecisions and suggesting several improvements.

References

[CZ92]   R. Caferra and N. Zabel A Method for Simultaneous Search for Refutations and Models by Equational Constraint Solving. *Journal of Logic and Computation*, 13:613-641, 1992.

[CF95]   D. Cantone and A. Ferro. Techniques of computable set theory with applications to proof verification. *Comm. Pure App. Math.*, 48(9-10):1–45, 1995. Special Issue in honor of J.T. Schwartz.

[CFO89]   D. Cantone, A. Ferro, and E. G. Omodeo. *Computable Set Theory. Vol. 1*. Oxford University Press, 1989. Int. Series of Monographs on Computer Science.

[CRC95]   D. Cantone and R. Ruggeri Cannata. Deciding set-theoretic formulae with the predicate *finite* by a tableau calculus. *Matematiche*, 50(2):99–118, 1995.

[CRC97]   D. Cantone and R. Ruggeri Cannata. A decidable tableau calculus for a fragment of set theory with iterated membership. To appear in Journal of Automated Reasoning, 1997.

[Fit90]   M.C. Fitting. *First-Order Logic and Automated Theorem Proving*. Springer-Verlag, 1990.

[FOS80]   A. Ferro, E. G. Omodeo, and J. T. Schwartz. Decision Procedures for Elementary Sublanguages of Set Theory I. Multilevel Syllogistic and Some Extensions. *Communications on Pure and Applied Mathematics*, 33:599–608, 1980.

[Jec78]   T.J. Jech. *Set theory*. Academic Press, New York, 1978.

Hintikka Multiplicities in Matrix Decision Methods for Some Propositional Modal Logics

Serenella Cerrito[1] and Marta Cialdea Mayer[2]*

[1] Université de Paris-Sud, LRI, Bât 490,
F-91405 Orsay Cedex, France.
e-mail: serena@lri.fr
[2] Dipartimento di Informatica e Automazione, Università di Roma Tre,
via Vasca Navale 84, 00146 Roma, Italia.
e-mail: cialdea@inf.uniroma3.it

Abstract. This work is a study of the inter-translatability of two closely related proof methods, i.e. tableau (or sequent) and connection based, in the case of the propositional modal logics K, $K4$, T, $S4$, paying particular attention to the relation between matrix *multiplicity* and multiple use of ν_0-formulae (contractions) in tableaux/sequent proofs.

The motivation of the work is the following. Since the role of a multiplicity in matrix methods is the encoding of the number of copies of a given formula that are needed in order to prove a valid formula, it is important to find upper bounds for multiplicities in order to reduce as much as possible the search space for proofs. Moreover, it is obviously a crucial issue if the matrix method is to be used as a decision method. We exploit previous results establishing upper bounds on the number of contractions in tableau/sequent proofs [4], in order to establish upper bounds for multiplicities in matrix systems.

We obtain two kinds of upper bounds: in function of the size of the formula to be proved and in function of the number of the atomic paths through the unindexed formula-tree. Such bounds may be non-optimal. However, the method used to establish them may be useful for obtaining finer upper bounds.

1 Introduction

Among the different proof methods that have been defined for modal logics, two classes of methods are clearly strongly related: sequent/tableau-based [5, 13] and connection-based methods [17, 18]. The efficiency of proof search strongly depends on features allowing the calculus to reduce the size of the search space. Recently, many improvements of tableau systems for modal logics have been proposed (see for example [8, 9, 12]). However, it has been argued that tableau and sequent-based proof search is intrinsically inadequate to deal with several

---

* This work has been partially supported by the Italian MURST project "Rappresentazione della Conoscenza e Meccanismi di Ragionamento" and by ASI – Agenzia Spaziale Italiana.

kinds of redundancies in the search space. By contrast, the *matrix proof method*, based on Bibel *connection method* [2], exploits the possibility of searching proofs of formulae by working on their parsing trees. It has been advocated as a very good approach to minimise redundancies, while, admittedly, sacrificing human readability (the problem of translating matrix proofs into "readable proofs" is addressed, for example, in [1, 3, 14, 15]).

From a proof-theoretical point of view, a main interesting aspect of matrix methods is that a matrix can be seen as uniquely encoding a set of tableau proofs which differ only with respect to "irrelevant" permutations of rule applications (pretty much as proof-nets give a unique representation of several "equivalent" sequent proofs in linear logic [7]).

In the matrix method, validity within a logic is characterised by the existence of a set of *connections* (i.e. pairs of complementary literals) within the formula. If every *atomic path* through the formula contains a connection from the set, such a set is said to *span* the formula. However, in modal propositional logics, like in classical first order logic, it may be necessary to use several copies of some subformulae in order to simultaneously "close" all of the atomic paths. This holds not only for logics whose tableau (or sequent) inference rules explicitly allow a reuse of the same formula (i.e. contain implicit or explicit contractions), such as, for example, T and the transitive logics, but also for the simplest normal modal logic K (see Example 1 below).

Given a modal propositional logic $\mathcal{L}$, an upper bound on the needed *multiplicity*, i.e. the number of copies of a given subformula that are needed in order to span a valid formula, is called an $\mathcal{L}$-*Hintikka multiplicity* in [18]. To the best of our knowledge, a Hintikka multiplicity has been so far explicitly established only in the case of $S5$ [18]. However, Wallen observes:

> We stress: defining an $\mathcal{L}$-Hintikka multiplicity for a given formula for the other logics is straightforward, since the number of subformulae of the formula is finite. This, essentially, is the approach taken in tableau methods, such as Fitting's [5] and Hughes and Cresswell's diagrammatic methods [10], to show that various modal logics are decidable in the first place. For the purposes of automated deduction, since the $\mathcal{L}$-Hintikka multiplicity defines the space that must be searched exhaustively to determine the non-validity of the formula, we are motivated by the desire to determine least such multiplicities. ([18], p.177).

Now, note that if *only* the fact that the number k of subformulae of a formula A is finite is used to determine $\mathcal{L}$-Hintikka multiplicities, without any further analysis, one gets, for instance, an $S4$-Hintikka multiplicity of the order of 2^{2^k} (2^{2^k} being the maximal number of leaves in a $S4$ loop-free tableau for A, such as can be determined by using only the subformula property). The aim of this work is to push the analysis further, so as to progress towards the definition of minimal $\mathcal{L}$-Hintikka multiplicities. Such a problem is obviously crucial if the matrix method has to be used as a *decision method*.

Wallen's presentation of modal matrix methods emphasises their relationship with tableau/sequent style systems. In particular, its definition of path mirrors the definition of expansion rules in tableau systems and the use of prefixes closely resembles the role of prefixes in prefixed tableau systems. Moreover, multiplicity corresponds to the possible multiple use of a ν_0-formula in a tableau proof. In effect, the possibility of translating a modal matrix proof into a sequent/tableau proof has been shown in [15], while [16] uses the inverse translation, in order to show how to define a matrix method starting from tableau systems.

In this work, we refine the analysis of the inter-translatability of tableau and matrix systems, paying particular attention to the relation between multiplicity and multiple use of ν_0-formulae. As far as transitive logics are concerned, the multiplicity of a ν_0-formula reflects also the number of times a contraction affects such a formula (or better, its ν-parent) in a tableau proof. Now, such a number can be bounded polynomially in the length of the initial formula when single branches of the tableau are considered [4, 8, 9]. The possibility of exploiting this result in order to establish an upper bound for the multiplicity in modal matrix methods is the original inspiration of this work.

However, the paths through a matrix do not preserve the tree structure of a tableau. Since a matrix is made of pointers to subformulae, multiple copies are needed also when an occurrence of a formula is used in different branches of the corresponding tableau. Intuitively, in a matrix proving A, the value $\mu(a_j)$, where μ is the multiplicity function and a_j the position of a ν_0-subformula B of $\langle A, 0\rangle$, gives a measure of the number n of distinct occurrences of B needed in a tableau $\mathcal{T}$ in order to refute FA. The number n is a function of two quantities:

1. the maximal number of copies of such a B needed in a branch of $\mathcal{T}$;
2. the number of branches of $\mathcal{T}$.

This remark is better illustrated by some examples (see Section 2 for a brief survey of the modal tableau and matrix methods).

Example 1. Consider the K-valid formula A_1: $\Diamond(\neg p \vee \neg q) \vee (\Box p \wedge \Box q)$. Any K-tableau proof of A_1 contains two branches. Each branch contains an application of the π_K-rule, "producing" two copies of the ν_0-formula $F(\neg p \vee \neg q)$: one in the first branch and the other in the second one. Let a_j be the unique position whose label is $\neg p \vee \neg q$ (with polarity 0) in the formula-tree for $\langle A_1, 0\rangle$. It is easy to see that no K-admissible substitution σ and no set of σ-complementary connections can span $\langle A_1, 0\rangle^\mu$ if $\mu(a_j)$ is strictly smaller than 2. In fact, the (variable) position a_j must be simultaneously unified with the (constant) positions associated to p (with polarity 0, sub-formula of $\Box p$) and q (with polarity 0, sub-formula of $\Box q$). Intuitively, the position corresponding to $\neg p \vee \neg q$ must be used twice in the matrix, in order to "match" p on a branch and q on the other.

Example 2. Consider the T-valid formula A_2: $\Diamond(q \to \Box q)$. Any T-tableau proof of A_2 has just one branch but needs at least two copies of the signed formula $F(q \to \Box q)$: the first one, generated by a ν_T-rule, yields $F(q \to \Box q)$, thus $F\Box q$, holding in the actual world. The second one is produced (as a "side effect") when the π_K-rule is applied to $F\Box q$. Let a_j be the (unique) position in the formula tree for $\langle A_2, 0\rangle$ whose label is $q \to \Box q$ (and polarity 0). It is easy to see that no T-admissible substitution σ and no set of σ-complementary connections can span $\langle A_2, 0\rangle^\mu$ if $\mu(a_j)$ is strictly smaller than 2.

Example 3. Consider again the formula A_2 of the previous example, but in the context of $S4$. Again, any $S4$-tableau $\mathcal{T}$ proving A_2 contains exactly one branch. A first application of the ν_T-rule generates a first copy of the ν_0-formula $F(q \to \Box q)$. If $\mathcal{T}$ is a proof of A_2, later on in the branch there is an application of the π_{S4} rule on $F\Box q$; this inference generates the π_0-formula Fq but does not produce any ν_0-formula. Differently from the case of T, a new application of the ν_T rule to $F\Diamond(q \to \Box q)$ is needed in order to close the branch (note the use of a contraction, in sequent terms). This application is responsible of the generation of a second copy of the ν_0-formula $F(q \to \Box q)$. Again, if a_j is the unique position whose label is $q \to \Box q$ (with polarity 0) in the formula-tree for $\langle A_2, 0 \rangle$, no multiplicity μ such that $\mu(a_j)$ is strictly smaller than 2 will do.

In this work we provide upper bounds for the multiplicity needed to span a valid formula in the logics K, T, $K4$ and $S4$, to be used when working with the matrix method as a *decision* method. The results are obtained via a translation of tableau proofs into matrices and the exploitation of upper bounds on the number of copies of a single formula needed in a tableau branch in order to obtain a closed tableau. Since the number of branches in tableaux must also be taken into account and such a number is, in general, exponential in the size of the initial formula, the bounds that we obtain in this way are exponential too. It seems plausible that much better cannot be done: in general, a given ν-formula may be used in all the branches of a tableau.

However, since the number of branches in a tableau proof bears some relation with the number a of atomic paths in the unindexed formula tree, it may be interesting in practical cases to limit the multiplicity in terms of a, that may be known in advance and is often far lower than its general upper bound. We have explored also this possibility and found that, while in the case of the logic K the value a itself is a Hintikka multiplicity, in the case of $K4$, T and $S4$ the Hintikka multiplicity is still an exponential function of a. In many cases however, it is a better bound to consider than the Hintikka multiplicity expressed in terms of the size of the formula. The possibility of finding a polynomial bound in terms of a remains an open issue.

This work is organised as follows. Next section is a brief overview of tableau and matrix proof systems for the modal logics we are considering. Section 3 contains the main results concerning tableau systems (3.1) and illustrates how to shift them in the context of matrix systems in order to obtain $\mathcal{L}$-Hintikka multiplicities (3.2). The proof passes through the provably correct construction of a matrix proof from a tableau proof of a given formula.

2 Tableau and Matrix Proof Methods for Modal Logic

The propositional modal language extends the classical language by means of the modal operators $\Box$ and $\Diamond$. In this work we consider the modal systems K, $K4$, T and $S4$. The system K is axiomatically characterisable by addition of the following schema and inference rule to classical propositional axioms and rules (in the axiomatic characterisation of modal systems, the modal operator $\Diamond$ is usually a defined symbol: $\Diamond A =_{Def} \neg\Box\neg A$):

$$\text{(K)} \quad \Box(A \to B) \to (\Box A \to \Box B) \qquad\qquad \text{(Nec)} \quad \frac{A}{\Box A}$$

The systems T, $K4$ and $S4$ are the extensions of K obtained by addition, respectively, of the schema (T), the schema (4), and both schemata (T) and (4):

$$\text{(T)} \quad \Box A \to A \qquad\qquad \text{(4)} \quad \Box A \to \Box\Box A$$

In the following subsections we outline the tableau and connection proof methods for these logics.

2.1 Signed Tableaux for Modal Logics

Tableau systems are definable for modal logics in different formats: either signed or unsigned tableaux, and either prefixed or unprefixed ones. In this work we refer to unprefixed signed tableaux [5]. Although the prefixed version is closer to the matrix method, the use of unprefixed tableaux is preferable in this setting. In fact it restricts the application of rules expanding formulae of possible force (when a rule is applied to one of such formulae, the others are "lost"), therefore allowing for a better analysis of the structure of a tableau proof.

A signed formula is either TX or FX, where X is a formula. A compact formulation of the modal expansion rules can be obtained by defining ν-formulae (formulae of necessary force) and π-formula (formulae of possible force), with their ν_0 and π_0, as follows.

ν	ν_0
$T\Box X$	TX
$F\Diamond X$	FX

π	π_0
$T\Diamond X$	TX
$F\Box X$	FX

The tableau systems for the logics $S4$, $K4$, T, K are obtained by the addition of specific modal rules to the classical propositional ones, as the following table shows.

K			$(\pi_K)\ \dfrac{\pi, S}{\pi_0, S\#}$	where $S\# = \{\nu_0 \mid \nu \in S\}$
$K4$			$(\pi_4)\ \dfrac{\pi, S}{\pi_0, S\#}$	where $S\# = \{\nu, \nu_0 \mid \nu \in S\}$
T	$(\nu_T)\ \dfrac{\nu, S}{\nu_0, \nu, S}$		$(\pi_K)\ \dfrac{\pi, S}{\pi_0, S\#}$	where $S\# = \{\nu_0 \mid \nu \in S\}$
$S4$	$(\nu_T)\ \dfrac{\nu, S}{\nu_0, \nu, S}$		$(\pi_{S4})\ \dfrac{\pi, S}{\pi_0, S\#}$	where $S\# = \{\nu \mid \nu \in S\}$

When tableau rules are compared with sequent calculi rules [5], it is apparent that a contraction is implicit in the rule ν_T, a sequence of weakenings is implicit in the rules π_K and π_{S4}, and both contractions and weakenings are implicit in π_4. Note that the π_4 rule is a derived expansion rule in $S4$, thanks to the presence of the ν_T rule. We prefer, however, not to have implicit applications of the latter rule, when possible.

Signed subformulae of a signed formula are defined as follows: if B occurs positively in A, then TB is a subformula of TA and FB is a subformula of FA; if B occurs negatively in A, then FB is a subformula of TA and TB is a subformula of FA. The calculi enjoy the *subformula property*: if $\mathcal{T}$ is a tableau for the set of modal formulae S, then every formula in a node in $\mathcal{T}$ is a (signed) subformula of a formula in S. As a consequence, every node in a tableau for S is a subset of the set of subformulae in S; hence, any refutable set S of formulae has a closed tableau whose depth does not exceed 2^n, where n is the number of subformulae in S.

We shall use the expressions $\mathcal{L}$-tableau, $\mathcal{L}$-contradictory, etc., where $\mathcal{L}$ is a specific modal system, in the obvious sense.

2.2 A Short Survey of the Matrix Method

We very briefly recall here the basic definitions of the matrix proof method, in the restricted context of the propositional modal logics K, $K4$ T, $S4$. The reader should consult [17] and [18] for more information.

The matrix method works on signed formulae, using the notation $\langle A, i \rangle$, where A is a formula and i, its *polarity*, is an element of $\{0, 1\}$, instead of FA and TA. Below we often write just X to mean a pair $\langle A, i \rangle$.

A *formula tree* for a (signed) formula X – also called *unindexed formula tree* for X – is the representation of X as a tree. Sometimes, if A is a formula, we shall use the expression *unindexed formula tree for A* to mean "unindexed formula tree for $\langle A, 0 \rangle$". Each node of the tree is said to be a *position* and its *label* is a (signed) subformula Y of X. Hence, different occurrences of Y in X correspond to different positions, i.e. distinct nodes, in the formula tree for X. The *tree ordering* $<$ on the positions is the partial ordering induced by the formula tree (a node is seen as "smaller" than any of its sons). Positions are classified as $\alpha, \beta, \nu, \nu_0, \pi, \pi_0$ according to the corresponding classifications of their labels. Let $\mathcal{N}_0$ be the set of ν_0 position of a formula tree for X. A *modal multiplicity* μ is a function from $\mathcal{N}_0$ to the natural integers; as Wallen says, " it serves to encode the number of instances of subformulae of X in the scope of a modal operator of necessary force considered within a putative proof" [18].

Given a multiplicity μ for a (signed) formula X, X^μ is said to be an *indexed formula*. The notion of formula tree for X is then extended in a natural way so as to get the notion of *indexed formula tree* for the indexed formula X^μ; such a tree is a tree of indexed positions. Roughly, an indexed formula tree for X^μ may be seen as obtained from the unindexed formula tree by replacing each subtree T_a rooted in a positions a by n copies of T_a if $\mu(a) = n$; the roots of these n subtrees may be denoted by $a^1, \cdots, a^n$.

Paths through X^μ are sets of indexed positions; the rigorous definition is a recursive one. Here, we give just the "flavour" of how paths are generated by traversing the indexed formula tree for X^μ. The singleton containing the root position is the initial path. New paths may be obtained by replacing positions in the tree by their sons, with a *caveat*: at a β position we split into two paths, one containing β_1, the other β_2. In the case of a ν position, such a position is preserved, but any of its ν_0 sons in the tree for X^μ can be added to get a new path. An *atomic path* is a path that cannot generate any new path.

A *connection* is a subset of an atomic path consisting of two positions of different polarities but labelled by the same (unsigned) atomic formula.

A *modal position* is either a ν_0 position or a π_0 position. The *modal prefix* $pre(a)$ of a position a in an indexed formula tree is the string of modal positions dominating a in the ordering $<$, to which a itself is appended, if it is a modal position. Those elements of $pre(a)$ which are ν_0 positions are seen as variables, the others as constants.

A *modal substitution* σ is essentially a function from the set of ν_0 positions (in the indexed formula tree for X^μ) to strings of modal positions. Given a modal logic $\mathcal{L}$, specific conditions, depending on $\mathcal{L}$, are imposed on σ to ensure its $\mathcal{L}$-*correctness* (our terminology). These conditions reflect the properties of the accessibility relation between words typical of each logic.

A modal substitution σ induces a relation $<<$ on positions such that, if $\sigma(a) = string$ and $string$ is not itself a position of ν_0 type, then, for any position a' occurring in $string$, $a' << a$. The transitive closure of the union of the partial ordering $<$ and $<<$ defines a binary relation $\lhd$ on positions called *reduction ordering*. Given a modal logic $\mathcal{L}$, a modal substitution σ is said to be $\mathcal{L}$-*admissible* if it is $\mathcal{L}$-correct and the induced reduction ordering is irreflexive.

If σ is a modal substitution for an indexed formula X^μ, a connection $\{a, a'\}$ is said to be σ-*complementary* iff $\sigma(pre(a)) = \sigma(pre(a'))$ (the substitution σ being extended to strings in the natural way). A set of σ-complementary connections *spans* X^μ when every atomic path through X^μ contains a σ-complementary connection.

The following theorem is the base of the matrix method.

Theorem 1 (Wallen [17]). *A propositional modal formula A is $\mathcal{L}$-valid iff there is a modal multiplicity μ, an $\mathcal{L}$-admissible modal substitution and a set of σ-complementary connections that spans the indexed formula $\langle A, 0 \rangle^\mu$.*

Example 4. As a short example, let us consider the K-valid formula $A_4 : \Box\neg p \lor \Diamond p$ and the constant multiplicity μ whose value is always 1. An unindexed formula tree for $\langle A_4, 0 \rangle$ is:

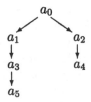

where $label(a_0) = \langle A_4, 0 \rangle$, $label(a_1) = \langle \Box\neg p, 0 \rangle$, $label(a_2) = \langle \Diamond p, 0 \rangle$, $label(a_3) = \langle \neg p, 0 \rangle$, $label(a_4) = \langle p, 0 \rangle$ and $label(a_5) = \langle p, 1 \rangle$. There exists a single atomic path through this tree, namely $\{a_2, a_4, a_5\}$. The prefix of a_5 is $a_0 a_3$, where both

a_0 and a_3 are constant, because their labels are π_0-formulae (the root of the tree is always treated as a π_0 position). The prefix of a_4 is $a_0 a_4$, where a_4 is a variable, because $label(a_4)$ is a ν_0-formula. Such an atomic path is σ-complementary for the K-admissible substitution σ such that $\sigma(a_3) = a_4$.

3 Hintikka Multiplicities in Matrix Proof Search

3.1 Some Useful Results about Tableau Systems

We noticed in Section 1 that the multiplicity $\mu(a_j)$ – where a_j is a position with $label(a_j) = X$ – needed to span a valid formula A is a function of both the maximal number of inferences "producing" X in a branch and the maximal number of branches in a tableau proof of A. This fact is formally stated and proved in next subsection. Here, we establish upper bounds to the number of inferences "producing" X in a branch and the number of branches in a tableau proof of A. The main results stated in this section are reformulations of results in [4], where they are used in order to give a polynomial translation of $S4$ into T in the style of [6] and to obtain a bound on the maximal number of ν-inferences needed in a branch of an $S4$ tableau proof of A. The latter result yields a bound to the number of inference steps per branch equal to the cube of the length of A and an upper bound to the number of contractions needed in $S4$ sequent style proofs.

Some definitions are used in the sequel:

Definition 2. 1. Let $\mathcal{T}$ be a tableau proof of A and X a (signed) subformula of FA. An inference $\mathcal{J}$ in $\mathcal{T}$ is said to *produce* X if X does not occur in the premise of $\mathcal{J}$ and it occurs in the conclusion. In particular, an application of the ν_T-rule $\dfrac{\nu, S}{\nu_0, \nu, S}$ produces ν_0 (if $\nu_0 \notin S$), and applications of the rules π_K or π_4, $\dfrac{\pi, S}{\pi_0, S\#}$ produce, besides the π_0 specific of the inference, also all the ν_0-formulae such that $\nu \in S$.

2. If $\mathcal{T}$ is a tableau, then $br(\mathcal{T})$ is the number of branches in $\mathcal{T}$ and, for any signed formula X, $k(\mathcal{T}, X)$ is the maximal number of rules producing X in any branch of $\mathcal{T}$.

3. If A is a signed formula and S a set of formulae, then:

   (a) If X a subformula of A, $occ(X, A)$ is the number of occurrences of X in A.

   (b) $\beta(A)$ denotes the number of *occurrences* of β-subformulae in A.

   (c) $\nu(A)$ is the number of ν-subformulae of A.

   (d) $\pi(A)$ is the number of π-subformulae of A.

   (e) If X a subformula of A, $deg(X, A)$ is the modal degree of X in A (i.e. the number of modal operators governing X in A) and $deg(A)$ is the maximal modal degree of a subformula in A.

   (f) $deg(S)$ is the maximal modal degree $deg(A)$ for $A \in S$.

If $\mathcal{L}$ is any of the logics K, $K4$, T, $S4$, A is an $\mathcal{L}$-valid formula and X a (signed) subformula of FA, then (abusing the functional notation) we shall use $k(X)$ and $br(A)$ to denote upper bounds for $k(\mathcal{T}, X)$ and $br(\mathcal{T})$, respectively, where $\mathcal{T}$ is a tableau proof of A, i.e. integers such that there exists a tableau proof $\mathcal{T}$ of A such that $k(\mathcal{T}, X) \leq k(X)$ and $br(\mathcal{T}) \leq br(A)$.

The following lemma is a preliminary result that will be used in the sequel.

Lemma 3. *If $\mathcal{L}$ is $K4$, T or $S4$ and A is an $\mathcal{L}$-valid formula, then there exists an $\mathcal{L}$-tableau proof $\mathcal{T}$ of A such that the number of applications of the π-rule (either the π_4 or the π_K or the π_{S4}-rule) in any branch is bounded by:*
1. *$deg(A)$, if $\mathcal{L} = T$;*
2. *$(\nu(A) + 1) \times \pi(A)$, if $\mathcal{L} = K4$ or $\mathcal{L} = S4$;*

Proof. 1. Since the modal degree of the conclusion of a π_K-rule application is smaller than the modal degree of its premise, after the n-th application of the π_K rule in a branch of a T-tableau there may occur only formulae Y such that $deg(Y, A) \geq n$.
2. It can be proved that if A is a $K4$- or $S4$-valid formula, then there is a $K4$-(resp. $S4$)-tableau proof of A where every branch contains at most $\nu(A) + 1$ applications of the π_4 (resp. π_{S4})-rule to the same formula (see [4] or [8], making ideas in [11] explicit). □

The next two results establish upper bounds for $k(X)$ and $br(A)$, respectively.

Lemma 4. *If $\mathcal{L}$ is K, $K4$, T or $S4$ and A is an $\mathcal{L}$-valid formula, then there exists an $\mathcal{L}$-tableau proof $\mathcal{T}$ of A such that, for every ν_0-subformula X of FA there are at most $k(X)$ applications of rules producing X in any branch of $\mathcal{T}$, where:*
1. *if $\mathcal{L} = K$, $k(X) = occ(X, A)$;*
2. *if $\mathcal{L} = T$, $k(X) = deg(X, A) + 1$;*
3. *if $\mathcal{L} = K4$ or $\mathcal{L} = S4$, $k(X) = 1 + ((\nu(A) + 1) \times \pi(A))$;*

This is summarised by the following table:

	K	T	$K4$ and $S4$
$k(X)$	$occ(X, A)$	$deg(X, A) + 1$	$1 + ((\nu(A) + 1) \times \pi(A))$

Proof. 1. Every inference in K "consumes" its principal formula and a consumed occurrence never produces copies of itself.
2. The ν_T-rule contains an implicit contraction, therefore a ν-formula can produce copies of itself in the same branch. However, let $\mathcal{B}$ be a branch in a tableau proof of A and $\mathcal{J}_1$, $\mathcal{J}_2$ two applications of rules producing a ν_0 formula X in $\mathcal{B}$. It can be shown that, if between the conclusion of $\mathcal{J}_1$ and the conclusion of $\mathcal{J}_2$ there is no application of the π_K-rule (hence, in particular, $\mathcal{J}_2$ is not an application of the π_K-rule), then the inference $\mathcal{J}_2$ can be eliminated and the resulting tableau is still a proof of A.

The same reasoning proving Item 1 in Lemma 3 shows that, if no wasteful application of the ν_T rule produces X, there may be at most $deg(X, A) + 1$ inferences producing X in a branch of a tableau proof of A (the "+1" is due to the fact that, possibly, a ν_T-rule produces X before any application of the π_K-rule).

3. Since in $K4$ the only rule producing ν_0-formulae is the π_4-rule, the result follows immediately from Lemma 3.

Similarly to the case of T, any $S4$-tableau proof of A can be transformed into a tableau proof of A such that for every branch B, if J_1 and J_2 are two applications of the ν_T rule to the same formula in B, then at least one application of the π_{S4} rule is between them. This fact and Lemma 3 clearly imply the desired result (again, the "+1" is due to possibility of a ν_T-rule producing X before any application of the π_K-rule). $\qquad\square$

Lemma 5. *If $\mathcal{L}$ is K, $K4$, T or $S4$ and A is an $\mathcal{L}$-valid formula, then there exists an $\mathcal{L}$-tableau proof $\mathcal{T}$ of A such that the number of branches of $\mathcal{T}$ is not greater than $br(A)$, where:*

1. *if $\mathcal{L} = K$, $br(A) = 2^{\beta(A)}$;*
2. *if $\mathcal{L} = T$, $br(A) = 2^{\beta(A) \times (deg(A)+1)}$;*
3. *if $\mathcal{L} = K4$ or $\mathcal{L} = S4$, $br(A) = 2^{\beta(A) \times (1+((\nu(A)+1) \times \pi(A)))}$.*

This is summarised by the following table:

	K	T	$K4$ and $S4$
$br(A)$	$2^{\beta(A)}$	$2^{\beta(A) \times (deg(A)+1)}$	$2^{\beta(A) \times (1+((\nu(A)+1) \times \pi(A)))}$

Proof. We preliminarily observe that if a tableau $\mathcal{T}$ contains no more than b applications of the β-rule in any branch, then $br(\mathcal{T}) \leq 2^b$.

1. Obvious, since any K-tableau has no more than one inference per formula occurrence in each branch.
2. We must set a limit to the number of times a β-formula can be produced, hence used for a β-inference, in a T-tableau. By Lemma 3, any β-subformula of A needs not be used more than $deg(A) + 1$ times in each branch (even if a β-formula occurs more than once in A, it needs not be expanded more than once in a branch segment between two consecutive applications of the π_K-rule).
3. Using Lemma 3 again and the fact that no formula need be used more then once in each branch segment between two applications of the π_4 (resp. π_{S4})-rule. $\qquad\square$

3.2 Hintikka Multiplicities for K, K4, T and S4

All the results of this section descend from a basic result, expressing a relation between the multiplicity, in a matrix proving A, for a position a whose label is X and the number of copies of X produced in a tableau proof of A.

Following [18], we say that a multiplicity μ is an $\mathcal{L}$-*Hintikka multiplicity* for a propositional formula A whenever the absence of an $\mathcal{L}$-admissible modal substitution σ and of a set of σ-complementary connections spanning $\langle A, 0 \rangle^\mu$ implies that A is not $\mathcal{L}$-valid. In other words, an $\mathcal{L}$-Hintikka multiplicity for a formula provides a bound on the increase of multiplicity needed to decide whether A is $\mathcal{L}$-valid or not. The result below provides a general method to compute $\mathcal{L}$-Hintikka multiplicities. Its proof constitutes also a syntactical completeness proof of the connection method.

Lemma 6. *Let FA be the root of a closed $\mathcal{L}$-tableau $\mathcal{T}$ whose number of branches is $br(\mathcal{T})$. For any occurrence of a ν_0-subformula X in FA, corresponding to position x in the unindexed formula tree for $\langle A, 0 \rangle$, let $prod(x)$ be the maximal number of copies of this occurrence produced in any branch of $\mathcal{T}$. Let μ be the multiplicity defined by: for any ν_0 position x in the unindexed formula tree for $\langle A, 0 \rangle$,*

$$\mu(x) = prod(x) \times br(\mathcal{T}).$$

The multiplicity μ is an $\mathcal{L}$-Hintikka multiplicity for A.

Proof. A proof of this result can be given either by a double step mapping (of signed tableau proofs into prefixed tableau proofs, preserving $k(\mathcal{T}, X)$ and $br(\mathcal{T})$, and then mapping the branches of a prefixed tableau proof to the atomic paths of a formula tree with given multiplicity), or else directly mapping signed tableau proofs to matrices. We give here a sketch of the latter proof, that does not require knowledge of prefixed tableau systems.

Let μ be the multiplicity defined above. Let $Tree_{\mathcal{T}}$ denote the indexed formula tree for A^μ (we use the indexed notation for such a tree in order to emphasise that it actually depends on $\mathcal{T}$, in that μ itself depends on $\mathcal{T}$).

Let $\mathcal{B}_1, ..., \mathcal{B}_n$ be the set of all the branches in $\mathcal{T}$. The multiplicity μ has been chosen "big enough" to allow the construction of a mapping φ from formula occurrences in any of the $\mathcal{B}_i$ to positions in $Tree_{\mathcal{T}}$ such that:

1. for every occurrence X_i of a signed formula X in $\mathcal{T}$, the label of $\varphi(X_i)$ is X;
2. given any ν_0-subformula X of FA:
   (a) if X_1 and X_2 are two occurrences of X produced by distinct inferences in $\mathcal{B}_i$, then $\varphi(X_1) \neq \varphi(X_2)$.
   (b) if two occurrences X_i and X_j of X are produced by two inferences occurring respectively in two branches $\mathcal{B}_i$ and $\mathcal{B}_j$ of $\mathcal{T}$, then $\varphi(X_i) \neq \varphi(X_j)$.
   (c) if X is produced by an inference $\mathcal{I}$ in the common part of $\mathcal{B}_i$ and $\mathcal{B}_j$, then two different occurrences of X are however considered: $X_i \in \mathcal{B}_i$ and $X_j \in \mathcal{B}_j$ and $\varphi(X_i) \neq \varphi(X_j)$. (This choice is perhaps superfluous but it does not affect the computation of the $\mathcal{L}$-Hintikka multiplicity and makes the proof easier).

The proof consists of the following steps.

1. Given any branch $\mathcal{B}_i$ of $\mathcal{T}$ whose leaf closes via a pair Tp, Fp of atoms, we show that there is a modal substitution $\sigma_{\mathcal{B}_i}$ which unifies the prefixes of the positions $\varphi(Tp)$ and $\varphi(Fp)$.
2. We define a substitution $\sigma_{\mathcal{T}}$ as the union of the $\sigma_{\mathcal{B}_i}$. (Note that such a substitution is well defined because of our choice 2c above).
3. We show that σ_T is $\mathcal{L}$-admissible.

4. We show that any atomic path through $Tree_{\mathcal{T}}$ contains a sub-path $\{a, a'\}$ such that the respective labels of a and a' are two signed atomic subformulae of FA which close some branch in $\mathcal{T}$ (i.e. the subformulae have the form Fp, Tp).

5. We conclude that μ is an $\mathcal{L}$-Hintikka multiplicity for A.

To prove step 1, we reason by induction on the length of the branch $\mathcal{B} = S_0, S_1 ..., S_n$ ending with a leaf closing via, say, Fp, Tp. Of course, the induction shows that the result holds for *sets* of formulae (the S_i's), so that actually formula trees for sets of formulae have to be considered (we do not give the tedious details here). Let p_i and p_j denote the occurrences of Fp, Tp in S_n. We show the existence of a substitution σ_B which unifies the prefixes of $\varphi(p_i)$ and $\varphi(p_j)$ (which are determined by the formula-tree $Tree_{\mathcal{T}}$, where $\mathcal{T}$ is a tableau for S_0).

Obviously, each specific logic $\mathcal{L}$ requires a different construction of σ_B. *We outline here the construction for* T. The cases where the top inference in $\mathcal{T}$ is not a π_K-inference are quite easy and we do not spell them out here. When $\mathcal{B}$ begins with an application of the π_K-rule, such an inference has the form:

$$\frac{\pi, \nu^1, \cdots, \nu^m, E}{\pi_0, \nu^1{}_0, \cdots, \nu^m{}_0}$$

where $\nu^1, \cdots, \nu^m$ are ν-formulae and no formula in E is a ν-formula. Say that the remaining segment of $\mathcal{B}$ is $\mathcal{B}' = S_1 ..., S_n$. Let $\mathcal{T}'$ be the subtree of $\mathcal{T}$ with root node $S_1 = \{\pi_0, \nu^1{}_0, \cdots, \nu^m{}_0\}$, a the position of π_0 in $Tree_{\mathcal{T}}$ (as determined by the mapping φ) and, for i s.t. $1 \le i \le m$, let x_i be the (variable) position of ν_0^i (as determined by φ).

Let v_1 and v_2 be the prefixes of $\varphi(p_1)$ and $\varphi(p_2)$ in $Tree_{\mathcal{T}}$ and u_1 and u_2 the prefixes of $\varphi(p_1)$ and $\varphi(p_2)$ in $Tree_{\mathcal{T}'}$.

Since Fp and Tp both occur in S_n, thus also in the conclusion of the π_K-rule, they are subformulae of either π or some of the ν^i in the premise. Therefore, the prefixes v_1 and v_2 have the form, respectively, $a_i u_1$ and $a_j u_2$, where a_i (resp. a_j) is the position associated (via φ) either to the occurrence of π in the displayed rule or to that of the ν^i's where the atom occurs.

Thus, three cases have to be considered:

1. both Tp and Fp are subformulae of π; therefore $a_i = a_j$;
2. both Tp and Fp are subformulae of some of the ν^i; therefore a_i and a_j are both variable positions, let us say $a_i = x_i$ and $a_j = x_j$
3. Tp is a subformula of some of the ν^i's and Fp is a subformula of π (or *vice versa*); therefore a_i is a variable position, let us say $a_i = x_i$.

In the the first case, it suffices to set σ_B equal to $\sigma_{B'}$. In the second and third case, it must be noted beforehand that $x_i \notin u_1$ and $x_i \notin u_2$; similarly for x_j. In the second case, it suffices to add the substitution x_i/x_j (or *vice versa*) to $\sigma_{B'}$. In the third one, it suffices to add the substitution x_i/a to $\sigma_{B'}$.

The heart of step 3 is the proof that the reduction ordering $\lhd$ induced by $\sigma_{\mathcal{T}}$ is irreflexive. We outline here the proof for the logic T. Suppose that we have a $\lhd$-cycle c: $a_1, ... a_i, ..., a_n$ (where $a_1 = a_n$). A $\Rightarrow$-*motif* in the cycle c is a *maximal* segment of c where any two consecutive positions, say a_j and a_{j+1} are such that $a_j \ll a_{j+1}$. By exploiting the definition of $\sigma_\mathcal{T}$, we show that:

1. if a putative cycle contains a number m of $\Rightarrow$-motifs, where $m \geq 2$, than it has a sub-cycle containing a number $m - 1$ of $\Rightarrow$-motifs.
2. there exists no cycle containing either one $\Rightarrow$-motif or no $\Rightarrow$-motif at all.

Finally, for step 4, it suffices to observe that the atomic paths through $Tree_\mathcal{T}$ can be generated in the same order in which the inference rules are applied in $\mathcal{T}$. □

By the above result and the already established lemmas we get the following theorem.

Theorem 7. *Let A be an $\mathcal{L}$-valid formula, X a subformula of A and $k(X)$, $br(A)$ as in Lemmas 4 and 5, respectively, for the logic $\mathcal{L}$. Let μ be the multiplicity such that, for any ν_0 position a, labelled by X, in the formula tree for A*
$$\mu(a) = k(X) \times br(A).$$
The multiplicity μ is an $\mathcal{L}$-Hintikka multiplicity for A.
Therefore, if $label(a) = X$:

	K	T
$\mu(a)$	$2^{\beta(A)}$	$(deg(X,A) + 1) \times 2^{\beta(A) \times (deg(A)+1)}$

	$K4$ and $S4$
$\mu(a)$	$(1 + ((\nu(A) + 1) \times \pi(A))) \times 2^{\beta(A) \times ((1+(\nu(A)+1) \times \pi(A)))}$

The above result gives exponential bounds for multiplicities and the reason for this is that the maximal number of branches in a tableau has to be taken into account. However, in many practical cases the actual number of branches in a tableau proof is far from reaching its upper bound. Therefore it can be useful to characterise $br(A)$ as a function of the number of atomic paths in the unindexed formula tree for A. This can be done, thanks to the following lemma.

Lemma 8. *Let A be an $\mathcal{L}$-valid formula, and let $atp(A)$ be the number of atomic paths in the unindexed formula tree for $\langle A, 0 \rangle$. Then $br(A)$ is not greater than $br^*(A)$, where:*

1. $br^*(A) = atp(A)$, *if $\mathcal{L} = K$;*
2. $br^*(A) = atp(A)^{1+deg(A)}$, *if $\mathcal{L} = T$;*
3. $br^*(A) = atp(A)^{1+((\nu(A)+1) \times \pi(A))}$, *if $\mathcal{L} = K4$ or $\mathcal{L} = S4$.*

Proof. The case of K is obvious. For the other logics, we note that a maybe wasteful but complete strategy for finding a tableau proof of A applies propositional expansion rules and ν-rules exactly once for each formula, then a π-rule, then again propositional expansion rules and ν-rules, etc. Each of such cycles produces a subtree with no more than $atp(A)$ branches. Therefore we can see the proof as a tree with branching factor equal to $atp(A)$ and whose depth is equal to $1 + m$ where m is the maximal number of applications of the π-rule in a branch. The result then follows from Lemma 3. □

The above lemma can be combined with Lemma 6, obtaining the following:

Theorem 9. *Let A be an $\mathcal{L}$-valid formula, X a subformula of A and $k(X)$ as established in Lemma 4. Let μ be the multiplicity such that, for any ν_0 position a, labelled by X, in the formula tree for A*
$$\mu(a) = k(X) \times br^*(A).$$
The multiplicity μ is an $\mathcal{L}$-Hintikka multiplicity for A.

Therefore, if $label(a) = X$:

	K	T
$\mu(a)$	$atp(A)$	$(deg(X, A) + 1) \times atp(A)^{1+deg(A)}$

	$K4$ and $S4$
$\mu(a)$	$(1 + ((\nu(A) + 1) \times \pi(A))) \times atp(A)^{1+((\nu(A)+1)\times\pi(A))}$

4 Concluding Remarks

In this work we have addressed the problem of finding $\mathcal{L}$-Hintikka multiplicities for the modal logics K, $K4$, T and $S4$. The problem is crucial if the matrix proof methods for such logics are to be used as decision methods. Upper bounds for multiplicity, ensuring completeness, have been established both in terms of the size of the formula A to be proved and the number of atomic paths in the unindexed formula tree for A. These bounds are obtained by exploiting the close relation of matrices to tableaux and previous results setting a bound on the number of "copies" of the same formula needed in a tableau proof.

The problem of establishing good bounds to the proof length in tableau or sequent based proof search has been addressed by several authors. With respect to *time* complexity, all of them consider only the complexity of the construction of a single branch; the number of branches in the tree, responsible of the exponential upper bounds in connection based decision methods, is not taken into account. The main results established in [4] have been already outlined in Section 3.1. Similar ideas have been nicely used in [8] to define a sequent calculus for $S4$ embedding a search strategy which prevents the construction of any looping branch. In [9], a rather different approach is taken to bound the length of $S4$ sequent proofs, based on the extension of the language by means of a new modal connective (and a significant reformulation of the rules of the calculus); this allows the definition of a measure that strictly decreases at any application of an inference rule. It results that not only the length of each branch can be polynomially bounded in the length of the proved formula, but also the space complexity for the associated decision procedure is improved w.r.t. [11]. It would be of interest to analyse how the new calculus relates to the matrix method and find out whether better bounds can be obtained.

In fact, the main issue that still remains to be addressed is which lower bounds can be established for $\mathcal{L}$-Hintikka multiplicities and, consequently, whether the upper bounds can be significantly improved or not.

Acknowledgements. The authors wish to thank the anonymous referees of previous works for their suggestion to work in this direction and Alain Heuerding for useful discussions on this topic.

References

1. P. Andrews. Transforming matings into natural deduction proofs. In *Proc. of CADE-5*, pages 281–292, 1980.
2. W. Bibel. *Automated Theorem Proving*. Vieweg Verlag, 1987.
3. W. Bibel, D. Korn, C. Kreitz, and S. Schmitt. Problem-oriented applications of automated theorem proving. In *Proc. of DISCO-96*, pages 1–21, 1996.
4. S. Cerrito and M. Cialdea Mayer. A polynomial translation of S4 into T and contraction free tableaux for S4. *Logic Journal of the IGPL*, 5(2):287–300, 1997.
5. M. Fitting. *Proof Methods for Modal and Intuitionistic Logics*. Reidel Publishing Company, 1983.
6. M. Fitting. First-order modal tableaux. *Journal of Automated Reasoning*, 4:191–213, 1988.
7. J.-Y. Girard. Linear logic. *Theoretical Computer Science*, 50, 1987.
8. A. Heuerding, M. Seyfried, and H. Zimmermann. Efficient loop-check for backward proof search in some non-classical propositional logics. In *Proc. of the 5th Int. Workshop on Theorem Proving with Analytic Tableaux and Related Methods (Tableaux '96)*, 1996.
9. J. Hudelmaier. Improved decision procedures for the modal logics K, T and S4. In *Proc. of the 9th International Workshop, CSL'95, Annual Conference of the EACSL*, pages 320–333. Springer Verlag, LNCS 1092, 1995.
10. G. E. Hughes and M. J. Cresswell. *An Introduction to Modal Logic*. Methuen & Co., 1968.
11. R. E. Ladner. The computational complexity of provability in systems of modal propositional logic. *SIAM Journal of Computation*, 6:467–480, 1977.
12. F. Massacci. Strongly analytic tableaux for normal modal logics. In *Proceedings of CADE-12*, pages 723–737, 1994.
13. M. Ohnishi and K. Matsumoto. Gentzen method in modal calculi. *Osaka Mathematical Journal*, 9:113–130, 1957.
14. S. Schmitt and C. Kreitz. On transforming intuitionistic matrix proofs into standard sequent proof. In *Proc. of the 4th Int. Workshop on Theorem Proving with Analytic Tableaux and Related Methods (Tableaux '95)*, pages 106–153, 1995.
15. S. Schmitt and C. Kreitz. Converting non-classical matrix proofs into sequent-style systems. In *Proc. of CADE-13*, 1996.
16. L. A. Wallen. Generating connection calculi from tableau- and sequent-based proof systems. In A. G. Cohn and J. R. Thomas, editors, *Artificial Intelligence and its Applications*, pages 35–50. John Wiley & Sons, 1986.
17. L. A. Wallen. Matrix proof methods for modal logic. In *Proc. of the 10th International Joint Conference on Artificial Intelligence (IJCAI-10)*, pages 917–923, 1987.
18. L. A. Wallen. *Automated Deduction in Nonclassical Logics: Efficient Matrix Proof Methods for Modal and Intuitionistic Logics*. MIT Press, 1990.

Automated Natural Deduction Prover and Experiments

Li Dafa
Dept. of Applied Mathematics
Tsinghua University
Beijing 100084, CHINA
Fax: (8610)62562768
email: dli@math.tsinghua.edu.cn

Jia Peifa
Dept. of Computer Science   Tsinghua University

Abstract

The paper presents the heuristics and techniques which are used to implement automated natural deduction prover (ANDP), the natural deduction was adapted from Gentzen system. There are 4 rules for quantifiers in Gentzen system and we have two unification algorithms in the system ANDP to handle quantifiers. Andrews' Challenge and Turing Halting Problem were solved using ANDP.

Keywords: Gentzen system, heuristics, natural deduction, quantifiers

Introduction

There are many rules in natural deduction system adapted from Gentzen system. The rules UG (Universal Generalization) and EG (Existential Generalization) are used to introduce quantifiers. The rules US (Universal Specialization) and ES (Existential Specialization) are used to eliminate quantifiers.

We have two algorithms to handle quantifiers, one is for introducing quantifiers, the other is for eliminating quantifiers[5]. In [2] we only described how our system ANDP works.

Given two formulas if they can become equal after applying the rules for introducing quantifiers, then our algorithm can find the series of operations UG and EG and decide which occurences of terms should be universally or existentially generalized upon[5].

Similarly given two formulas if they can become equal after applying the rules for eliminating quantifiers, then our algorithm can find the series of operations US and ES and decide what terms are used to replace the bound variables without using Skolem functions.

*The project supported by NSFC, partially by Intelligence Technology and System Lab.

The paper presents some heuristics and techniques used to implement the automated Gentzen system. We omitted the two unification algorithms here. Please see [5] for detail.

1. Subsumption strategy

2. How To Treat The Goal Being Of The Form $\exists x_n Q_{n-1} x_{n-1} \ldots\ldots Q_1 x_1 A$, where Q_i is quantifier $\forall$ or $\exists$.

It is hard to treat the goal with top existential quantifier. We can try to test if the goal can be infered from antecedence lines by using the rules for quantifiers. It is more difficult to process existential conclusion when we programme an automated natural deduction proving system.

We have two cases.

Case 1, in the case a goal is of the form $\exists x A$, where A is any formula. By using our algorithms for quantifiers we can try to infer the goal from an antecedence line, that is, apply a series of operations UG and EG to an antecedence line.

Case 2, When conclusion is of the form $\exists x_n Q_{n-1} x_{n-1} \ldots Q_1 x_1 [B \wedge C]$, where B and C are any formulas. Usually people have to construct its indirect proofs, that is, negate it, then infer a contradiction. By using our algorithm we can try to justify the goal from **antecedence lines.**Usually it is difficult to process the goal directly, we can process $\exists x_n Q_{n-1} x_{n-1} \ldots Q_1 x_1 B$ and $\exists x_n Q_{n-1} x_{n-1} \ldots Q_1 x_1 C$, respectively. By using our algorithm if we can infer $\exists x_n Q_{n-1} x_{n-1} \ldots Q_1 x_1 B$ from an antecedence line B' and $\exists x_n (Q_{n-1} x_{n-1}) \ldots (Q_1 x_1) C$ from an antecedence line C' by case 1, respectively, let x_i correspond to t_i in B' and to s_i in C' which are the terms existentially generalized upon to x_i of $\exists x_n Q_{n-1} x_{n-1} \ldots Q_1 x_1 B$ and $\exists x_n Q_{n-1} x_{n_1} \ldots Q_1 x_1 C$, respectively. If t_i and s_i are unifiable (otherwise it fails to do it), let σ be the MGU, then $t_i \sigma = s_i \sigma$, $Q_i x_i \ldots Q_1 x_1 [B \wedge C]$ can be obtained from new goal $\sigma Q_{i-1} x_{i-1} \ldots Q_1 x_1 [B \wedge C]$ by universally or existentially generalizing upon $t_i \sigma$, $i = n, n-1, \ldots, 1$.

3. The ordering of the inference rules and lengths of proofs

A length of a proof of a logic formula clearly depends on the structure of the formula. Usually the simpler the structure of a formula is, the shorter the length of its proof is. >From eight-year-experence we realize that the following ordering of the rules of inference can usually find short proofs.

Step 1, it reasons from bottom to top, that is, from conclusion to axioms and premises.

For the goal being of the form $\Gamma \vdash A \vee B$ it can be justified using the implication rule IMP, the new goal $\Gamma \vdash \sim A \to B$ is established. If a goal line is of the form $\Gamma \vdash A \to B$, it can be justified using the conditional rule CP, the new goal $\Gamma, A \vdash B$ is established. For the goal being of the form $\Gamma \vdash A \wedge B$, it can be infered from two new goal lines $\Gamma \vdash A$ and $\Gamma \vdash B$ using conjunction rule CONJUN. For the goal being of the form $\Gamma \vdash A \leftrightarrow B$, it can be infered from the new goal $\Gamma \vdash [A \to B] \wedge [B \to A]$ using the EQUIV rule. If the goal is of the form $\sim A$, the new goal is obtained by pushing $\sim$ into A if possible.

For the goal being of the form $\Gamma \vdash \exists x A$, it is hard to treat it. We can try to push quantifiers inside to find the minimal scopes of quantifiers, thus the structure of the goal may become simple so that the methods above can be used. Therefore it always try to reason from bottom to top except that the goal is an atomic formula or the wff with top existential quantifier.

Note that for the goal being of the form $\Gamma \vdash \forall x A$, it will not be treated in the step, otherwise it will produce many redundant formulas.

Step 2, when it fails to reason from bottom to top, it turns to do from top to bottom. It prefers using propositional rules to quantifier rules. The rules of inference for propositional logic are used by the following ordering: MP(modus ponens), MT(modus tollens), DS(disjunctive syllogism), SIMP(simplication), PUSH-NEGATION(push negation inward using De.Morgan rule), EQUIVALENCE. If it fails to use a rule, then it tries using the next rule in turn. If it succeeds in using a rule then it will return the top of the order list above and try the rules one by one. If it fails to use propositional rules then it will try using following rules of quantifiers.

4. The order of US and ES

If it eliminates the top universal quantifiers in precdence to the top existential quantifiers, sometimes it fails to prove some formulas. By our experience a length of a proof is much imposed on by following three rules: universal specialization US, existential specialization ES and CASES.

5. The rule UG and lengths of proofs

6. How To Eliminate The Top Existential Quantifier Without Using Skolem Functions In An Antecedence Line

Top existential quantifier is eliminated in Gentzen system like that given $\Gamma \vdash \exists x A(x)$, if $\{\Gamma, A(a)\} \vdash C$ then $\Gamma \vdash C$ where a is not free in Γ or $A(x)$ or C.

Case 1, if no free variables occur in $\exists x A(x)$, then the top existential quantifier can be eliminated casually, that is, if $\Gamma, A(a) \vdash C$, then $\Gamma \vdash C$, it does not need an algorithm to do it.

Case 2, if there are some free variables such as y occuring in the formula $\exists x A(x, y)$, after the existential quantifier $\exists x$ is eliminated, $\Gamma, A(a, y) \vdash A(a, y)$ is obtained, clearly the variable y occurs in hypothesis line, it is not possible to universally generalize the variable y or substitute any term for the variable y because y occurs in hypothesis line. Therefore for the case it needs a complicated algorithm to treat it.

7. How To Use The Rule US

For rule US it means that $\Gamma \vdash A(t)$ is infered from that $\Gamma \vdash \forall x A(x)$, the problem is what term is substituted for the individual variable x in $A(x)$. It is why it is hard to programme an automated natural deduction system. Usually the purpose for using rule US is to test if rules MP, MT, DS can be used next step or test if A and A' in $\sim A'$ can become equal after using rule US so that rule $\sim$ elimination or rule IP can be used or the goal can be justified immediately after using rule US.

For detail please refer the algorithm for eliminating quantifiers[6,10].

8. The strategy for minmal scopes of quantifiers

9. How To Use The Rule CASES

If the rule CASES is applied to a disjunction, then two disjuncts of it will be used as new hypotheses. Maybe it produces new constants from the new hypotheses, then many new Herbrand terms and irrelevant and redundant formulas. Now we present the following strategies to prevent the case above from happening as possible.

1: The rule CASES is first applied to premises.

2: The rule CASES is then applied to the disjunctions from which it will not produce new constants.

3: The rule CASES is then applied to short formulas.

The strategies will not affect the completeness. A lot of experiments proved that the strategies were general. Using the strategies ANDP did not only find a mechanical proof in natural deduction style of the new formalization of the halting problem but also produced the small search spaces for Burkholder's original formalization of the Halting problem in [1,3,6] and Pelletier's 75-problems.

Experiments

(1). The Mechanical Proofs Of Andrews' Challenge In Natural Deduction style

Andrews Challenge is logically simple, but its size makes it difficult to prove it.

A natural deduction proof of the challenge was obtained with our ANDP system, the system is loaded in SUN 4-286 Work-Station, it is written in Common LISP. The length of the proof of the challenge constructed by ANDP is 522 steps. The CPU time used is 8 seconds. ANDP is for a general purpose, we did not use special technique to prove the challege. We realized that when equivallence formulas are used it becomes easy to prove a logic theorem.

Other Andrews' problems and Pelletier's problems and other complicated problems are proved by our ANDP system.

(2). The Mechanical Proof Of Halting Problem IN Natural Deduction Style

L.Burkholder presented Turing halting problem [1] as 76th automated theorem proving problem, halting problem is important theorem in computer science. The English statement for halting problem is omitted, please refer [1].

It Failed To Prove Halting Problem By Resolution

Massimo Bruschi reported the following facts in [1]:

"No mention of (automated solutions of) this problem has been made in subsequent issues of the Newsletter" "I was unable to obtain a mechanical proof of the theorem simply by applying ENprover". "Applications of OTTER, Argonne's theorem prover, also were unsuccesseful. The problem seems to lie in the number and the length of the clauses derived from the transformation of the given input: 84 from the the axiom and 2 from the negation of the thesis, some of them with 7 literals."

The Natural Deduction Proof Of Halting Problem Is Obtained

ANDP system is loaded in SUN 4-286 workstation in AI Lab. of Tsinghua University. The natural deduction proof of the problem is obtained using ANDP [3,4,6].

References

[1] Bruschi, M., The Halting Problem, AAR Newsletter 17, March 1991.

[2] Li Dafa, A Natural Deduction Automated Theorem Proving System, Proc. of CADE-11: 11th international Conference On Automated Deduction, June 15-18, 1992, NY, USA, P668-672.

[3] Li Dafa, A Mechanical Proof of the Halting Problem, AAR Newsletter, No.23, June 1993.

[4] Li Dafa, The Formalization of the Halting Problem Is Not Suitable for Describing the Halting Problem, AAR Newsletter 27, October 1994.

[5] Li Dafa, The Algorithms For Eliminating and Introducing Quantifiers , Proc. of PRICAI'94 : The 3rd Pacific Rim International Conference On Artificial Intelligence, Aug. 15-18, 1994, Beijing, P208-214.

[6] Li Dafa , An Automated Natural Deduction Proof of the Formalization of the Halting Problem, AAR NEWSLetter, No. 32, March 1996.

Non-elementary Speed-ups in Proof Length by Different Variants of Classical Analytic Calculi*

Uwe Egly

Institut für Informationssysteme E184.3
Technische Universität Wien
Treitlstraße 3, A–1040 Wien, Austria
e-mail: uwe@kr.tuwien.ac.at

Abstract. In this paper, different variants of classical analytic calculi for first-order logic are compared with respect to the length of proofs possible in such calculi. A cut-free sequent calculus is used as a prototype for different other analytic calculi like analytic tableau or various connection calculi. With modified branching rules (β–rules), non-elementary shorter minimal proofs can be obtained for a class of formulae. Moreover, by a simple translation technique and a standard sequent calculus, analytic cuts, i.e., cuts where the cut formulae occur as subformulae in the input formula, can be polynomially simulated.

1 Introduction

Analytic first-order calculi like (free-variable) analytic tableaux [4, 15, 23] or various connection calculi [5] are well suited for implementing automated deduction on a computer. In order to search for a proof in such calculi, only subformulae of the input formula have to be considered. This property, often referred to as the subformula property, makes such calculi highly attractive for automated proof search in classical as well as in non-classical logics. Although the subformula property does not imply cut-free calculi, most of the aforementioned calculi are cut-free. The price that we have to pay for the simplicity introduced by these restrictions is the weakness of these calculi with respect to proof length. Short and structured proofs should be preferred because they are easier to understand and easier to check.

In this paper, we show that this weakness can be eliminated to an enormous extend without giving up these restrictions. In what follows, we will use different variants of the cut-free LK-calculus [16] because this calculus can be considered as a prototype for many other analytic calculi including the ones mentioned above (see [23] or [7] for details). We stress that our results also hold for free-variable tableaux which are often used in implementations. The idea on which the improvement is based is the introduction of a subformula in both polarities

---

* The author would like to thank Hans Tompits and the referees for their useful comments on an earlier version of this paper.

although the subformula may occur in one polarity only in the input formula, i.e., the formula to be proven.

Our observation was that different translations to definitional form, namely Eder's variant [12] and the variant of Plaisted and Greenbaum [19], yield different normal forms, where, for some classes of formulae, the lengths of shortest proofs are non-elementarily related [14]. Both translations introduce labels abbreviating subformulae but the former translation introduces equivalences, whereas the latter one introduces implications whenever possible. Roughly speaking, if equivalences are used then not only the subformula is available in the resulting formula but also the negation of the subformula. It is well known from the literature that resolution combined with the latter translation polynomially simulates the cut-free LK-calculus but resolution combined with the former translation yields a stronger calculus with respect to proof length (see [13] for the propositional case). Therefore, we introduce formulae in different polarities by two mechanisms, namely by using modified branching rules[2] and by a translation of the input formula. As an example, consider the following LK-rule for $\to l$.

$$\frac{\Gamma \vdash \Delta_1, A, \Delta_2 \quad \Gamma_1, B, \Gamma_2 \vdash \Pi}{\Gamma, \Gamma_1, (A \to B), \Gamma_2 \vdash \Delta_1, \Delta_2, \Pi} \to l$$

This rule is replaced by the following two rules.

$$\frac{\Gamma \vdash \Delta_1, A, B, \Delta_2 \quad \Gamma_1, B, \Gamma_2 \vdash \Pi}{\Gamma, \Gamma_1, (A \to B), \Gamma_2 \vdash \Delta_1, \Delta_2, \Pi} \to l_1 \qquad \frac{\Gamma \vdash \Delta_1, A, \Delta_2 \quad \Gamma_1, B, A, \Gamma_2 \vdash \Pi}{\Gamma, \Gamma_1, (A \to B), \Gamma_2 \vdash \Delta_1, \Delta_2, \Pi} \to l_2$$

With these new rules, some restricted forms of analytic cuts can be simulated resulting in a non-elementary speed-up in minimal proof length for some classes of formulae.[3]

The second possibility to introduce formulae in both polarities is a translation of the input formula F. Instead of proving F, the equivalent formula $(\bigwedge_{A \in \Sigma(F)} \forall (A \to A)) \to F$ is proved, where $\Sigma(F)$ denotes the set of all subformulae of F. Observe that the length of the resulting formula[4] is at most quadratic in the length of F. With a standard cut-free sequent calculus and this translation scheme, analytic cuts can be simulated with only a moderate increase of proof length. This is important for existing implementations because sequent- or tableau-based theorem provers can be extended by analytic cut without modifying them. Generating a slightly extended formula can be done in an inexpensive preprocessing step.

One may object that the introduction of additional formulae increases the branching degree of nodes in the search space. This is a correct observation but we reply with the following three arguments. First, it is common mathematical practise to use lemmata (and, therefore, cuts) in proofs. These lemmata usually improve the structure and readability of proofs. Second, weaker forms of

---

[2] I.e., the β–rule in tableau notation.

[3] In [2], non-elementary speed-ups are obtained by an optimized δ-rule. Roughly speaking, antiprenexing is incorporated into the δ-rule, such that the non-elementary speed-up described in [3] can be reproduced in the tableau format.

[4] I.e., the number of symbol occurrences.

asymmetric β-rules (e.g., the folding-down rule in SETHEO or KOMET) and lemma mechanisms have been proved useful in practical applications. It is immediately apparent that such extensions do not improve the search for a proof for all formulae, but some of them can only be proved with these extensions, even if they increase the branching degree. A third argument concerns tableaux with lean induction. In [1], we introduce induction principles into tableau calculi with equality in a simple and elegant way. Induction is implemented by modified closure conditions for branches. Such tableau calculi can be strengthened[5] by allowing analytic cuts or modified branching rules. For instance, if the axiom E is deleted from Example 2 (a weak form of arithmetic) in [1], then the resulting formula is not provable in our tableau calculus with lean induction, but it becomes provable if analytic cuts are allowed.

The structure of the paper is as follows. In Section 2, definitions and notations are introduced. In Section 3, we present improved branching rules for LK. In Section 4, we show that these modified branching rules can enable nonelementary speed-ups of the length of shortest proofs. Section 5 is devoted to a simple translation of the input formula such that analytic cuts can be polynomially simulated by proving the result of the translation in LK without cut. Finally, we discuss related concepts relevant for theorem proving systems.

2   Definitions and Notations

We consider a usual first-order language with function symbols.

Definition 2.1
Given a formula F, we call G an immediate s-subformula *(structural subformula) of F if either*

1. *F is an atom and $G = F$, or*
2. *$F = (\neg G_1)$ and $G = G_1$, or*
3. *$F = (G_1 \circ G_2)$ ($\circ \in \{\wedge, \vee, \rightarrow\}$) and $G = G_1$ or $G = G_2$, or*
4. *$F = Qx\, G_1(x)$ ($Q \in \{\forall, \exists\}$) and $G = G_1(x)$.*

The relation "is s-subformula of" is the reflexive and transitive closure of the relation "is an immediate s-subformula of". We call G an immediate subformula *of F if either one of 1.–3. holds, or*

4'. *$F = Qx\, G_1(x)$ ($Q \in \{\forall, \exists\}$) and $G = G_1(t)$ for an arbitrary term t.*

The relation "is subformula of" is the reflexive and transitive closure of the relation "is an immediate subformula of".

Definition 2.2
Let F be a formula. $\Sigma(F)$ denotes the set of all s-subformulae of F and $\forall F$ denotes the universal closure of F.

---

[5] "Strengthening" here does not mean "proving with a shorter proof", but proving formulae which are unprovable in the weaker calculus.

Definition 2.3
Let $A_1, \ldots, A_n$ and $B_1, \ldots, B_m$ be first-order formulae. Then,

$$A_1, \ldots, A_n \vdash B_1, \ldots, B_m$$

is called a sequent. *The informal meaning of the sequent is the same as the informal meaning of the formula* $(\bigwedge_{i=1}^{n} A_i) \to (\bigvee_{i=1}^{m} B_i)$.

Let S be either a formula or a sequent. An occurrence of a formula G occurs positively (negatively) in S if the number of explicit ($\neg$) or implicit ($\to$, $\vdash$) negation signs of G is even (odd). For instance, A and C occur positively in $A \to B \vdash C$, whereas B occurs negatively in the same sequent.[6]

We use the following "multiplicative" sequent calculus. In contrast to Gentzen's original formulation in [16], rule applications are allowed at arbitrary places in the sequent. As a consequence, the exchange rule can be omitted. Let Γ, Δ, and Λ (possibly subscripted) denote sequences of formulae and let A, B, and F denote formulae.

The inference rules for LK are the *logical rules*, the *quantifier rules*, and the *structural rules* without cut. $\mathsf{LK_{cut}}$ is the calculus LK extended by the cut rule. For all sequent calculi in this paper, the *initial sequents* (or the *axioms*) are of the form $F \vdash F$ for a formula F.

LOGICAL RULES

$$\frac{\Delta_1, A, \Delta_2, B, \Delta_3 \vdash \Gamma}{\Delta_1, (A \wedge B), \Delta_2, \Delta_3 \vdash \Gamma} \wedge l \qquad \frac{\Gamma \vdash \Delta_1, A, \Delta_2 \quad \Lambda \vdash \Pi_1, B, \Pi_2}{\Gamma, \Lambda \vdash \Delta_1, \Pi_1, (A \wedge B), \Delta_2, \Pi_2} \wedge r$$

$$\frac{\Gamma_1, A, \Gamma_2 \vdash \Delta_1 \quad \Pi_1, B, \Pi_2 \vdash \Delta_2}{\Gamma_1, \Pi_1, (A \vee B), \Gamma_2, \Pi_2 \vdash \Delta_1, \Delta_2} \vee l \qquad \frac{\Gamma \vdash \Delta_1, A, \Delta_2, B, \Delta_3}{\Gamma \vdash \Delta_1, (A \vee B), \Delta_2, \Delta_3} \vee r$$

$$\frac{\Gamma \vdash \Delta_1, A, \Delta_2 \quad \Gamma_1, B, \Gamma_2 \vdash \Pi}{\Gamma, \Gamma_1, (A \to B), \Gamma_2 \vdash \Delta_1, \Delta_2, \Pi} \to l \qquad \frac{\Gamma_1, A, \Gamma_2 \vdash \Delta_1, B, \Delta_2}{\Gamma_1, \Gamma_2 \vdash \Delta_1, (A \to B), \Delta_2} \to r$$

$$\frac{\Delta \vdash \Gamma_1, A, \Gamma_2}{\neg A, \Delta \vdash \Gamma_1, \Gamma_2} \neg l \qquad \frac{\Gamma_1, A, \Gamma_2 \vdash \Delta}{\Gamma_1, \Gamma_2 \vdash \Delta, \neg A} \neg r$$

For all logical rules, A and B are called *auxiliary formulae* and $(A \wedge B)$, $(A \vee B)$, $(A \to B)$, and $\neg A$ are called *principal formulae* of the corresponding $\wedge$, $\vee$, $\to$, and $\neg$ rules. The inference rules $\vee l$, $\to l$, and $\wedge r$ are called *branching rules*.

QUANTIFIER RULES

$$\frac{\Delta_1, A(t), \Delta_2 \vdash \Gamma}{\Delta_1, \forall x\, A(x), \Delta_2 \vdash \Gamma} \forall l \qquad \frac{\Gamma \vdash \Delta_1, A(y), \Delta_2}{\Gamma \vdash \Delta_1, \forall x\, A(x), \Delta_2} \forall r$$

---

[6] Positive resp. negative occurrences coincide with positive resp. negative parts in [22].

$$\frac{\Delta_1, A(y), \Delta_2 \vdash \Gamma}{\Delta_1, \exists x\, A(x), \Delta_2 \vdash \Gamma} \; \exists l \qquad\qquad \frac{\Gamma \vdash \Delta_1, A(t), \Delta_2}{\Gamma \vdash \Delta_1, \exists x\, A(x), \Delta_2} \; \exists r$$

$\forall r$ and $\exists l$ must fulfill the eigenvariable condition, i.e., the (free) variable y does not occur in $\Gamma, \Delta_1, \Delta_2$, or $A(x)$. The term t is any term not containing a bound variable. $A(t)$ and $A(y)$ are auxiliary formulae and $\forall x A(x)$ and $\exists x A(x)$ are principal formulae.

STRUCTURAL RULES

WEAKENING

$$\frac{\Gamma_1, \Gamma_2 \vdash \Delta}{\Gamma_1, A, \Gamma_2 \vdash \Delta} \; wl \qquad\qquad \frac{\Delta \vdash \Gamma_1, \Gamma_2}{\Delta \vdash \Gamma_1, A, \Gamma_2} \; wr$$

A is called the *weakening formula*.

CONTRACTION

$$\frac{\Gamma_1, A, \Gamma_2, A, \Gamma_3 \vdash \Delta}{\Gamma_1, A, \Gamma_2, \Gamma_3 \vdash \Delta} \; cl \qquad\qquad \frac{\Delta \vdash \Gamma_1, A, \Gamma_2, A, \Gamma_3}{\Delta \vdash \Gamma_1, A, \Gamma_2, \Gamma_3} \; cr$$

CUT

$$\frac{\Gamma \vdash \Delta_1, A, \Delta_2 \quad \Lambda_1, A, \Lambda_2 \vdash \Pi}{\Gamma, \Lambda_1, \Lambda_2 \vdash \Delta_1, \Delta_2, \Pi} \; cut$$

A is called the *cut formula*.

We say that inference I *introduces* F if F is the principal or weakening formula of I.

Let $\mathcal{LK}$ be a meta-variable for sequent calculi. A *derivation* (in tree form) of a sequent S in $\mathcal{LK}$ is defined as usual.

Definition 2.4
An application of the cut rule in an $\mathsf{LK_{cut}}$*-derivation of* $\vdash F$ *is called* analytic *if the cut formula is a subformula of* F. $\mathsf{LK_{acut}}$ *is the calculus* $\mathsf{LK_{cut}}$, *where all cuts are analytic.*

Analytic cuts preserve the subformula property because the cut formula occurs (as a subformula) in the formula of the end sequent.

Definition 2.5
A sequence of sequents in an $\mathcal{LK}$*-derivation* α *of* S *is called a* branch *of* α *if the following conditions are satisfied.*

1. *The sequence begins with an initial sequent and ends with* S.
2. *Every sequent in the sequence except* S *is an upper sequent of an inference* I, *and is immediately followed by the lower sequent of* I.

A path *is a partial branch beginning in a sequent (not necessarily an initial sequent) and ending with* S.

Definition 2.6
The length *of a formula* F, *denoted by* $|F|$, *is the number of symbol occurrences
in the string representation of* F. *If* $\Delta = F_1, \ldots, F_n$ *or* $\Delta = \{F_1, \ldots, F_n\}$, *then*
$|\Delta| = \sum_{i=1}^{n} |F_i|$. *For a sequent* S *of the form* $\Gamma \vdash \Delta$, $|S| = |\Gamma| + |\Delta|$. *The length
of an* $\mathcal{LK}$-*derivation* α, *denoted by* $|\alpha|$, *is* $\sum_{S \in \mathcal{T}} |S|$, *where* $\mathcal{T}$ *is the multiset of
sequences occurring in* α *The number of sequents occurring in* α *is denoted by*
$\#seq(\alpha)$. *The* height $h(\alpha)$ *of* α *is the number of sequents occurring on the longest
branch.*

Let s be the non-elementary function with $s(0) = 1$ and $s(n+1) = 2^{s(n)}$ for all
$n \in \mathbb{N}$. The following definition of a polynomial simulation (resp. an elementary
simulation) is adapted from [12] and restricted to the case that the connectives
in both calculi are identical.

Definition 2.7
A calculus P_1 *can* polynomially simulate (elementarily simulate) *a calculus* P_2 *if
there is a polynomial* p (*an elementary function* e) *such that the following holds.
For every proof of a formula* F *in* P_2 *of length* n, *there is a proof of* F *in* P_1,
whose length is not greater than $p(n)$ $(e(n))$.

3 Alternative Branching Rules

In this section, we define LK^a, an asymmetric variant of LK for first-order logic by
replacing each of the branching rules $\vee l$, $\wedge r$, and $\rightarrow l$ by two variants. Moreover,
we show that LK^a is sound and complete. In the next section, we demonstrate
that these modified branching rules are strongly related with the analytic cut
rule.

We replace $\vee l$ by $\vee l_1$ and $\vee l_2$, $\wedge r$ by $\wedge r_1$ and $\wedge r_2$, and $\rightarrow l$ by $\rightarrow l_1$ and $\rightarrow l_2$
and obtain the calculus LK^a from LK. The new branching rules are as follows.

$$\frac{\Gamma_1, A, \Gamma_2 \vdash \Delta_1, B \quad \Pi_1, B, \Pi_2 \vdash \Delta_2}{\Gamma_1, \Pi_1, (A \vee B), \Gamma_2, \Pi_2 \vdash \Delta_1, \Delta_2} \vee l_1 \qquad \frac{\Gamma_1, A, \Gamma_2 \vdash \Delta_1 \quad \Pi_1, B, \Pi_2 \vdash \Delta_2, A}{\Gamma_1, \Pi_1, (A \vee B), \Gamma_2, \Pi_2 \vdash \Delta_1, \Delta_2} \vee l_2$$

$$\frac{\Gamma, B \vdash \Delta_1, A, \Delta_2 \quad \Lambda \vdash \Pi_1, B, \Pi_2}{\Gamma, \Lambda \vdash \Delta_1, \Pi_1, (A \wedge B), \Delta_2, \Pi_2} \wedge r_1 \qquad \frac{\Gamma \vdash \Delta_1, A, \Delta_2 \quad \Lambda, A \vdash \Pi_1, B, \Pi_2}{\Gamma, \Lambda \vdash \Delta_1, \Pi_1, (A \wedge B), \Delta_2, \Pi_2} \wedge r_2$$

$$\frac{\Gamma \vdash \Delta_1, A, B, \Delta_2 \quad \Gamma_1, B, \Gamma_2 \vdash \Pi}{\Gamma, \Gamma_1, (A \rightarrow B), \Gamma_2 \vdash \Delta_1, \Delta_2, \Pi} \rightarrow l_1 \qquad \frac{\Gamma \vdash \Delta_1, A, \Delta_2 \quad \Gamma_1, B, A, \Gamma_2 \vdash \Pi}{\Gamma, \Gamma_1, (A \rightarrow B), \Gamma_2 \vdash \Delta_1, \Delta_2, \Pi} \rightarrow l_2$$

Obviously, LK^a is an analytic calculus and the subformula property is retained
by these replacements of inference rules. What is new is the possibility to get
a subformula occurring in one polarity only in the lower sequent in different
polarities in the upper sequents of the new branching rules. It is immediately
apparent by the translations below that, for any LK-derivation α of $\vdash F$, there

exists an LK^a-derivation α^a of the same end sequent and $|\alpha^a| \leq 2 \cdot |\alpha|$. Hence, LK^a polynomially simulates LK but the reverse simulation is not elementary as we will demonstrate in the next section.

We consider a transformation of LK-derivations into LK^a-derivations of the same end sequent. An application of $\vee l$ of the form

$$\frac{\Gamma_1, A, \Gamma_2 \vdash \Delta_1 \qquad \Pi_1, B, \Pi_2 \vdash \Delta_2}{\Gamma_1, \Pi_1, (A \vee B), \Gamma_2, \Pi_2 \vdash \Delta_1, \Delta_2} \ \vee l$$

is replaced by

$$\frac{\dfrac{\Gamma_1, A, \Gamma_2 \vdash \Delta_1}{\Gamma_1, A, \Gamma_2 \vdash \Delta_1, B} \ wr \qquad \Pi_1, B, \Pi_2 \vdash \Delta_2}{\Gamma_1, \Pi_1, (A \vee B), \Gamma_2, \Pi_2 \vdash \Delta_1, \Delta_2} \ \vee l_1$$

The replacements for the rules $\to l$ and $\wedge r$ are similar to the one presented above, i.e., the necessary formula is introduced by weakening above the new branching rule. We get the following theorem.

Theorem 3.1
LK^a *polynomially simulates* LK.

We consider a transformation of LK^a-derivations into LK_{acut}-derivations of the same end sequent. An application of $\vee l_1$ of the form

$$\frac{\Gamma_1, A, \Gamma_2 \vdash \Delta_1, B \qquad \Pi_1, B, \Pi_2 \vdash \Delta_2}{\Gamma_1, \Pi_1, (A \vee B), \Gamma_2, \Pi_2 \vdash \Delta_1, \Delta_2} \ \vee l_1$$

is replaced by

$$\frac{\dfrac{\dfrac{\dfrac{A \vdash A \qquad B \vdash B}{(A \vee B) \vdash A, B} \ \vee l \qquad \Gamma_1, A, \Gamma_2 \vdash \Delta_1, B}{\Gamma_1, (A \vee B), \Gamma_2 \vdash \Delta_1, B, B} \ cut}{\Gamma_1, (A \vee B), \Gamma_2 \vdash \Delta_1, B} \ cr \qquad \Pi_1, B, \Pi_2 \vdash \Delta_2}{\Gamma_1, \Pi_1, (A \vee B), \Gamma_2, \Pi_2 \vdash \Delta_1, \Delta_2} \ cut$$

The replacements for the new rules $\vee l_2$, $\to l_1$, $\to l_2$, $\wedge r_1$, and $\wedge r_2$ are similar to the one presented above. We get the following theorem.

Theorem 3.2
LK_{acut} *polynomially simulates* LK^a.

Theorems 3.1 and 3.2 imply the following one.

Theorem 3.3
LK^a *is sound and complete.*

We will demonstrate in the next section that the new branching rules indeed have the power of cut for some class of formulae. Observe that the cuts, introduced by the translation, are analytic cuts.

4 A Non-Elementary Speed-Up Result

In this section, we compare LK with LK^a. Our discussion is based on a sequence $H_1, H_2, \ldots$ of formulae which are modified and extended variants of formulae $F_1, F_2, \ldots$ presented in [18].

Definition 4.1
Let F_k occur in the infinite sequence of formulae $(F_k)_{k \in \mathbb{N}}$ where

$$F_k = \forall b \, ((\forall w_0 \exists v_0 \, p(w_0, b, v_0) \wedge$$
$$\forall uvw \, (\exists y \, (p(y, b, u) \wedge \exists z \, (p(v, y, z) \wedge p(z, y, w)))) \to p(v, u, w)))$$
$$\to \exists v_k \, (p(b, b, v_k) \wedge \exists v_{k-1} \, (p(b, v_k, v_{k-1}) \wedge \ldots \wedge \exists v_0 \, p(b, v_1, v_0)) \ldots).$$

Using this sequence $(F_k)_{k \in \mathbb{N}}$, Orevkov showed that cut elimination can tremendously affect proof length. More precisely, he proved that there exists an LK_{cut}-derivation of $\vdash F_k$ with a single occurrence of the cut rule and the number of sequents in this derivation is linear in k, but any cut-free LK-derivation of $\vdash F_k$ has height $\geq 2 \cdot s(k) + 1$. We first sketch a slightly modified version of Orevkov's LK_{cut}-derivation with one application of the cut rule. Let ψ_k be this LK_{cut}-derivation. The cut is then changed to an analytic cut by extending F_k by $(q \vee \neg q) \vee A$, where A is the cut formula and q is an atom neither occurring in F_k nor in A. The remaining derivation is adjusted accordingly. Instead of F_k, we get an extended formula H_k. Then we show that any LK-derivation of $\vdash H_k$ has length $> (2 \cdot s(k))^{1/2}$ and that we can simulate the cut in the derivation of $\vdash H_k$ by the modified branching rules. Hence, we obtain a short cut-free derivation of $\vdash H_k$ in LK^a.

Abbreviations shown in Figure 1 are used in the following in order to simplify the notation. We do not present the LK_{cut}-derivation of $\vdash F_k$ in all details but sketch the proof stressing the relevant details. Slightly differing LK_{cut}-derivations of F_k can be found in [18] or in [14].

There are two kinds of LK-derivations, namely β_k and $\delta_k(b_0)$, which are relevant in the following for $k > 0$. For $k = 0$, only $\delta_0(b_0)$ is relevant, which is an LK-derivation of the sequent $A_0(b_0), C \vdash B_0(b_0)$. Then, the LK-derivation of $\vdash F_0$, denoted by ψ_0, is as follows.

$$\frac{\delta_0(b_0)}{\vdash \forall b \, ((A_0(b) \wedge C) \to B_0(b))} \; \wedge l, \to r, \forall r$$

For $k > 0$, β_k is an LK-derivation of $A_0(b_0), C \vdash A_k(b_0)$ and $\delta_k(b_0)$ is an LK-derivation of $A_0(b_0), C, A_k(b_0) \vdash B_k(b_0)$. Then, ψ_k is as follows.

$$\frac{\dfrac{\beta_k \qquad \delta_k(b_0)}{A_0(b_0), C \vdash B_k(b_0)} \; cut, cl, cl}{\vdash \forall b \, ((A_0(b) \wedge C) \to B_k(b))} \; \wedge l, \to r, \forall r$$

The derivation ψ_k of $\vdash F_k$ in LK_{cut} discussed so far has one application of the cut rule where the cut formula has a free variable. We transform ψ_k into ϕ_k and make the application of the cut rule analytic.

$$C_1(\alpha, \beta, \gamma) = \exists z \, (p(\alpha, \beta, z) \wedge p(z, \beta, \gamma))$$
$$C_2(\alpha, \beta, \gamma) = \exists y \, (p(y, b_0, \alpha) \wedge C_1(\beta, y, \gamma))$$
$$C = \forall u \forall v \forall w \, (C_2(u, v, w) \rightarrow p(v, u, w))$$
$$B_0(\alpha) = \exists v_0 \, p(b_0, \alpha, v_0)$$
$$B_{i+1}(\alpha) = \exists v_{i+1} \, (p(b_0, \alpha, v_{i+1}) \wedge B_i(v_{i+1}))$$
$$A_0(\alpha) = \forall w_0 \exists v_0 \, p(w_0, \alpha, v_0)$$
$$A_{i+1}(\alpha) = \forall w_{i+1} \, (A_i(w_{i+1}) \rightarrow \overline{A}_{i+1}(w_{i+1}, \alpha))$$
$$\overline{A}_0(\alpha, \delta) = \exists v_0 \, p(\alpha, \delta, v_0)$$
$$\overline{A}_{i+1}(\alpha, \delta) = \exists v_{i+1} \, (A_i(v_{i+1}) \wedge p(\alpha, \delta, v_{i+1}))$$

Fig. 1. Abbreviations used in the following.

Definition 4.2
Let H_k ($k \in \mathbb{N}$) be a formula of the form

$$\forall b \, ((((q \vee \neg q) \vee A_k(b)) \wedge (A_0(b) \wedge C)) \rightarrow B_k(b))$$

where q is a predicate with a predicate symbol not occurring elsewhere in $A_i(b)$, $B_i(b)$ ($0 \leq i \leq k$), and C.

An $\mathsf{LK_{cut}}$-derivation of $\vdash H_k$ is obtained from the derivation of $\vdash F_k$ in $\mathsf{LK_{cut}}$ presented above by simply adding a wl-inference with weakening formula $(q \vee \neg q) \vee A_k(b)$ directly below β_k ($k \geq 1$) or directly below $\delta_0(b_0)$ ($k = 0$). Then ϕ_0 and ϕ_k ($k \geq 1$) are as follows.

$$\cfrac{\cfrac{\delta_0(b_0)}{\cfrac{(q \vee \neg q) \vee A_0(b_0), A_0(b_0), C \vdash B_0(b_0)}{\cfrac{((q \vee \neg q) \vee A_0(b_0)) \wedge (A_0(b_0) \wedge C) \vdash B_0(b_0)}{\vdash \forall b \, ((((q \vee \neg q) \vee A_0(b)) \wedge (A_0(b) \wedge C)) \rightarrow B_0(b))} \rightarrow r, \forall r} \wedge l, \wedge l} wl}$$

$$\cfrac{\cfrac{\cfrac{\beta_k}{(q \vee \neg q) \vee A_k(b_0), A_0(b_0), C \vdash A_k(b_0)} wl \qquad \delta_k(b_0)}{\cfrac{(q \vee \neg q) \vee A_k(b_0), A_0(b_0), C \vdash B_k(b_0)}{\cfrac{((q \vee \neg q) \vee A_k(b_0)) \wedge (A_0(b_0) \wedge C) \vdash B_k(b_0)}{\vdash \forall b \, ((((q \vee \neg q) \vee A_k(b)) \wedge (A_0(b) \wedge C)) \rightarrow B_k(b))} \rightarrow r, \forall r} \wedge l, \wedge l} cut, cl, cl}$$

The $\mathsf{LK_{cut}}$-derivation ϕ_k has only one application of the analytic cut rule. Hence, ϕ_k is an $\mathsf{LK_{acut}}$-derivation of $\vdash H_k$.

Lemma 4.1
Let ϕ_k be the $\mathsf{LK_{acut}}$-derivation of $\vdash H_k$ with one application of an analytic cut described above. Then, the length of ϕ_k is $\leq c \cdot 2^{d \cdot k}$ for constants c, d.

Proof. By Theorem 2 in [18], the number of sequents in the $\mathsf{LK_{acut}}$-derivation of $\vdash F_k$ is less than $e \cdot k$, where e is a constant. Since $|(q \vee \neg q) \vee A_k| \leq a \cdot 2^{b \cdot k}$ for constants a, b, there exist constants c, d such that the length of ϕ_k is $\leq c \cdot 2^{d \cdot k}$. $\quad\square$

Next, we transform the $\mathsf{LK_{acut}}$-derivation ϕ_k with one occurrence of an analytic cut into a cut-free $\mathsf{LK^a}$-derivation of the same end sequent.

Lemma 4.2
There exists an $\mathsf{LK^a}$-derivation ρ_k of $\vdash H_k$ such that the length of ρ_k is $\leq c \cdot 2^{d \cdot k}$ for constants c, d.

Proof. The derivations ρ_0 and ρ_k $(k \geq 1)$ are as follows.

$$
\dfrac{
\dfrac{
\dfrac{
\dfrac{A_0(b_0) \vdash A_0(b_0)}{q \vee \neg q, A_0(b_0) \vdash A_0(b_0)} \; wl \qquad \delta_0(b_0)
}{(q \vee \neg q) \vee A_0(b_0), A_0(b_0), C \vdash B_0(b_0)} \; \vee l_1
}{((q \vee \neg q) \vee A_0(b_0)) \wedge (A_0(b_0) \wedge C) \vdash B_0(b_0)} \; \wedge l, \wedge l
}{\vdash \forall b\, ((((q \vee \neg q) \vee A_0(b)) \wedge (A_0(b) \wedge C)) \to B_0(b))} \; \to r, \forall r
$$

$$
\dfrac{
\dfrac{
\dfrac{
\dfrac{\beta_k}{q \vee \neg q, A_0(b_0), C \vdash A_k(b_0)} \; wl \qquad \delta_k(b_0)
}{(q \vee \neg q) \vee A_k(b_0), A_0(b_0), C \vdash B_k(b_0)} \; \vee l_1, cl, cl
}{((q \vee \neg q) \vee A_k(b_0)) \wedge (A_0(b_0) \wedge C) \vdash B_k(b_0)} \; \wedge l, \wedge l
}{\vdash \forall b\, ((((q \vee \neg q) \vee A_k(b)) \wedge (A_0(b) \wedge C)) \to B_k(b))} \; \to r, \forall r
$$

Hence, we have an $\mathsf{LK^a}$-derivation ρ_k of $\vdash H_k$ and the length of ρ_k is $\leq c \cdot 2^{d \cdot k}$ for constants c, d. $\quad\square$

In contrast to the short $\mathsf{LK_{cut}}$-derivation of $\vdash F_k$, any derivation of the same end sequent in LK has length non-elementary in k. The following lemma is a corollary of Theorem 1 in [18].

Lemma 4.3
Let ψ_k be an LK-derivation of $\vdash F_k$. Then, $h(\psi_k) \geq 2 \cdot \mathsf{s}(k) + 1$.

As a consequence, any derivation of $\vdash F_k$ in LK has length $\geq 2 \cdot \mathsf{s}(k) + 1$. In the remainder of this section, we prove a non-elementary lower bound on the length of any LK-derivation of $\vdash H_k$.

Lemma 4.4
Let ϕ be an LK-derivation of $A_1, \ldots, A_n \vdash B_1, \ldots, B_m$. Then, $\#seq(\phi) \geq n + m - 1$.

Proof. By induction on the structure of ϕ. $\quad\square$

Lemma 4.5
Let ϕ_k be an LK-derivation of $\vdash H_k$. Then, $|\phi_k| > (2 \cdot \mathsf{s}(k))^{1/2}$.

Proof. There are two possible inferences by which the formula $Q = (q \lor \neg q) \lor A_k(b_0)$ can be introduced into ϕ_k, namely wl and $\lor l$. We first eliminate all occurrences of Q introduced by $\lor l$. Then, all occurrences of Q introduced by wl are eliminated. The resulting derivation ϕ'_k is an LK-derivation of $\vdash F_k$ and $|\phi'_k| > 2 \cdot s(k)$. A simple calculation yields the desired result.

Step 1. Q is introduced by $\lor l$. Select the first $\lor l$-inference (with respect to some tree ordering) such that α_1 and α_2 do not have an $\lor l$-inference with principal formula Q. If there is no such inference, then goto Step 2. Otherwise, this first inference has the following form (I_1 and I_2 are LK-inferences.).

$$
\cfrac{
  \cfrac{\alpha_1}{\Gamma_1, q \lor \neg q, \Gamma_2 \vdash \Delta_1}\ I_1
  \qquad
  \cfrac{\alpha_2}{\Pi_1, A_k(b_0), \Pi_2 \vdash \Delta_2}\ I_2
}{\Gamma_1, \Pi_1, ((q \lor \neg q) \lor A_k(b_0)), \Gamma_2, \Pi_2 \vdash \Delta_1, \Delta_2}\ \lor l
$$
$$\beta$$

Construct an LK-derivation of $\vdash H_k$ of the form

$$
\cfrac{
  \cfrac{\alpha'_1}{\Gamma'_1, \Gamma'_2 \vdash \Delta'_1}\ I'_1 \\[2pt]
  \vdots \ \ wl, wr \ \ (*) \\[2pt]
  \Gamma_1, \Gamma_2 \vdash \Delta_1 \\[2pt]
  \vdots \ \ wl, wr \ \ (**) \\[2pt]
  \Gamma_1, \Pi_1, \Gamma_2, \Pi_2 \vdash \Delta_1, \Delta_2
}{\Gamma_1, \Pi_1, (q \lor \neg q) \lor A_k(b_0), \Gamma_2, \Pi_2 \vdash \Delta_1, \Delta_2}\ wl
$$
$$\beta$$

where α'_1 is obtained from α_1 by omitting all weakenings introducing formulae q, $\neg q$, or $q \lor \neg q$ and by omitting contractions upon such formulae. Moreover, all $\lor l$-inferences with principal formula $q \lor \neg q$ are omitted. I'_1 is either I_1 or an inference occurring in α_1. The number of sequents occurring in the resulting derivation is not greater than the number of sequents in ϕ_k because q, $\neg q$, and $q \lor \neg q$ occurring in Γ_1, Γ_2, or Δ_1 are re-introduced at $(*)$, and $\#seq(\alpha_2)$ is not less than the number of wl and wr in $(**)$. Replace all such $\lor l$-inferences without increasing the number of sequents resulting in an LK-derivation where all occurrences of Q are introduced by wl.

Step 2. Omit all wl introducing q, $\neg q$, $q \lor \neg q$, or Q and adjust the derivation by omitting all contractions upon these formulae and all inferences with auxiliary formulae $q, \neg q, q \lor \neg q$ or Q. Since all occurrences of Q are introduced by wl, $\vdash F_k$ is derived.

Now, we have an LK-derivation ϕ'_k of $\vdash F_k$, and $\#seq(\phi'_k) \leq \#seq(\phi_k)$. Since any sequent occurring in ϕ'_k has length less than $|\phi_k|$, the increase of length is at most quadratic. Now, $2 \cdot s(k) < |\phi'_k| \leq |\phi_k|^2$ and $|\phi_k|$ is greater than $(2 \cdot s(k))^{1/2}$.

$\square$

Combining Lemma 4.2 and Lemma 4.5 yields the following theorem.

Theorem 4.1
LK *cannot elementary simulate* LK^a.

Observe that similar results also hold for closely related calculi including free-variable tableaux. Moreover, the search space decreases non-elementarily (for some classes of formulae) if LK^a is applied instead of LK.

5   A Simple Transformation of the Input Formula

In this section, we introduce a simple translation scheme based on the introduction of closed formulae $\forall (A \rightarrow A)$ for any s-subformula A of the given formula F. We show that LK$_{\text{acut}}$ can be polynomially simulated by this transformation in combination with LK. Moreover, the reverse simulation is also polynomial. We start with the definition of the implicational form of a formula.

Definition 5.1
Let $\Sigma(F)$ denote the set of all s-subformulae of a given formula F. Then, the implicational form *of F, denoted by $\iota(F)$, is defined as follows.*

$$\iota(F) = \bigwedge_{A \in \Sigma(F)} \forall (A \rightarrow A)$$

Instead of proving $\vdash F$, the sequent $\vdash \iota(F) \rightarrow F$ is considered. This is possible because F is equivalent to $\iota(F) \rightarrow F$. The length of $\iota(F)$ is estimated as follows. Since the number of s-subformulae of F is $\leq |F|$,

$$|\iota(F)| \leq |F| \cdot |F \rightarrow F| \leq |F| \cdot (2 \cdot |F| + 1).$$

Definition 5.2
A sequent $\vdash F$ is derivable in LK$_{\rightarrow}$ if $\vdash \iota(F) \rightarrow F$ is derivable in LK.

Theorem 5.1
LK$_{\rightarrow}$ *polynomially simulates* LK$_{\text{acut}}$.

Proof Idea. Let ϕ be a derivation of $\vdash F$ in LK$_{\text{acut}}$ and let A be an s-subformula of F. We show that replacing an application of *cut* with cut formula A by an application of $\rightarrow l$ and $\forall l$-inferences to close the resulting formula $A \rightarrow A$ yield an LK$_{\text{acut}}$-derivation with only a moderate increase of derivation length. After the replacements of all analytic cuts, we eventually get an LK$_{\rightarrow}$-derivation of $\vdash F$ after some additional inferences. □

Theorem 5.2
LK$_{\text{acut}}$ *polynomially simulates* LK$_{\rightarrow}$.

Proof Idea. Let ϕ be a derivation of $\vdash F$ in $\mathsf{LK}_\to$, i.e., an LK-derivation of $\vdash \iota(F) \to F$. The idea is to replace some $\to l$-inferences with principal formula $A \to A$ by an analytic cut and wl.

Consider an application of $\to l$ with lower sequent S in ϕ (shown on the left below) such that the length of the path β is minimal (I_1 and I_2 are LK-inferences).

$$
\cfrac{\cfrac{\alpha_1}{\Gamma \vdash \Delta_1, A, \Delta_2}\;I_1 \quad \cfrac{\alpha_2}{\Gamma_1, A, \Gamma_2 \vdash \Pi}\;I_2}{\underset{\beta}{(A \to A), \Gamma, \Gamma_1, \Gamma_2 \vdash \Delta_1, \Delta_2, \Pi}}\;\to l
\qquad
\cfrac{\cfrac{\alpha_1}{\Gamma \vdash \Delta_1, A, \Delta_2}\;I_1 \quad \cfrac{\alpha_2}{\Gamma_1, A, \Gamma_2 \vdash \Pi}\;I_2}{\underset{\beta'}{\Gamma, \Gamma_1, \Gamma_2 \vdash \Delta_1, \Delta_2, \Pi}}\;cut
$$

Replace the left inference figure by the right inference figure, where sequents in β' are adjusted accordingly. Iterating this elimination of $\to l$, deleting some weakenings and contractions yields an $\mathsf{LK_{acut}}$-derivation ψ of $\vdash F$, because only such formula occurrences are deleted which occur in $\iota(F)$. The length of ψ is $\leq |\phi|$. $\qquad\qquad\Box$

Theorem 5.1 and Theorem 3.2 imply

Theorem 5.3
$\mathsf{LK}_\to$ *and* $\mathsf{LK_{acut}}$ *polynomially simulate* $\mathsf{LK^a}$.

This theorem, together with Lemma 4.2 imply the following

Corollary 5.1
There exists an $\mathsf{LK}_\to$ *-derivation of* $\vdash H_k$ *with length exponential in* k.

6 Conclusion and Discussion

We have shown that a slight modification of the branching rules of an LK-calculus yields a non-elementary decrease of proof length for a sequence of formulae. The reason for such a tremendous "speed-up" by the new rules is the utilization of both polarities of a subformula of F, which occurs only in one polarity in $\vdash F$.

The idea to use asymmetric branching rules in classical propositional sequent calculi is not new. Prawitz [21] (see also [5, 6]) used a similar idea in propositional logic. He introduced a reduction which can be simulated by (a sequence of) asymmetric branching rules and contractions.

More recently, D'Agostino showed in [11] that a class $(T_n)_{n\in\mathbb{N}}$ of propositional formulae, which was used by Cook and Reckhow [10] to indicate that analytic tableaux are not super, possess short proofs (in the length of the input formula) in a tableau system with modified branching rules, but any proof of a formula from this class in a standard tableau system has exponential length. He observed that, for $(T_n)_{n\in\mathbb{N}}$, the truth table method is computationally superior to analytic semantic tableaux. Semantically, a β-rule represents a case analysis, e.g., $A \lor B$ is true if either A is true or B is true. The observation resulting in asymmetric β-rules is the possibility to assume the negation of one case in the other without violating correctness. Hence, we can additionally assume $\neg A$ in case B, or $\neg B$ in case A. The formulae $(T_n)_{n \in \mathbb{N}}$ have short resolution proofs (even in tree resolution) and short tableau proofs if a linear (sequence) form is allowed, i.e., the tableau derivation has dag[7] form instead of tree form. Bibel [6] showed that T_n, which he calls "complete matrix in n variables", possess short proofs in connection calculi allowing a kind of factorization.

In contrast to the propositional case and $(T_n)_{n \in \mathbb{N}}$, where resolution yields short proofs, it can be shown that any resolution proof of H_k must have length non-elementary in the length of the input formula if the formula is translated to clause form using the standard method (without introducing additional definitions for subformulae). Hence, we get a non-elementary speed-up of LK^a over resolution in the first-order case. Observe that not only proof length decreases non-elementarily, but also the size of the search space if, for instance, breadth-first search is assumed. The reason is that the increase of the size of the search space by the additional formulae introduced by the modified rules is only elementary, which is more than compensated by the much shorter proof length. Similar results for the search space hold for $\mathsf{LK}_\to$ and $\mathsf{LK}_{\mathrm{acut}}$.

Another phenomenon has been observed in [20]. Shannon graphs (or *Binary Decision Diagrams*) use essentially the idea of the modified β-rules. In some sense, BDDs mimic the behavior of tableau systems with modified β-rules.

Finally, the folding-down operation in clausal connection tableaux implements the modified β-rules restricted to clauses. It is shown in [17] that folding-down can be viewed as an implementation technique for factorization in these tableaux.

References

1. M. Baaz, U. Egly, and C. G. Fermüller. Lean Induction Principles for Tableaux. In *Proceedings of the International Conference on Theorem Proving with Analytic Tableaux and Related Methods*. Springer Verlag, 1997.
2. M. Baaz and C. G. Fermüler. Non-elementary Speedups between Different Versions of Tableaux. In P. Baumgartner, R. Hähnle, and J. Posegga, editors, *Proceedings of the Fourth Workshop on Theorem Proving with Analytic Tableaux and Related Methods*, pages 217–230. Springer Verlag, 1995.
3. M. Baaz and A. Leitsch. On Skolemization and Proof Complexity. *Fundamenta Informaticae*, 20:353–379, 1994.
4. E. W. Beth. Semantic Entailment and Formal Derivability. *Mededlingen der Koninklijke Nederlandse Akademie van Wetenschappen*, 18(13), 1955.
5. W. Bibel. *Automated Theorem Proving*. Vieweg, Braunschweig, second edition, 1987.
6. W. Bibel. *Deduction: Automated Logic*. Academic Press, London, 1993.
7. W. Bibel and E. Eder. Methods and Calculi for Deduction. In D. M. Gabbay, C. J. Hogger, and J. A. Robinson, editors, *Handbook of Logic in Artificial Intelligence and Logic Programming*, volume 1, chapter 3, pages 67–182. Oxford University Press, Oxford, 1993.

---

[7] directed acyclic graph

8. G. Boolos. Don't Eliminate Cut. *J. Philosophical Logic*, 13:373–378, 1984.

9. G. Boolos. A Curious Inference. *J. Philosophical Logic*, 16:1–12, 1987.

10. S. Cook and R. Reckhow. On the Lenghts of Proofs in the Propositional Calculus. In *Proceedings of the 5th Annual ACM Symposium on Theory of Computing*, pages 135–148. University of Toronto, 1974.

11. M. D'Agostino. Are Tableaux an Improvement on Truth-Tables? Cut-Free Proofs and Bivalence. *J. Logic, Language and Information*, 1:235–252, 1992.

12. E. Eder.. *Relative Complexities of First Order Calculi*. Vieweg, Braunschweig, 1992.

13. U. Egly. An Answer to an Open Problem of Urquhart, 1996. Submitted.

14. U. Egly. On Different Structure-preserving Translations to Normal Form. *J. Symbolic Computation*, 22:121–142, 1996.

15. M. Fitting. *First-Order Logic and Automated Theorem Proving*. Springer Verlag, second edition, 1996.

16. G. Gentzen. Untersuchungen über das logische Schließen. *Mathematische Zeitschrift*, 39:176–210, 405–431, 1935. English translation in [24].

17. R. Letz, K. Mayr, and C. Goller. Controlled Integration of the Cut Rule into Connection Tableau Calculi. *J. Automated Reasoning*, 13:297–337, 1994.

18. V. P. Orevkov. Lower Bounds for Increasing Complexity of Derivations after Cut Elimination. *Zapiski Nauchnykh Seminarov Leningradskogo Otdeleniya Matematicheskogo Instituta im V. A. Steklova AN SSSR*, 88:137–161, 1979. English translation in *J. Soviet Mathematics*, 2337–2350, 1982.

19. D. A. Plaisted and S. Greenbaum. A Structure-Preserving Clause Form Translation. *J. Symbolic Computation*, 2:293–304, 1986.

20. J. Posegga and P. H. Schmitt. Automated Deduction with Shannon Graphs. *J. Logic and Computation*, 5(6):697–729, 1995.

21. D. Prawitz. A Proof Procedure with Matrix Reduction. In M. Laudet, D. Lacombe, L. Nolin, and M. Schützenberger, editors, *Symposium on Automatic Demonstration*, volume 125 of *Lecture Notes in Mathematics*, pages 207–214. Springer Verlag, 1970.

22. K. Schütte. *Proof Theory*. Springer Verlag, 1977.

23. R. Smullyan. *First-Order Logic*. Springer Verlag, 1968.

24. M. E. Szabo, editor. *The Collected Papers of Gerhard Gentzen*. Studies in Logic and the Foundations of Mathematics. North-Holland Publishing Company, 1969.

25. G. Takeuti. *Proof Theory*, volume 81 of *Studies in Logic and the Foundations of Mathematics*. North-Holland Publishing Company, 1975.

Ordered Tableaux: Extensions and Applications*

Reiner Hähnle and Christian Pape

Universität Karlsruhe, Institut für Logik, Komplexität und Deduktionssysteme
76128 Karlsruhe, Germany, {reiner,pape}@ira.uka.de

Abstract. In this paper several conceptual extensions to the theory of order-restricted free variable clausal tableaux which was initiated in [9, 8] are presented: atom orderings are replaced by the more general concept of a selection function, the substitutivity condition required for lifting is for certain variants of the calculus replaced by a much weaker assumption, and a first version of order-restricted tableaux with theories is introduced. The resulting calculi are shown to be sound and complete. We report on first experiments made with a prototypical implementation and indicate for which classes of problems order-restricted tableaux calculi are likely to be beneficial.

1 Introduction

In this paper we continue to develop the theory of order-restricted free variable clausal tableaux which was initiated in [9, 8].

A-ordered tableaux (full definitions of all notions are given in Section 2 and 3) are regular clause tableaux with two different kinds of extension steps: given an A-ordering [6] $\prec$ on literals, a clause C can be used to extend a tableau branch **B** iff (i) C has a maximal connection into **B**, i.e. the connection literal of C is $\prec$-maximal in C *or* (ii) C has a maximal connection into another clause D, i.e. the connection literals of both clauses are $\prec$-maximal in the clause, where they occur.

On the one hand we present several conceptual extensions of ordered tableaux: in Section 4 atom orderings are replaced by the more general concept of a selection function; in Section 5 we demonstrate how such calculi can be implemented with the help of constraints; for a somewhat less restrictive class of calculi we show that the substitutivity condition required for lifting can be replaced by a much weaker assumption: *stability wrt variable renaming* is already sufficient for completeness. The resulting calculus, called *tableaux with input selection function* is shown to be complete.

Finally, a first version of tableaux with selection function and theories is presented (Section 6.1).

On the other hand, in Section 7 we report our experiences made with a prototypical implementation. We indicate for which classes of problems order-restricted tableaux calculi are likely to be beneficial.

---

* This work has been partially supported by the Deutsche Forschungsgemeinschaft, Schwerpunktprogramm "Deduktion".

On a methodological level we compare tableaux with selection function to restart model elimination, recently developed by Baumgartner & Furbach [3] (Section 6.2).

2 Preliminaries

Given a signature Σ, i.e. a set of predicate, function, constant and variable symbols, then **atoms** and **literals** are constructed from Σ and the negation sign $\neg$ as usual. The set of all literals for Σ is denoted by $\mathbf{L}_\Sigma$. We omit the index Σ if no confusion can arise.

We denote substitutions by σ, τ, or write them explicitly as a (finite) set $\sigma = \{x_1 \leftarrow t_1, \ldots, x_n \leftarrow t_n\}$ with the meaning $\sigma(x_i) = t_i$ and $\sigma(x) = x$ for all $x \neq x_i$. The special case of a variable renaming (all t_i are distinct variables) is denoted by θ. We use such substitutions only for replacing the variables of a clause by new distinct variables.

A **clause** is a sequence $L_1 \vee \ldots \vee L_n, n \geq 1$ of disjunctively connected literals. The variables in such clauses are assumed to be (implicitly) universally quantified. An **instance** of a clause C is a sequence of literals $C\theta$ such that θ replaces the variables of C by new variables. The variables of instances of clauses are only placeholders for terms, they are *not* assumed to be universally quantified. $\mathbf{C}$ is the set of all clauses. We write $L \in C$ for short if a literal L occurs in a clause C. $\overline{L}$ is the **complement** of a literal L, i.e. $\overline{A} = \neg A$ and $\overline{\neg A} = A$ if A is an atom. *Atom* selects the atomic part of literals that is $Atom(A) = Atom(\neg A) = A$ for every atom A.

A **clause tableau** $\mathfrak{T}$ is an ordered tree where the root node is labeled with *true* or a literal and all other nodes are labeled with literals. For a node n of $\mathfrak{T}$ the clause $ClauseOf(n)$ is constructed from the literals of the children of n in the order from left to right. $Predecessor(n)$ denotes the parent node of node n. A path from the root node to a leaf literal of $\mathfrak{T}$ is called a **branch** of $\mathfrak{T}$. A tableau is **closed** if every branch contains (at least) two complementary literals. We sometimes describe a tableau as a finite set of branches and a branch as a finite set of literals. We also often identify branches with the set of literals on them. A branch $\mathbf{B}$ is said to be **regular** if (i) every literal of a node of $\mathbf{B}$ occurs only once on $\mathbf{B}$ and (ii) $ClauseOf(n)$ is not a tautology for every node n of $\mathbf{B}$. A tableau $\mathfrak{T}$ is **regular** if all branches of $\mathfrak{T}$ are regular.

Partial Interpretations are associated with a consistent set of ground literals. An interpretation I **satisfies** a ground clause C iff there exists $L \in C$ with $L \in I$. I is said to be a **model** for a set S of first order clauses iff I satisfies all clauses of every ground instance of S.

3 A-Ordered Tableaux

To ease comparison with previous results in this section we briefly rehash A-ordered clausal ground tableaux as defined in [8].

Definition 1. An *A*-ordering is a binary relation $\prec_A$ on atoms, such that for all atoms A, B, C:

1. $A \not\prec A$ (**irreflexivity**),
2. $A \prec_A B$ and $B \prec_A C$ implies $A \prec_A C$ (**transitivity**), and
3. $A \prec_A B$ implies $A\sigma \prec_A B\sigma$ for all substitutions σ (**substitutivity**).

For sake of readability we focus on the ground version of *A*-ordered tableaux. Lifting to first order logic is handled in Section 5 in the context of tableaux with selection function. The results established there hold as well for *A*-ordered tableaux. As in ordered resolution connections between clauses are restricted to literals that occur $\prec_A$-maximally.

Definition 2. A literal L_j occurs $\prec_A$-**maximally** in a clause $L_1 \vee \ldots \vee L_n$ iff $Atom(L_j) \not\prec_A Atom(L_i)$ for all $i = 1, \ldots, n$.

A clause $C = L_1 \vee \ldots \vee L_n$ possesses a $\prec_A$-**maximal connection** to a clause $C' = L'_1 \vee \ldots \vee L'_{n'}$ iff $L_i = \overline{L'_j}$ for some $1 \leq i \leq n, 1 \leq j \leq n'$, L_i occurs maximally in C, and L'_j occurs maximally in C'. If, moreover, $C, C' \in S$ then C is called a **restart clause** of S.

A clause C has a **maximal connection into a set of literals B** iff C has a maximal connection to a literal of **B**.

A-ordered tableaux are regular[2] clause tableaux with the additional restriction that a clause $C \in S$ can be used to extend a branch **B** only if C has a maximal connection into **B** or to another clause of S:

Definition 3. An *A*-**ordered ground clause tableau** is a regular ground clause tableau $\mathfrak{T}$ such that

1. $C = ClauseOf(n)$ has a maximal connection into the branch ending in n *or*
2. C is a restart clause of S.

At the start of a refutation the initial tableau is empty and only the second extension rule (called **restart rule**) can be used to expand it. But even if a relevant clause for the initial step is used, it is still necessary to allow restarts later on to obtain a complete calculus, as the following example shows:

Example 1.
Take the unsatisfiable clause set $M = \{A \vee \underline{B},\ A \vee \neg \underline{B},\ \neg A \vee \underline{C},\ \neg A \vee \neg \underline{C}\}$ and *A*-ordering $A < B < C$ (maximal literals are underlined). Each clause of M is a restart clause. Fig. 1 shows a closed *A*-ordered tableau for M. The solid lines correspond to a (maximal)

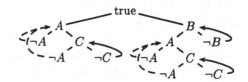

Fig. 1. Restarts are necessary

extension step: these branches can be closed immediately with the maximal literal of the maximal connected clause. The dashed arrows indicate a **reduction step**, i.e. the branch is closed by non maximal literals of a clause.

---

[2] Regularity and ordering restrictions are orthogonal concepts, but we choose to start out from regular tableaux, because regularity fits naturally into our completeness proof below. All results do still hold, of course, if regularity is dropped.

Note that without a second application of the restart rule no closed A-ordered tableau for M can be constructed, independently of the choice of clause for the initial step.

Theorem 4 [9]. *For any unsatisfiable set S of ground clauses and A-ordering $\prec_A$ exists a closed $\prec_A$-ordered tableau for S.*

In the ground case the procedure stays complete even if restart steps are delayed until no extension steps are possible.

The first order case is handled as usual with Herbrand's Theorem and a lifting lemma which, by substitutivity of $\prec_A$, is straightforward. In Section 5 we will see that lifting is even possible under a weaker assumption.

L-orderings (and L-ordered tableaux) are defined exactly as A-orderings, but on literals instead of atoms. Hence, each A-ordering is also an L-ordering, but not vice versa. It is easy to show that the previous theorem holds for L-orderings as well.

4 Tableaux with Selection Function

In this section we still work with ground clauses. The first order case is considered in the following section.

As shown in the previous section, in ordered tableaux only the literals which are maximal wrt the clause in which they occur need to be considered for an extension step. This kind of restriction can be generalized with the help of certain functions.

Each A-ordering $\prec_A$ induces a function $f_{\prec_A} : \mathbf{C} \to (2^{\mathbf{L}} - \{\emptyset\})$ by stipulating $f_{\prec_A}(C) = \{L \mid L \text{ is } \prec_A\text{-maximal in } C\}$. On the other hand, not every function f from $\mathbf{C}$ to $(2^{\mathbf{L}} - \emptyset)$ can be realized with an A-ordering.[3]

Definition 5. A **selection function** is a total function $f : \mathbf{C} \to (2^{\mathbf{L}} - \{\emptyset\})$ such that all literals in $f(C)$ occur also in C for all $C \in \mathbf{C}$.

A selection function f is **deterministic** iff $|f(C)| = 1$ for all $C \in \mathbf{C}$.

Example 2. Consider a *total L-ordering* $\prec_L$ on ground literals. Then exactly one literal in each ground clause is maximal. Thus $\prec_L$ defines a deterministic selection function.

A total A-ordering on ground literals does not define a deterministic selection function in general as, for example, in $L \vee \overline{L}$ both literals are maximal.

We define particular deterministic selection functions f_{last} and f_{first} which select the exactly last, respectively, the first literal of a clause.

In tableaux with selection function the selected literals play the rôle of maximal literals in ordered tableaux. Accordingly, connections between clauses are restricted to selected literals.

---

[3] Consider the clauses $C_1 = L_1 \vee L_2$, $C_2 = L_2 \vee L_1$, and assume $f(C_i) = \{L_i\}$ for $i = 1, 2$. Any ordering $\prec_A$ such that $f_{\prec_A} = f$ must be cyclic which contradicts irreflexivity.

Definition 6. Given a selection function f, two ground clauses C and D have a **connection via** f iff there are $L \in f(C)$, $M \in f(D)$ such that L is complementary to M. If, moreover, $C, D \in S$ then we say that C is a **restart clause of** S **(wrt** f**).**

A clause C has a **weak connection into a set of literals B via** f iff some $L \in f(C)$ is complementary to a literal of **D**.

In tableaux with selection function the only admissible extension steps are with clauses that have a weak connection into the branch they extend or if they are restart clauses wrt f (see Definition 7). In the latter case $C = D$ is not excluded, however, such clauses are tautologies and, by regularity, are not allowed to be considered for extension steps.

Note that the set of restart clauses can easily be computed in a pre-processing step and thus causes no additional cost during proof search.

Definition 7. Given a selection function f and a set of ground clauses S, a **tableau with selection function** f **for** S is a regular ground clause tableau $\mathfrak{T}$ such that

1. $C = \mathit{ClauseOf}(n)$ has a weak connection via f into the branch ending in n
   or
2. C is a restart clause of S wrt f.

Example 3. Reconsider the set M of Example 1. The selection function f_{last} selects the underlined literals and thus the ordered tableau displayed in Fig. 1 is also a tableaux with selection function f_{last} for M.

Because each restart clause can extend any tableau branch at any time (provided that it was not used before in this branch which would violate regularity), one should be very careful in the choice of a selection function. To gain a maximum of search space restriction for a given set of clauses, one should choose a selection function for which the number of restart clauses in S is minimal. This prohibits extensive use of restarts and leads to stronger connected proofs.

By virtue of regularity of tableaux with selection function it is not possible to build infinite branches in an attempt to refute a finite set S of ground clauses. In our completeness proof this property ensures that a model of S can be constructed from an open *finite* tableau branch which cannot be extended further. As a consequence, ground tableaux with selection function are proof confluent.

Definition 8. Let S be a set of ground clauses, f a selection function, and $\mathfrak{T}$ a tableau with selection function f for S. $\mathfrak{T}$ is **saturated** iff there is no tableau $\mathfrak{T}'$ with selection function f for S such that $\mathfrak{T}$ is a proper subtree of $\mathfrak{T}'$.

Theorem 9. *For any finite unsatisfiable set S of ground clauses and selection function f there exists a closed tableau with selection function f for S.*

Proof. Assume there were no closed tableaux with selection function f for S. We construct a model of S.

Regularity does not permit to extend tableau branches with tautologies, hence assume wlog that S does not contain any tautologies.

Let $\mathfrak{T}$ be any saturated tableau with selection function f for S which is finite by regularity and finiteness of S. $\mathfrak{T}$ is not closed, so it has a finite open branch **B**. The literals on **B** form a partial interpretation $I_\mathbf{B}$ which satisfies at least the clauses of S that were used to extend **B**.

Let $S' \subseteq S$ be the set of clauses not satisfied by $I_\mathbf{B}$. The clauses in S' have the following properties:

1. There are no clauses $C, D \in S'$ such that C is connected to D via f. Otherwise, C and D were restart clauses and thus were used to extend **B**, because $\mathfrak{T}$ is saturated. But then C and D are satisfied by definition of $I_\mathbf{B}$.
2. There is no clause $C \in S'$ such that C has a weak connection via f into **B**. Otherwise, C was used to extend **B** and, as before, is satisfied by definition of $I_\mathbf{B}$.

By 1. $J = \bigcup_{C \in S'} f(C)$ is a well-defined partial interpretation which is trivially also a model of S'.

By 2. $I_\mathbf{B} \cup J$ is a well-defined interpretation and thus a model of S, because $I_\mathbf{B}$ is a model of $S - S'$ and J is a model of S'. $\qquad\square$

This proof can be adapted to formulas in negation normal form (as done for A-ordered tableaux in [8]) or even to arbitrary first order formulas, see [15] for details. Likewise, the proof works as well for infinite sets of ground clauses provided that the clauses are selected in a fair manner for extension.

It is also easy to see that the proof goes through unaltered for a generalized procedure in which before each restart or extension step the selection function f can be changed arbitrarily.

In the case of A-orderings the proof can be made even shorter by making use of a result by Bachmair & Ganzinger [1] that any set of clauses which is saturated wrt ordered resolution and does not contain the empty clause is satisfiable. If this is assumed then the model building part in our proof can be omitted and it simply remains to show that if **B** is a saturated ordered open tableau branch for S, then $\mathbf{B} \cup S$ is saturated wrt ordered resolution, an observation due to Bachmair (personal communication).

We refrain from using the latter insight, because the results in [1] (i) are only for the clausal case, whereas our approach can be generalized to non-normal-form [15], and (ii) are only for the theory of equality, whereas we employ other theories as well.

5 Lifting

As usual in semantic tableaux, there are (at least) two different ways for lifting ground tableaux with selection function to first order logic.

One can enforce a fair selection of all the ground instances $\hat{S}$ of a first order clause set S to build a (possibly infinite) ground tableau with selection function for $\hat{S}$ which in turn gives a first-order tableau with selection function for S. Completeness of this calculus follows directly from Herbrand's Theorem and Theorem 9. This gives a calculus in the spirit of Smullyan [17].

For efficiency reasons we favor the so-called *free variable* approach (cf., for example, [7]): rather than guessing the "right" instantiation of a universally quantified formula, one uses free variables and unification to search for a closing substitution. For this purpose we generalize the notions of connection and of selection function.

Definition 10. A clause C has a **connection** with a clause C' iff there exist literals $L \in C$, $L' \in C'$, such that $L\sigma = \overline{L'}\sigma$ with mgu σ. L and L' are called **connection literals**.

Given a selection function f, C and C' have a **connection via** f iff they have a connection with literals L and L' and mgu σ such that $L\sigma \in f(C\sigma)$ and $L'\sigma \in f(C'\sigma)$. If, moreover, $C, C' \in S$ we say that C is a **restart clause** in S (with connection literal L and mgu σ).

By Herbrand's Theorem we know that for any unsatisfiable set S of first order clauses there is a finite unsatisfiable set $\hat{S}$ of ground instances of S. For $\hat{S}$ there is a closed ground tableau with selection function by Theorem 9. It is straightforward to lift this tableau to a closed first order tableau provided that the selection function used for the latter has the following property:

Definition 11. A selection function f is **stable wrt substitutions** iff $L\sigma \in f(C\sigma)$ implies $L \in f(C)$ for all substitutions σ and clauses C.

Example 4. The functions f_{last} and f_{first} defined in Example 2 are stable wrt substitutions.

Theorem 12. *For any unsatisfiable set S of first order clauses and selection function f which is stable wrt substitutions exists a closed tableau with selection function f for S.*

In general it is not possible to extend a deterministic selection function on ground clauses to a deterministic selection function on first order clauses which is stable wrt substitutions as the following simple example shows (one can obtain, however, a slightly different notion of selection function by using literal *occurrences* in the definition of stability which admits deterministic extension):

Example 5. Consider any deterministic selection function f and the first order clause $C = p(x) \vee p(f(y))$. For the substitution $\sigma = \{x \leftarrow f(a), y \leftarrow a\}$ obviously $f(C\sigma) = \{p(x)\sigma, p(f(y))\sigma\} = \{p(f(a))\}$ holds. Then every extension $\hat{f}$ of f to first order clauses must, by the substitutivity condition (Def. 11), select both literals $p(x)$ and $p(f(y))$. Therefore $\hat{f}$ is not deterministic.

Table 1 shows a first order proof calculus based on Definition 7. The additional set $\mathfrak{C}$ is a constraint which guarantees that every ground instance of the resulting tableau is also a ground tableau with selection function. More precisely, $\mathfrak{C}$ is a set of literal/clause pairs $\langle L, C \rangle$ which record the so far selected literals so one can check whether the substitution associated with a connection respects the selection function. Accordingly, one defines $\langle L, C \rangle \sigma = \langle L\sigma, C\sigma \rangle$ and $\mathfrak{C}$ to be **satisfiable** iff $L \in f(C)$ for all $\langle L, C \rangle \in \mathfrak{C}$. The idea of using constraints to express global restrictions on tableau search is due to [10], where constraints were used to enforce regularity. To increase readability these regularity constraints are omitted in Table 1.

$$\frac{\mathfrak{T} \cup \{\mathbf{B}\} \parallel \mathfrak{C}}{(\mathfrak{T} \cup \bigcup\limits_{\substack{L \neq L' \\ L \in C}} \{\mathbf{B} \cup \{L\theta\}\})\sigma \parallel (\mathfrak{C} \cup \{\langle L', C \rangle \theta\})\sigma}$$

where C has a weak connection via f into $\mathbf{B}$ with connection literal L' and mgu σ, $\mathfrak{C}\sigma$ is satisfiable, and θ is a variable renaming of C.

$$\frac{\mathfrak{T} \cup \{\mathbf{B}\} \parallel \mathfrak{C}}{(\mathfrak{T} \cup \bigcup\limits_{L \in C} \{\mathbf{B} \cup \{L\theta\}\})\sigma \parallel (\mathfrak{C} \cup \{\langle L', C \rangle \theta\})\sigma}$$

where C is a restart clause in S with connection literal L' and mgu σ, $\mathfrak{C}\sigma$ is satisfiable, and θ is a variable renaming of C.

$$\frac{\mathfrak{T} \cup \{\mathbf{B}\} \parallel \mathfrak{C}}{(\mathfrak{T} - \{\mathbf{B}\})\sigma \parallel \mathfrak{C}\sigma}$$

if σ is an mgu of $\{L, \overline{L'}\} \subseteq \mathbf{B}$ and $\mathfrak{C}\sigma$ is satisfiable.

Table 1. First order tableaux with selection function

Checking validity of a constraint $\mathfrak{C}$ can be expensive. For example, the total A-ordering based on the lexicographical path ordering (LPO) leads to a satisfiability test for $\mathfrak{C}$ which is NP-complete [13].

A compromise is to simply omit the constraints and apply the selection function on uninstantiated clauses. The resulting calculus is still complete, but somewhat less restrictive. It corresponds to the rules in Table 1 if constraints and the conditions imposed on them are removed. In other words, one constructs a tableau in which each extension step *at the time when it is performed* is restricted by a selection function f, but this is not necessarily the case for the final closed tableau.

Such an intermediate calculus is possible, because in tableau calculi each clause used for an extension step, by definition, is an *input clause*. Let us, therefore, call the resulting tableau calculi—in analogy to input resolution—**tableaux with input selection function.**

It turns out that for tableaux with input selection function substitutive selection functions are not required: below we show completeness when the selection function merely is stable wrt variable renaming:

Definition 13. A selection function f is **stable wrt variable renaming** iff $L \in f(C)$ implies $L\theta \in f(C\theta)$ for all variable renamings θ and clauses C.

Obviously, each selection function stable wrt substitutions is also a selection function stable wrt variable renaming. Moreover, the inclusion is proper:

Example 6. Let $f_\#$ be the function which selects the first literal (from the left) among the literals containing a maximal number of (constant and function) symbols.

It is easy to show that $f_\#$ is stable wrt variable renaming. $f_\#$ is not stable wrt substitutions: let $C = P(f(x)) \vee Q(y)$, then $f_\#(C) = P(f(x))$, but $f_\#(C\{y \leftarrow f(f(u))\}) = Q(f(f(u)))$.

Obviously, one has much greater flexibility in the choice of a suitable selection function for a given problem in this larger class.

For the following completeness proof we have to modify some of our notions. A ground clause is considered to be a sequence of disjunctive connected *indexed literals* $L : I$, where the additional index I can be chosen from any set. It is straightforward to adapt the definition of ground tableau with selection function and the proof of Theorem 9 to indexed clauses.

Theorem 14. *Given an unsatisfiable set S of first order clauses and a selection function f which is stable wrt variable renaming. Then there exists a closed tableau with input selection function f for S.*

For the proof of this theorem we need the following technical lemma. Its easy proof is based on the fact that for unifiable terms idempotent mgus always exist.

Lemma 15. *Given a tableau $\mathfrak{T}$ and a ground substitution ν for $\mathfrak{T}$ such that $\mathfrak{T}\nu$ is closed. Let $\mathbf{B}$ be a branch of $\mathfrak{T}$ closed by two complementary literals $L\nu$ and $L'\nu$ and let σ be the mgu of $L, \overline{L'} \in \mathbf{B}$.*

Then $\mathfrak{T}\sigma\nu = \mathfrak{T}\nu$, and hence: $\mathfrak{T}$ can be closed by applying the mgus of all complementary literals in arbitrary order.

Proof of Theorem 14. By Herbrand's Theorem there exists a finite unsatisfiable set $\widehat{S}$ of ground instances of S.

From S and $\widehat{S}$ we construct the set $\widehat{S}{:}S$ of *indexed* ground clauses as follows: for each $(L_1 \vee \cdots \vee L_n)\sigma \in \widehat{S}$ with $L_1 \vee \cdots \vee L_n \in S$ let $L_1\sigma{:}L_1\theta \vee \cdots \vee L_n\sigma{:}L_n\theta$ be in $\widehat{S} : S$, where θ is a renaming of the variables in $L_1 \vee \cdots \vee L_n$.

We extend f from S to $\widehat{S}{:}S$ by stipulating $\widehat{L}{:}L \in \widehat{f}(\widehat{C}{:}C)$ iff $L \in f(C)$.

$\widehat{f}$ is well-defined, because f is stable wrt variable renaming. $\widehat{S}{:}S$ is a set of indexed ground clauses. From Theorem 9 we know that a closed tableau $\widehat{\mathfrak{T}}$ with selection function $\widehat{f}$ for $\widehat{S}{:}S$ exists.

From $\widehat{\mathfrak{T}}$ we build a tableau $\mathfrak{T}$ with selection function f for S:

1. Obtain $\mathfrak{T}$ from $\widehat{\mathfrak{T}}$ by replacing each literal $\widehat{L} : L$ in $\widehat{\mathfrak{T}}$ by L.
2. If two complementary literals $\widehat{L} : L$ and $\widehat{L'}{:}L'$ close a branch in $\widehat{\mathfrak{T}}$ with mgu σ, then apply σ to $\mathfrak{T}$. Lemma 15 guarantees that the corresponding branch in $\mathfrak{T}\sigma$ is closed, too.

The result is a closed tableau that can be constructed with the rules of Table 1, neglecting constraints. Thus it is a tableau with input selection function f.

□

6 Extensions and Related Calculi

6.1 Theory Connections

With theory reasoning we mean a general method to integrate theories such as the equality theory (efficiently) into deductive systems. Various sound and complete methods have been developed for several calculi including resolution, the connection method, and tableaux. All of the more efficient methods for tableau-like calculi share two main ideas: (i) an extension of the notion of connection and connection unifiers to theory connections and theory unifiers and (ii) the partition of the automated deduction system into a general purpose foreground reasoner and a theory-specific background reasoner.

In the case of **total theory reasoning** the background reasoner calculates theory unifiers for a given input set of formulas given by the foreground reasoner, whereas in **partial theory reasoning** the background reasoner in addition derives new formulas, so-called *residues*, that have to be used by the foreground reasoner. In our context we are only interested in total theory reasoning and we give only a very brief overview of the necessary constituents of theory reasoning. [4] is a detailed survey of theory reasoning in tableau calculi.

In general, every satisfiable set of first order clauses defines a theory $\mathcal{T}$. The notions and semantics of an interpretation, the satisfiability of formulas, etc. are restricted relative to a given theory.

Definition 16. A **theory** $\mathcal{T}$ is a satisfiable set of first order clauses.

An interpretation I is a $\mathcal{T}$-**interpretation** iff I satisfies $\mathcal{T}$.

A formula ϕ is $\mathcal{T}$-**satisfiable** iff ϕ is satisfied by a $\mathcal{T}$-interpretation; ϕ is $\mathcal{T}$-**unsatisfiable** otherwise.

Example 7. The clause set $\mathcal{O} = \{\neg x < x,\ \neg x < y \vee \neg y < z \vee x < z\}$ defines the theory of strict orderings.

For a given signature Σ the equality theory $\mathcal{E}_\Sigma$ can be defined by the axioms of reflexivity, symmetry, transitivity, and monotonicity for function and predicate symbols of the equality predicate $\approx$. $\mathcal{E}_\Sigma$ is finite iff Σ is.

The notions of complementary literal and connection unifier are relativized to theories as well.

Definition 17. Let **B** be a set of literals. **B** is $\mathcal{T}$-**complementary** iff the existential closure of the conjunction of the members of **B** is $\mathcal{T}$-unsatisfiable.

A substitution σ is a $\mathcal{T}$-**unifier** for **B** iff **B**σ is $\mathcal{T}$-complementary. If, moreover, **B**σ is a minimal set of $\mathcal{T}$-complementary literals, i.e. no proper subset of it is $\mathcal{T}$-complementary, then **B** is a $\mathcal{T}$-**connection** and σ is a $\mathcal{T}$-**connection unifier for B**. As usual, a $\mathcal{T}$-**mgu** is a most general $\mathcal{T}$-unifier.

The minimality condition for theory connections is due to [2], where a connection calculus with theories is presented for the first time. In [16] $\mathcal{T}$-connections are used without minimality condition and the rôle of $\mathcal{T}$-connections is taken by a so-called set of complete $\mathcal{T}$-connections. The latter gives more flexible control over the $\mathcal{T}$-unifiers that have to be calculated to obtain a complete calculus. This can be of importance in theories where decidability is not guaranteed for $\mathcal{T}$-connections but for complete sets of $\mathcal{T}$-connections. Our notion of $\mathcal{T}$-connections has the advantage that it leads straightforwardly to a theory version of tableaux with selection function. It can be adapted easily to the terminology of [16].

Example 8. In the equality theory $\mathcal{E}$ the literal set $\mathbf{B} = \{a{\approx}b,\ p(a),\ \neg p(b),\ p(x)\}$ is $\mathcal{E}$-complementary, but it is no $\mathcal{E}$-connection. The substitution $\sigma = \{x \leftarrow a\}$ is an $\mathcal{E}$-unifier for $\mathbf{B}' = \{a \approx b,\ \neg p(b),\ p(x)\}$ which is an $\mathcal{E}$-connection.

If $\mathcal{T} = \emptyset$, then the standard notions of complementary literals, connection, and connection unifier are instances of the above definitions.

To build theories into tableaux with selection function we must modify our definition of connection via a selection function:

Definition 18. A set S of instances of first order clauses has a $\mathcal{T}$-**connection via** f iff there is a $\mathcal{T}$-connection $\mathbf{B}$ with unifier σ such that $L \in f(C)$ with $C \in S$ for all $L \in \mathbf{B}$.

Depending on the given theory, more than two clauses may be involved in a connection via f. Instead of defining $\mathcal{T}$-restart clauses directly, we generalize the notion of a connection into a tableau branch, i.e., into a set of literals.

Definition 19. Given a set of literals $\mathbf{B}$, a selection function f, a theory $\mathcal{T}$, and a set S of first order clauses, then a clause $C \in S$ has a $\mathcal{T}$-**connection into** $\mathbf{B}$ iff there is a set $S' \subseteq \{C | C$ is an instance of a clause in S or $C \in \mathbf{B}\}$ with $C \in S'$ such that S' has a $\mathcal{T}$-connection via f.

$f(C)$ is a **connection literal** of S'.

If S' contains only clauses of S then every $C \in S'$ is (an instance of) a **restart clause** and if $S' \cap \mathbf{B} \neq \emptyset$ then every $C \in (S' - \mathbf{B})$ has a **weak connection** into $\mathbf{B}$.

Table 2 shows $\mathcal{T}$-tableaux with selection function. The main difference to the previous version is the uniform handling of extension steps and, as a consequence, there is no closure involved in any extension step. In theory reasoning it does not make sense to separate out the case, where an immediate closure is possible, because in general a tableau branch may have to be extended more than once before any new open branch can be closed with a theory unifier.

Theorem 20. *Given a theory $\mathcal{T}$, a $\mathcal{T}$-unsatisfiable set S of first order clauses and a selection function f which is stable wrt substitutions. Then there exists a closed $\mathcal{T}$-tableau with selection function f for S.*

It is possible to modify the previous completeness proofs. Lifting to first order logic is possible with the help of a theory version of Herbrand's Theorem [15, 16].

$$\frac{\mathfrak{T} \cup \{\mathbf{B}\} \parallel \mathfrak{C}}{(\mathfrak{T} \cup \bigcup_{L \in C} \{\mathbf{B} \cup \{L\theta\}\})\sigma \parallel (\mathfrak{C} \cup \{\langle L', C \rangle \theta\})\sigma}$$

if C has a $\mathcal{T}$-connection via f into $\mathbf{B}$ with connection literal L' and $\mathcal{T}$-mgu σ, $\mathfrak{C}\sigma$ is satisfiable, and θ is a variable renaming of C.

$$\frac{\mathfrak{T} \cup \{\mathbf{B}\} \parallel \mathfrak{C}}{(\mathfrak{T} - \{\mathbf{B}\})\sigma \parallel \mathfrak{C}\sigma}$$

if there is a $\mathcal{T}$-connection $\mathbf{B}' \subseteq \mathbf{B}$ with mgu σ and $\mathfrak{C}\sigma$ is satisfiable.

Table 2. $\mathcal{T}$-tableaux with selection function

The background reasoner is used for two tasks in this calculus: First, it is used to calculate theory connections and, second, it is used to check for closure. Both can be done by calculating theory unifiers.

The calculation of theory unifiers can be very expensive and even undecidable. Without any restriction one has to consider each set $\mathbf{B}$ with $C \in S$ for all $L \in \mathbf{B}$. Therefore, a restriction on certain literals as, for example, in tableaux with selection function, is extremely useful.

6.2 Restart Model Elimination

In [3] a modification of Model Elimination [11] is introduced which bears some similarities to tableaux with selection function. This calculus, called Restart Model Elimination, also uses a selection function f to restrict connections to those clauses which contain literals selected by f.

In contrast to tableaux with selection function, (i) only *positive* literals of a clause are selected by f and are considered as connection literals and (ii) a clause can extend a branch $\mathbf{B}$ only if it has a **strong connection via f into $\mathbf{B}$**, i.e. if it has a connection to the *leaf*-literal of $\mathbf{B}$. Furthermore strong connections are restricted to *negative* leaf literals.

Like in tableaux with selection function [8] imposing the strong connection condition leads to an incomplete calculus. To avoid this negative clauses are admitted for restarts.

The strong connection condition also implies that regularity in restart model elimination has to be relaxed in order to obtain a complete calculus. A branch $\mathbf{B}$ needs only be regular wrt all *positive* literals and within each *block* of $\mathbf{B}$, where a block is defined as a sequence of literals in $\mathbf{B}$ which lies totally between two subsequent applications of the restart rule.

The completeness proof for restart model elimination is based on the literal-excess method. Unfortunately, it gives no clue on how the calculus can be modified such that the selection function does yield negative literals as well.

7 Implementation and Application

$\mathcal{T}$-Tableaux with (input) selection function were implemented in $_3\mathcal{T}^A\!P$ [5], an automated theorem prover for sorted multiple-valued full first order logic with equality.

(Equality) theory reasoning is used to close a branch if the problem contains equalities, but is not used to calculate theory connections. Instead, the test is performed with a weaker condition (giving more candidates for theory connections than necessary) that can be computed in polynomial time [18]. This latter method is, of course, restricted to selected literals.

Completely unrestricted extensions of tableau branches with restart clauses is problematic, because such clauses might be totally independent from the branch on which they are used. It turns out that a good heuristic for avoiding extensions with potentially useless restart clauses is to prefer such restart clauses that have a connection, not necessarily via the selection function, into the current branch.

$_3\mathcal{T}^A\!P$ uses a further technique due to [14] to avoid redundant extension steps in hindsight: if a subtableau below a node n can be closed without using the literal of this node, then all open branches produced by *ClauseOf(Predecessor(n))* are *pruned*, i.e. closed immediately.

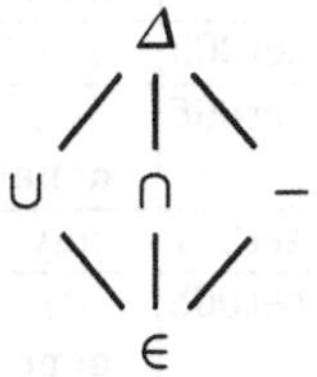

$$\Delta$$
$$/\ |\ \backslash$$
$$\cup \quad \cap \quad -$$
$$\backslash\ |\ /$$
$$\in$$

Fig. 2. Hierarchy of predicate symbols

We have found that tableaux with selection function work quite well when the formulas of a problem to be solved are in some sense hierarchical. Mathematical theories, for example, are based on definitions which themselves are based on definitions of more primitive notions. Most of these definitions introduce new predicate or function symbols. Thus, these symbols form a hierarchy.

Let us consider a first order axiomatization of a fragment of set theory. Symmetrical difference Δ between two sets can be defined by the union $\cup$, intersection $\cap$, and set difference $-$, all of which are based on the membership predicate $\in$.

Fig. 2 shows the hierarchy induced by these definitions. Formally, such a hierarchy is an irreflexive, partial ordering $<_H$ on predicate, function and constant symbols. In our example we have $\in <_H \cup, \cap, - <_H \Delta$.

Definition 21. Let $<$ be any irreflexive, partial ordering on function and constant symbols. Then for any terms $s = f(s_1, \ldots, s_n)$, $t = g(t_1, \ldots, t_m)$ we define $s \prec_L t$ to hold iff both of the following conditions hold:

1. $f < g$ or $f = g$, $n = m$, and $t_i \prec_L s_i$ for $i = 1, \ldots, n$;
2. the variables of s are a subset of the variables of t.

$\prec_L$ can be extended to an A-$(L$-$)$ordering in an obvious way provided that $<$ is an ordering on predicate symbols as well.

In the set theory example the ordering $<_H$ gives a good correspondence between the literals ordered by $\prec_L$ and the hierarchy of set theory. $\prec_L$ often prevents literals from unrelated hierarchies to be unifiable and, as a consequence, a clause C has no $\prec_L$-connection into a branch which only contains literals from a hierarchy that is not reachable from a hierarchy of the maximal literals of C.

Note that even if the ordering does not perfectly reflect the hierarchy within a problem (or if there is no real hierarchy), completeness of the calculus still guarantees a proof can be found, though proof search might not be influenced as favorably.

name	ordering	R	CB	$_3\mathcal{T}^A\mathcal{P}$[s]	Otter[s]
set001	$\prec_L$	30	11	1.18	0.06
	none	10	4	0.12	
set002	$\prec_L$	546	206	27.82	2.32
	none			> 500	
set003	$\prec_L$	12	5	0.26	0.08
	none	48	8	0.66	
set004	$\prec_L$	12	5	6.12	0.05
	none	48	8	0.72	
set005	any			> 500	> 300
set006	$\prec_L$	650	40	52.15	0.05
	none			> 500	
set007	any			> 500	> 300
set008	$\prec_L$	1406	536	102.29	0.96
	none			> 500	
set009	$\prec_L$	22	10	0.27	0.47
	none	22	10	0.17	

Fig. 3. Results for TPTP-problem SET

Fig. 3 shows some results for a few problems from the TPTP problem class SET [19]. The calculus used are tableaux with input selection function. None of the problems is formulated with equality.

As the time needed for refutation is only a rough measure of the complexity of a proof, we give the number of rule applications (R)[4] and of branches (CB) in the closed tableau. Lines corresponding to the smallest proof are darkly shaded. The final column shows the times Otter [12] needs for a proof.[5] Otter is usually faster, but the proofs it generates are typically longer, sometimes drastically. Also the basic speed of the experimental system $_3\mathcal{T}^A\mathcal{P}$ is slower than that of Otter by a factor of several hundred.

8 Conclusion

We introduced tableaux with selection function which are a generalization of A-ordered tableaux. In addition we showed that lifting is possible with respect to a large class of selection functions which makes it easy to find a suitable selection function for certain problems.

It turns out that theories can be build into tableaux with selection function using the concepts of theory connection and theory reasoning.

Our first experiments with an implementation of this calculus show promising results with problems that have a hierarchical structure.

In [15] tableaux with selection function are extended to the non-clausal case and also an implementation with constraints is discussed.

Acknowledgements

We owe Peter Baumgartner and an anonymous referee many useful comments on a previous version of this paper.

---

[4] In $_3\mathcal{T}^A\mathcal{P}$ clauses are quantified explicitly. To extend a tableau branch with a clause several γ and β-rule applications have to be made. Because of this, R is usually several times larger than the actual number of extension steps.

[5] The figures were obtained with Otter version 3.0.4 in June 1996. They are available via ftp://info.mcs.anl.gov/pub/Otter/www-misc/otter304-tptp121.txt.gz. Both systems were run on roughly comparable machines.

References

1. L. Bachmair and H. Ganzinger. Rewrite-based equational theorem proving with selection and simplification. *J. of Logic and Computation*, 4(3):217–247, 1994.
2. P. Baumgartner. A Model Elimination Calculus with Built-in Theories. In H.-J. Ohlbach, editor, *Proceedings of the 16-th German AI-Conference (GWAI-92)*, pages 30–42. Springer-Verlag, 1992. LNAI 671.
3. P. Baumgartner and U. Furbach. Model Elimination without Contrapositives and its Application to PTTP. *J. of Automated Reasoning*, 13:339–359, 1994.
4. B. Beckert. Equality and other theory inferences. In M. D'Agostino, D. Gabbay, R. Hähnle, and J. Posegga, editors, *Handbook of Tableau Methods*. Kluwer, Dordrecht, to appear, 1997.
5. B. Beckert, R. Hähnle, P. Oel, and M. Sulzmann. The tableau-based theorem prover $_3T^AP$, version 4.0. In *Proceedings, 13th International Conf. on Automated Deduction (CADE), New Brunswick, NJ, USA*, LNCS 1104, pages 303–307. Springer, 1996.
6. C. Fermüller, A. Leitsch, T. Tammet, and N. Zamov. *Resolution Methods for the Decision Problem*. LNAI 679. Springer-Verlag, 1993.
7. M. C. Fitting. *First-Order Logic and Automated Theorem Proving*. Springer-Verlag, New York, second edition, 1996.
8. R. Hähnle and S. Klingenbeck. A-ordered tableaux. *J. of Logic and Computation*, 6(6):819–834, 1996.
9. S. Klingenbeck and R. Hähnle. Semantic tableaux with ordering restrictions. In A. Bundy, editor, *Proc. 12th Conf. on Automated Deduction CADE, Nancy*, volume 814 of *LNCS*, pages 708–722. Springer-Verlag, 1994.
10. R. Letz, J. Schumann, S. Bayerl, and W. Bibel. SETHEO: A high-perfomance theorem prover. *J. of Automated Reasoning*, 8(2):183–212, 1992.
11. D. Loveland. Mechanical theorem proving by model elimination. *Journal of the ACM*, 15(2), 1968.
12. W. W. McCune. *OTTER 3.0 Reference Manual and Guide*. Argonne National Laboratory/IL, USA, 1994.
13. R. Nieuwenhuis. Simple LPO constraint solving methods. *Information Processing Letters*, 47:65–69, 1993.
14. F. Oppacher and E. Suen. HARP: A tableau-based theorem prover. *J. of Automated Reasoning*, 4:69–100, 1988.
15. C. Pape. Vergleich und Analyse von Ordnungseinschränkungen für freie Variablen Tableau. Diplomarbeit, Fakultät für Informatik, Universität Karlsruhe, May 1996. In German. Available via anonymous ftp to ftp.ira.uka.de under pub/uni-karlsruhe/papers/techreports/1996/1996-30.ps.gz.
16. U. Petermann. How to build in an open theory into connection calculi. *Journal on Computers and Artificial Intelligence*, 2:105–142, 1992.
17. R. M. Smullyan. *First-Order Logic*. Dover Publications, New York, second corrected edition, 1995. First published 1968 by Springer-Verlag.
18. M. Sulzmann. Ausnutzung von KIV-Spezifikationen in $_3T^AP$. Studienarbeit, Fakultät für Informatik, Universität Karlsruhe, July 1995.
19. G. Sutcliffe, C. Suttner, and T. Yemenis. The TPTP problem library. In A. Bundy, editor, *Proc. 12th Conf. on Automated Deduction CADE, Nancy*, LNAI, pages 252–266. Springer, 1994.

Two Loop Detection Mechanisms: A Comparison

Jacob M. Howe

Computer Science Division
University of St Andrews, Scotland, KY16 9SS
jacob@dcs.st-and.ac.uk

Abstract. In order to compare two loop detection mechanisms we describe two calculi for theorem proving in intuitionistic propositional logic. We call them both MJ^{Hist}, and distinguish between them by description as 'Swiss' or 'Scottish'. These calculi combine in different ways the ideas on focused proof search of Herbelin and Dyckhoff & Pinto with the work of Heuerding *et al* on loop detection. The Scottish calculus detects loops earlier than the Swiss calculus but at the expense of modest extra storage in the history. A comparison of the two approaches is then given, both on a theoretic and on an implementational level.

1 Introduction

The main interest of this paper is the comparison of the two loop detection mechanisms described below. In order to do this we illustrate their use on the permutation-free sequent calculus MJ for the propositional fragment of intuitionistic logic. This gives calculi whose implementations are suitable for theorem proving.

Backwards proof search and theorem proving with a standard cut-free sequent calculus, Gentzen's LJ, for the propositional fragment of intuitionistic logic is inefficient because of three problems. Firstly, the proof search is not in general terminating, due to the possibility of looping. Secondly, it will produce proofs which are essentially the same; they are permutations of each other, and correspond to the same natural deduction. Thirdly, there are choice points where it has to be decided which of several rules to apply and where to apply them.

The sequent calculus MJ for intuitionistic logic was introduced (with another name, LJT) by Herbelin in [7]. The propositional fragment of the calculus MJ is displayed in Figure 1. This uses Girard's idea of a special place for formulae in the antecedent, the stoup first seen in [6]. The calculus was developed by Dyckhoff and Pinto [3] because it has the property that proofs are in 1–1 correspondence with the normal natural deductions of NJ. MJ is a permutation-free sequent calculus; it avoids the problems of permutations in the cut-free sequent calculus of Gentzen. This removes the second of the problems. In this paper we are more interested in theorem proving than in proof search, hence the second problem is not directly relevant. But notice that permutations are avoided in MJ by a focusing method — several choice points are removed. That is, MJ partly addresses the third problem and hence is advantageous as a calculus for theorem proving. However, the naïve implementation of MJ will lead to the possibility of looping.

Looping may easily be removed by checking whether a sequent has already occurred in a branch. Implementation of this is inefficient as it requires much information to be stored. Recent work by Heuerding *et al* [9] (the intuitionistic case of which is closely related to that of Gabbay in [5]) shows how to use a 'history' to prevent looping in a far more efficient way.

In this paper the history mechanism is developed in two ways and applied to MJ. Both the resulting calculi have advantages and disadvantages. These are discussed theoretically and also pragmatically (in terms of the speed with which Prolog implementations give proofs). We call the new calculus MJ^{Hist}, the two varieties 'Swiss' and 'Scottish'.

$$\frac{A, \Gamma \Rightarrow B}{\Gamma \Rightarrow A \supset B} \; (\supset_{\mathcal{R}}) \qquad \frac{\Gamma \Rightarrow A \quad \Gamma \xrightarrow{B} C}{\Gamma \xrightarrow{A \supset B} C} \; (\supset_{\mathcal{L}})$$

$$\frac{\Gamma \xrightarrow{A} C}{\Gamma \xrightarrow{A \wedge B} C} \; (\wedge_{\mathcal{L}1}) \qquad \frac{\Gamma \xrightarrow{B} C}{\Gamma \xrightarrow{A \wedge B} C} \; (\wedge_{\mathcal{L}2})$$

$$\frac{\Gamma \Rightarrow A \quad \Gamma \Rightarrow B}{\Gamma \Rightarrow A \wedge B} \; (\wedge_{\mathcal{R}}) \qquad \frac{A, \Gamma \Rightarrow C \quad B, \Gamma \Rightarrow C}{\Gamma \xrightarrow{A \vee B} C} \; (\vee_{\mathcal{L}})$$

$$\frac{\Gamma \Rightarrow A}{\Gamma \Rightarrow A \vee B} \; (\vee_{\mathcal{R}1}) \qquad \frac{\Gamma \Rightarrow B}{\Gamma \Rightarrow A \vee B} \; (\vee_{\mathcal{R}2})$$

$$\frac{A, \Gamma \xrightarrow{A} B}{A, \Gamma \Rightarrow B} \; (C) \qquad \frac{}{\Gamma \xrightarrow{A} A} \; (ax) \qquad \frac{}{\Gamma \xrightarrow{\perp} A} \; (\perp)$$

Define $\neg A \equiv A \supset \perp$.
A, B, C are formulae. Γ is a multiset of formulae.
B, Γ is shorthand for $\{B\} \cup \Gamma$, where $\cup$ is multiset union.
Sequent $\Gamma \Rightarrow C$ has context Γ, goal C and no stoup.
Sequent $\Gamma \xrightarrow{A} C$ has context Γ, goal C and a single formula, A, in the stoup.

Fig. 1. Propositional Fragment of the calculus MJ.

2 Calculi With Histories

In this section we first discuss the idea of the history mechanism, and then describe the two calculi. We shall conclude with a comparison of the two calculi.

2.1 The Use of Histories to Prevent Looping

Looping can very easily occur in MJ, for example:

$$\frac{\begin{array}{cc}\vdots & \vdots\\(p \wedge p) \supset p \Rightarrow p & (p \wedge p) \supset p \Rightarrow p\end{array}}{(p \wedge p) \supset p \Rightarrow p \wedge p} \; (\wedge_{\mathcal{R}}) \qquad \frac{}{(p \wedge p) \supset p \xrightarrow{p} p} \; (ax)$$

$$\frac{(p \wedge p) \supset p \xrightarrow{(p \wedge p) \supset p} p}{(p \wedge p) \supset p \Rightarrow p} \; (C) \qquad (\supset_{\mathcal{L}})$$

The sequent $(p \wedge p) \supset p \Rightarrow p$ may continue to occur in the proof tree for this sequent using the MJ calculus. We can see that there is a loop: we need a mechanical way to detect such loops.

One way to do this is to add a *history* to each sequent. The history is the set of all sequents that have occurred so far on the branch of a proof tree. After each backwards inference the new sequent (without its history) is checked to see whether it is a member of this set. If it is we have looping and backtrack. If not the new history is the extension of the old history by the old sequent (without its history), and we try to prove the new sequent, and so on. Unfortunately this method is inefficient as it requires long lists of sequents to be stored by the computer, and all of this list has to be checked at each stage. When the sequents are stored we are keeping far more information than is necessary. Efficiency would be improved by cutting down the amount of storage and checks to the bare minimum needed to prevent looping.

The basis of the reduced history is the realisation (as in [9]) that one need only store goal formulae in order to loop-check. The rules of MJ are such that the context cannot decrease; once a formula is in the context it will be in the context of all sequents above it in the proof tree. For two sequents to be the same they obviously need to have the same context. Therefore we may empty the history every time the context is extended. All we need store in the history are goal formulae. If we have a sequent whose goal is already in the history, then we have the same goal and the same context as another sequent, that is, a loop.

There are two slightly different ways of doing this. There is the straightforward extension and modification of the calculus described in [9] (which we call a 'Swiss history'). The other approach involves storing slightly more formulae in the history, but detects loops more quickly. This we describe as the 'Scottish history'; it can, in many cases, be much more efficient than the Swiss method.

2.2 The Swiss History

Before continuing, we should point out that the calculus we describe here as Swiss is significantly different from the one in [9]. This is partly due to our use of MJ as a base calculus, and partly because we are trying to focus on the history mechanism, hence we have not included the subsumption checks that the calculus in [9] uses.

The Swiss-style calculus MJ^{Hist} is displayed in Figure 2. Let us make some general points about it (which will apply to the Scottish MJ^{Hist} as well). We have given explicit rules for negation (which are just special cases of the rules for implication) for the sake of completeness of connectives. Also, notice that there are two rules for $(\supset_{\mathcal{R}})$.

These correspond to the two cases where the new formula is or is not in the context. As noted above (in §2.1) this is very important for MJ^{Hist}. Also note that the number of formulae in the history is at most equal to the length of the formula we check for provability.

The loop checking due to the history in the calculus works in a similar way to that of $\mathsf{IPC}^{RP}{}_{\wedge, \rightarrow}{}^{SU}$ in [9]. A sequent is matched against first the conclusions of right rules until the goal formula is either a propositional variable, falsum, or a disjunction (note that disjunction isn't covered in [9], and requires special treatment). This is ensured by condition $\star$ on rule (C). Then a formula from the context is selected and placed in the stoup by the (C) rule, the sequent is then matched against stoup formulae of left rules (this focusing does not occur in [9]). The history mechanism is used to prevent looping in the $(\supset_{\mathcal{L}})$ rule (and similarly in the $(\neg_{\mathcal{L}})$ rule). The left premiss of the rule has the same context as the conclusion, but the goal is generally different. If the goal, C, of the conclusion is not in the history, $\mathcal{H}$, we store C in $\mathcal{H}$ and continue backwards proof search on the left premiss. Alternatively, C might already be in $\mathcal{H}$. In this case there is a loop, and so this branch is not pursued. We backtrack and look for a proof in a different way.

There is another place where the rules are restricted in order to prevent looping. This is the condition placed on the $(\vee_{\mathcal{L}})$ rule. For the $(\supset_{\mathcal{R}})$ rule (which attempts to extend the context) there are two cases corresponding to when the context is and when it is not extended. Something similar is happening in the $(\vee_{\mathcal{L}})$ rule. In both the premisses of the rule a formula may be added to the context. If both contexts really are extended, then we continue building the proof tree. If one or both contexts are not extended then the sequent with the non-extended context, S, will be the same as some sequent at a lesser height in the proof tree — there is a loop. This is easy to see: since the context and goal of S are the same as that of the conclusion, the sequent before the stoup formula (or a formula containing it as a subformula) was selected into the stoup must be the same as S.

We now state the equivalence theorem. This is done in two stages.

Theorem 1 *The calculi MJ and MJ^{Hist} (without $\star$) are equivalent. That is, a sequent S is provable in MJ if and only if $S; \phi$ (the sequent with the empty history) is provable in MJ^{Hist} (without $\star$).*

PROOF:(Sketch) The $\Leftarrow$ direction is straightforward.

To prove the $\Rightarrow$ direction we take an MJ proof tree and use it to build an MJ^{Hist} proof tree.

We start at the root, $\Gamma \Rightarrow A$ in MJ and we have root $\Gamma \Rightarrow A; \{A\}$ in MJ^{Hist}.

Given a fragment of MJ proof tree with corresponding fragment of MJ^{Hist} proof tree, we look at the next inference in the MJ tree. We have a recipe which we can use to build a fragment of MJ^{Hist} proof tree corresponding to a strictly larger fragment of the MJ proof tree.

As proof trees are finite, this process must be terminating.

For full details see [10]. ∎

$$\frac{A, \Gamma \Rightarrow B; \phi}{\Gamma \Rightarrow A \supset B; \mathcal{H}} \; (\supset_{\mathcal{R}1}) \quad \text{if } A \notin \Gamma \qquad \frac{\Gamma \Rightarrow B; \mathcal{H}}{\Gamma \Rightarrow A \supset B; \mathcal{H}} \; (\supset_{\mathcal{R}2}) \quad \text{if } A \in \Gamma$$

$$\frac{A, \Gamma \Rightarrow \bot; \phi}{\Gamma \Rightarrow \neg A; \mathcal{H}} \; (\neg_{\mathcal{R}1}) \quad \text{if } A \notin \Gamma \qquad \frac{\Gamma \Rightarrow \bot; \mathcal{H}}{\Gamma \Rightarrow \neg A; \mathcal{H}} \; (\neg_{\mathcal{R}2}) \quad \text{if } A \in \Gamma$$

$$\frac{\Gamma \Rightarrow A; (C, \mathcal{H}) \quad \Gamma \xrightarrow{B} C; \mathcal{H}}{\Gamma \xrightarrow{A \supset B} C; \mathcal{H}} \; (\supset_{\mathcal{L}}) \quad \text{if } C \notin \mathcal{H}$$

$$\frac{\Gamma \Rightarrow A; (C, \mathcal{H})}{\Gamma \xrightarrow{\neg A} C; \mathcal{H}} \; (\neg_{\mathcal{L}}) \quad \text{if } C \notin \mathcal{H}$$

$$\frac{\Gamma \xrightarrow{A} C; \mathcal{H}}{\Gamma \xrightarrow{A \wedge B} C; \mathcal{H}} \; (\wedge_{\mathcal{L}1}) \qquad \frac{\Gamma \xrightarrow{B} C; \mathcal{H}}{\Gamma \xrightarrow{A \wedge B} C; \mathcal{H}} \; (\wedge_{\mathcal{L}2})$$

$$\frac{\Gamma \Rightarrow A; \mathcal{H} \quad \Gamma \Rightarrow B; \mathcal{H}}{\Gamma \Rightarrow A \wedge B; \mathcal{H}} \; (\wedge_{\mathcal{R}})$$

$$\frac{A, \Gamma \Rightarrow C; \phi \quad B, \Gamma \Rightarrow C; \phi}{\Gamma \xrightarrow{A \vee B} C; \mathcal{H}} \; (\vee_{\mathcal{L}}) \quad \text{if } A \notin \Gamma \text{ and } B \notin \Gamma$$

$$\frac{\Gamma \Rightarrow A; \mathcal{H}}{\Gamma \Rightarrow A \vee B; \mathcal{H}} \; (\vee_{\mathcal{R}1}) \qquad \frac{\Gamma \Rightarrow B; \mathcal{H}}{\Gamma \Rightarrow A \vee B; \mathcal{H}} \; (\vee_{\mathcal{R}2})$$

$$\frac{A, \Gamma \xrightarrow{A} B; \mathcal{H}}{A, \Gamma \Rightarrow B; \mathcal{H}} \; (C)^{\star} \qquad \frac{}{\Gamma \xrightarrow{A} A; \mathcal{H}} \; (ax) \qquad \frac{}{\Gamma \xrightarrow{\bot} A; \mathcal{H}} \; (\bot)$$

$\star$ B is either a propositional variable, $\bot$ or a disjunction.

A, B, C are formulae, Γ is a multiset of formulae, and $\mathcal{H}$ is a set of formulae.

B, Γ is shorthand for $\{B\} \cup \Gamma$.

Sequent $\Gamma \Rightarrow C; \mathcal{H}$ has context Γ, goal C, history $\mathcal{H}$ and no stoup.

Sequent $\Gamma \xrightarrow{A} C; \mathcal{H}$ has context Γ, goal C, history $\mathcal{H}$ and stoup A.

When the history has been extended we have parenthesised $(C, \mathcal{H})$ for emphasis.

Fig. 2. The propositional calculus MJ^{Hist}, Swiss style

$$\frac{A, \Gamma \Rightarrow B; \{B\}}{\Gamma \Rightarrow A \supset B; \mathcal{H}} \ (\supset_{\mathcal{R}1}) \quad \text{if } A \notin \Gamma \qquad \frac{A, \Gamma \Rightarrow \bot; \{\bot\}}{\Gamma \Rightarrow \neg A; \mathcal{H}} \ (\neg_{\mathcal{R}1}) \quad \text{if } A \notin \Gamma$$

$$\frac{\Gamma \Rightarrow B; (B, \mathcal{H})}{\Gamma \Rightarrow A \supset B; \mathcal{H}} \ (\supset_{\mathcal{R}2}) \quad \text{if } A \in \Gamma, \quad \text{if } B \notin \mathcal{H}$$

$$\frac{\Gamma \Rightarrow \bot; (\bot, \mathcal{H})}{\Gamma \Rightarrow \neg A; \mathcal{H}} \ (\neg_{\mathcal{R}2}) \quad \text{if } A \in \Gamma, \quad \text{if } \bot \notin \mathcal{H}$$

$$\frac{\Gamma \Rightarrow A; (A, \mathcal{H})}{\Gamma \xrightarrow{\neg A} C; \mathcal{H}} \ (\neg_{\mathcal{L}}) \quad \text{if } A \notin \mathcal{H}$$

$$\frac{\Gamma \Rightarrow A; (A, \mathcal{H}) \quad \Gamma \xrightarrow{B} C; \mathcal{H}}{\Gamma \xrightarrow{A \supset B} C; \mathcal{H}} \ (\supset_{\mathcal{L}}) \quad \text{if } A \notin \mathcal{H}$$

$$\frac{\Gamma \xrightarrow{A} C; \mathcal{H}}{\Gamma \xrightarrow{A \wedge B} C; \mathcal{H}} \ (\wedge_{\mathcal{L}1}) \qquad \frac{\Gamma \xrightarrow{B} C; \mathcal{H}}{\Gamma \xrightarrow{A \wedge B} C; \mathcal{H}} \ (\wedge_{\mathcal{L}2})$$

$$\frac{\Gamma \Rightarrow A; (A, \mathcal{H}) \quad \Gamma \Rightarrow B; (B, \mathcal{H})}{\Gamma \Rightarrow A \wedge B; \mathcal{H}} \ (\wedge_{\mathcal{R}}) \quad \text{if } A \notin \mathcal{H} \text{ and } B \notin \mathcal{H}$$

$$\frac{A, \Gamma \Rightarrow C; \{C\} \quad B, \Gamma \Rightarrow C; \{C\}}{\Gamma \xrightarrow{A \vee B} C; \mathcal{H}} \ (\vee_{\mathcal{L}}) \quad \text{if } A \notin \Gamma \text{ and } B \notin \Gamma$$

$$\frac{\Gamma \Rightarrow A; (A, \mathcal{H})}{\Gamma \Rightarrow A \vee B; \mathcal{H}} \ (\vee_{\mathcal{R}1}) \quad \text{if } A \notin \mathcal{H} \qquad \frac{\Gamma \Rightarrow B; (B, \mathcal{H})}{\Gamma \Rightarrow A \vee B; \mathcal{H}} \ (\vee_{\mathcal{R}2}) \quad \text{if } B \notin \mathcal{H}$$

$$\frac{A, \Gamma \xrightarrow{A} B; \mathcal{H}}{A, \Gamma \Rightarrow B; \mathcal{H}} \ (C)^{\star} \qquad \frac{}{\Gamma \xrightarrow{A} A; \mathcal{H}} \ (ax) \qquad \frac{}{\Gamma \xrightarrow{\bot} A; \mathcal{H}} \ (\bot)$$

$\star$ B is either a propositional variable, $\bot$ or a disjunction.

A, B, C are formulae, Γ is a multiset of formulae, $\mathcal{H}$ is a set of formulae.

B, Γ is shorthand for $\{B\} \cup \Gamma$.

Sequent $\Gamma \Rightarrow C; \mathcal{H}$ has context Γ, goal C, history $\mathcal{H}$ and no stoup.

Sequent $\Gamma \xrightarrow{A} C; \mathcal{H}$ has context Γ, goal C, history $\mathcal{H}$ and stoup A.

When the history has been extended we have parenthesised $(C, \mathcal{H})$ for emphasis.

Fig. 3. The propositional calculus MJ^{Hist}, Scottish style

Theorem 2 *The calculus MJ^{Hist} with condition $\star$ placed on rule (C) is equivalent to MJ^{Hist} without the extra condition.*

PROOF:(Sketch) The $\Leftarrow$ direction is trivial.

To prove the $\Rightarrow$ direction, we first prove that MJ and MJ with condition $\star$ on (C) are equivalent. This is done by a simple induction on the depth of the proof and on complexity of formulae.

For any MJ^{Hist} (without $\star$) proof that doesn't satisfy $\star$, we can consider it as an MJ proof. Then we can find an MJ proof satisfying $\star$. Using the procedure in the proof of theorem 1, we can build an MJ^{Hist} (with $\star$) proof tree.

For full details see [10] ∎

2.3 The Scottish History

In this section we discuss the Scottish MJ^{Hist}. We go through its theory where it is different from the Swiss style calculus and explaining the motivations for the alternative approach. The Scottish MJ^{Hist} is given in Figure 3.

We said earlier that when using a history mechanism to prevent looping it would be good to cut down the amount of storage and checking needed to a bare minimum. This was done in the Swiss MJ^{Hist} — the history mechanism operates in one place only and other restrictions for loop prevention involve no storage. However it is not clear that this is the best and most attractive approach. There is a tradeoff between these advantages and the obvious disadvantage of not looking for loops very often. We will find loops more quickly if we look for them at more points. That is, we might continue building a tree needlessly, when a loop might already have been spotted. The Scottish MJ^{Hist} has larger histories, but this allows us to check for loops more often, and in certain situations this is advantageous.

As in the Swiss history, when attempting to prove a sequent, right rules are applied first, then (C), then left rules. Also, looping is prevented by the $(\vee_{\mathcal{L}})$ rule in the same way. The difference between the two calculi is in the way that the history mechanism works.

Whereas the Swiss calculus only places formulae in the history which have been the goal of the conclusion of a $(\supset_{\mathcal{L}})$ (or $(\neg_{\mathcal{L}})$) rule, the Scottish calculus keeps as the history a complete record of the goal formulae of sequents between context extensions. At each of the places where the history might be extended, the new goal is checked against the history. If it is in the history, then there is a loop. The heart of the difference between the two calculi is that in the Swiss calculus loop checking is done when a formula leaves the goal, whereas in the Scottish calculus it is done when it becomes the goal.

We have the same equivalence theorems as for the Swiss calculus. These are proved in a similar manner. For details again see [10].

Theorem 3 *The calculi MJ and MJ^{Hist} (without $\star$) are equivalent. That is, a sequent $\Gamma \Rightarrow A$ is provable in MJ if and only if $\Gamma \Rightarrow A; \{A\}$ (the sequent with its trivial history) is provable in MJ^{Hist} (without $\star$).*

PROOF: Similar to that of theorem 1. ■

Theorem 4 *The calculus MJ^{Hist} with condition $\star$ placed on rule (C) is equivalent to MJ^{Hist} without the extra condition.*

PROOF: Similar to that of theorem 2. ■

2.4 Comparison of the Two Calculi

Because of the way that the Swiss history works, loop detection is delayed. Let us illustrate this with an example. Consider the sequent:

$$p, q, (p \supset q \supset r) \supset r \Rightarrow p \supset q \supset r$$

In the Swiss style MJ^{Hist} (where $\Gamma = p, q, (p \supset q \supset r) \supset r$, and $G = p \supset q \supset r$) this gives the following:

$$\frac{\dfrac{\dfrac{\dfrac{\dfrac{\Gamma \Rightarrow G; \{r\} \quad \overline{\Gamma \xrightarrow{r} r; \phi}\ (ax)}{\Gamma \xrightarrow{(p \supset q \supset r) \supset r} r; \phi}\ (\supset_{\mathcal{L}})}{\Gamma \Rightarrow r; \phi}\ (C)}{\Gamma \Rightarrow q \supset r; \phi}\ (\supset_{\mathcal{R}2})}{\Gamma \Rightarrow G; \phi}\ (\supset_{\mathcal{R}2})$$

We have to go through all the inference steps again (in the branch above the left premiss) before the loop is detected. However, in the Scottish calculus we get:

$$\frac{\dfrac{\dfrac{\dfrac{\Gamma \Rightarrow G; \{G, r, q \supset r, G\} \quad \overline{\Gamma \xrightarrow{r} r; \{r, q \supset r, G\}}\ (ax)}{\Gamma \xrightarrow{(p \supset q \supset r) \supset r} r; \{r, q \supset r, G\}}\ (\supset_{\mathcal{L}})}{\Gamma \Rightarrow r; \{r, q \supset r, G\}}\ (C)}{\Gamma \Rightarrow q \supset r; \{q \supset r, G\}}\ (\supset_{\mathcal{R}2})}{\Gamma \Rightarrow G; \{G\}}\ (\supset_{\mathcal{R}2})$$

The topmost inference, $(\supset_{\mathcal{L}})$, is not valid, because the left premiss has goal formula, G, which is already in the history. That is, the loop is detected, and is detected lower in the proof tree than in the Swiss style calculus.

Spotting the loop as it occurs is not only theoretically more attractive, but could also prevent a lot of costly extra computation.

The two calculi both have their good points. The Swiss calculus is efficient from the point of view that its history mechanism requires little storage and checking. The Scottish calculus is efficient in that it detects loops as they occur, avoiding unnecessary computations.

The question is whether or not in general an overhead in storage and checking of the history (which shouldn't be too great due to regular resetting) is preferable to the larger proof trees which are the result delaying checking. Perhaps the best way to decide this is to look at empirical results in the form of timings for implementations of the calculi. Note that as the two calculi are rather similar it is more than likely that any optimisation that can be applied to the one can be applied to the other.

3 Implementation of the Decision Procedure

Our implementation of the calculus is syntax directed. A sequent $\Gamma \Rightarrow A; \phi$ for the Swiss calculus, or $\Gamma \Rightarrow A; \{A\}$ for the Scottish, is passed to the theorem prover. For a sequent with an empty stoup, the next inference is determined by the goal. If the goal is an implication, negation or conjunction, then the appropriate rule on the right is applied. If an instance of one of these rules fails, then we have to backtrack as no other rule is applicable. If the goal is a propositional variable, falsum or a disjunction, the contraction rule is applied, selecting a formula and placing it in the stoup. If a contraction fails, then a different formula is placed in the stoup. If the goal is a propositional variable or falsum, and contraction has failed for all possible stoup formulae, we backtrack. If the goal is a disjunction and contraction has failed for all possible stoup formulae, then we may apply disjunction on the right. If this fails we have to backtrack. For a sequent with a stoup formula, the next inference is decided upon by the stoup formula. The next inference must be an instance of the appropriate rule on the left. If such an inference fails, then we have to backtrack. Note that in $(\supset_{\mathcal{L}})$ we check the right branch, the one with the stoup formula, first. We get failure if at any point no rule instance can be applied. We give an example of failure due to the history:

$$\frac{}{p, \Gamma \Rightarrow p \supset q; \{p, q\}} \ (\supset_{\mathcal{R}2})$$

fails due to $q \notin \{p, q\}$ not being satisfied. Because of condition $\star$, no other rule instances are applicable to this sequent and so we must backtrack.

For this implementation we do not need to know anything about the invertibility of any of the rules. However, it may be of some independent interest to point out rules which are invertible and those which are not. For all three calculi - MJ, MJ^{Hist} (Swiss) and MJ^{Hist} (Scottish) - all rules are fully invertible with the exception of $(\wedge_{\mathcal{L}})$, $(\vee_{\mathcal{R}})$ and (C).

4 Results

The issue we are concerned with here is that of speed: how quickly we find out whether or not a certain sequent or formula is provable. We tested the two theorem provers on a sample of problems, some easy, some more problematic.

The calculi were implemented in prolog (naïvely, code can be found in [10]). The programs were run using SICStus Prolog2.1 on a SUN SparcStation 10. The times given are runtimes (in milliseconds), i.e. "CPU time used whilst executing, excluding time spent garbage collecting, stack shifting or in system calls" [15]. In Figure 4 we present the formulae we gave to the theorem prover (the quantified formulae were instantiated over finite universes). In Table 1 we give the results and average timings (where NR means that the machine had not proved the example after running overnight).

The results indicate that although the Swiss calculus can be quicker on some examples, this advantage is less significant than the disadvantage of the several examples where the Swiss calculus is several orders of magnitude slower than the Scottish calculus. It should also be added that the times for the calculi implemented compare poorly with our implementation of the single succedant LJT calculus of [1].

1. $((A \lor B) \land (D \lor E) \land (G \lor H)) \supset ((A \land D) \lor (A \land G) \lor (D \land G) \lor (B \land E) \lor (B \land H) \lor (E \land H))$

2. $((A \lor B \lor C) \land (D \lor E \lor F) \land (G \lor H \lor J) \land (K \lor L \lor M)) \supset (A \land D) \lor (A \land G) \lor (A \land K) \lor (D \land G) \lor (D \land K) \lor (G \land K) \lor (B \land E) \lor (B \land H) \lor (B \land L) \lor (E \land H) \lor (E \land L) \lor (H \land L) \lor (C \land F) \lor (C \land J) \lor (C \land M) \lor (F \land J) \lor (F \land M) \lor (J \land M)$

3. $((A \lor B \lor C) \land (D \lor E \lor F)) \supset ((A \land B) \lor (B \land E) \lor (C \land F))$

4. $(A \supset B) \supset (A \supset C) \supset (A \supset (B \land C))$

5. $(A \land \neg A) \supset B$

6. $(A \lor C) \supset (A \supset B) \supset (B \lor C)$

7. $((((A \supset B) \land (B \supset A)) \supset (A \land B \land C)) \land (((B \supset C) \land (C \supset B)) \supset (A \land B \land C)) \land (((C \supset A) \land (A \supset C)) \supset (A \land B \land C))) \supset (A \land B \land C)$

8. $((\neg\neg P \supset P) \supset P) \lor (\neg P \supset \neg P) \lor (\neg\neg P \supset \neg\neg P) \lor (\neg\neg P \supset P)$

9. $(((G \supset A) \supset J) \supset D \supset E) \supset (((H \supset B) \supset I) \supset C \supset J) \supset (A \supset H) \supset F \supset G \supset ((((C \supset B) \supset I) \supset D) \supset (A \supset C) \supset (((F \supset A) \supset B) \supset I) \supset E$

10. $A \supset B \supset ((A \supset B \supset C) \supset C) \supset (A \supset B \supset C)$

11. $((\neg\neg(\neg A \lor \neg B) \supset (\neg A \lor \neg B)) \supset (\neg\neg(\neg A \lor \neg B) \lor \neg(\neg A \lor \neg B))) \supset (\neg\neg(\neg A \lor \neg B) \lor \neg(\neg A \lor \neg B))$

12. $B \supset (A \supset (((A \land B) \supset C_1) \supset (((A \land B) \supset C_2) \supset (((A \land B) \supset C_3) \supset (((A \land B) \supset (B \supset C_1 \supset C_2 \supset C_3 \supset B)) \supset (A \land B))))))$

13. $((A \land B \lor C) \supset (C \lor (C \land D))) \supset (\neg A \lor ((A \lor B) \supset C))$

14. $\neg\neg((\neg A \supset B) \supset (\neg A \supset \neg B) \supset A)$

15. $\neg\neg(((A \leftrightarrow B) \leftrightarrow C) \leftrightarrow (A \leftrightarrow (B \leftrightarrow C)))$

16. $\forall x \exists y \forall z (p(x) \land q(y) \land r(z)) \leftrightarrow \forall z \exists y \forall x (p(x) \land q(y) \land r(z))$

17. $\exists x_1 \forall y_1 \exists x_2 \forall y_2 \exists x_3 \forall y_3 (p(x_1, y_1) \land q(x_2, y_2) \land r(x_3, y_3)) \supset \forall y_3 \exists x_3 \forall y_2 \exists x_2 \forall y_1 \exists x_1 (p(x_1, y_1) \land q(x_2, y_2) \land r(x_3, y_3))$

18. $\neg \exists x \forall y (mem(y, x) \leftrightarrow \neg mem(x, x))$

19. $\neg \exists x \forall y (q(y) \supset r(x, y)) \land \exists x \forall y (s(y) \supset r(x, y)) \supset \neg \forall x (q(x) \supset s(x))$

20. $\forall z_1 \forall z_2 \forall z_3 (q(z_1, z_2, z_3, z_1, z_2, z_3)) \supset \exists x_1 \exists x_2 \exists x_3 \exists y_1 \exists y_2 \exists y_3 ((p(x_1) \land p(x_2) \land p(x_3) \leftrightarrow p(y_1) \land p(y_2) \land p(y_3)) \land q(x_1, x_2, x_3, y_1, y_2, y_3))$

21. $((\exists x (p \supset f(x))) \land (\exists x_1 (f(x_1) \supset p))) \supset (\exists x_2 ((p \supset f(x_2)) \land (f(x_2) \supset p)))$

22. $(\exists x (p(x)) \land (\forall x_1 (f(x_1) \supset (\neg g(x_1) \land r(x_1))) \land (\forall x_2 (p(x_2) \supset (g(x_2) \land f(x_2))) \land (\forall x_3 (p(x_3) \supset q(x_3)) \lor \exists x_4 (p(x_4) \land r(x_4)))))) \supset \exists x_5 (q(x_5) \land p(x_5))$

23. $((\exists x (p(x)) \leftrightarrow \exists x_1 (q(x_1))) \land \forall x_2 \forall y ((p(x_2) \land q(y)) \supset (r(x_2) \leftrightarrow s(y)))) \supset (\forall x_3 (p(x_3) \supset r(x_3)) \leftrightarrow \forall x_4 (q(x_4) \supset s(x_4)))$

24. $(\forall x ((f(x) \lor g(x)) \supset \neg h(x)) \land \forall x_1 ((g(x_1) \supset \neg i(x_1)) \supset (f(x_1) \land h(x_1)))) \supset \forall x_2 (i(x_2))$

25. $(\neg \exists x (f(x) \land (g(x) \lor h(x))) \land (\exists x_1 (i(x_1) \land f(x_1)) \land \forall x_2 (\neg h(x_2) \supset j(x_2)))) \supset \exists x_3 (i(x_3) \land j(x_3))$

26. $(\forall x ((f(x) \land (g(x) \lor h(x))) \supset i(x)) \land (\forall x_1 ((i(x_1) \land h(x_1)) \supset j(x_1)) \land \forall x_2 (k(x_2) \supset h(x_2)))) \supset \forall x_3 ((f(x_3) \land k(x_3)) \supset j(x_3))$

27. $\neg \exists y \forall x (f(x, y) \leftrightarrow \neg \exists z (f(x, z) \land f(z, x)))$

Fig. 4. Example Formulae

Example	Universe	Result	Swiss Time	Scottish Time
1.		Provable	14	18
2.		Provable	1388	1701
3.		Unprovable	15	21
4.		Provable	0.2	0.2
5.		Provable	0.1	0.1
6.		Provable	0.6	0.8
7.		Provable	11	14
8.		Provable	0.5	0.5
9.		Provable	4.3	4.3
10.		Unprovable	0.4	0.5
11.		Unprovable	24	10
12.		Provable	0.7	1.0
13.		Unprovable	4.5	3.2
14.		Provable	3.5	2.7
15.		Provable	50	57
16.	3	Provable	803	961
17.	2	Provable	7497	8450
18.	4	Provable	63	8.5
18.	5	Provable	146	15
19.	2	Provable	7.8	8.1
19.	3	Provable	18420	27
20.	2	Provable	1.1	2.1
20.	4	Provable	5.3	6.6
21.	2	Unprovable	8.6	10
21.	3	Unprovable	27	33
22.	2	Provable	366	22
22.	3	Provable	12320	514
23.	2	Provable	35	45
23.	3	Provable	2186	1407
24.	2	Unprovable	49	31
25.	2	Provable	10790	20
25.	4	Provable	NR	365
26.	2	Provable	3.4	5.8
26.	5	Provable	17	30
27.	2	Provable	10082	47

Table 1. Results and Timings (averages in milliseconds)

5   Conclusion

The use of a pared down history makes for a seemingly efficient means of loop detection for a theorem prover. However, as other intuitionistic theorem provers are written in different languages, are run on different machines and (in most cases) deal with first-order formulae, comparison is hard. An (incomplete) list of other intuitionistic theorem provers is: [2], [4], [8], [11], [12], [13], [14]. Of the two calculi given here, the one with the smallest history and the least checking (the Swiss one) can become inefficient (see example 27.) when delay in loop checking allows many extra branches to be pursued. In the Scottish style calculus the inefficiency of the increased history is more than counterbalanced by the early loop detection.

We have illustrated the use of the two history mechanisms on a particular calculus for intuitionistic propositional logic - one in which we are particularly interested, rather than because it is the best illustration. We anticipate similar advantages of the Scottish history mechanism in treatment of modal logics.

A final issue to be addressed is that of proof search. For this neither calculus is really suited, as they only find loop free proofs (plus a few more in the Swiss case). For details see [10].

References

1. Dyckhoff, R.: Contraction-free Sequent Calculi for Intuitionistic Logic. Journal of Symbolic Logic **57(3)** (1992) 795–807
2. Dyckhoff, R.: MacLogic implementation. Available from URL http://www-theory.dcs.st-and.ac.uk/~rd/logic/soft.html
3. Dyckhoff, R., Pinto, L.: A Permutation-free Sequent Calculus for Intuitionistic Logic. University of St Andrews Research Report CS/96/9 (1996)
4. Dyckhoff, R., Pinto, L.: Implementation of a Loop-free Method for Construction of Countermodels for Intuitionistic Propositional Logic. University of St Andrews Research Report CS/96/8 (1996)
5. Gabbay, D.: Algorithmic Proof With Diminishing Resources, Part 1. Proceedings of the 1990 workshop Computer Science Logic, eds. Börger, E., Kleine Büning, H., Richter, M. M., Schönfeld, W.; Springer LNCS **533** (1991) 156–173
6. Girard, J.-Y.: A New Constructive Logic: Classical Logic. Mathematical Structures in Computer Science **1** (1991) 255–296
7. Herbelin, H.: A λ-calculus Structure Isomorphic to Gentzen-style Sequent Calculus Structure. Proceeding of the 1994 workshop Computer Science Logic, eds. Pacholski, L., Tiuryn, J.; Springer LNCS **933** (1995) 61–75
8. Heuerding, A., Jäger, G., Schwendimann, S., Seyfried, M.: Propositional Logics on the Computer. Proceedings of the 1995 international workshop on Theorem Proving with Analytic Tableaux and Related Methods (TABLEAUX '95), eds. Baumgartner, P., Hähnle, R., Posegga, J.; Springer LNAI **918** (1995) 310-323
9. Heuerding, A., Seyfried, M., Zimmermann, H.: Efficient Loop-Check for Backward Proof Search in Some Non-classical Propositional Logics. Proceedings of the 1996 international workshop on Theorem Proving with Analytic Tableaux and Related Methods (TABLEAUX '96), eds. Miglioli, P., Moscato, U., Mundici, D., Ornaghi, M.; Springer LNAI **1071** (1996) 210–225
10. Howe, J.M.: Theorem Proving and Partial Proof Search for Intuitionistic Propositional Logic Using a Permutation-free Calculus with Loop Checking. University of St Andrews Research Report CS/96/12 http://www-theory.cs.st-and.ac.uk/~jacob/papers/tpil.html (1996)
11. Sahlin, D., Franzén, T., Haridi, S.: An Intuitionistic Predicate Logic Theorem Prover. Journal of Logic and Computation **2(5)** (1992) 619–656

12. Shankar, N.: Proof Search in the Intuitionistic Sequent Calculus. Proceedings of the 1992 international conference on Automated Deduction (CADE-13), ed., Kupar, D.; Springer LNAI **607** (1992) 522–536
13. Stoughton, A.: porgi:a Proof-Or-Refutation Generator for Intuitionistic propositional logic. http://www.cis.ksu.edu/~allen/home.html
14. Tammet, T.: A Resolution Theorem Prover for Intuitionistic Logic. Available from the URL http://www.cs.chalmers.se/tammet/ (1996). This is a longer version of the paper in Proceedings of the 1996 international conference on Automated Deduction (CADE-13), eds. McRobbie, M. A., Slaney, J. K.; Springer LNAI **1104** (1996) 2-16
15. SICStus Prolog User's Manual. Swedish Institute of Computer Science (1993)

Acknowledgements

The author is indebted to Roy Dyckhoff for many useful discussions. The comments of the anonymous referees have also been very helpful and were greatly appreciated.

Subgoal Alternation in Model Elimination*

Ortrun Ibens and Reinhold Letz

Institut für Informatik
Technische Universität München
80290 München, Germany
{ibens,letz}@informatik.tu-muenchen.de
phone +49-(0)89-521097

Abstract. The order in which subgoals are selected and solved has a strong influence on the search space of model elimination procedures. A general principle is to prefer subgoals with few solutions over subgoals with many solutions. In this paper we show that the standard selection methods are not flexible enough to satisfy this principle. As a generalization of the standard paradigm the new method of subgoal *alternation* is presented and integrated into the theorem prover SETHEO. Among other advantages, subgoal alternation also provides more look-ahead information about the needed proof resources than the standard method; this information can be used for search pruning. The evaluation of the new technique on a large number of formulae shows a significant improvement in performance.

1 Introduction

The model elimination calculus [Lov78] can be viewed as a refined tableau method, namely the *free-variable* tableau system [Fit90] with the *connection* condition [LSBB92, LMG94]. The particular suitability of model elimination for automated deduction results from its *goal-directedness*, which is ensured by the connection condition. The same condition, however, is responsible for the main disadvantage of the calculus, its non-confluence. In order to find a proof in model elimination, one has to enumerate *all* possible deductions, in contrast to general tableaux, where it is sufficient to work on *one* tableau.

The search space of model elimination can be seen as a complex form of an and-or-tree in which tableaux are and-nodes (connecting the contained subgoals) and subgoals are or-nodes (connecting the possible unification partners). While, for any subgoal, all possible unification partners have to be tried in order to guarantee completeness, soundness requires to solve all subgoals. The order of subgoal selection strongly influences the size of the search space. This is because subgoals normally share variables and thus the solution substitutions of one subgoal have an influence on the solution substitutions of the other subgoals.

---

* Work funded within the project 'SETHEO II' (Je112/5) of the Deutsche Forschungsgemeinschaft (DFG).

A general *least commitment* paradigm is to prefer subgoals that produce fewer solutions.

This has been recognized, for example, in the logic programming community, and respective heuristics have been developed for subgoal selection. One common feature of all standard subgoal selection methods, however, is that, once a subgoal has been selected, one sticks to that preference for all its unification partners, which is also a basic search paradigm of Prolog. In this paper, we show that this paradigm has a fundamental search-theoretic weakness. It is demonstrated that one can profit from changing the subgoal preference dynamically, depending on the remaining unification partners in the or-nodes. We develop a new method of *subgoal alternation* which is more flexible than the standard approach and can better satisfy the aforementioned fewest-solution principle.

The integration of subgoal alternation into model elimination also benefits from a number of synergetic effects with other search properties. On the one hand, one can profit from *look-ahead* information concerning the minimal number of inferences needed for closing a tableau. On the other hand, local failure caching [LMG94] is supported by subgoal alternation. The new method has been implemented within the model elimination theorem prover SETHEO. In combination with other techniques contained in the system, a significant performance gain could be obtained, particularly for the special equality handling in SETHEO. The use of subgoal alternation is one of the main reasons why SETHEO was a winner in the first world-wide theorem prover competition at the Conference on Automated Deduction 1996 [MIL+97].

The paper is organized as follows. In Section 2 we give a short introduction to the underlying proof procedure. Section 3 explains the general selection principles for subgoal selection. In Section 4 we illustrate the weakness of standard subgoal selection and present our new approach. In Section 5 the integration of the subgoal selection strategies in SETHEO are described. Section 6 reports on the performance gains achieved by subgoal alternation. We conclude then with a summary of the advantages of subgoal alternation and we describe our future work.

2   Model Elimination and Connection Tableaux

Historically, model elimination was introduced as a two-sorted variant of resolution (see, e.g., [Lov78]). Proof-theoretically, however, model elimination can also be viewed as a specialized tableau method, the so-called *connection tableau calculus* [LSBB92, LMG94].[2] A connection tableau for a set of clauses is generated by first applying the *start rule* and then repeatedly applying either the *reduction* or the *extension rule*. Employment of the start rule means selecting a clause from a set of possible start clauses and attaching its literals to the root node; the start rule is the standard tableau expansion rule. The reduction rule permits

---

[2] One major argument for this interpretation is that, when viewed as a tableau method, model elimination (in its weak variant) is *cut-free*.

the closing of a branch (or *path*) by unifying a subgoal[3] K with the complement of a path literal L—L is called a *connected path literal*; the reduction rule is the standard closure rule of free-variable tableaux. The extension rule requires the performing of a unification step by attaching the (instantiated) literals of a *connected clause* to the current subgoal K, i.e., a clause that contains a literal which can be unified with the complement of K. The extension rule, which is obviously a specific combination of tableau expansion and reduction, implements the connectedness condition.[4]

The connectedness condition leads to the *goal-directedness* of the calculus, since every literal in the tableau bears a relation to the start clause, which need not be the case for general tableaux. On the other hand, however, the connectedness condition causes the *non-confluence* of the calculus, i.e., if an open tableau branch cannot be expanded, it does not guarantee the satisfiability of the input clauses.[5] As a consequence, when searching for a proof, one has to enumerate *all* connection tableaux. The most efficient way of tableau enumeration is by using iterative deepening search procedures with backtracking (see [Sti88]).

Pure connection tableaux are not successful in automated deduction, since too many tableaux have to be inspected. Important *structural refinements* of clausal tableaux are the *regularity* condition, which is a variant of tableau *strictness*, and the avoidance of *tautological* tableau clauses [LSBB92]. With these techniques many tableaux can be ignored simply because of the violation of certain structural properties. It can be shown, however, that a lot of redundancy is still contained in the search space, since many tableaux are redundant in the presence of *other* tableaux. One such phenomenon is *tableau subsumption* [LMG94]. As an attempt to eliminate redundancy of this type, *failure caching* was introduced in [AS92]. Both of its variants, *solution caching* and *path caching*, turned out not to be successful in practice, for the following reasons. Solution caching is not compatible with regularity (and tautology-freeness) and incomplete in the non-Horn case. Path caching is inefficient and not very effective. As a remedy, a *local* version of failure caching has been developed in [LMG94].[6] *Local failure caching* is a particular tableau subsumption technique which overcomes the mentioned weaknesses of *global* failure caching. In local failure caching the cache is used to store path-dependent solutions of subgoal *occurrences* in the tableau. Because for each cached solution the path dependence is the same, local failure caching is complete even in the non-Horn case. Moreover, it can be efficiently implemented, because the path itself need not be stored, but only its variables. Finally, local failure caching is compatible with the structural refinements of connection tableaux (regularity, tautology-freeness), provided these are confined

---

[3] A *subgoal* is the literal at the leaf of an open branch.

[4] SLD-resolution can be viewed simply as the connection tableau calculus without reduction steps. This shows that SLD-resolution is inherently cut-free and not a cut calculus like general resolution.

[5] For this reason, in contrast to general tableaux, connection tableaux (and hence model elimination) are not suited for model *generation*.

[6] There it was termed as *anti-lemmata*.

to the *open part* of the tableau (see [LMG94]).

A general weakness of tableaux is that even the shortest proofs can be long, which is due to the cut-freeness of the calculus. The additional *folding up* rule, which generalizes Shostak's *c-reduction* [Sho76], provides a controlled integration of the atomic cut rule [LMG94]. Briefly, folding up permits the use of the complement of a solved subgoal L as an additional assumption. This assumption is made available to all subgoals which are located below the lowest path literal that is situated above L and was used for solving L. Folding up can also be viewed as a restricted but highly efficient kind of lemma generation.[7]

3   Subgoal Selection

If an open connection tableau is constructed during proof search, then *all* its subgoals have to be solved. The selection of the *next* subgoal can have a strong influence on the size of the search space. In order to identify a non-closable connection tableau as early as possible, the solutions of a subgoal should be exhausted as early as possible. Therefore, subgoals for which probably only few solutions exist should be selected earlier than subgoals for which many solutions exist. This results in the *fewest-solutions* principle for subgoal selection.[8]

Fortunately, the subgoal selection has no influence on the proof size, i.e., the number of symbols occurring in the proof. Even more, connection tableaux are *strongly independent* of the subgoal selection, that is, regardless in which order subgoals are selected, one can always generate the same tableaux, modulo variable renaming [Let93].[9]

One particular useful form of choosing subgoals is *depth-first* selection, i.e., always selecting a subgoal with maximal depth in the tableau. Model elimination can just be *defined* as the connection tableau calculus restricted to *depth-first* subgoal selection. Depth-first selection has a number of advantages, the most important being that the search is kept relatively local (which involves that failure caching techniques are very effective). Furthermore, efficient implementations are possible. One should note, however, that there exist formulae, where depth-first subgoal selection performs much worse than free subgoal selection (see [Let93]).

Depth-first selection means that all subgoal alternatives stem from one clause of the input set. Therefore, the selection order of the literals in a clause can be determined *statically*, i.e., once and for all before starting the proof search, as

---

[7] Note that, in connection tableaux, folding up is superior to the asymmetric form of the β-rule with cut, which is currently discussed in the tableau community (see [LMG94]).

[8] The special case of the fewest-solutions principle without solutions is the *first-fail* principle.

[9] This strong form of independence does not hold if the folding up rule is added. Although the proof size is preserved, the *depth* of the tableau may change. This becomes important when search strategies based on tableau depth are used, as in 6.2.

in [LSBB92]. Subgoal selection can also be performed *dynamically*, whenever the literals of the clause are handled in a tableau. The static version is cheaper (in terms of performed comparisons), but often an optimal subgoal selection cannot be determined statically, as can be seen, for example, when considering the transitivity clause $p(X, Z) \lor \neg p(X, Y) \lor \neg p(Y, Z)$. Statically, none of the literals can be preferred. Dynamically, however, when performing an extension step entering the transitivity clause from a subgoal $\neg p(a, Z)$, the first subgoal $\neg p(X, Y)$ is instantiated to $\neg p(a, Y)$. Since now it contains only one variable, is should be preferred according to the fewest-solutions principle. Entering the transitivity clause from a subgoal $\neg p(X, a)$ leads to preference of the second subgoal $\neg p(Y, a)$.

4 Subgoal Alternation

When a subgoal in a tableau has been selected for solution, a number of complementary unification partners are available, viz. the connected path literals and the connected literals in the input clauses. Together they form the so-called *choice point* of the subgoal. One common principle of standard backtracking search procedures in model elimination (and in Prolog) is that, whenever a subgoal has been selected, its choice point must be completely finished, i.e., when retracting an alternative in the choice point of a subgoal, one has to stick to the subgoal and try another alternative in its choice point. This standard methodology has a deep search-theoretic weakness that has not been recognized so far.

The weakness of remaining in the same choice point can be illustrated with the following generic example, variants of which often occur in practice. Given the subgoals $\neg p(X, Y)$ and $\neg q(X, Y)$ in a tableau, assume the following clauses be in the input.

(1) $p(a, a)$,
(2) $p(X, Y) \lor \neg p'(X, Z) \lor \neg p'(Y, Z)$,
(3) $p'(a_i, a)$, $\quad 1 \le i \le n$,
(4) $q(a_i, b)$, $\quad 1 \le i \le n$.

Suppose further we have decided to select the first subgoal and perform depth-first subgoal selection. The critical point, say at time t, is after unit clause (1) in the choice point was tried and no compatible solution instance for the other subgoal was found. Now we are forced to enter clause (2). Obviously, there are n^2 solution substitutions (unifications) for solving clause (2) (the product of the solutions of its subgoals). For each of those solutions we have to perform n unifications with the q-subgoal, which all fail. Including the unifications spent in clause (2), this amounts to a total effort of $1 + n + n^2 + n^3$ unifications (see Figure 1). Observe now what would happen when at time t we would not have entered clause (2), but would switch to the q-subgoal instead. Then, for each of the n solution substitutions $q(a_i, b)$, one would jump to the p-subgoal, enter clause (2) and perform just n failing unifications for its first subgoal. This sums up to a total of just $n + n(1 + n) = 2n + n^2$ unifications (see Figure 2).

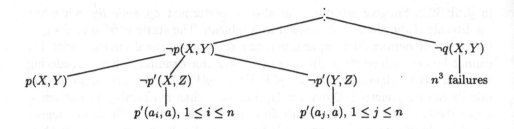

Fig. 1. Effort in case of standard subgoal processing.

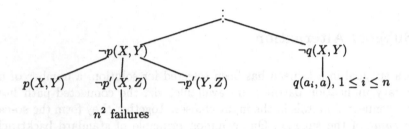

Fig. 2. Effort when switching to another subgoal.

It is apparent that this phenomenon has to do with the fewest-solutions principle. Clause (2) generates more solutions for the subgoal $\neg p(X, Y)$ than the clauses in the choice point of the subgoal $\neg q(X, Y)$. This shows that taking the remaining alternatives of *all* subgoals into account provides a choice which can better satisfy the fewest-solution principle. The general principle of *subgoal alternation* is that one always switches to that subgoal with a next clause that produces the fewest solutions.

One might object that with a different subgoal selection, selecting the q-subgoal first, one also could avoid the cubic effort. But it is apparent that the example could be extended such that the q-subgoal would additionally have a longer clause as alternative, so that the total number of its solutions would be even larger than that of the p-subgoal. In this case, with subgoal alternation one could jump back to the p-subgoal and try clause (2) next, in contrast to standard subgoal selection. Another possibility of jumping to the q-subgoal *after* having entered clause (2) would be free subgoal selection. In fact, subgoal alternation under depth-first subgoal selection comes closer to standard free subgoal selection, but both methods are not identical.

The question is, when it is worthwhile to stop the processing of a choice point and switch to another subgoal? As a matter of fact, it cannot be determined in advance, how many solutions a clause in the choice point of a subgoal produces for that subgoal. A useful criterion, however, is the *shortest-clause* principle, since, in the worst case, the number of subgoal solutions coming from a clause is the product of the numbers of solutions of its subgoals.[10]

In summary, subgoal (or *choice point*) alternation works as follows. The standard subgoal selection and clause selection phases are combined and result in a single selection phase that is performed before each derivation step. The selection yields the subgoal for which the most suitable unification partner exists wrt. the number of solutions probably produced. For this, the unification partners of all subgoals are compared with each other using, for instance, the shortest-clause principle. If more than one unification partner is given the mark of 'best', their corresponding subgoals have to be compared due to the principles for standard subgoal selection, namely the first-fail principle and the fewest-solutions principle.

standard backtracking	subgoal alternation
A1 B2	A1 B2
A1 B4	A1 B4
A1 B6	A1 B6
A3 B2	↻ B2 A3
A3 B4	B2 A5
A3 B6	↻ A3 B4
A5 B2	A3 B6
A5 B4	↻ B4 A5
A5 B6	↻ A5 B6

Table 1. Order of tried clauses for subgoals A and B with clauses of lengths 1,3,5 and 2,4,6 in their choice points, respectively. ↻ indicates choice point alternations.

In order to compare the working of choice point alternation (using the shortest-clause principle) with the standard non-alternating variant, consider two subgoals A and B with clauses of lengths 1,3,5 and 2,4,6 in their choice points, respectively. Table 1 illustrates the order in which clauses are tried.

Subgoal alternation has a number of interesting other effects on the proof search, which we will mention next.

4.1 Early Instantiation and Preference of Small Subproofs

A particularly beneficial effect of preferring short clauses, especially the preference of unit clauses, is the early instantiation of variables. Unit clauses are usually more instantiated than longer clauses, because they represent the 'facts' of the input problem, whereas longer clauses in general represent the axioms of the underlying theory. Since normally variables are shared between several

---

[10] Also, the number of variables in the *calling* subgoal and in the *head* literal of a clause matter for the number of solutions produced.

subgoals, the solution of a subgoal by a unit clause usually leads to instantiating variables in other subgoals. These instantiations reduce the number of solutions of the other subgoals and thus reduce the search space to be explored when selecting them. Advantage is also taken from choice point alternation when combined with local failure caching. Failure caching can only exploit information from *closed* sub-tableaux, thus a large number of small subproofs provides more information for caching than a small number of large sub-tableaux that cannot be closed. Since choice point alternation prefers short clause and hence small subproofs, the local failure caching mechanism is supported.

4.2 Computing Look-Ahead Information

Choice point alternation leads to simultaneously processing several choice points. This provides the possibility of computing *look-ahead information* concerning the minimal number of inferences still needed for closing a tableau. A simple estimation of this inference value is the number of subgoals plus the number of all subgoals in the shortest alternative of each subgoal. In general, when using standard subgoal selection, every choice point except the current one contains connected path literals and connected unit clauses, that is, the number of subgoals within the shortest alternative for each subgoal equals zero.

Using choice point alternation, at several choice points the reduction steps and the extension steps with unit clauses have already been tried, so that only the unification partners with subgoals are left in the choice points of several subgoals. Thus, one obtains more information about the needed inference resources than in the standard procedure. This *look-ahead information* can be used for search pruning, whenever the number of inferences has an influence on the search bound (see Section 6).

4.3 Disadvantages of Subgoal Alternation

If a subgoal cannot be solved at all, switching to another subgoal may be worse than sticking to the current choice point, since this may earlier lead to the retraction of the whole clause. Obviously, this is important for ground subgoals, because they have maximally one solution substitution. Since ground subgoals do not contain free variables, they cannot profit from early instantiations achieved by choice point alternation, i.e., switching to brother subgoals and instantiating their free variables cannot lead to instantiations within a ground subgoal. Therefore, when processing a ground subgoal, the fewest-solutions principle for subgoal selection becomes more important than the shortest-clause principle for subgoal alternation. For this reason, choice point alternation should not be performed when the current subgoal is ground.

5 Integration of Subgoal Alternation into SETHEO

The mentioned method of subgoal selection and alternation has been implemented as an extension of the theorem prover SETHEO [LSBB92, GLMS94, MIL+97], which is a model elimination prover for first order clause logic.

5.1 Standard Clause Ordering and Subgoal Selection

The ordering of the input clauses is performed as follows in SETHEO. Statically, for each subgoal K and each connected literal in the input clauses, a so-called *contrapositive* of this input clause is inserted into the choice point of K, i.e., a variant of the clause with the connected literal as *head* and the other literals as subgoals. The contrapositives that are connected to a subgoal are ordered by their lengths, due to the shortest-clause principle. If several contrapositives have the same length, the ones with lower complexity and more free variables are preferred. In the proof search, connected path literals are preferred to contrapositves, i.e., reduction steps are tried before extension steps.

SETHEO performs a static and a dynamic ordering of the subgoals in a contrapositive. The static selection criteria are described in [LSBB92]. Currently, in SETHEO dynamic subgoal selection is performed in a depth-first manner only, i.e., between the literals of the contrapositive attached in the last extension step.[11] Although some other measures may be used in SETHEO, we favour the following criterion: generally, ground subgoals are preferred; then, from the subgoals with the highest complexity one with a minimal number of variables is selected. Optionally, we can use an additional proviso, namely, not to switch to a subgoal with a different predicate symbol, according to the order of subgoals computed statically.

5.2 Subgoal Alternation

In general, SETHEO's method of choice point alternation subscribes to the shortest-clause principle as follows. The (remaining) choice points of the subgoals in a contrapositive are inspected and only the ones with an alternative of minimal length are considered. After this pre-selection the standard subgoal selection methods from above are applied. Optionally, we offer a restricted version of subgoal alternation, called *single-alternation*, which permits the switching away from a subgoal only once, namely, before its first non-unit clause is tried. Accordingly, the general choice point alternation strategy is called *multi-alternation*. Furthermore, subgoal alternation may be switched off completely, depending on the resources left for proving a subgoal. This is sensible since the resources also have a strong influence on the number of solutions of a subgoal.

5.3 Implementation

Standard dynamic subgoal selection is realized by *reordering* the initial sequence of the subgoals (generated by static subgoal reordering). Choice point alternation, however, is implemented as a *delay* mechanism for selected subgoals. Processing the choice point of a selected subgoal has to be interrupted and later

---

[11] Therefore, currently SETHEO is really a model elimination prover.

finished; meanwhile, choice points of other subgoals are processed. In order to save storage capacity, the data structure used for choice points is shared between all simultaneously processed subgoals.

6 Experiments and Results

In order to evalute whether subgoal alternation is really beneficial on a large number of examples, we have tested the new technique on the TPTP problem library [SSY94] (version 1.2.1), which is *the* standard collection of benchmarks for theorem provers. The experiments have been performed on HPUX computers with a maximal runtime of 1000 seconds CPU time per problem.[12] Since the effect of subgoal alternation strongly depends on the used search method, we must briefly review the search bounds that are available in SETHEO.

6.1 The Inference Bound

The *inference bound* was the first search bound used for depth-first iterative deepening search in model elimination [Sti88]. Experiments have demonstrated, however, that the inference bound is not the most successful strategy (consult [LSBB92] and below). The main weaknesses are the following. Firstly, the bound is too optimistic, since it implicitly assumes that subgoals which are not yet processed may be solved with just one inference step (but cf. [Har96] for a slight improvement). Secondly, the inference bound is *global* in the sense that brother subgoals share their inference resources, which has disadvantages concerning local failure caching (see, e.g., [MIL+97]).

An improvement with regard to the first weakness, however, could be expected from the employment of look-ahead information, as it is delivered by subgoal alternation. Since the look-ahead information provides a more realistic estimation of the minimal number of inferences that still have to be performed, the search procedure reaches the current inference limit more quickly.

6.2 The Weighted-Depth Bound

A standard bound that performs better in practice than the inference bound is the *depth bound* [LSBB92]. This limits the *depth*[13] of the tableaux considered in the current search round. The weakness of the depth bound, which is *local* wrt. resource sharing, is that it is too coarse and does not permit a smooth increase of the iterative deepening levels. A certain limited advance could be achieved by using the inference and the depth bound together [LMG94].

In order to overcome the limitations of the above bounds, the *weighted-depth bound* has been developed and integrated into SETHEO. The main idea of the

---

[12] This amounts to about 200 seconds on a Sun Ultra 2.

[13] Actually, only the depth of non-unit clauses is limited in SETHEO. This implements a certain unit preference strategy.

weighted-depth bound is to take the inferences into account when modifying the depth bound. Concretely, when computing the available depth resource for the solution of a subgoal, the number Δi of inferences consumed since entering the contrapositive is used. The modifications of the depth bound can be controlled by parameters (see [MIL+97] for a more detailed description). Depending on the parameter choices, the weighted-depth bound can simulate the depth bound, the inference bound, or any combination of them; even finer bounds than the inference bound may be defined. Again, the look-ahead information can be employed. To Δi a value Δl is added, which is computed as follows. Let l_0 be the look-ahead value when entering the clause and l_c the current look-ahead value. Then $\Delta l = l_c - l_0$. Adding Δl to Δi leads to a stronger pruning of the search.

6.3 Adaptations to Problem Characteristics

The tuning of the search parameters depending on a given problem can significantly improve the system performance. Since a manual tuning contradicts the paradigm of automated deduction, we use a very simple heuristic but fully automatic parameter selection that turns out to be very effective in practice.

In general, dynamic subgoal selection is performed using the proviso of not switching to a different predicate symbol, according to the order determined by static subgoal reordering. Our main distinction is whether or not a problem is ground. If it is we perform no iterative deepening at all and no subgoal alternation (see Section 4.3). All other problems are run with subgoal alternation. For non-ground problems that have no function symbols (of arity > 0), the standard depth bound is by far the most successful strategy. On the remaining problems the weighted-depth bound is the most successful bound. Another important criterion is whether the input formula contains equality or not. If not, we apply multi-alternation or single-alternation depending on whether or not the formula contains clauses of length > 3. For problems with equality axioms, we use single-alternation uniformly, but drop the proviso.[14] Finally, folding up is applied only in the non-Horn case.

resource bound	subgoal alternation	solved problems
inference	no	597
inference	yes	613
weighted-depth	no	758
weighted-depth	yes	794

Table 2. Performance results for the 2402 TPTP problems with function symbols and variables

---

[14] This is reasonable, since equality *connects* all literals and hence switching between different predicate symbols should be permitted.

In Table 2 we have listed the results of running the 2402 problems in the TPTP that are non-ground and contain proper function symbols with and without subgoal alternation. For those problems subgoal alternation turned out to be most beneficial. Although the weighted-depth bound is superior to the inference bound, we have added the figures for the inference bound, since a relative speed-up can also be of interest. The performance increase gained from choice point alternation is 2.7% for the inference bound and 4.7% for the weighted-depth bound.

6.4 Equality Handling

A fundamental weakness of model elimination is the handling of equality. The naïve approach, which is to simply add the congruence axioms of equality, as in the test runs above, suffers from the severe deficiency that equality specific redundancy elimination techniques cannot be applied. Furthermore, using the axiomatic approach, equality inferences on the deeper term level are much more expensive than on the top level, which is intuitively unacceptable. Therefore, we have integrated an alternative equality treatment into SETHEO which is based on *lazy basic paramodulation* [MLS95, Mos96]. Lazy basic paramodulation is a variant of paramodulation [RW69] which is compatible with the goal-directedness of the model elimination calculus, but which provides, e.g., no ordering restrictions on the rule applications. The method has been implemented within SETHEO by means of a preprocessing step. During this preprocessing step an input formula with equality axioms is transformed into an E-equivalent[15] formula without equality axioms (for details on the transformation read [Mos97]). Model elimination on the output formula then behaves like model elimination with lazy basic paramodulation on the input formula. The transformation can also be viewed as a variant of the STE-modification introduced by Brand [Bra75].

The structure of the problems is completely different in both approaches. In the axiomatic approach, equality is represented by short clauses (the reflexivity, symmetry, transitivity axioms yield clauses of length one, two, three, respectively), whereas during STE-modification the equality axioms are removed; instead the respective properties are compiled directly into the remaining clauses. This leads to a flattening of the literals[16] and an enlarging of the clause lengths (see [Mos97] for a detailed description). As an adaptation to the specific structure of the STE-modified problems, we have applied the multi-alternation variant of choice point alternation.

Table 3 shows the results of experiments performed with the axiomatic approach and the equality handling via STE-modification. For both approaches the inference bound and the weigted-depth bound have been investigated. The improvement obtained by subgoal alternation for the axiomatic approach was 7.5%, when using the inference bound, and 6.4% for the weighted-depth bound.

---

[15] I.e., equivalent wrt. the standard interpretation of the equality axioms.
[16] In the modified formula, all terms have a maximal depth of 2.

For equality handling by STE-modification we could achieve a 14.6% improvement for the inference bound, and 17.5% for the weighted-depth bound. Before the development of choice point alternation there was no use in applying STE-modification, since it performed worse than the axiomatic approach. In combination with choice point alternation, however, STE-modification becomes a promising paradigm for equality treatment, as already shown in the ATP system competition of CADE-13 [MIL+97].

equality handling	resource bound	subgoal alternation	solved problems
axiomatic	inference	no	347
axiomatic	inference	yes	373
axiomatic	weighted-depth	no	470
axiomatic	weighted-depth	yes	500
STE-modification	inference	no	431
STE-modification	inference	yes	494
STE-modification	weighted-depth	no	456
STE-modification	weighted-depth	yes	536

Table 3. Performance results for the 1920 TPTP problems containing equality

Interestingly, the sets of problems solved by the axiomatic and the STE-modification method differ significantly. The number of problems solved with one *or* the other methods is 600, when using the weighted-depth bound. This motivates a competitive use of both equality handling approaches, which, on sequential hardware, means a doubling of the run time limit.

6.5 Summarized Results

Table 4 shows SETHEO's performance on the whole TPTP (version 1.2.1), which consists of 2752 problems. Here only the results for the preferred strategy are given, i.e., no iterative deepening for ground problems, depth bound for problems without function symbols, and weighted-depth bound otherwise. The use of subgoal alternation is as described in Sections 6.3 and 6.4. With subgoal alternation we could achieve the following improvements: 3.3% for axiomatic equality handling, 7.9% for STE-modification, and 5.5% for the competitive variant. The 1232 problems solved with the competitive variant correspond to 45% of the TPTP.

7 Conclusion

We have presented the new method of subgoal alternation for search procedures of the model elimination type. This is a generalization of standard subgoal

equality handling	subgoal alternation	solved problems
axiomatic	no	1096
axiomatic	yes	1132
STE-modification	no	1082
STE-modification	yes	1168
competitive	no	1168
competitive	yes	1232

Table 4. Performance results for the 2752 TPTP problems

processing and provides more flexibility. With subgoal alternation some search-theoretic disadvantages of the standard approach can be overcome. The theoretical considerations are supported by experimental results on a large number of problems. The highest increase in performance is achieved for problems containing equality, particularly when the equality handling is carried out using STE-modification [Bra75, Mos97]. In fact, if subgoal alternation is not applied, the equality handling by STE-modification is less successful than the axiomatic approach (see Section 6.4). Furthermore, subgoal alternation supports a number of other methods for search pruning. It achieves search space reductions through early instantiations and provides look-ahead information concerning the search resources.

For the future, we envisage the following promising research perspectives. On the one hand, subgoal alternation may be combined with free subgoal selection, which is not restricted to subgoals in the current clause. On the other hand, the criteria for subgoal alternation can be refined. Firstly, subgoal alternation should not be adapted to the characteristics of a whole formula, but to each clause individually. Secondly, in order to improve the detection of forthcoming failures, the number of connected unification partners could additionally be used as an alternation criterion. Finally, specific variable dependencies between subgoals may be taken into account.

Acknowledgements

We want to thank Klaus Mayr for lots of fruitful discussions, Christoph Goller for checking the specification of choice point alternation, and Max Moser for providing the STE-modification.

References

[AS92]   O. L. Astrachan, M. E. Stickel. Caching and Lemmaizing in Model Elimination Theorem Provers. *CADE-11*, LNAI 607, p. 224–238, Springer, 1992.

[Bra75] D. Brand. Proving theorems with the modification method. *SIAM Journal of Computing*, 4(4):412–430, 1975.

[Fit90]   M. Fitting. *First-Order Logic and Automated Theorem Proving*, Springer, 1990.

[GLMS94] C. Goller, R. Letz, K. Mayr, J. Schumann. SETHEO V3.2: Recent Developments – System Abstract –. *CADE-12*, LNAI 814, p. 778–782, Springer, 1994.

[Har96]   J. Harrison. Optimizing proof search in model elimination. *CADE-13*, LNAI 1104, p. 313–327, Springer, 1996.

[Let93]   R. Letz. *First-Order Calculi and Proof Procedures for Automated Deduction*. Dissertation, Technische Hochschule Darmstadt, 1993.

[LSBB92] R. Letz, J. Schumann, S. Bayerl, W. Bibel. SETHEO: A High-Performance Theorem Prover. *Journal of Automated Reasoning*, 8:183–212, 1992.

[LMG94]   R. Letz, K. Mayr, C. Goller. Controlled Integration of the Cut Rule into Connection Tableau Calculi. *Journal of Automated Reasoning*, 13:297–337, 1994.

[Lov78]   D. W. Loveland. *Automated theorem proving: A logical basis*. North Holland, New York, 1978.

[MLS95]   M. Moser, C. Lynch, J. Steinbach. Model elimination with basic ordered paramodulation. Technical Report AR-95-11, Institut für Informatik, Technische Universität München, 1995.

[Mos96]   M. Moser. *Goal-directed reasoning in clausal logic with equality*. Dissertation, Institut für Informatik, TU München, 1996.

[Mos97]   M. Moser. Compiling basic paramodulation to logic. Technical report, Institut für Informatik, TU München, 1997.

[MIL+97] M. Moser, O. Ibens, R. Letz, J. Steinbach, C. Goller, J. Schumann, K. Mayr. SETHEO and E-SETHEO. *Special issue of the Journal of Automated Reasoning*, to appear 1997.

[RW69]   G. A. Robinson, L. Wos. Paramodulation and theorem proving in first-order theories with equality. *Machine Intelligence*, 4:135–150, 1969.

[Sho76]   R. E. Shostak. Refutation graphs. *Artificial Intelligence*, 7:51–64, 1976.

[Sti88]   M. E. Stickel. A Prolog technology theorem prover: Implementation by an extended Prolog compiler. *Journal of Automated Reasoning*, 4:353–380, 1988.

[SSY94]   G. Sutcliffe, C. B. Suttner, T. Yemenis. The TPTP problem library. *CADE-12*, LNAI 814 of, p. 778–782, Springer, 1994.

Projection: A Unification Procedure for Tableaux in Conceptual Graphs

Gwen Kerdiles

Faculty of Philosophy (University of Amsterdam) and LIRMM (University of Montpellier), 161 Rue Ada, 34392 Montpellier Cedex 5, France
E-mail: kerdiles@lirmm.fr

Abstract. Conceptual Graphs offer a formalism for knowledge representation in Artificial Intelligence, inspired by both order-sorted logic and Peirce's Existential Graphs. These graphical structures provide an attractive and intuitive representation of information and are particularly suitable for human-machine interfaces. Conceptual Graphs borrow from order-sorted logic the notion of sort. Sorting not only provides an intuitive classification of objects of the language, but also an efficient way of restricting search spaces (for example, in unification).

The formalism calls for efficient systems of reasoning in order to compete with logical programming. Projection is one such tool for a language limited to conjunction and existential quantification (Simple Conceptual Graphs). Projection is very efficient for certain classes of Conceptual Graphs and offers an original approach to deduction: the perspective of graph matching.

The aim of this paper is twofold: enrich the language of Simple Conceptual Graphs with implication and negation, and propose an efficient analytic deduction system that combines analytic tableaux with projection.

1 Introduction

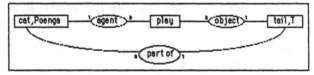

Fig. 1. About Poenga's favourite occupation.

Conceptual Graphs are a simple and expressive knowledge representation formalism introduced by Sowa in [Sow84]. They combine in a natural graphical presentation[1] Semantic Networks of Artificial Intelligence with the logical graphs of Peirce and order-sorted logic. Knowledge is represented in a simple way by bipartite graphs: *concept nodes* alternate with *relation nodes*. For instance, the information "the cat, Poenga, plays with her tail" may be paraphrased by "the cat, Poenga is the agent of playing with the object, the tail T, which is part of Poenga". In a simplified Conceptual graph notation, this can be represented as in Fig. 1.

---

[1] This feature is particularly adapted to web interfaces: there are several sites "in construction" to provide knowledge storage and retrieval, in the Conceptual Graph formalism, via Internet.

Poenga and T are called individual markers; *cat* and *tail*, concept types; *agent*, *object* and *part of*, relations. There is a special individual marker, *, which plays the role of an indefinite marker, e.g., in Fig. 1, if *Poenga* is replaced by *, then the information: "there is a cat who plays with its tail, T" is represented. Conceptual Graphs adopt from order-sorted logic a hierarchy of concept types. The possibility of partitioning the universe of a formal language has been fruitfully investigated since the pioneering work of Herbrand. Oberschelp and Mahr give an overview of the advantages of sorting in [BHR90]. Some work in Automated Deduction has highlighted the benefits brought by sorting in terms of limitation of search space, e.g., [SS89] studies some computational aspects of different kinds of order-sorted logics with or without equational theories. Conceptual Graphs also introduce an order over relations of common arity. For instance, a ternary relation *to be in the middle of* can be defined as a subtype of a ternary relation *to be between*.

Projection is an inference mechanism for the language of Simple Conceptual Graphs, informally presented so far. This tool was investigated in graph theory by Chein and Mugnier [CM92]. They also proved the completeness of projection and compared it to some Constraint Satisfaction Problems. Projection is a labeled graph morphism which can be efficiently implemented (e.g., the Corali system, LIRMM, University of Montpellier). Projection yields linear complexity for certain classes of graphs and is therefore be the cornerstone of the proposed proof procedure.

Section 3 presents an extension of the Simple Conceptual Graph language which includes implication and negation. In Section 4, a tableau procedure for the extended language is defined. It decomposes the sentences with analytic tableau rules (e.g., [Smu68]) and conclude proofs by using projection. Satchmo, [MB88], a theorem prover based on a model-generation paradigm and implemented in Prolog takes benefit from range-restricted clauses which are inherent to many-sorted logic (and therefore, order-sorted logic). This approach is very close to a tableau calculus for order-sorted logic, e.g., [SW90], but in both methods, conclusions are drawn by matching at the atomic level, whereas we propose to match whole Simple Conceptual Graphs corresponding to existentially closed conjunctions of atoms.

Structural Skolemisation [BL94] is an efficient tool in Resolution, however it modifies the information represented. As a human oriented theorem prover is desired (all steps of a proof should be kept readable and intuitive), and as our language does not contain functional terms, the advantages of structural Skolemisation are obtained without representing the Skolemised terms: Skolem functions can be replaced by accessibility relations in the proof structure, as suggested by Reyle and Gabbay (Appendix of [RG94]).

2 Simple Conceptual Graphs

Let us recall some slightly modified basic definitions which are fully developed, for instance in [Sow84] or [CM92]. These definitions are adopted to a large extent by more recent work, e.g., [PMC95], [SM96], or The Peirce-project (http:// www.cs.adelaide.edu.au/ users/ peirce).

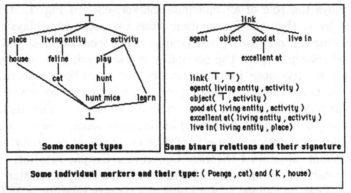

Fig. 2. A partial definition of a support.

Definition 1 (Supports). A support represents an ontology of a specific application domain (for example, Fig. 2 pictures partially a support). Briefly, it is a 5-tuple $\Sigma = (T_C, T_R, \sigma, M, \tau)$ where:

- T_C is a poset of concept types with a supremum $\top$ and an infimum $\bot$ which is also called *the absurd type*.
- T_R is a partition of posets of relation names.
- M is an infinite set of markers partitioned into a set of individual markers I and a set of variable markers V. There are infinitely many individual markers of each concept type except for $\bot$: *false* is the only marker of the absurd type. τ associates to any individual marker the type of the individual represented.
- σ associates to any relation, its arity and the maximal type of each of its arguments (also called, *signature* of the relation).

Definition 2 (SCGs). A Simple Conceptual Graph (SCG) related to a support Σ, is a bipartite, finite and not necessarily connected multigraph $G = (R, C, U, label)$ such that:

- R and C denote the two classes of relation and concept vertices.
- *label* is a mapping respecting σ which associates to a relation vertex its name, and to a concept vertex $c \in C$ a pair $(type(c), ref(c))$ where:
  1. $ref(c) \in (M \cup \{*x/x \in V\})$. If $ref(c) = *x$ then c is called *a declaration of x* and a function, *marker*, is defined by $marker(c) = x$, otherwise $marker(c) = ref(c)$. If $ref(c) = x \in M \setminus \{false\}$, c is said to be a *free occurrence of x*.
  2. if $ref(c) \in I$ then $type(c) = \tau(ref(c))$, else $type(c) \in T_C \setminus \{\bot\}$.
  3. For any two distinct concept nodes $c1$ and $c2$ in C, it holds that $marker(c1) \neq marker(c2)$.
- U is the set of edges such that the edges incident to each relation vertex are totally ordered and a concept vertex with $marker = false$ cannot be connected to an edge.

Simple Conceptual Graphs are sometimes defined as oriented graphs but the total order on the neighbours of a relation vertex provides a representation closer to predicate logic formulae.

Dec(G) (respect. **Free(G)**) is the set of pairs $(type(c), marker(c))$ such that c is a declaration (respect. a free occurrence) in G.

For any support, two Simple Conceptual Graphs have a special feature: **False** $= (\emptyset, \{c\}, \emptyset, label(c) = (\bot, false))$ and **Emptygraph** $= (\emptyset, \emptyset, \emptyset, \emptyset)$.

A SCG is **closed** if and only if every marker that occurs free in it, is an individual marker.

Merging of two SCGs: $G \uplus H$, the merging of two SCGs, G and H, is obtained by first juxtaposing the two graphs and then, merging all concept vertices with a common *marker*. If one of the concept vertices of a merging class is a declaration then the resulting vertex is a declaration too. Note that up to a renaming of variable markers (indeed, a variable marker may be associated to a different concept type in each of the two graphs), the merging of two SCGs is a SCG and that the merging of two closed SCGs is closed.

Definition 3 (Projections). A projection from a closed SCG, $G = (R_G, C_G, U_G, label_G)$, to a closed SCG, $H = (R_H, C_H, U_H, label_H)$, is either the *empty function* if there exist a concept node c in C_H such that $label(c) = (\bot, false)$ or a total function π from $(R_G \cup C_G)$ to $(R_H \cup C_H)$ such that:

- $\forall r \in R_G, \pi(r) \in R_H$ and $label(\pi(r)) \leq_{T_R} label(r)$
- $\forall c \in C_G, \pi(c) \in C_H, type(\pi(c)) \leq_{T_C} type(c)$ and if $ref(c) \in I$ then $ref(\pi(c)) = ref(c)^2$
- $\forall r \in R_G$ and $\forall 1 \leq i \leq arity(r), \pi(G_i(r)) = H_i(\pi(r))$ where $X_i(r)$ is the ith-neighbour of r in X.

Example 4. Suppose that we have a support which is partially pictured in Fig. 2. The closed SCG, H, in Fig. 3 is a representation of "The cat, Poenga, who lives in the house, K, learns to hunt mice and is excellent at hunting mice". G is a representation of "There is a feline who learns to hunt and who is good at playing, and there is a cat who lives in a house". There is a projection from G to H (in other words, G follows from H and the information conveyed by the support) which is represented in Fig. 3 by arrows from nodes of G to nodes of H.

Projection can be efficiently implemented. Indeed, despite the fact that in general the problem of determining whether there is a projection from a closed

---

2 It is important to note that for the language of Simple Conceptual Graphs, variable markers are not primordial; Indeed, for the deduction process, projection, it is sufficient to have in the referent field of a concept node either an individual marker or a '*' (without any explicit variable name) because a concept node with an individual marker is projected on a concept node with the same marker and a declaration is projected on any concept node of a subtype.

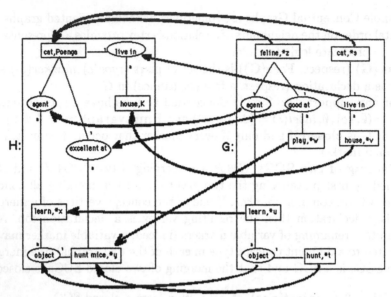

Fig. 3. A projection between two closed SCGs.

SCG to another closed SCG is an NP-complete problem (see [MC92]), when restricted to certain classes of graphs, the problem is polynomial. Mugnier and Chein present in [MC92], a polynomial algorithm for computing a projection from a closed SCG, which is a tree, to another closed SCG (of any form) and for computing the number of such projections. Their algorithm is exploited as a preprocessing routine in the general case.

3 Conceptual Graphs

[SM96] and [Gho96] study rules of the form "a Simple Conceptual Graph implies a Simple Conceptual Graph". These rules can be associated to SCG bases to generate SCGs which follow from a base and some rules (Forward Chaining) or prove that some SCGs can be derived (Backward Chaining), however, in the CG rule language, information such as (i) "The cat, Poenga, lives in a house and every feline who lives in that house learns to hunt mice" cannot be represented. In order to reach the expressivity of full predicate logic (without functional terms), the language of SCGs is enriched with a binary symbol ($\Rightarrow$) for implication, e.g., Fig. 4 is a representation of (i). The referent field of concept nodes provides the key for the binding of free occurrences of variable markers by declarations and the interlocking structure defines the scope of those declarations.

Conceptual Graph languages as expressive as the one presented are proposed for instance in [Sow84], [Ber93] and [Wer95]. These languages are very close to Peirce's notation, using the notions of sheet of assertion and negative box. Although, readability of both notations (negative boxes vs implicative forms) is defensible, the graphical deduction systems proposed in the quoted papers, i.e., adaptations of Peirce's graphical deduction rules, are not analytic and therefore are not easily implementable. We could have adopted a notation with negative boxes (it would have even simplified the tableau rules and projection could have been used similarly) but implicative forms seem to be more intuitive and are closer to the syntax of spread languages, e.g., Prolog style languages.

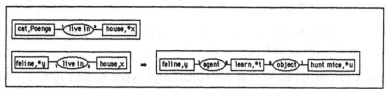

Fig. 4. A Conceptual Graph.

Definition 5 (CGs). A Conceptual Graph(CG), G, is a set of entries containing exactly one atomic entry, where an entry is either

- atomic if it is a SCG
- or complex: $G1 \Rightarrow G2$ where $G1$ and $G2$ are CGs

Notation: A Conceptual Graph which is a singleton (its only entry is a SCG) is called atomic. Sets of entries are represented by juxtaposing them inside a frame (see Fig. 4). To lighten the representation, a single external frame is drawn for an atomic CG. When there is no ambiguity, we omit to draw the *Emptygraph* and a frame around the SCG *False*. For example, (i) represents the negation of "there is a cat who lives in the house K", whereas (ii) should be drawn.

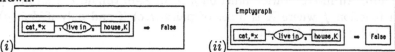

Definition 6. For any Conceptual Graph G, we define (or extend some definitions of Sect. 2):

1. **atomic** is a function which takes a CG and returns its atomic entry and **complex** is a function which returns the set of complex entries of a CG.
2. **Degree** of G: if G is atomic then $degree(G) = 0$, else $degree(G)$ is the sum of the degrees of the complex entries in G.
   The degree of a complex entry $G1 \Rightarrow G2$ is $1 + degree(G1) + degree(G2)$.
3. $Dec(G) = Dec(atomic(G))$ is the set of **declared** variable markers in the atomic entry of G. For instance, let G be the Conceptual Graph represented in Fig. 4, $Dec(G) = \{(house, x)\}$.
4. $Free(G)$ is the set of **free** markers in G:
   $Free(G) = (\bigcup_{E \in G} Free(E)) \setminus Dec(G)$ where for a complex entry E of the form $G1 \Rightarrow G2$, $Free(E) = Free(G1) \cup Free(G2) \setminus Dec(G1)$.

5. A CG G is called **closed** if and only if every marker that occurs free in G is an individual marker.

6. A CG is **pure** if it does not contain a variable marker declared twice and if none of its free variable markers occurs in a declaration. A CG which is not pure can easily be transformed into an equivalent pure CG by proceeding outside-in and renaming re-declared variable markers.
   In the rest of the paper, only pure Conceptual Graphs are considered.

7. The **merging** of two CGs, G and H, is the CG: $G \uplus H = \{atomic(G) \uplus atomic(H)\} \cup complex(G) \cup complex(H)$.

Definition 7 (Models). A model, $M_\Sigma = ((Dom, \subseteq), F)$ with respect to a support $\Sigma = (T_C, T_R, \sigma, M, \tau)$ consists of

- a poset $(Dom, \subseteq)$ such that for all concept types t_1 and t_2 in T_C which are not the absurd type, it holds that Dom_{t_1} and Dom_{t_2} are elements of Dom and if $t_1 \leq_{T_C} t_2$ then $Dom_{t_1} \subseteq Dom_{t_2}$. These sub-domains of interpretation respect the ordering conveyed by the ontology (support) and the property of non-empty universe(s): $\forall Dom_t \in Dom, Dom_t \neq \emptyset$.
- an interpretation function F on $(I \setminus \{false\}) \cup T_R$ (individual markers and relation types of the support) verifies:
  - if $c \in (I \setminus \{false\})$ then $F(c) \in Dom_{\tau(c)}$,
  - if $R \in T_R$ and $arity(R) = n$ then $F(R) \subseteq Dom_{\sigma_1(R)} \times \ldots \times Dom_{\sigma_n(R)}$ where $\sigma_i(R)$ is the maximal type of the ith-argument of R,
  - $\forall R, R' \in T_R, R \leq_{T_R} R'$ implies that $F(R) \subseteq F(R')$.

  The interpretation function also respects the chosen ontology. F_I is the interpretation function of the individual markers: the restriction of F to the set of individual markers (without $false$).

Notation: An interpretation function g is an **X-extension** of an interpretation function f where X is a set of pairs (t, c) if and only if $dom(g) = dom(f) \cup \{c/(t, c) \in X\}$ and $g \supseteq f$.

Definition 8 (Truth of a CG). The truth of a CG, G, in a model $M_\Sigma = ((Dom, \subseteq), F)$ under an interpretation f (a partial function from $M \setminus \{false\}$ to $Dom_\top$), noted $M_\Sigma \models_f G$, is defined by:

1. $M_\Sigma \models_f G$ if and only if $\exists g$, $Dec(G)$-extension of f, such that for every entry E in G, it holds that $M_\Sigma \models_g E$

2. $M_\Sigma \models_g G1 \Rightarrow G2$ if and only if $\forall h$, $Dec(G1)$-extension of g, $M_\Sigma \models_h G1$ implies $M_\Sigma \models_h G2$

3. for the case of an atomic entry E: $M_\Sigma \models_g E$ if and only if
   (a) there is no concept node c in E such that $marker(c) = false$ and
   (b) for every concept node c in E, it holds that $g(marker(c)) \in Dom_{type(c)}$ and
   (c) for every relation node r of arity n in E, $< g(marker(c_1)), \ldots, g(marker(c_n)) > \in F(label(r))$, where c_i is the ith-neighbour of r in E.

1. E is atomic:
   (a) To each concept node c such that $ref(c) = false$, associate an atom $\neg X$ where X is some tautology.
   (b) To each relation node r in E such that $label(r) = R$, associate an atom
   $R((type(c_1), marker(c_1)), \ldots, (type(c_n), marker(c_n)))$ where n is the arity of R and c_i the ith-neighbour of r in E
   (c) $\Phi(E)$ is the conjunction of these atoms.
2. E has the form $G1 \Rightarrow G2$: let $< x1, \ldots, xn >$ be any ordering of $Dec(G1)$ (n=0 if empty), if $G1$ contains neither relation node nor $False$ then $\Phi(E) = \forall x1, \ldots, xn(\Phi(G2))$, otherwise $\Phi(E) = \forall x1, \ldots, xn((\Phi(E'_1) \wedge \ldots \wedge \Phi(E'_m)) \rightarrow \Phi(G2))$ where $< E'_1, \ldots, E'_m >$ is any ordering of $G1$.

For example, a translation of the CG in Fig. 4 can be:
$\exists(house, x)[live\ in((cat, Poenga), (house, x)) \wedge$
$\forall(feline, y)[live\ in((feline, y), (house, x)) \rightarrow \exists(learn, t)\exists(hunt\ mice, u)$
$[agent((feline, y), (learn, t)) \wedge object((hunt\ mice, u), (learn, t))]]]$

The translation into a CG related to a support Σ, $\Upsilon(\phi)$, of any formula ϕ of PL_Σ is given by:

1. $\Upsilon(r(t_1, \ldots, t_n)) = \boxed{(R, C, U, label)}$ where
   $R = \{r_1\}$, $U = \{(r_1, c_1), \ldots, (r_1, c_n)\}$,
   $C = \{c_1, \ldots, c_n\}$, such that $c_i = c_j$ $(1 \le i \ne j \le n)$ if and only if $t_i = t_j$
   $label(r_1) = r$ and $label(c_i) = t_i$.
2. $\Upsilon(\phi \vee \psi) = \Upsilon(\neg(\neg\phi \wedge \neg\psi))$
3. $\Upsilon(\phi \wedge \psi) = \Upsilon(\phi) \uplus \boxed{Emptygraph} \Rightarrow \Upsilon(\psi)$
4. $\Upsilon(\phi \rightarrow \psi) = \boxed{\Upsilon(\phi) \Rightarrow \Upsilon(\psi)}$
5. $\Upsilon(\neg\phi) = \boxed{\Upsilon(\phi) \Rightarrow False}$
6. $\Upsilon(\forall(t, x)\phi) = \boxed{\boxed{t,^*x} \Rightarrow \Upsilon(\phi(x))}$
7. $\Upsilon(\exists(t, x)\phi) = \boxed{t,^*x} \uplus \boxed{Emptygraph} \Rightarrow \Upsilon(\phi)$

4 Combining tableaux with projections

Five disjoint classes of signed Conceptual Graphs are distinguished:

1. We call α a signed CG of the form $+ \boxed{\begin{array}{c}\alpha_1 \\ \ldots \\ \alpha_n\end{array}}$, $1 \le n$ such that if $n = 2$ then $atomic(\alpha) \ne Emptygraph$ and if $n = 1$ then $Dec(\alpha) \ne \emptyset$.

2. We call β a signed CG of the form $- \boxed{\begin{array}{c}\beta_1 = atomic(\beta) \\ \ldots \\ \beta_n\end{array}}$, $1 < n$ such that if $n = 2$ then $atomic(\beta) \ne Emptygraph$.

- A CG, G, related to a support Σ, is **satisfiable** if and only if there exists a model M_Σ and an interpretation function f, which is a $Free(G)$-extension of F_I, and such that $M_\Sigma \models_f G$. G is also called *satisfiable in* M_Σ, if the model is given.
- A finite set of CGs, S, related to a support Σ, is satisfiable if and only if there exists a model M_Σ and an interpretation function f, which is a $Free(S)$-extension of F_I, and such that, for every G in S, it holds that $M_\Sigma \models_f G$. $Free(S)$ is the union of all $Free(G)$, $G \in S$.
- A CG, G, related to a support Σ, is **valid** if and only if for every model M_Σ and for every interpretation function, f, which is a $Free(G)$-extension of F_I, it holds that $M_\Sigma \models_f G$.

Theorem 9 (Soundness and Completeness of projection). *It follows from [Sow84] and [CM92] that projection is sound and complete for closed SCGs, i.e., given a support Σ, there is a projection from a closed SCG, G, to a closed SCG, H, if and only if for any model M_Σ, H is true in M_Σ implies that G is true in M_Σ.*

Definition 10 (Signed Conceptual Graphs). Under any interpretation, the truth value of a signed CG, $+G$ (G is called positive), is the same as that of G. The one of $-G$ (G is called negative) is the same as that of $\boxed{G \Rightarrow False}$. $Dec(\pm G) = Dec(G)$, $Free(\pm G) = Free(G)$ and $atomic(\pm G) = atomic(G)$.

Definition 11 (Substitution functions). For two sets of pairs (t, m), X and Y, such that t is a concept type, m is a marker different from $false$, and if m is an individual marker, then $t = \tau(m)$, $f : X \to Y$ is a substitution function if and only if $f((t, m)) = (t', m')$ implies (i) $t' \leq_{T_C} t$ and (ii) If m is an individual marker then $t' = t$ and $m' = m$.

Abusing notations, substitution functions are also applied to Conceptual Graphs, sets of Conceptual Graphs or labels of concept nodes. In Fig. 3, the substitution function which would correspond to the pictured projection is the following function π': $\pi'((feline, z)) = \pi'((cat, s)) = (cat, Poenga)$, $\pi'((learn, u)) = (learn, x)$, $\pi'((hunt, t)) = \pi'((play, w)) = (hunt\ mice, y)$ and $\pi'((house, v)) = (house, K)$.

Definition 12 (Translations). A translation of any CG (not necessarily pure), G, into PL_Σ (a language of order-sorted logic which has a common vocabulary with a support Σ and which respects the same constraints on types) is $\Phi(G) = \exists x1, \ldots, xn(\Phi(E_1) \wedge \ldots \wedge \Phi(E_m))$ where $< E_1, \ldots, E_m >$ is any ordering of the entries of G and $< x1, \ldots, xn >$ is any ordering of $Dec(G)$ (if $Dec(G)$ is empty then n=0). If G contains neither relation node nor $False$ then $(\Phi(E_1) \wedge \ldots \wedge \Phi(E_m))$ is some tautology.

Let E be an entry, $\Phi(E)$ is defined by:

3. We call γ a signed CG of the form $+$ $\boxed{\gamma_1 \Rightarrow \gamma_2}$.
4. We call δ a signed CG of the form $-$ $\boxed{\delta_1 \Rightarrow \delta_2}$.
5. We call χ a negative atomic CG which does not contain any relation node.

For convenience, we talk about $\alpha_1, \ldots, \alpha_n$ as the entries of a node α (Idem for β, γ or δ).

Definition 13 (Tableaux). A tableau is a pair, $(\mathcal{T}, \Pi)$, composed of an ordered (the successors of a node are ordered) tree $\mathcal{T}$ whose nodes are occurrences of signed CGs and *a substitution function* Π. $(\mathcal{T}_0, \Pi_0)$ is a tableau for a CG G where $\mathcal{T}_0$ is a single node $-G$ and Π_0 is the identity function on the set of all pairs $(type(c), marker(c))$ where c is a concept node in G. Suppose $(\mathcal{T}, \Pi)$ is a tableau for G, let H be a leave of $\mathcal{T}$, then the tableau may be extended by either of the following rules:

Rule for α's: If an α occurs on the path P_H, then let Θ_α be a substitution bijection which associates to every marker in $Dec(\alpha)$ **a new individual marker** of the same type ("new" in the sense that it does not appear so far in $\mathcal{T}$) and define $\alpha'_{1 \leq i \leq n} = +\Theta_\alpha(\boxed{\alpha_i})$. We may adjoin successively $\alpha'_1, \ldots, \alpha'_n$ such that α'_1 is the sole successor of H and $\alpha'_{2 \leq i \leq n}$ is the sole successor of α'_{i-1} (if $n = 1$ then α'_1 is the sole successor of H).

Rule for β's: If a β occurs on the path P_H, then let Θ_β be a substitution bijection which associates to every marker in $Dec(\beta)$, **a new variable marker** of the same type and define $\beta'_{1 \leq i \leq n} = -\Theta_\beta(\boxed{\beta_i})$. We may simultaneously adjoin β'_1 (as leftmost successor of H) to β'_n as successors of H.

Rule for γ's: If a γ occurs on the path P_H, then let Θ_γ be a substitution bijection which associates to every marker in $Dec(\gamma_1)$, **a new variable marker** of the same type, and to every marker in $Dec(\gamma_2)$, **a new individual marker** of the same type. $\gamma'_1 = -\Theta_\gamma(\gamma_1)$ and $\gamma'_2 = +\Theta_\gamma(\gamma_2)$. We may simultaneously adjoin γ'_1 as the left successor of H and γ'_2 as the right successor of H.

Rule for δ's: If a δ occurs on the path P_H, then let Θ_δ be a substitution bijection which associates to every marker in $Dec(\delta_1)$, **a new individual marker** of the same type, and to every marker in $Dec(\delta_2)$, **a new variable marker** of the same type. $\delta'_1 = +\Theta_\delta(\delta_1)$ and $\delta'_2 = -\Theta_\delta(\delta_2)$. We may adjoin successively δ'_1 as the sole successor of H and δ'_2 as the sole successor of δ'_1.

Rule for χ's: If some χ occurs on the path P_H, then let Θ_χ be a substitution function, which associates to every referent in $Dec(\chi)$ **any individual marker** of a subtype and define $\chi' = +\Theta_\chi(\chi)$. We may adjoin χ' as the sole successor of H.

If one of the preceding rules, $x \in \{\alpha, \beta, \gamma, \delta, \chi\}$, is applied, then Π must be adapted to the newly introduced markers: Π becomes $(Id_{codom(\Theta_x)} \cup \Pi)$.

Definition 14 (Proofs). A proof of G is a tableau for G which has all of its branches closed. A branch B of a tableau, $(\mathcal{T},\Pi)$, **closes** if there are in B a closed atomic node $-\boxed{H}$, and some closed atomic nodes $+\boxed{G_1}$, $\dots$, $+\boxed{G_n}$ such that there exists a projection from $\Pi(H)$ to $(\uplus_{1 \le i \le n}\Pi(G_i))$ with associated substitution function π. After closing the branch B, Π becomes $(\pi \circ \Pi)$.

Theorem 15 (Completeness). *If a CG, G, related to a support Σ is valid if and only if there exists a closed tableau for G.[3]*

Simple Conceptual Graphs could have been decomposed into elementary graphs, i.e., simple graphs containing at most one relation node, as is done in Prolog, [Gho96] or in the DRS-calculus of [RG94], but to take benefit from projection, it is more efficient to gather all these atoms into SCGs which correspond to their conjunction. Indeed, performing a complete decomposition would correspond to a classical tableau calculus for a CG language. The gain brought by the graphical representation (graph matching) would be lost and it would then be equivalent to use the translations of Def. 12 and a tableau calculus for order-sorted logic, e.g., the one presented in [SW90].

Example 16 (A closed tableau). Suppose that we want to prove that a graph (4) representing the information "there is a cat of a certain colour who learns to hunt mice" follows from the graphs representing the information "The cat, Poenga, is white", "Poenga lives in the house, K" and "Every feline who lives in a house learns to hunt mice". We may start a tableau, $(\mathcal{T},\Pi)$, where $\mathcal{T}$ is a single branch with the nodes (1), (2), (3) and (4), Π is the identity function on the set of pairs $(type(c), marker(c))$ such that c is a concept node in $\mathcal{T}$ (for conciseness, we omit in the construction of the tableau, the first two steps which only decompose the theorem to be proved).

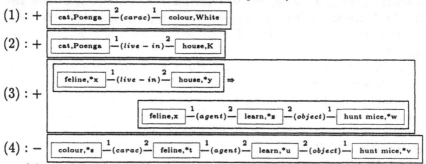

(3) is a δ, thus we may extend the tableau by adding two direct successors to (4): (5) and (6) where x' and y' are new variable markers, a and b are new individual markers of the respective types *learn* and *hunt mice*. Π remains the identity function but extended to the newly introduced markers.

---

[3] The proof of this theorem proposed in [Ker97] is a classical model-generation proof using the completeness of the projection procedure.

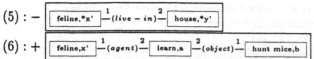

$(5) : -$ | feline,*x' $\overset{1}{-}(live-in)\overset{2}{-}$ house,*y'

$(6) : +$ | feline,x' $\overset{1}{-}(agent)\overset{2}{-}$ learn,a $\overset{2}{-}(object)\overset{1}{-}$ hunt mice,b

Note that (6) cannot yet play a role in closing a branch because projection is defined on closed graphs and x' occurs free in (6). The underlying notion is Skolem dependencies. There is a projection from (5) to (2) which closes the left branch and Π remains the identity function except for x' and y': $\Pi((feline, x')) = (cat, Poenga)$ and $\Pi((house, y')) = (house, K)$. There is now another projection which closes the tableau, from (4) to the merging of $\Pi(1)$ and $\Pi(6)$. The theorem is proved and the final value of Π tells that "the white cat, Poenga" is a solution to the request.

The language of Conceptual Graphs does not include functional terms, thus, Skolem dependencies have to be expressed in a different way. Reyle and Gabbay propose in appendix of [RG94], a deduction system for Discourse Representation Structures where Skolem dependencies are controlled by a notion of accessibility relation in a proof structure. The tree structure of a tableau is appropriate for obtaining similar properties in our proof system: a branch closes only if there is a projection between nodes of that branch, but, as projection is defined only on closed atomic CGs, the branches are forced to close in a certain order which is decided by the Skolem dependencies. There are essentially two methods of Skolemising a formula: structural Skolemisation, i.e., elimination of strong quantifiers at their position, and prefix Skolemisation where the formula is first transformed in prenex form. It is shown in [BL94] that compared to structural Skolemisation, prefix Skolemisation may result in a nonelementary increase of computational complexity. Although, the concept of Skolemisation adopted is the structural one, proofs are often more complex than they would be if we had functional terms: a greater number of individual markers are introduced within the proofs than if we could transform the graphs into a structural Skolemised form before applying the tableau rules.

Example 17. Suppose that we have a concept type t, two unary relations P and Q and a binary one R. We omit to write concept types because every marker in the example is of type t. Show that the following closed formula, F, is not valid.

$$F \equiv \exists a((P(a) \wedge \forall u(P(u) \to \exists y(Q(y) \wedge R(u,y)))) \to (\exists w(Q(w) \wedge \forall x(P(x) \to R(x,w)))))$$

A structural Skolemisation of F is the following formula:

$$(P(c_1) \wedge \forall u(P(u) \to (Q(f_1(u)) \wedge R(u, f_1(u))))) \to (\exists w(Q(w) \wedge (P(f_2(w)) \to R(f_2(w), w))))$$

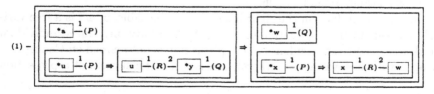

$\delta(1)$: (2), (3) where c_1 is a new individual marker and w' a new variable marker.

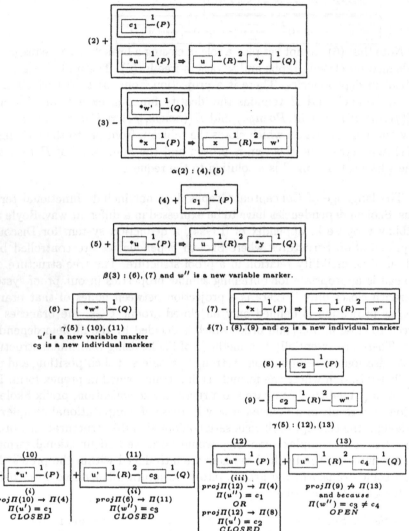

Continuing (13) with $\gamma(5)$ does not help because y is always replaced by a new individual marker and therefore, we cannot project (9) on the resulting graph. Continuing (13) with $\beta(3)$ introduces two branches. On the left one, we have a new declared variable w''' which by projection unifies with c_4. On the right branch, x is replaced by a new individual marker c_5 and therefore, there is no projection from new nodes to (13) because $c_5 \neq c_1$ and $c_5 \neq c_2$. The right branch remains open.

In (7), and in fact on the right hand side of the tableau, w'' is a free variable marker. In the application of δ on (7), the newly introduced individual marker c_2 becomes dependent on w'' (projection is not defined on Conceptual Graphs with free variable markers), precisely on the success of a projection of the graph (which is on the left hand side of the tableau) containing the declaration of w''. Therefore, there is no need of explicitly expressing this dependence (by introducing $f_1(w'')$ in place of c_2 as it is done in the Skolemised for of F). The same phenomena occur with c_3 and c_4 that are dependent on respectively u' and u''. From an implementation perspective, separating $\mathcal{T}$, the information obtained while constructing a tableau $(\mathcal{T}, \Pi)$, from Π, the substitutions associated to projections which close branches, enables backtracking in cases were different projections could close a single branch. For instance, in the branch containing node (12) there are two possible projections that close the branch. When trying one of them does not help, we may backtrack and try the other one (In this example, it does not help neither).

5 Conclusion

An extension of the formalism of Simple Conceptual Graphs and a sound and complete deduction system are proposed. The tableau rules decompose a complex graph into *smaller* ones and projection is the mechanism for closing a branch. Together, the tableau structure and projection provide a control on the choices of instantiations and therefore, on the soundness of the whole deduction procedure. Contrary to other complete systems for Conceptual Graphs (e.g., [Sow84], [Ber93] and [Wer95] present adaptations of Peirce's deduction rules), the proposed deduction system has an essential property in the perspective of an implementation: it is analytic. Nevertheless, the possibility of restricting the range of declarations by means of shifting should be considered. It was proved in [BL94] that this pretreatment never increases the complexity of proofs. Interlacing operations in Graph Theory (projection) and a classical method of Logic (analytic tableaux) offers a novel perspective on deduction (graph matching). It also provides a way to combine results from these two intensively studied fields.

Further work will study the inclusion of Nested Conceptual Graphs (see [PMC95]) in the proposed system. These graphs extend the formalism of Simple Conceptual Graphs by adding a *description* field (which is itself a Nested Conceptual Graph) in the concept nodes. *Zoom in* on a concept node is possible. In that formalism, it is easy and intuitive to represent the information "there is a photo hanging on the wall and this photo is one of Poenga playing with her tail": the second part of the sentence is represented by a graph which occurs as the description of the concept node representing "there is a photo". [Ber93] has presented rules (in Peirce's style) for different modal logics in the framework of Existential Graphs. Tableaux for modal logics may also be combined with projection.

Acknowledgment: This paper was written in the framework of a cooperation program between the University of Montpellier and the University of Amsterdam. I am very grateful to Anne Preller, Frank Veltman and Dick de Jongh. I would like to thank the anonymous referees for their valuable comments. Many thanks to Bridget and Hervé for turning into English so many of those French flavoured sentences.

References

[Ber93]   H. v.d. Berg. *Knowledge Graphs and Logic: One of two kinds*. PhD thesis, Universiteit Twente, September 1993.

[BHR90]   K.H. Bläsius, U. Hedtstück, and C.-R. Rollinger, editors. *Sorts and Types in Artificial Intelligence*, volume 418 of *LNAI*. Springer-Verlag, 1990. Proceedings of Workshop Eringerfeld, FRG, April 1989.

[BL94]   M. Baaz and A. Leitsch. On Skolemization and Proof Complexity. *Fundamenta Informaticae*, 20:353–379, 1994.

[CM92]   M. Chein and M.L. Mugnier. Conceptual Graphs, Fundamental Notions. *RIA*, 6.4:365–406, 1992.

[Gho96]   B.C. Ghosh. *Conceptual Graph Language : A Language of Logic and Information in Conceptual Structures*. PhD thesis, Asian Institute of Technology, Bangkok, Thailand, February 1996.

[Ker97]   G. Kerdiles. Analytic Tableaux for an extended language of Conceptual Graphs. *RR LIRMM*, 97002, 1997.

[MB88]   R. Manthey and F. Bry. SATCHMO: a theorem prover implemented in Prolog. *LNCS*, 310:415–434, 1988. CADE 9.

[MC92]   M.L. Mugnier and M. Chein. Polynomial algorithms for projection and matching. In *Proceedings of the 7th Workshop on Conceptual Graphs*, 1992. New Mexico State University, Las Cruces, New Mexico.

[PMC95]   A. Preller, M.L. Mugnier, and M. Chein. Logic for Nested Graphs. *Computational Intelligence Journal*, 95-02-558, 1995.

[RG94]   U. Reyle and D.M. Gabbay. Direct Deductive Computation on Discourse Representation Structures. *Linguistics and Philosophy*, 17:343–390, August 1994.

[SM96]   E. Salvat and M.L. Mugnier. Sound and Complete Forward and Backward Chaining of Graph Rules. In *proceedings of ICCS'96*, volume 1115 of *LNAI*, pages 248–262. Springer-Verlag, 1996.

[Smu68]   R.M. Smullyan. *First-Order Logic*. Springer-Verlag, 1968.

[Sow84]   J.F. Sowa. *Conceptual Structures, Information Processing in Mind and Machine*. Addison Wesley, 1984.

[SS89]   M. Schmidt-Schauß. *Computational Aspects of an Order-Sorted Logic with Term Declarations*, volume 395 of *LNAI*. Springer-Verlag, Edited by J.Siekmann, 1989.

[SW90]   P.H. Schmitt and W. Wernecke. Tableau Calculus for Order Sorted Logic. In *Sorts and Types in Artificial Intelligence*, volume 418 of *LNAI*, pages 49–60. Springer-Verlag, 1990.

[Wer95]   M. Wermelinger. Conceptual Graphs and First-Order Logic. In *proceedings of ICCS'95, Santa Cruz, USA*, volume 954 of *LNAI*, pages 323–337. Springer-Verlag, 1995.

On Quasitautologies

Ján Komara and Paul J. Voda

Institute of Informatics, Faculty of Mathematics and Physics,
Comenius University, Mlýnská dolina, 842 15 Bratislava, Slovakia.
e-mail: {komara,voda}@fmph.uniba.sk
http://www.fmph.uniba.sk/~{komara,voda}

Abstract. Quasitautologies are formulas valid solely by the properties of identity and of propositional connectives. In connection with Herbrand's theorem the quasitautologies are central to the automated theorem proving with identity. It is a matter of folklore in logic that the predicate of being a quasitautology is decidable. We present a finitary proof of this fact by a powerful method of transformation of tableaux with identity rules. This should shed some light on the subtleties of tableaux with identity. We have extended this method in a separate paper to a much harder finitary proof of the conservativity of Skolem axioms. The question of why we should prefer finitary over model-theoretic proofs occurs frequently in logic. The answer is always simple: we obtain more information from a finitary proof than from a model-theoretic one. In our case, we get that quasitautologies can be proved by tableaux with a subterm property.

1 Introduction

Following Shoenfield [Sho67] we say that a formula of first-order logic with identity is a *quasitautology* if it is a tautological (i.e. propositional) consequence of the *axioms of identity* **Eq**. These are the *symmetry, transitive, function substitution,* and *predicate substitution* axioms listed in that order:

$$a \doteq b \to b \doteq a \tag{1}$$

$$a \doteq b \to b \doteq c \to a \doteq c \tag{2}$$

$$a_1 \doteq b_1 \to \cdots \to a_n \doteq b_n \to f(a_1, \ldots, a_n) \doteq f(b_1, \ldots, b_n) \tag{3}$$

$$a_1 \doteq b_1 \to \cdots \to a_n \doteq b_n \to p(a_1, \ldots, a_n) \to p(b_1, \ldots, b_n). \tag{4}$$

We do not exclude in (3) and (4) the case when $n = 0$; in such a case (3) is also called the *reflexivity* axiom for the constant f.

Quasitautologies are very important in the automated theorem proving (ATP). Namely, by Herbrand's theorem we have for an existential formula of the form $\exists x_1 \ldots \exists x_k A[x_1, \ldots, x_k]$ with a quantifier-free matrix A:

the formula $\exists x_1 \ldots \exists x_k A[x_1, \ldots, x_k]$ is valid iff there is a number n and terms $a_1^1 \ldots a_i^j \ldots a_k^n$ such that the following is valid

$$A[a_1^1, \ldots, a_k^1] \lor A[a_1^2, \ldots, a_k^2] \lor \cdots \lor A[a_1^n, \ldots, a_k^n]. \tag{5}$$

The disjunction (5), being a quantifier-free formula, is valid iff it is a quasitautology by the Hilbert-Ackermann's theorem (see [Sho67]). Thus Herbrand's theorem reduces the problem of validity of existential formulas in first-order logic with identity to the problem of generation of quasitautologies.

The importance of Herbrand's theorem in ATP lies in a suggestion of a method for the recognition of validity of existential formulas. Given an existential formula $\exists x_1 \ldots \exists x_k A[x_1, \ldots, x_k]$ the method first guesses a number $n \geq 1$ (called the *multiciplity*) and then tries to find terms $a_1^1 \ldots a_i^j \ldots a_k^n$ solely from the shape of the matrix A which turns the disjunction (5) into a quasitautology. For arbitrary but fixed number n the last step is called the problem of *Herbrand skeletons of size n*.

It is a matter of folklore in logic that the predicate of being a quasitautology is decidable. See for instance [BJ80] for a proof relying on finite models. Decidability of quasitautologies and related problems were proved by similar techniques in [Sho78,Sho84], [NO79,NO80], and [CLS96].

Nevertheless, we feel that the proof of the decidability of quasitautologies should be expressed in the well-developed apparatus of predicate calculus by finitary means. From the finitary proof we get a strong property of tableaux: the so-called subterm property. At the heart of our proof lies the main lemma 4.5 where we present a constructive transformation of an identity tableau for a formula A which removes from it all terms not occurring already in the formula A. As a consequence A is a quasitautology iff A has a tableau proof not introducing any terms not already in A. We will prove in Theorem 5.2 that this is equivalent to the existence of identity axioms $E_1, \ldots, E_n$ containing only terms from A such that $E_1 \rightarrow \cdots \rightarrow E_n \rightarrow A$ is a tautology. The last is decidable as there are only finitely many such axioms.

The technique used in the proof of our main lemma introduces a highly complex method of transformation of tableaux. We believe that it sheds a light on the extremely subtle properties of tableaux with identity which are not yet sufficiently appreciated. We have used an extension of the method in a far more complex situation. In [KV96] we give a finitary (i.e. non model-theoretic) proof of the conservativity of Skolem axioms in first-order logic with identity.

In contrast to the decidability of quasitautologies there is no recursive bound on the value n and the size of terms $a_1^1 \ldots a_i^j \ldots a_k^n$ turning the disjunction (5) into a quasitautology. This is a consequence of the well known fact that the problem of validity in first-order logic is undecidable. The undecidability of Herbrand skeletons of size $n \geq 1$ is a stronger result which has been proved only recently in [VK95] although the undecidability of the so-called simultaneous rigid E-unification proved by Degtyarev and Voronkov [DV95,DV96] (see also [DGV96]) can be used to show the undecidability for the case $n = 1$. This is a kind of a setback for ATP where it was long believed that it is sufficient to specify the size of a Herbrand skeleton, i.e. the multiciplity. From the undecidability of Herbrand skeletons we can see that, in addition to the multiplicity, we also need a bound on the size of terms.

The above form of Herbrand's theorem can be strengthened and used in many ways (see [Bus96]). For instance, let T be an open theory with the language $L(T)$ and let $\Delta(T)$ be the set of all formulas of $L(T)$ which are the (closed) instances of axioms of T. Then:

the formula $\exists x_1 \ldots \exists x_k A[x_1, \ldots, x_k]$ of $L(T)$ with quantifier-free matrix A is valid in T iff there is a number n and terms $a_1^1 \ldots a_i^j \ldots a_k^n$ of $L(T)$ such that the disjunction (5) is valid in $\Delta(T)$.

Noting that $\Delta(T)$ and (5) are quantifier-free, it is easy to see by the Hilbert-Ackermann's theorem that the disjunction (5) is valid in $\Delta(T)$ iff it is a *quasitautological consequence* of $\Delta(T)$, i.e. it is a tautological consequence of $\Delta(T)$ and the axioms of identity **Eq**.

This paper is structured as follows. In Sect. 2 we give our notation. In Sect. 3 we introduce quasitautologies and quasitautological consequence, and show how to reduce them to propositional logic. In Sect. 4 we explain our basic machinery — the identity tableaux — and prove the subterm property for them. In Sect. 5 we prove the main result of the paper, i.e. the decidability of quasitautologies.

2 Notation and Logical Background

2.1 Object language.

We assume that our object language FOL (First Order Language) has denumerable many function f_i^n and predicate p_i^n symbols of every arity $n \geq 0$. Note that this includes constants and propositional variables as they have the arity 0.

Semiterms are formed from variables by means of application of function symbols in the usual way. *Terms* are *closed*, i.e. without variables, semiterms. *Semiformulas* are formed from *atomic* semiformulas $a \doteq b$ and $p(a_1, \ldots, a_n)$ by the propositional connectives $\bot$ (false), $\top$ (true), $\neg, \wedge, \vee, \rightarrow$ (which are right associative and listed in the order of decreasing precedence) and quantifiers $\exists, \forall$ in the usual way. *Formulas* are *closed*, i.e. with all variables bound, semiformulas.

We use the syntactic variables f, g, etc. for function symbols, p, q, etc. for predicate symbols, a, b, etc. for (semi)terms, A, B, etc. for (semi)formulas, and x, y, etc. for variables. To every syntactic variable we can affix the suffix s to obtain a syntactic variable which ranges over finite (possibly empty) sequences of the corresponding objects. The sequence notation a, as stands for a sequence starting with the (semi)term a followed by the sequence of (semi)terms as. The notation As, Bs stands for the concatenation of two sequences As and Bs of (semi)formulas. We extend the set-theoretic predicates $\in$ and $\subset$ to include finite sequences (eg: $A \in As$, $As \subset T$). In this case a finite sequence stands for the corresponding finite set formed from the elements of the sequence. Occasionally, we will write T, S as an abbreviation for $T \cup S$.

We generalize some of the propositional connectives to for finite sequences. The expression $As \rightarrow_s A$ stands for A if As is empty and for $B \rightarrow (Bs \rightarrow_s A)$ if the sequence $As = B, Bs$. We define $As \vee_s A$ similarly. The generalized connectives $\rightarrow_s$ and $\vee_s$ are right associative and have respectively the same precedence as $\to$ and $\lor$. The sequence $as \doteq, bs$ $(as \not\doteq, bs)$ stands for the sequence of equalities (inequalities) between the corresponding elements of the two sequences which are assumed to be of the same length.

In order to be able to operate on semiterms, semiformulas, proofs (tableaux), and finite sequences by primitive recursive functions we identify these objects with their codes as natural numbers. We leave it to the reader to choose his own favorite *gödelization* of the object language (see for instance [HP93]). From now on we will refer to the codes of finite sequences (which are numbers) as *lists*.

2.2 Remark. We wish to stress here that under 'terms' and 'formulas' we understand 'closed terms' and 'closed formulas' (sentences). Our use of 'semiterms' and 'semiformulas' comes from Takeuti [Tak75]. The set of axioms of identity **Eq** is a set of formulas and it is an infinite decidable set.

2.3 Semantics. In order to establish the terminology we start with a quick review of some semantic notions of first-order logic. The reader is referred to [Bar77] for more details.

A first-order formula A is a *tautology* if it is true only on account of the standard properties of propositional constants and connectives regardless of what truth assignment is given to its *propositional atoms* (*p-atoms* for short) which are the atomic and quantifier formulas. This is clearly a primitive recursive predicate. Its definition can be given via the usual method of truth tables.

The following formulas are tautologies:

$$\forall x\, p(x) \to \forall x\, p(x) \tag{1}$$
$$(\forall x\, p(x) \to \exists y\, q(y)) \to \neg \exists y\, q(y) \to \neg \forall x\, p(x)\,, \tag{2}$$

but the following, although they are valid, are not tautologies:

$$a \doteq a \tag{3}$$
$$\forall x\, (p(x) \to p(x)) \tag{4}$$
$$\forall x\, p(x) \to \neg \exists x\, \neg p(x)\,. \tag{5}$$

Formulas (3) and (4) are p-atoms. Formula (5) has the form $A \to \neg B$ for some p-atoms A and B.

We write $T \models_p A$ when the formula A is a *tautological (propositional) consequence* of a set of formulas T. Intuitively, $T \models_p A$ holds when from the assumption that the formulas of T are true we can determine the truth of A solely by the properties of propositional connectives. The precise meaning can be given by propositional truth sets in the usual way. We abbreviate $\emptyset \models_p A$ to $\models_p A$. This abbreviation is used also with the other forms of consequence and/or provability defined below.

The relation between tautologies and tautological consequence is captured by the following theorem which is also known as the *compactness theorem for propositional logic*. The theorem can be proved either directly or indirectly from Thm. 4.2 in [Bar77] which is based on the dual notion of propositional satisfiability.

2.4 Theorem (Tautological reduction). $T \models_p A$ *iff there is a list of formulas* $As \subset T$ *such that the formula* $As \rightarrow_s A$ *is a tautology.*

3 Quasitautologies

3.1 Definition. We say that a formula A is a *quasitautological consequence* of a set of formulas T, written as $T \models_i A$, if A is a tautological consequence of the set $T \cup \mathbf{Eq}$, i.e. $T, \mathbf{Eq} \models_p A$. A formula A is a *quasitautology* if $\models_i A$.

We say that a list of identity axioms Es is an *identity associate* (*i-associate* for short) of a formula A *in* a set of formulas T if there is a list $As \subset T$ such that the formula $Es \rightarrow_s As \rightarrow_s A$ is a tautology. For $T = \emptyset$ we say that Es is an *i-associate* of A. We say that Es has the *subterm property* if there is $As \subset T$ such that $Es \rightarrow_s As \rightarrow_s A$ is a tautology and all terms of Es occur in As or A.

3.2 Remark. Since the set of identity axioms $\mathbf{Eq}$ is primitive recursive, the predicate of being an *i*-associate in T of A is semidecidable in T and the predicate of being an *i*-associate of A is primitive recursive.

Note also that *i*-associates have the following monotone property. If Es_1 is an *i*-associate of A in T and $Es_1 \subset Es_2 \subset \mathbf{Eq}$ then Es_2 is also an *i*-associate of A in T. Consequently, if a formula has an *i*-associate then it has *i*-associates of arbitrary length and size of terms.

3.3 Theorem. *A quantifier-free formula is valid iff it is a quasitautology.*

Proof. This follows from the Hilbert-Ackermann's theorem (see [Sho67]) or, equivalently, from cut-free proofs of quantifier-free formulas. □

3.4 Theorem. *A formula* A *is a quasitautological consequence of* T *iff it has an i-associate in* T. *In particular,* A *is a quasitautology iff it has an i-associate.*

Proof. The formula A is a quasitautological consequence of T iff $T, \mathbf{Eq} \models_p A$ iff, by the tautological reduction, there are lists of formulas $Es \subset \mathbf{Eq}$ and $As \subset T$ such that the formula $Es \rightarrow_s As \rightarrow_s A$ is a tautology iff the formula A has an *i*-associate in T. □

3.5 Discussion. As a consequence of Thm. 3.4, the predicate $T \models_i A$ in A is semidecidable in T. In particular, the predicate of being a quasitautology is semidecidable. The main goal of this paper is to give a constructive proof that the problem of the recognition of quasitautologies is decidable (actually primitive recursive). In Sect. 5 we will strengthen Thm. 3.4 by proving the following claim:

'a formula is a quasitautology iff it has an *i*-associate

with the subterm property'. (1)

Note that there are only finitely many distinct terms in a formula A and so there is only a finite number of identity axioms containing only those terms. Let us denote by Es_A the list of *all* such identity axioms. As a consequence of (1) and of the monotonicity of i-associates we will get:

'the formula A is a quasitautology iff Es_A is i-associate of A'. (2)

Since Es_A is a primitive recursive function in A, the predicate of being a quasitautology will become primitive recursive. Thus, it suffices to consider only those identity axioms the terms of which are in the formula to be proved.

The proof of (1) is based on the following observation:

- tautologies are closed under the replacement of its terms;
- if E is an identity axiom which is not a function substitution axiom then the replacement of a subterm of E yields an identity axiom of the same kind.

Thus, if we have an i-associate Es of A which does not contain any function substitution axioms then the desired i-associate of A with the subterm property is obtained by replacing in Es every term $f(as)$ which does not occur in A by some constant of A. If the i-associate Es contains function substitution axioms this transformation is no longer correct since function substitution axioms are not invariant under the replacement of terms. This means that the above replacement applied to a function substitution axiom may yield a formula which is not an identity axiom. The core of the proof of (1) will consist of finding an invariant method of replacing terms in function substitution axioms. This will be done by tableaux in the following section.

3.6 Example. Let a, b, and c be distinct constants. The formula A of the form

$$b \doteq a \rightarrow a \doteq c \rightarrow f(b) \doteq f(c)$$ (1)

is a quasitautology as the list $Es_1 \subset Eq$ consisting of the following formulas

$$b \doteq a \rightarrow f(b) \doteq f(a)$$ (2)
$$a \doteq c \rightarrow f(a) \doteq f(c)$$ (3)
$$f(b) \doteq f(a) \rightarrow f(a) \doteq f(c) \rightarrow f(b) \doteq f(c)$$ (4)

is its i-associate. Since the term $f(a)$ does not occur in A, the list Es_1 does not have the subterm property. The replacement of the term $f(a)$ by a constant k of A yields a list Es_2 consisting of:

$$b \doteq a \rightarrow f(b) \doteq k$$ (5)
$$a \doteq c \rightarrow k \doteq f(c)$$ (6)
$$f(b) \doteq k \rightarrow k \doteq f(c) \rightarrow f(b) \doteq f(c).$$ (7)

Note that while the formula $Es_2 \rightarrow_s A$ is a tautology, Es_2 is not an i-associate of A since neither (5) nor (6) are identity axioms.

However, the formula A has an i-associate with the subterm property, for instance the list Es_3 consisting of

$$b \doteq a \to a \doteq c \to b \doteq c \tag{8}$$

$$b \doteq c \to f(b) \doteq f(c). \tag{9}$$

4 Identity Tableaux with the Subterm Property

The elimination of redundant function substitution axioms from i-associates in order to get an associate with the subterm property is difficult to manage while working on the level of i-associates. The elimination process is still quite complex (see the main lemma 4.5) but manageable when it is done on the level of tableaux.

4.1 Identity tableaux. We now briefly describe the method of identity tableaux. The reader may refer for details to [KV95]. Our tableaux are dual to the tableaux of Smullyan [Smu68]. Smullyan's tableaux are based on refutation as they show unsatisfiability. Our tableaux demonstrate logical consequence and in the case of identity tableaux, quasitautological consequence. In order to assure the synthetic consistency property we add to our tableaux *cut rules* which are eliminable.

An *identity tableau* X *for As in* T, written as $X : T \vdash_i^+ As$, establishes $T \models_i As \vee_s \bot$. It can be visualized as a downward growing dyadic tree whose nodes are formulas. We start the tree with a branch where we write down the formulas from As one after another and then append X to it. The tableau is formed from the initial branch As called the *theorem list* by expanding the branches according to the *rules of expansion* until all branches are *closed*. The list As may be empty. The formulas of T are called the *axioms* of the tableau. The leaf at every branch is the *empty tableau* $\bot$.

A branch of a tableau is closed if it contains $\top$ or $\neg\bot$ or else it contains a pair of formulas B and $\neg B$. This is called the *closure rule*. The expansion rules for identity tableaux consist of *propositional*, *identity*, *axiom*, and *cut* rules. The propositional rules are *negation* rules $\frac{\neg\neg B}{B}$, *disjunction* rules:

$$\frac{B \vee C}{B} \quad \frac{B \vee C}{C} \quad \frac{B \to C}{\neg B} \quad \frac{B \to C}{C} \quad \frac{\neg(B \wedge C)}{\neg B} \quad \frac{\neg(B \wedge C)}{\neg C},$$

and *conjunction* rules:

$$\frac{B \wedge C}{B \mid C} \quad \frac{\neg(B \to C)}{B \mid \neg C} \quad \frac{\neg(B \vee C)}{\neg B \mid \neg C}.$$

The identity rules correspond to the identity axioms **Eq**:

$$\frac{a \not\doteq b}{b \not\doteq a} \quad \frac{a \not\doteq b \quad b \not\doteq c}{a \not\doteq c} \quad \frac{as \not\doteq_s bs}{f(as) \not\doteq f(bs)} \quad \frac{as \not\doteq_s bs \quad \neg p(as)}{\neg p(bs)}.$$

We do not exclude the case when the list as is empty; in such case a function substitution rule is also called the *reflexivity* rule for the constant f. The axiom rules are $\frac{}{\neg A}$ for every axiom A. The cut rules are $\frac{}{A \mid \neg A}$.

We use the syntactic variables X, Y, Xt, etc. to range over tableaux. If the tableau $X : T \vdash_i^+ As$ is not expanded by identity rules we will qualify the provability sign $\vdash$ by p: $\vdash_p^+$ and say that the tableau is a *propositional* tableau. In both forms we will omit the qualifier $+$ if the tableau is a *cut free* tableau, i.e. a tableau without any cut rules. We will write $T \vdash_{q_1}^{q_2} As$ in all qualified forms if there is a corresponding tableau. If the set T is empty then we say that a tableau in T is *axiom-free*.

For the proof of the following theorem see Thm(s). 7 and 12 in [KV95].

4.2 Theorem (Soundness and completeness of tableaux). *We have:*

(i) A *is a tautology iff* $\vdash_p^+ A$ *iff* $\vdash_p A$;

(ii) $T \models_p A$ *iff* $T \vdash_p^+ A$;

(iii) $T \models_i A$ *iff* $T \vdash_i^+ A$.

4.3 Subterm property for identity tableaux.

We say that an expansion rule of an identity tableau has the *subterm property* if either the rule is an axiom rule or else every term occurring in the conclusion of the rule occurs also in the branch above. An identity tableau has the *subterm property* if all its expansion rules have the subterm property.

It is easy to see that the propositional, symmetry, transitivity and predicate substitution rules have the subterm property. The expansion rules which may violate the subterm condition are the function substitution and cut rules. Note that if $X : \vdash_i^+ As$ is an identity tableau with the subterm property then every term of X occurs also in As because axiom rules are not applicable.

All identity tableau expansion and closure rules except axiom and function substitution rules are invariant under the replacement of all occurrences of a term by another term. Replacement applied to an invariant rule yields a rule of the same kind.

4.4 Example (continued).

Consider the formula A from Par. 3.6. Figure 1 shows two identity tableau proofs of A. The tableau (a) uses identity rules corresponding to the i-associate Es_1 and has not the subterm property because the function substitution rule α_1 introduces a new term $f(a)$. The tableau (b) uses identity rules corresponding to Es_3 and has the subterm property. The proof of the main lemma 4.5 shows how to get from the tableau (a) to the tableau (b) by constructive means.

4.5 Main lemma.

Let $f(as)$ be a maximal term (i.e. a term not occurring as a proper subterm of any other term) of the tableau

$$X : \vdash_i^+ As \tag{1}$$

which does not occur in the theorem list As. Let As contain at least one constant k. Then there is a tableau $Y : \vdash_i^+ As$ which contains the same terms as (1) except for the term $f(as)$.

$$b \doteq a \to a \doteq c \to f(b) \doteq f(c)$$
$$a \doteq c \to f(b) \doteq f(c)$$
$$f(b) \doteq f(c)$$
$$b \not\equiv a$$
$$a \not\equiv c$$
$$f(b) \not\equiv f(a)$$
$$f(a) \not\equiv f(c)$$
$$f(b) \not\equiv f(c)$$
$$\perp$$

(a)

$$b \doteq a \to a \doteq c \to f(b) \doteq f(c)$$
$$a \doteq c \to f(b) \doteq f(c)$$
$$f(b) \doteq f(c)$$
$$b \not\equiv a$$
$$a \not\equiv c$$
$$b \not\equiv c$$
$$f(b) \not\equiv f(c)$$
$$\perp$$

(b)

Fig. 1. Examples of identity tableaux without (a) and with (b) the subterm property

Proof. Let us temporarily call a term $f(bs)$ of a tableau *critical* if the term $f(bs)$ is distinct from $f(as)$ and the tableau contains either $f(as) \not\equiv f(bs)$ or $f(bs) \not\equiv f(as)$ as the conclusion of an f-substitution rule. The proof of the lemma is by complete induction on the number of different critical terms in the tableau X. By the maximality of $f(as)$ none of the critical terms of X contains the term $f(as)$. We consider two cases.

If there are no critical terms in X then we obtain the tableau Y from X by replacing every occurrence of $f(as)$ by the constant k. Note that every f-substitution rule of the form $\frac{as \not\equiv_s as}{f(as) \not\equiv f(as)}$ is transformed into the function substitution rule without premises, i.e. into the reflexivity rule $\frac{}{k \not\equiv k}$ for the constant k.

If there is at least one critical term $f(bs)$ in X then the direct replacement of $f(as)$ by k is no longer possible. Suppose that $as = a_1, \ldots, a_n$ and $bs = b_1, \ldots, b_n$ for some $n > 0$ (as $f(bs) \neq f(as)$). We intend to obtain the desired tableau Y by n cut rules as shown on the top of Fig. 2. The subtableaux

$$Z_i : \vdash_i^+ a_i \doteq b_i, As \tag{2}$$

for $1 \leq i \leq n$ contain at most the terms from X but not $f(as)$. The tableau

$$Z : \vdash_i^+ as \not\equiv_s bs, As \tag{3}$$

contains exactly the same terms as X except for $f(as)$.

Each subtableau Z_i is formed from X by a transformation performed on all topmost occurrences of function substitution rules with the critical term $f(bs)$. The middle of Fig. 2 shows two such rules α_1 and α_2. The bottom of the figure shows the transformation leading to a tableau

$$X_i : \vdash_i^+ a_i \doteq b_i, As . \tag{4}$$

The left subtableau of X starting with $f(as) \not\equiv f(bs)$ is replaced by the empty tableau because the branch closes on the pair of formulas $a_i \doteq b_i$ and $a_i \not\equiv b_i$

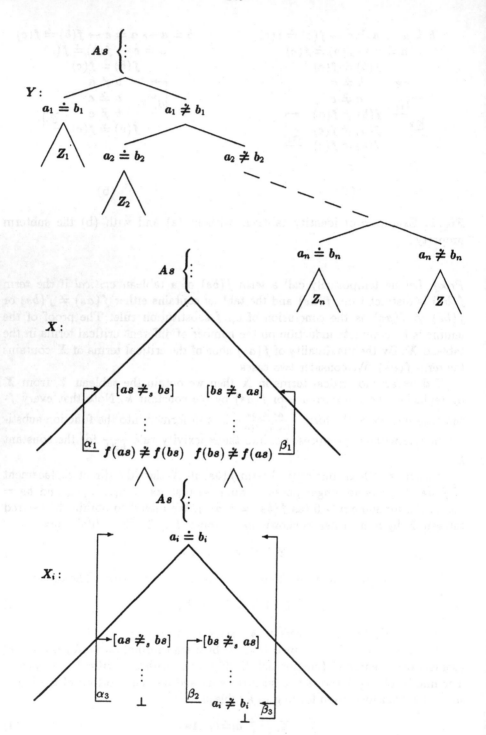

Fig. 2. Main lemma — elimination of critical terms

(α_3). The right subtableau of X starting with $f(bs) \not\doteq f(as)$ is replaced by a subtableau expanded first by the symmetry rule $\beta_2 : \frac{b_i \not\doteq a_i}{a_i \not\doteq b_i}$ and then closed on the pair of formulas $a_i \doteq b_i$ and $a_i \not\doteq b_i$ (β_3). Now the term $f(bs)$ cannot occur critically in the tableau X_i, so X_i has a lesser number of critical terms and the desired tableau Z_i is obtained from it by inductive hypothesis.

We intend to replace $f(as)$ by $f(bs)$ in the tableau X in order to obtain the tableau Z. This may cause problems if the first tableau contains f-substitution rules non-invariant under this replacement. The top of Fig. 3 shows three such typical rules α_1, β_1, and γ_1 which may occur in the tableau X. Here it is assumed that both $f(cs)$ and $f(ds)$ are distinct from $f(as)$. By the maximality assumption the terms $f(cs)$ and $f(ds)$ do not contain $f(as)$. The transformation leading to Z consists of two steps.

In the first step we modify all f-substitution rules non-invariant under the replacement of $f(as)$ by $f(bs)$ as shown in the middle of Fig. 3. This is done by a series of symmetry (α_2, β_2) and transitive rules (α_3, β_3, γ_3). We obtain a tableau

$$Z' : \vdash_i^+ \ as \not\doteq_s bs, As\,, \tag{5}$$

where the branches above the conclusions of the non-invariant rules α_1, β_1, and γ_1 contain the lists $bs \not\doteq_s cs$, $bs \not\doteq_s bs$, and $ds \not\doteq_s bs$ respectively.

In the second step we replace every occurrence of $f(as)$ in the tableau Z' by $f(bs)$ as shown at the bottom of the figure. The formulas $f(bs) \not\doteq f(cs)$, $f(bs) \not\doteq f(bs)$, and $f(ds) \not\doteq f(bs)$ are now conclusions of f-substitution rules α_4, β_4, and γ_4 respectively. This yields the desired tableau Z. □

4.6 Theorem (Subterm property for identity tableaux). *If $\vdash_i^+ A$ then there is a tableau $X : \vdash_i^+ A$ with the subterm property.*

Proof. Take $Y : \vdash_i^+ A$ and consider two cases. If the formula A contains at least one constant k then the theorem is proved by complete induction on the number of different terms in Y which do not occur in A. If there are none then we set $X = Y$. Otherwise, there is a maximal term $f(as)$ of Y not occurring in A and by Lemma 4.5 there is a tableau $Z : \vdash_i^+ A$ without $f(as)$ and without any new terms. The desired tableau X is now obtained from Z by inductive hypothesis.

If the formula A is without constants then every p-atom of A is either a propositional variable or a quantifier formula. From the cut elimination theorem as proved in [KV95] we obtain a cut-free identity tableau $Z : \vdash_i A$. Terms can occur in Z only in formulas $a \not\doteq a$ which can be removed whereby we get a cut-free propositional tableau $X : \vdash_p A$ which is trivially with the subterm property. We note that by the soundness of axiom-free propositional tableaux 4.2(i) the formula A is a tautology similar to 2.3(1) or 2.3(2). □

4.7 Remark. We can strengthen Thm. 4.6 to assert that the tableau X with the subterm property is cut free: $X : \vdash_i A$. This follows from the above theorem and from the fact that the cut elimination theorem as proved in [KV95] preserves the subterm property.

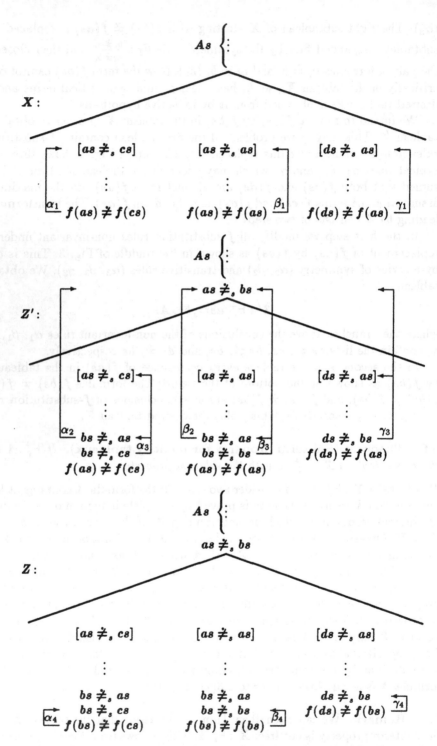

Fig. 3. Main lemma — elimination of non-invariant f-substitution rules

5 Decidability of Quasitautologies

5.1 Remark. In Thm. 11 of [KV95] we proved a theorem on the *elimination of identity rules*:

$$\text{if } T \vdash_i^+ A \text{ then } T, \mathbf{Eq} \vdash_p^+ A. \qquad (1)$$

The proof uses the correspondence between the identity axioms and identity rules. The transformation going from $X : T \vdash_i^+ A$ to $Y : T, \mathbf{Eq} \vdash_p^+ A$ eliminates every identity rule by the corresponding axiom. Figure 4 shows the elimination of a function substitution rule as an example. Note that if the rule α has the subterm property then the corresponding axiom does not introduce any new terms. The same holds for the remaining identity rules.

Thus, if the tableau $X : \vdash_i^+ A$ is with the subterm property and $\mathbf{Es}$ is the list of identity axioms corresponding to the identity rules used in X then every term of $\mathbf{Es}$ occurs in the formula A and the transformation (1) yields a propositional tableau $Y : \mathbf{Es} \vdash_p^+ A$.

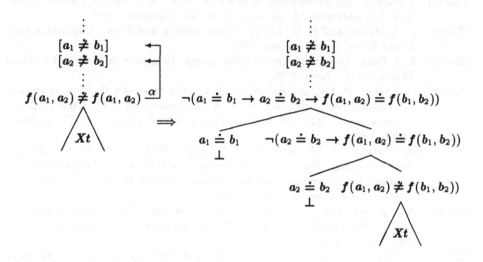

Fig. 4. Elimination of a function substitution rule

We have the *deduction theorem for identity tableaux*:

$$T \vdash_i^+ A \text{ iff there is a list of formulas } \mathbf{As} \subset T \text{ such that } \vdash_i^+ \mathbf{As} \rightarrow_s A. \qquad (2)$$

The proof is the same as that for the propositional tableaux (see Thm. 10 in [KV95]).

5.2 Theorem (Subterm property). $T \models_i A$ *iff there is an i-associate of A in T with the subterm property. In particular, a formula is a quasitautology iff it has an i-associate with the subterm property.*

Proof. The implication ($\Leftarrow$) follows from Thm. 3.4. For the implication ($\Rightarrow$) assume that $T \models_i A$. By the completeness of identity tableaux 4.2(iii) we have $T \vdash_i^+ A$. By the deduction theorem 5.1(2) there is a list $As \subset T$ such that $\vdash_i^+ As \rightarrow_s A$. By Thm. 4.6 there is a tableau $X : \vdash_i^+ As \rightarrow_s A$ with the subterm property. By the elimination of identity rules (see Par. 5.1), there is a list Es of identity axioms containing only the terms of A and As such that $Es \vdash_p^+ As \rightarrow_s A$. By the deduction theorem for propositional tableaux (see Thm. 10 in [KV95]) $\vdash_p^+ Es \rightarrow_s As \rightarrow_s A$. By the soundness of axiom-free propositional tableaux 4.2(i) the formula $Es \rightarrow_s As \rightarrow_s A$ is a tautology. The list Es is thus an i-associate of A in T with the subterm property. $\square$

5.3 Theorem. *The predicate of being a quasitautology is primitive recursive.*

Proof. By Thm. 5.2 and the argument in Par. 3.5. $\square$

References

[Bar77] J. Barwise. An introduction to first-order logic. In J. Barwise, editor, *Handbook of Mathematical Logic*, pages 5–46. North-Holland, 1977.

[BJ80] G. S. Boolos and R. C. Jeffrey. *Computability and Logic*. Cambridge University Press, second edition, 1980.

[Bus96] S. R. Buss. An introduction to proof theory. To appear in *Handbook of Proof Theory* (ed. S. Buss), 1996.

[CLS96] D. Cyrluk, P. Lincoln, and N. Shankar. On Shostak's decision procedure for combinations of theories. In M. A. McRobbie and J. K. Slaney, editors, *Proceedings of CADE-13*, number 1104 in LNAI, pages 463–477. Springer Verlag, 1996.

[DGV96] A. Degtyarev, Y. Gurevich, and A. Voronkov. Herbrand's theorem and equational reasoning: Problems and solutions. In *Bulletin of the European Association for Theoretical Computer Science*, volume 60, October 1996. The "Logic in Computer Science" column.

[DV95] A. Degtyarev and A. Voronkov. Simultaneous rigid *E*-unification is undecidable. UPMAIL Technical Report 105, Uppsala University, Computing Science Department, May 1995.

[DV96] A. Degtyarev and A. Voronkov. The undecidability of simultaneous rigid *E*-unification. *Theoretical Computer Science*, 166:291–300, 1996.

[HP93] P. Hájek and P. Pudlák. *Metamathematics of First-Order Arithmetic*. Springer Verlag, 1993.

[KV95] J. Komara and P. J. Voda. Syntactic reduction of predicate tableaux to propositional tableaux. In P. Baumgartner, R. Haehnle, and J. Posegga, editors, *Proceedings of TABLEAUX '95*, number 918 in LNAI, pages 231–246. Springer Verlag, 1995.

[KV96] J. Komara and P. J. Voda. On Skolem axioms. Technical report, Institute of Informatics, Faculty of Mathematics and Physics, Comenius University, Bratislava, November 1996.

[NO79] G. Nelson and D. C. Oppen. Simplification by cooperating decision procedures. *ACM Transactions on Programming Languages and Systems*, 1(2):245–257, 1979.

[NO80]   G. Nelson and D. C. Oppen. Fast decision procedures based on congruence closure. *Journal of the ACM*, 27(2):356–364, 1980.

[Sho67] J. R. Shoenfield. *Mathematical Logic*. Addison-Wesley, 1967.

[Sho78] R. E. Shostak. An algorithm for reasoning about equality. *Communications of the ACM*, 21(7):583–585, July 1978.

[Sho84] R. E. Shostak. Deciding combinations of theories. *Journal of the ACM*, 31(1):1–12, January 1984.

[Smu68] R. Smullyan. *First Order Logic*. Springer Verlag, 1968.

[Tak75] G. Takeuti. *Proof Theory*. North-Holland, 1975.

[VK95]   P. J. Voda and J. Komara. On Herbrand skeletons. Technical report, Institute of Informatics, Faculty of Mathematics and Physics, Comenius University, Bratislava, July 1995. Revised January 1996. Submitted for publication.

Tableaux Methods for Access Control in Distributed Systems

Fabio Massacci*

Computer Laboratory
University of Cambridge, England (UK)
e-mail: Fabio.Massacci@cl.cam.ac.uk

Abstract. The aim of access control is to limit what users of distributed systems can do directly or through their programs. As the size of the systems and the sensitivity of data increase formal methods of analysis are often required.

This paper presents a prefixed tableaux method for the calculus of access control in distributed system developed at DEC-SRC by Abadi, Lampson et. al. Beside the applicative interest, the calculus poses interesting technical challenges, since it has not the tree-model property, introduces relations between modalities which cannot be compiled into axiom schemas, and has some features of the universal modality.

As a side-effect we show a tableaux calculus for the universal modality which distinguishes it from $S5$ (via satisfiability on non tree-models).

1 Introduction

Access control is a key issue for the security of computer systems (see [25] for an introduction). Its main purpose is to restrain the actions which legitimate (or malicious) users may perform, either directly or indirectly (through their programs). Its need arises in any system with multiple users and sensitive information or shared resources such as the military [4], banking and commerce [7] or health care services [2].

Distributed systems face additional challenges (e.g. large scale, insecure communications, delegation of management etc.) which require new modelling and reasoning techniques. For instance, access control must be combined with authentication [31], and policies must be refined at the various levels of delegation [22]. The definition of the jurisdiction capabilities of communicating agents plays also a key role in the analysis of security protocols [5, 28]. Indeed access control is just a problem of jurisdiction in complex and distributed systems.

As systems become more complex, human (and informal) verification becomes infeasible. Hence formal methods, logics and automated reasoning techniques can be useful tools for the verification of security policies and access control procedures (see e.g. [5, 20, 19, 22, 31]). Our target is the development of (tableaux based) automated reasoning techniques for access control.

---

* Current address: Dip. Informatica e Sistemistica, Universitá di Roma "La Sapienza", via Salaria 113, I-00198 Roma (I), e-mail massacci@dis.uniroma1.it.

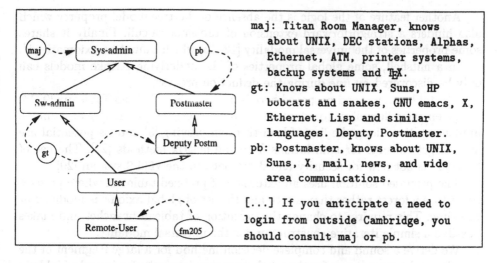

maj: Titan Room Manager, knows about UNIX, DEC stations, Alphas, Ethernet, ATM, printer systems, backup systems and TEX.

gt: Knows about UNIX, Suns, HP bobcats and snakes, GNU emacs, X, Ethernet, Lisp and similar languages. Deputy Postmaster.

pb: Postmaster, knows about UNIX, Suns, X, mail, news, and wide area communications.

[...] If you anticipate a need to login from outside Cambridge, you should consult maj or pb.

Fig. 1. From "Computing Facilities at the Computer Laboratory"

The principles of access control can be described with few abstractions: *subjects* (humans, programs etc.), *objects* (data, other programs etc.) and *privileges* which subjects detain on objects (e.g. read, write and execute in UNIX). The use of these abstractions is the basis of most formal models proposed in the literature, starting from the classical access matrix [18, 24] to more advanced systems [10, 19, 22, 23]. A key feature of the new approaches is the attempt to model more closely the (hierarchical) relationships between the various subjects, where some privileges can be inherited along the chains (e.g. Fig. 1).

The use of these abstractions leads naturally towards a formalisation of the problem with multi modal logics: one subject, one modality. There has a been a number of works on modelling security and obligations in a multiagents setting, e.g. [8, 17, 30], and in particular we focus on the expressive calculus developed at DEC-SRC [1]. This calculus is interesting for a number of reasons:

- it provides a uniform framework for reasoning about access control in presence of delegation and has a simple semantics [1];
- it constitutes the basis of a real system [19, 31];
- its features pose interesting technical challenges for deduction.

One of the characteristics, which challenge "standard" tableaux calculi, is the presence of formulae used for modelling delegation certificates and hierarchical relationships between subjects (i.e. modalities). Those relations have the same force of axiom schemas and are close to role-value-map constructs of AI languages [26]. The key difficulty is that we cannot "compile" them into tableaux rules (nor axiom schemas) since their presence depends on the particular non logical axioms *and* the particular theorem we want to prove. Two different theorems (i.e. access requests with different delegation certificates) may impose totally different relations between subjects.

Another feature of the logic is the absence of the tree model property which also hinders a straightforward extension of tableaux calculi. Finally it shares some properties of the universal modality [12, 13] which cannot be axiomatised[2].

In a nutshell, some global properties of the underlying Kripke models can only be discovered *on-line* during the deduction process.

The tableaux method proposed here is based on Simple Step Tableaux [9, 14, 21] given their flexibility to adapt to various logics, the possibility of a simple implementation along the line of lean theorem proving [3] and its potential use as a target calculus for translations from the matrix methods [27]. The use of tableaux is not new for security and dates back to the VERUS system [20].

The proposed solution uses an extension of prefixed tableaux where prefixes constitute a forest rather than a tree and the set of global axioms is modified at run time. This requires to change the definition of tableau branches and makes possible a simple iterative construction for the universal modality.

We derive a sound and complete decision method for a large fragment of the calculus and a correct one for the whole language (the calculus is undecidable in general). This method extends the deduction capabilities of [1] as we can prove important properties which must be added as non-logical axioms in [1, 19].

An important side-effect of these techniques, shown in the appendix, is an EXPTIME tableau for multi-modal logics with the universal modality [12, 13].

In what follows, we present the DEC-SRC calculus (§2) and analyse its semantical features (§3). We discuss the tableaux calculus (§4) with some examples (§5) and sketch its soundness and completeness proofs (§6). Conclusions (§7) are followed by an appendix on the universal modality.

2 The DEC-SRC Calculus for Access Control

To make the paper self-contained we sketch the intuitions behind the calculus and refer to [1] for a formal treatment and to [19, 31] for its applications.

Users, roles, groups and cryptographic keys are represented by atomic subjects, or *principals*, and denoted by A, B, K_A etc. Complex principals (P or Q) are built by conjunction " & " and quoting " | ". The intuition is that $P \& Q$ is a principal with the privileges of both P and Q, whereas $P \mid Q$ corresponds to the principal P claiming to quote a request from Q. Notice that P may claim to quote Q even when Q never said anything.

Other operators are possible i.e. A for B and A as R [1]: the former is used when A is claiming to act as a delegate for B; the latter when A speaks using a role[3] R. Since they can be encoded using $\mid$ and $\&$, we do not use them. For instance A for $B \doteq (A \& D) \mid B$ where D is a delegation server [1].

Operations over objects are represented by *statements* denoted by s. Atomic statements (r) are uninterpreted operations or requests [1, 23] i.e. propositional letters which can be true in a particular state of the system (request granted)

---

[2] This may also explain why only a sound axiomatisation has been devised in [1].

[3] For a distinction between the security concepts of role and group see [10, 23].

$$
\begin{array}{ll}
SysAdm \Rightarrow SwAdm \wedge & maj \Rightarrow SysAdm \wedge \\
SysAdm \Rightarrow PostMaster \wedge & gt \Rightarrow SwAdm \wedge \\
SwAdm \Rightarrow Usr \wedge & gt \Rightarrow DepPostM \wedge \\
PostMaster \Rightarrow Usr \wedge & pb \Rightarrow PostMaster \wedge \\
Usr \Rightarrow RemUsr & fm205 \rightarrow RemUsr \\
\wedge & \wedge \\
SysAdm \text{ controls } (fm205 \Rightarrow Usr) \wedge & maj \text{ says } (fm205 \Rightarrow Usr) \wedge \\
Usr \text{ controls } login(telnet) \wedge & fm205 \text{ says } login(telnet) \wedge \\
RemUsr \text{ controls } login(ftp) \wedge & fm205 \text{ says } read(mail) \\
(fm205 \mid Usr) \text{ controls } read(mail) &
\end{array}
$$

Fig. 2. A Logical Formalisation of Fig. 1

or false (not granted). Complex statements are built with boolean connectives $\wedge, \neg, \supset$ etc.: for instance $login\_telnet \supset login\_ftp$ whose intuitive (and formal) interpretation is "if telnet has been granted so has ftp".

To represent *user requests* we use the modal statement "P says s": principal P requests s to be granted. If P is a group then we follow [1, 19] and interpret it as "somebody in group P says s".

Hierarchical relations between principals are constructed with the *speaks-for* statement $P \Rightarrow Q$. The intuition is that P has at least all the privileges of Q i.e. P can speak for Q. If P says s this would be as Q itself said s. It is also used for group membership: $P \Rightarrow G$ means that P has at least all privileges of group G.

Principals and statements are linked by *privileges attributions* [23]: the statement "P controls s" captures the intuition that principal P has access control over s. In the literature on authentication this is called *jurisdiction* of a principal [5, 28] and axiomatised as A says $s \wedge A$ controls $s \supset s$.

The aim is to to replace it by a more complex but more realistic axiom, where the relation between two principals A and B is expressed with the $\Rightarrow$ operator:

$$A \text{ says } s \wedge B \text{ controls } s \wedge \text{ "some relation between } A \text{ and } B\text{" } \supset s.$$

A particular $P \Rightarrow Q$ may depends on the statements of other principals. For instance $fm205 \Rightarrow Usr$ in Fig. 2 depends on maj's statements.

3 Formal Syntax and Semantics

The language (described informally in §2) is the following, where A is an atomic principal and r an atomic propositional request:

$$P, Q ::= A \mid P \& Q \mid (P \mid Q) \qquad s, s' ::= r \mid \neg s \mid s \wedge s' \mid P \text{ says } s \mid (P \Rightarrow Q)$$

Other connectives are abbreviations, e.g. $s \supset s' \equiv \neg(s \wedge \neg s')$. Also P controls s is a shortcut for $(P \text{ says } s) \supset s$.

In the sequel we assume that in $P{\Rightarrow}Q$ either P or Q is an atomic principal, w.l.o.g. since $P{\Rightarrow}Q$ is equivalent to $P{\Rightarrow}A \wedge A{\Rightarrow}Q$ for a new atomic A.

A statement is *left (right) restricted* when speaks-for subformulae have the form $A{\Rightarrow}Q$ (respectively $P{\Rightarrow}A$) i.e. the left (right) principal is atomic. It is *weakly left (right) restricted* when statements $P{\Rightarrow}Q$ are admitted if they occur under the scope of an odd number of negations[4]. It is *request restricted* when in each statement of the form P **says** s, the statement s is either an atomic request or a group membership (both possibly negated). For instance the formalisation in Fig. 2 is left, right and request restricted.

In practice statements are right and request restricted. If $\Rightarrow$ is used for hierarchies and group and role membership, as in Fig. 1, the rightmost principal is atomic. Moreover, in almost all systems [25], privileges attributions are represented by ACL (Access Control Lists). In the DEC-SRC language, an ACL for a request r is simply the conjunctions of statements $\bigwedge_i P_i$ **controls** r, where r is uninterpreted [19, 23]. If we add, among the possible privileges, the possibility to hand over delegation to other principals such as P_i **controls** $(Q_j{\Rightarrow}A_k)$, then we still have a right restricted language.

The semantics is based on Kripke models [1, 11, 16]: a relation models the compatibility of a state with the requests made by a principal in the real world.

A model is a pair $\langle W, \mathcal{I} \rangle$, where W is a non empty set of states and $\mathcal{I}$ an interpretation such that for every atomic principal A it is $A^{\mathcal{I}} \subseteq W \times W$ and for every propositional letter r it is $r^{\mathcal{I}} \subseteq W$. Then $\mathcal{I}$ is extended as follows:

$$
\begin{aligned}
(\neg s)^{\mathcal{I}} &= W - s^{\mathcal{I}} \\
(s \wedge s')^{\mathcal{I}} &= s^{\mathcal{I}} \cap s'^{\mathcal{I}} \\
(P{\Rightarrow}Q)^{\mathcal{I}} &= \text{if } Q^{\mathcal{I}} \subseteq P^{\mathcal{I}} \text{ then } W \text{ else } \emptyset \\
(P \text{ \bf says } s)^{\mathcal{I}} &= \{ w \mid \forall w^* \in W \text{ if } \langle w, w^* \rangle \in P^{\mathcal{I}} \text{ then } w^* \in s^{\mathcal{I}} \} \\
(P \text{ \& } Q)^{\mathcal{I}} &= P^{\mathcal{I}} \cup Q^{\mathcal{I}} \\
(P \mid Q)^{\mathcal{I}} &= \{ \langle w, w^* \rangle \mid \exists w^{**} \langle w, w^{**} \rangle \in P^{\mathcal{I}} \text{ and } \langle w^{**}, w^* \rangle \in Q^{\mathcal{I}} \}
\end{aligned}
$$

Definition 1. A statement s is *satisfiable* iff there is a model $\langle W, \mathcal{I} \rangle$ where $(s)^{\mathcal{I}}$ is not empty. A statement s is *valid* iff for every model $\langle W, \mathcal{I} \rangle$ it is $(s)^{\mathcal{I}} = W$.

For simplicity, we write $w \| {-} s$ for $w \in s^{\mathcal{I}}$ and interchange a set of statements with their conjunction. Next we introduce the set of *global axioms* G which holds in every possible world [11, 21]. They are non-logical axioms describing the access control system: groups membership, privileges attributions etc.

Definition 2. A statement s is a *logical consequence* of G, i.e. $G \models s$, iff for every $\langle W, \mathcal{I} \rangle$ if $\forall w \in W, w \| {-} G$, then $\forall w \in W, w \| {-} s$.

Global axioms can be incorporated in the axiomatization of [1] with the modal deduction theorem [16, 11], but their explicit representation is more effective because the modal deduction theorem leads to an exponential blow up [15].

---

[4] For instance the formula $\neg A$ **says** $((B \text{ \& } C){\Rightarrow}D)$ is weakly left restricted since the group membership $(B \text{ \& } C){\Rightarrow}D$ is under the scope of one negation.

Remark. The semantics of $P \Rightarrow Q$ reflects global properties of the model, is close to the universal modality [12, 13], and can introduce axiom schemas on the fly.

We can represent explicity the relation with the universal modality $[u]$ as:

$$P \Rightarrow Q \equiv [u](P \Rightarrow_{loc} Q)$$

where we have the following two conditions:

$$(P \Rightarrow_{loc} Q)^\mathcal{I} = \{w \mid \forall w^* \in W \text{ if } \langle w, w^* \rangle \in Q^\mathcal{I} \text{ then } \langle w, w^* \rangle \in P^\mathcal{I}\}$$
$$([u]s)^\mathcal{I} = \text{if } \forall v \in Wv\|{-}s \text{ then } W \text{ else } \emptyset$$

A key property is the possibility of *introducing axiom schemas "on the fly"*. For instance $P \Rightarrow P \mid P$ forces the transitivity of relation $P^\mathcal{I}$, where P may be a complex principal. Yet, these global properties may or may not be present. As an example, suppose we have $A \, \mathsf{says} \, (B \Rightarrow B \mid B)$. Transitivity of B will follow only if $\neg A \, \mathsf{says} \, \bot$ is the case. So B's properties depend on the particular global axioms and theorems we are trying to prove.

Another feature is the *absence of the tree-model property* [29, 16]:

$$A \Rightarrow A \mid A \wedge \neg(A \Rightarrow B) \wedge B \, \mathsf{says} \, \bot \wedge \neg(A \, \mathsf{says} \, \bot) \wedge A \, \mathsf{says} \, (B \, \mathsf{says} \, \bot \wedge \neg A \, \mathsf{says} \, \bot)$$

has no tree model at all, although it is satisfiable in the world 1 of the model with $W = \{1, 11, 2\}$ and $A^\mathcal{I} = \{\langle 1, 11 \rangle, \langle 11, 11 \rangle\}$ and $B^\mathcal{I} = \{\langle 2, 2 \rangle\}$. The key point is that this model has *two* clusters (connected components) so that 1 satisfies the (local) says statements and 2 satisfies the global $\neg(A \Rightarrow B)$.

In model theoretic terms, (un)satisfiability is not preserved under disjoint union as in "traditional" modal logics [29]. This is due to the "hidden" presence of the universal modality which makes impossible the complete characterisation of the logic with an Hilbert system [12, 13, 29].

The logic is also *undecidable* and a reduction to pushdown automata (without details) is pointed out in [1].

A simpler proof uses the techniques of [26] and reduces validity to the world problem of (semi)groups: map elements to atomic principals, composition "∘" to quoting "|" and equations between words $p \cong q$ to statements $P \Rightarrow Q \wedge Q \Rightarrow P$ (for short $P = Q$) for the corresponding P and Q. For groups introduce a principal I for identity, one A^c for the converse of each atomic A, and the relative equations. Next use a new principal Gr with global axioms $Gr \mid A = Gr$ for every atomic A and the statement $Gr \, \mathsf{says} \, (P = Q)$ for every $p \cong q$ characterising the group. Then one can prove that $Gr \, \mathsf{says} \, (P' = Q')$ is valid with those assumptions iff the equation $p' \cong q'$ holds for the group [26].

4 A Tableaux Calculus

Prefixed tableaux use *prefixed statements*, i.e. pairs $\langle \sigma : s \rangle$ where s is a statement and σ is an alternating sequence of integers n and atomic principals A called *prefix* and defined as $\sigma ::= n \mid \sigma.A.n$. A key difference from "standard" prefixed

$$\langle quote\rangle : \quad \frac{\sigma : \neg(P\,|\,Q \text{ says } s)}{\sigma : \neg P \text{ says } (Q \text{ says } s)} \qquad [quote] : \quad \frac{\sigma : P\,|\,Q \text{ says } s}{\sigma : P \text{ says } (Q \text{ says } s)}$$

$$\langle and\rangle : \quad \frac{\sigma : \neg(P\,\&\,Q \text{ says } s)}{\sigma : \neg P \text{ says } s \mid \sigma : \neg Q \text{ says } s} \qquad [and] : \quad \frac{\sigma : P\,\&\,Q \text{ says } s}{\sigma : P \text{ says } s} \\ \sigma : Q \text{ says } s$$

$$\text{Glob} : \quad \frac{\vdots}{\sigma : s} \quad \text{if } \sigma \text{ is present in } \mathcal{B} \text{ and } s \in G_{\mathcal{B}}$$

$$\langle A\rangle : \quad \frac{\sigma : \neg(A \text{ says } s)}{\sigma.A.m : \neg s} \quad \sigma.A.m \text{ new} \qquad [A] : \quad \frac{\sigma : A \text{ says } s}{\sigma.A.n : s} \quad \sigma.A.n \text{ present}$$

$$D(A) : \quad \frac{\sigma : A \text{ says } s}{\sigma : \neg A \text{ says } \neg s} \quad \text{with some } \sigma.A.n \text{ already present in the branch}$$

Fig. 3. PDL-like Rules for Modal Connectives

tableaux [11, 14] is that a set of prefixes now describes a *forest of trees*, where arcs are labelled with atomic principals and node with integers. With k different initial prefixes we have, in graph-theoretic terminology, k connected components or clusters [16]. With global axioms and the operator $\Rightarrow$ we can impose an euclidean or transitive closures on a cluster but we cannot collapse two clusters.

Still the definition of tableau is similar to prefixed tableaux for modal logics [11, 14, 21]: a *tableau* $\mathcal{T}$ is a rooted (binary) tree where nodes are labelled with prefixed statements in the usual fashion.

Definition 3. A *branch* of a tableau $\mathcal{T}$ is a pair $\langle \mathcal{B}, G_{\mathcal{B}}\rangle$ where $\mathcal{B}$ is a path from the root to a leaf of $\mathcal{T}$ and $G_{\mathcal{B}}$ is a set of global axioms.

Thus, each time we branch the tree we should also duplicate (in theory) the set of global axioms. This definition is essential because we need to modify the set of global axioms during the deduction process and therefore different branches may end up with different global axioms.

A prefix is *present* in a branch $\langle \mathcal{B}, G_{\mathcal{B}}\rangle$, if there is a prefixed statement with that prefix already in $\mathcal{B}$, and it is *new* if it is not already present.

The rules for propositional connectives are standard [11, 14, 21] and omitted. The rules for conjunction, quoting, the use of global axioms and the transitional rules for atomic principals are in Fig. 3.

To cope with $\Rightarrow$ we introduce a new set of propositional atoms x_i (distinct from r) to mark unsaid statements as in Fig. 4. Since $P \Rightarrow Q$ implies that if P says s then Q says s for all s, its negation means that there is "something" (an unknown x_i) which P said but Q didn't. The first two rules correspond to

$$\langle R_{gr}\rangle : \frac{\sigma : P \Rightarrow A \quad \sigma : \neg(A \text{ says } s)}{\sigma : \neg(P \text{ says } s)} \qquad [L_{gr}] : \frac{\sigma : A \text{ says } s \quad \sigma : A \Rightarrow Q}{\sigma : Q \text{ says } s}$$

$$\langle U_{gr}\rangle : \frac{\sigma : \neg(P \Rightarrow Q)}{n : P \text{ says } x_i} \quad x_i \text{ and } n \text{ new} \qquad [U_{gr}] : \frac{\sigma : P \Rightarrow Q}{G_B := G_B \cup \{P \Rightarrow Q\}}$$
$$n : \neg(Q \text{ says } x_i)$$

Fig. 4. Rules for the speaks-for operator

the local features of the $\Rightarrow$ operator, whereas the last is due to its "universal" flavour. The $\langle U_{gr}\rangle$-rule combines both aspects.

Remark. Weakly left restricted statements do *not* need rule $\langle R_{gr}\rangle$ and $D(A)$ while right restricted languages do not need rule $[L_{gr}]$.

Further simplifications are possible: $P \Rightarrow (A \& B) \mid Q$ is logically equivalent to $P \Rightarrow A \mid Q \wedge P \Rightarrow B \mid Q$. This rule can be added when Q is empty since it may lead to right-restricted formulae. Rule $\langle U_{gr}\rangle$ must be applied only once for each subformula $\neg(P \Rightarrow Q)$, no matter its prefix. In a similar way rule $\langle A \rangle$ can be skipped if a prefixed formula $\sigma.A.n : \neg s$ is already present etc.

Definition 4. A branch $\langle \mathcal{B}, G_B \rangle$ is *closed* if $\mathcal{B}$ contains both $\sigma : s$ and $\sigma : \neg s$, for some s and σ. It is *open* if all possible rules have been applied and it is not closed. A tableau is *closed* if all branches are closed; it is *open* if at least one branch is open.

Definition 5. A *validity tableau proof* for statement s with global axioms G is a closed tableau starting with the branch $\langle \{1 : \neg s\}, G \rangle$.

In a dual way a satisfiability witness is any open branch of the tableau starting with $\langle \{1 : s\}, G \rangle$, when the calculus is complete for the fragment at hand.

Theorem 6 (Strong Soundness). *If s has a tableau proof with global axioms G then s is a logical consequence of G.*

We give a completeness result only for two main fragments (as noted in §3 the second is the most important from an applicative point of view).

Theorem 7 (Strong WL-Completeness). *If s is a logical consequence of G and $G \cup \{\neg s\}$ are weakly left restricted then s has a proof.*

Theorem 8 (Strong WRR-Completeness). *If s is a logical consequence of G and $G \cup \{\neg s\}$ are weakly right and request restricted then s has a proof.*

Remark. A decision method (rather than a semidecidable procedure which could be based on first order translations) is important for security analysis because satisfiability gives information on security weaknesses.

$$
\begin{array}{lll}
(a) & 1 : \neg(\neg(A \text{ says } \bot) \supset (A \text{ controls } P \Rightarrow A)) & \\
(b) & 1 : \neg(A \text{ says } \bot) & \text{by } \alpha \text{ rules from } (a) \\
(c) & 1 : \neg A \text{ controls } (P \Rightarrow A) & \\
(d) & 1 : A \text{ says } (P \Rightarrow A) & \text{by reducing } \textbf{controls} \text{ from } (c) \\
(e) & 1 : \neg(P \Rightarrow A) & \\
(f) & 2 : P \text{ says } x_1 & \text{by } \langle U_{gr} \rangle \text{ from } (e) \\
(g) & 2 : \neg(A \text{ says } x_1) & \\
(h) & 1.A.3 : \neg\bot & \text{by } \langle A \rangle \text{ from } (b) \\
(i) & 1.A.3 : P \Rightarrow A & \text{by } [A] \text{ from } (d) \\
(l) & G_1 := \{P \Rightarrow A\} & \text{by } [U_{gr}] \text{ from } (i) \\
(m) & 2 : P \Rightarrow A & \text{by } Glob \text{ from } G_1 \\
& \times & \text{contradiction between } (e, m)
\end{array}
$$

Fig. 5. Tableaux Proof of an Hand-off Axiom

For a *decision method* a simple condition, checkable in polynomial time, can be imposed on the global axioms and the consequence s. Associate a graph to the global axioms and the negation of the formula to be proved: each atomic principal is represented by a node and for every $P \Rightarrow Q$, under the scope of an even number of negation, draw an arc from the atomic principals in P to those in Q. If this graph is *acyclic* then the tableau construction terminates by using loop checking with an extended notion of the Fisher-Ladner closure. Notice that, in the embedding of the word problem (§3), the principal Gr creates cycles.

In access control, acyclicity is not a restriction but rather a requirement [10, 23]: if $\Rightarrow$ is used for hierarchies of groups/roles then cycles are not allowed.

5 Examples

For sake of simplicity, we assume that we have a direct (obvious) rule for controls rather than translating it back to $\wedge$ and $\neg$. A first example is the derivation of an *hand-off axiom* [1, 19].

$$\neg(A \text{ says } \bot) \supset (A \text{ controls } (P \Rightarrow A)).$$

Such axioms are used by principals to hand-over their privileges in [1, 19]. Notice that, although valid, they cannot be proved within the Hilbert system developed by [1] and are added as axioms. The tableau derivation is shown in Fig. 5. The key step is rule $[U_{gr}]$ which cannot be axiomatised.

To check that global axioms must be associated to a branch, try the (satisfiable) formula below (or in §3) with only one set of global axioms as in [11, 21].

$$\neg(A \,\&\, B \text{ says } \bot) \wedge A \text{ says } (B \Rightarrow A) \wedge A \text{ says } r \wedge B \text{ says } \neg r$$

For a "real-life" deduction (Fig. 6) we take delegation without certificates from [1, page 719] where a careful (and non trivial) Hilbert proof is given.

(a) $1: \neg (K_B \mathbf{\,says\,} (Sc \mathbf{\,says\,} r) \wedge K_S \mathbf{\,says\,} (K_B{\Rightarrow}B) \wedge (B\,|\,A) \mathbf{\,controls\,} r \supset r)$
(b) $1: K_B \mathbf{\,says\,} (Sc \mathbf{\,says\,} r)$ $\quad\quad \alpha$ rules from (a)
(c) $1: K_S \mathbf{\,says\,} (K_B{\Rightarrow}B)$
(d) $1: B\,|\,A \mathbf{\,controls\,} r$
(e) $1: \neg r$
(f) $1: K_S{\Rightarrow}S$ $\quad\quad\quad\quad\quad$ by $Glob$
(g) $1: S \mathbf{\,says\,} (K_B{\Rightarrow}B)$ $\quad\quad$ by $[L_{gr}]$ from $(c),(f)$
(h) $1: S \mathbf{\,controls\,} (K_B{\Rightarrow}B)$ $\quad$ by $Glob$

(h) $1: \neg S \mathbf{\,says\,} (K_B{\Rightarrow}B)$ $\quad$ (i) $1: K_B{\Rightarrow}B$ $\quad\quad\quad\quad$ red $\mathbf{controls}$ from (h)
$\quad\quad\quad\quad\times$ $\quad\quad\quad\quad\quad\quad\quad$ (l) $1: B \mathbf{\,says\,} (Sc \mathbf{\,says\,} r)$ by $[L_{gr}]$ from $(b),(i)$

$\quad\quad\quad\quad\quad\quad$ (m) $1: r$ $\quad$ (n) $1: \neg (B\,|\,A \mathbf{\,says\,} r)$ $\quad\quad$ red $\mathbf{controls}$ from (d)
$\quad\quad\quad\quad\quad\quad\quad\quad\quad\times$ $\quad\quad$ (o) $1: \neg B \mathbf{\,says\,} (A \mathbf{\,says\,} r)$ by $\langle quote\rangle$ from (n)
$\quad\quad\quad\quad\quad\quad\quad\quad\quad\quad\quad\quad$ (p) $1.B.2: \neg A \mathbf{\,says\,} r$ $\quad$ by $\langle B\rangle$ from (o)
$\quad\quad\quad\quad\quad\quad\quad\quad\quad\quad\quad\quad$ (q) $1.B.2: Sc \mathbf{\,says\,} r$ $\quad$ by $[B]$ from (l)
$\quad\quad\quad\quad\quad\quad\quad\quad\quad\quad\quad\quad$ (r) $1.B.2: Sc{\Rightarrow}A$ $\quad\quad$ by $Glob$
$\quad\quad\quad\quad\quad\quad\quad\quad\quad\quad\quad\quad$ (s) $1.B.2: A \mathbf{\,says\,} r$ $\quad\quad$ by $[L_{gr}]$ from $(q),(r)$
$\quad\quad\quad\quad\quad\quad\quad\quad\quad\quad\quad\quad\quad\quad\quad\times$

Fig. 6. Tableau Proof of Delegation without Certificates

Example. "A delegates to B who makes requests to C. For instance A may be a user with a sufficiently powerful smartcard, B a workstation and C a file server. [...] When B wishes to make a request r on A's behalf, B sends the signed requests along with A's name... in the format $K_B \mathbf{\,says\,} (A \mathbf{\,says\,} r)$... When C receives the request r he has evidence that B has said A has requested r but not that A has delegated to B; then C consults the ACL [Access Control List] for request r and determines whether the request should be granted. [...] A certification authority provides the certificates for the principals' public keys as needed. The necessary certificates are $K_S \mathbf{\,says\,} (K_A{\Rightarrow}A)$ and $K_S \mathbf{\,says\,} (K_B{\Rightarrow}B)$, where K_S is S's public key."

We add a level of indirection to the original problem (by modelling explicitly the smartcard Sc) and use the logic for the reasoning of the server C. The set of global axioms and the statement to be proven are:

$$G \doteq \{K_S{\Rightarrow}S, S \mathbf{\,controls\,} (K_B{\Rightarrow}B), Sc{\Rightarrow}A\}$$
$$s \doteq (K_B \mathbf{\,says\,} (Sc \mathbf{\,says\,} r) \wedge K_S \mathbf{\,says\,} (K_B{\Rightarrow}B) \wedge (B\,|\,A) \mathbf{\,controls\,} r) \supset r$$

In Fig. 6 only $[L_{gr}]$-rule is used. A derivation with only $\langle R_{gr}\rangle$-rule is possible.

This is an example of the "incompilability" of $P{\Rightarrow}Q$ into axiom schemas (or rules). Indeed $K_B{\Rightarrow}B$ corresponds to $B^{\mathcal{I}} \subseteq K_B^{\mathcal{I}}$, i.e. $K_B \mathbf{\,says\,} s \supset B \mathbf{\,says\,} s$ but it is not always valid! It depends on the server's statements i.e. $S \mathbf{\,says\,} (K_B{\Rightarrow}B)$.

Without the server's certificate i.e. with a different theorem, it doesn't hold. The possibility of adding "on-line" properties is critical here since delegations and groups membership depend on the security policy and the current certificates.

6 Soundness and (Partial) Completeness

To prove soundness we map prefixes to states and show that satisfiability is preserved by tableaux rules. After this key lemma, the rest is standard [11].

Definition 9. Let B be a set of prefixed formulae and $\langle W, \mathcal{I} \rangle$ a model, a *mapping* is a function $\imath()$ from prefixes to states s.t. for all σ and $\sigma.A.n$ present in B it is $\langle \imath(\sigma), \imath(\sigma.A.n) \rangle \in A^{\mathcal{I}}$.

Definition 10. A tableau branch $\langle B, G_B \rangle$ is *satisfiable* (SAT for short) in the model $\langle W, \mathcal{I} \rangle$ if for every $s_g \in G_B$ and every $w \in W$ it is $w \| \! \! - s_g$ and there is a mapping $\imath()$ such that for every $\langle \sigma : s_b \rangle$ present in B it is $\imath(\sigma) \| \! \! - s_b$. A tableau is SAT if one branch is such.

Theorem 11 (Safe Extension). *If $\mathcal{T}$ is a SAT tableau, then the tableau $\mathcal{T}'$ obtained by an application of a tableau rule is also SAT.*

Proof. By induction on the rules applied as in [11, Chapter 8] or [9, 14, 21]. The $\Rightarrow$ operator is the only new case.

Suppose that $\langle \sigma : P \Rightarrow Q \rangle$ occurs in some SAT branch. Then there must be a model $\langle W, \mathcal{I} \rangle$ and a mapping $\imath()$ such that $\imath(\sigma) \| \! \! - P \Rightarrow Q$. Hence $(P \Rightarrow Q)^{\mathcal{I}} = W$ for the global property of $\Rightarrow$ (it is not empty as it contains $\imath(\sigma)$). Therefore adding it to G_B as done by the $[U_{gr}]$ does not change the satisfiability of the branch wrt the G_B condition (and future applications of the *Glob*-rule).

If $\langle \sigma : \neg(P \Rightarrow Q) \rangle$ is present then by hp $\imath(\sigma) \| \! \! - \neg(P \Rightarrow Q)$. Hence there is a $\langle w, w^* \rangle \in Q^{\mathcal{I}}$ with $\langle w, w^* \rangle \notin P^{\mathcal{I}}$. Set $\imath(n) = w$ for the new prefix n and $(x_i)^{\mathcal{I}} = W - \{w^*\}$ for the new x_i. Clearly $\imath(n) \| \! \! - P \text{ says } x_i$ but $\imath(n) \| \! \! \not\vdash (Q \text{ says } x_i)$. $\square$

For completeness we apply a systematic and fair procedure (see [11, 14] for a definition) and use an open branch to construct a model. A key property the procedure must guarantee is *downward saturation* [11, 21]: all applicable rules must have been applied. The proof is given for weakly left restricted statements and its extension is discussed.

Theorem 12 (Model Existence). *If $\langle B, G_B \rangle$ is an open branch with weakly left restricted statements only, then there is a model $\langle W, \mathcal{I} \rangle$ on which it is SAT.*

Proof. Construct a pre-model $\langle W, \mathcal{I}_0 \rangle$ as follows:

$$W \doteq \{\sigma \mid \sigma \text{ is present in } B\}$$
$$A^{\mathcal{I}_0} \doteq \{\langle \sigma, \sigma.A.n \rangle \mid \sigma \text{ and } \sigma.A.n \text{ are present in } B\}$$
$$r^{\mathcal{I}_0} \doteq \{\sigma \mid \sigma : r \in B\}$$

Incorporate the constraints due to $\Rightarrow$ and build $\mathcal{I}$ from $\mathcal{I}_0$ as follows:

– for every formula $\sigma : A \Rightarrow P$ occurring in $\mathcal{B}$
  • compute $P^{\mathcal{I}}$;
  • if $\langle \sigma, \sigma^* \rangle \in P^{\mathcal{I}}$ then add $\langle \sigma, \sigma^* \rangle$ to $A^{\mathcal{I}}$;
– repeat until a fix-point is reached.

After this *closure phase* we must prove that if $\langle \sigma : s \rangle \in \mathcal{B}$ then $\imath(\sigma) \Vdash s$ by induction on the construction of s, where $\imath(\sigma) = \sigma$. The proof is similar to those used for PDL [9] or modal logics [11, 14, 21].

The difficult case is $\langle \sigma : A \text{ says } s \rangle$ since we must prove that for all prefixes $\langle \sigma, \sigma^* \rangle \in A^{\mathcal{I}}$ it is $\langle \sigma^* : s \rangle$ in $\mathcal{B}$ so that we can apply induction to $\langle \sigma^* : s \rangle$, get $\sigma^* \Vdash s$ and then the claim.

The difference with "traditional" proofs [11, 21] is that some prefixes σ^* are introduced in $A^{\mathcal{I}}$ during the closure phase. Hence we can have (i) $\sigma^* = \sigma.A.n$ for some n or (ii) $\langle \sigma, \sigma^* \rangle \in P^{\mathcal{I}}$ before the closure phase and $P^{\mathcal{I}} \subseteq A^{\mathcal{I}}$ afterwards.

The first case is standard [11, 21] whereas for the second case we use the saturation of the branch and a double induction: on the formula size and on the iterations of the closure phase needed to enter $\langle \sigma, \sigma^* \rangle$ into $A^{\mathcal{I}}$. For the base case we use the following result, proven by induction on P and s as in [9]:

Proposition 13. *Before each iteration step, if $\langle \sigma : P \text{ says } s \rangle$ is present in $\mathcal{B}$ and $\langle \sigma, \sigma^* \rangle \in P^{\mathcal{I}}$ then $\langle \sigma^* : s \rangle$ is present in $\mathcal{B}$*

For the induction step observe that whenever $\langle \sigma : A \Rightarrow P \rangle$ is present then by saturation $\langle \sigma : A \text{ says } s \rangle$ implies $\langle \sigma : P \text{ says } s \rangle$. So apply Prop. 13 to get $\langle \sigma^* : s \rangle$ and therefore when σ^* was added in $A^{\mathcal{I}}$ in the closure phase also $\langle \sigma^* : s \rangle$ was present. Now apply the induction hypothesis.

The local condition for $\Rightarrow$ is satisfied by construction. For the global condition we use mutual saturation between *Glob* and $[U_{gr}]$ rules. If $\sigma : A \Rightarrow P$ occurs in $\mathcal{B}$ then $[U_{gr}]$ implies that $A \Rightarrow P \in G_{\mathcal{B}}$. By *Glob* we have that for all $s \in G_{\mathcal{B}}$ and all prefixes σ in $\mathcal{B}$ it is $\langle \sigma : s \rangle$. Hence every σ satisfies the local condition i.e. W satisfies the global condition and $P^{\mathcal{I}} \subseteq A^{\mathcal{I}}$. $\square$

For the right and request restricted fragment of the language the key point is that we only have literals or statements of the form $P \Rightarrow A$ under the scope of **says**. The operator $\Rightarrow$ does not create problems given its global nature and the only difficult part is due to literals l (r or $\neg r$).

The previous proof for $\langle \sigma : A \text{ says } s \rangle$ does not work since $\langle \sigma : P \text{ says } s \rangle$ for non atomic P cannot propagate over $\langle \sigma : P \Rightarrow A \rangle$. However we can prove the dual of the induction step: if $\langle \sigma : \neg(A \text{ says } l) \rangle$ and $P \Rightarrow A$ is also present then there is is a $\langle \sigma, \sigma^* \rangle \in P^{\mathcal{I}}$ such that $\sigma^* : \neg l$. This means that all $\langle \sigma : P \text{ says } l' \rangle$ are consistent with each $\langle \sigma : \neg(A \text{ says } l) \rangle$. At this stage we need to use the $D(A)$-rule, to prove that those P statements are consistent also with each $\langle \sigma : A \text{ says } l'' \rangle$. By $D(A)$ we obtain $\langle \sigma : \neg(A \text{ says } \neg l'') \rangle$ and then apply the dual property.

Since all l, l', l'' are literals this is enough: all A **says** l are consistent among themselves, and each of them with all P **says** l'. This means that when we add a $\langle \sigma, \sigma^* \rangle$ from $A^{\mathcal{I}}$ to $P^{\mathcal{I}}$ in the closure phase we can always extend the valuation of the unspecified l' or l'' in σ^* so that the result is still a model. Again, this only works for *request restricted* statements.

7 Conclusion

The major contribution of this paper is the development of a tableaux method for the calculus of access control of Abadi, Lampson et al. [1, 19, 31]. We have also clarified some model theoretic features of the calculus that makes difficult its axiomatic characterisation. The completeness results presented here extend those in [1] and provide the basis for a full fledged automatisation.

This tableaux method requires novel techniques such as passing from a tree-like tableau to a forest of prefixes, a run time update of global axioms and the corresponding modification of the notion of branch. Future research is in the direction of providing a fully automated verifier, possibly using the results of [3].

As an aside, these tableaux techniques can be used for multi modal logics with the universal modality [13] and a sound and complete calculus is presented in the appendix. Therefore we can distinguish, in proof theoretic terms, between $S5$ and the universal modality (on non tree models).

A claim that we do *not* make is that logic and semantic tableaux should be used for run-time decisions on access control. Although possible, this may lead to unacceptable slow-downs. Logic and tableaux (or similar logic-based methods) should be used for *verification* and prototyping, for checking that access protocols respect security policies. This work is a step in this direction.

Acknowledgements

I would like to thank L. Paulson and the Computer Laboratory for their hospitality in Cambridge, M. Abadi, the Computer Security group (Cambridge), the Applied Logic group (IRIT) and the anonymous referees for many suggestions which helped to improve this paper. This research has been partly supported by ASI, CNR and MURST 40% and 60% grants and by EPSRC grant GR/K77051 "Authentication Logics".

References

1. M. Abadi, M. Burrows, B. Lampson, and G. Plotkin. A calculus for access control in distributed systems. *ACM Trans. on Prog. Lang. and Sys.*, 15(4):706–734, 1993.
2. R. Anderson. A security policy model for clinical information systems. In *Proc. of the 15th IEEE Symp. on Security and Privacy*. IEEE Comp. Society Press, 1996.
3. B. Beckert and R. Goré. Free variable tableaux for propositional modal logics. In THESE PROCEEDINGS, 1997.
4. D. Bell and L. La Padula. Secure computer systems: unified exposition and MULTICS. Report ESD-TR-75-306, The MITRE Corporation, March 1976.
5. M. Burrows, M. Abadi, and R. Needham. A logic for authentication. *ACM Trans. on Comp. Sys.*, 8(1):18–36, 1990. Also available as Res. Rep. SRC-39, DEC - System Research Center, 1989.
6. M. Castilho and A. Herzig. An alternative to the iteration operator of propositional dynamic logic. Tech. Rep. 96-05-R, IRIT (Toulouse), Univ. Paul Sabatier, jan 1996.

7. D. Clark and D. Wilson. A comparison of commercial and military computer security policies. In *Proc. of the 6th IEEE Symp. on Security and Privacy*, pp. 184–194. 1987.

8. F. Cuppens and R. Demolombe. A deontic logic for reasoning about confidentiality. In *3rd Int. Workshop on Deontic Logic in Computer Science*, Sesimbra, Portugal, 1996.

9. G. De Giacomo and F. Massacci. Tableaux and algorithms for propositional dynamic logic with converse. In *Proc. of the 13th Int. Conf. on Automated Deduction (CADE-96)*, LNAI 1104, pp. 613–628, 1996.

10. D. Ferraiolo, J. Cugini, and K. Richard. Role-based access control (rbac): Features and motivations. In *Proc. of the Annual (Computer Security Applications Conf.*, 1995.

11. M. Fitting. *Proof Methods for Modal and Intuitionistic Logics*. Reidel, 1983.

12. V. Goranko. Modal definability in enriched languages. *Notre Dame J. of Formal Logic*, 31(1), 1990.

13. V. Goranko and S. Passy. Using the universal modality: Gains and questions. *J. of Logic and Computation*, 2(1):5–30, 1992.

14. R. Goré. Tableaux method for modal and temporal logics. Tech. Rep. TR-ARP-15-5, Australian National Univ., 1995.

15. J. Halpern and Y. Moses. A guide to completeness and complexity for modal logics of knowledge and belief. *Artificial Intelligence*, 54:319–379, 1992.

16. G. Hughes and M. Cresswell. *a Companion to Modal Logic*. Methuen, 1984.

17. C. Krogh. Obligations in multiagent systems. In *5th Scandinavian Conference on Artificial Intelligence (SCAI-95)*, pp. 29–31. ISO Press, 1995.

18. B. Lampson. Protection. *ACM Operating Sys. Reviews*, 8(1):18–24, 1974.

19. B. Lampson, M. Abadi, M. Burrows, and E. P. Wobber. Authentication in distributed systems: Theory and practice. *ACM Trans. on Comp. Sys.*, 10(4):265–310, 1992.

20. B. Marick. The **verus** design verification system. In *Proc. of the 2nd IEEE Symp. on Security and Privacy*, pp. 150–157, 1983.

21. F. Massacci. Strongly analytic tableaux for normal modal logics. In *Proc. of the 12th Int. Conf. on Automated Deduction (CADE-94)*, LNAI 814, pp. 723–737, 1994.

22. J. Moffet and M. Sloman. Policy hierarchies for distributed systems management. *IEEE J. on Selected Areas in Communications*, 11(9), 1993.

23. R. Sandhu, E. Coyne, H. Feinstein, and C. Youman. Role-based access controls models. *IEEE Computer*, 29(2), February 1996.

24. R. Sandhu. The typed access matrix model. In *Proc. of the 11th IEEE Symp. on Security and Privacy*, pp. 122–136, 1992.

25. R. Sandhu and P. Samarati. Access control: Principles and practice. *IEEE Communications Magazine*, pp. 40–48, September 1994.

26. M. Schmidt-Schauss. Subsumption in KL-ONE is undecidable. In *Proc. of the 1st Int. Conf. on the Principles of Knowledge Representation and Reasoning (KR-89)*, pp. 421–431, 1989.

27. S. Schmitt and C. Kreitz. Converting non-classical matrix proofs into sequent-style systems. In *Proc. of the 13th Int. Conf. on Automated Deduction (CADE-96)*, LNAI 1104, pp. 418–432, 1996.

28. P. F. Syverson and P. C. van Oorschot. On unifying some cryptographic protocols logics. In *Proc. of the 13th IEEE Symp. on Security and Privacy*. IEEE Comp. Society Press, 1994.

29. J. van Benthem. Correspondence theory. In *Handbook of Philosophical Logic*, volume II. Reidel, 1986.
30. R. van der Meyden. The dynamic logic of permission. *J. of Logic and Computation*, 6(3):465–479, 1996.
31. E. Wobber, M. Abadi, and M. Burrows. Authentication in the Taos operating system. *ACM Trans. on Comp. Sys.*, 12(1):3–32, 1994.

A   Tableaux for the Universal Modality

We can easily define a tableaux calculus for multi modal logics with the universal modality: use A says s as $[A]s$ to get the logic K_n or more precisely an almost PDL variant (close to [6] although they have a weaker common knowledge modality); add the Single Step Tableaux (SST) rules for the others modal logics of knowledge and belief between K_n and $S5_n$ [14, 21]; finally use a modified version of rule $\langle U_{gr} \rangle$ and $[U_{gr}]$ described below.

$$\langle u \rangle : \frac{\sigma : \neg[u]s}{n : \neg s} \; n \text{ new} \qquad [u] : \frac{\sigma : [u]s}{G_B := G_B \cup \{s\}}$$

Theorem 14 (Universal Modality). *If $\mathcal{R}_1, \ldots, \mathcal{R}_n$ are sound and complete SST rules for the (multi) modal logics $\mathcal{L}_1, \ldots, \mathcal{L}_n$ [14, 21] then the tableaux calculus enhanced with rules $[u]$ and $\langle u \rangle$ is sound and complete for multi modal logic $\mathcal{L}_1 \ldots \mathcal{L}_n$ with the universal modality.*

For the soundness part we replace $\sigma : P \Rightarrow Q$ with $\sigma : [u]s$ and $P \Rightarrow Q$ with s in the corresponding argument of Thm. 11. For completeness, the mutual induction between the application of the *Glob*-rule and the $[u]$-rule is identical to Thm. 12.

It is possible to "distinguish the (axiomatically) indistinguishable" i.e. $S5$ from the universal modality in proof theoretic terms (closed vs open branches for some formulae). Of course this distinction can only be done as satisfiability on k-clusters models for $k \geq 2$, since $S5$ and $[u]$ cannot be distinguished by validity [12, 13, 29] and thus by traditional tableaux (and 1-cluster models).

For instance, with the $S5$-rules for $[A]$ and $[B]$ given in [14, 21]:

$$\Phi_{SAT} \doteq \langle B \rangle [A]r \wedge \langle B \rangle [A]\neg r \qquad \Phi_{UNSAT} \doteq \langle B \rangle [A]r \wedge \langle B \rangle [u]\neg r$$

have different tableaux: one open, and the second closed. So replacing an occurrence of $[A]$ with $[u]$ changes the satisfiability of a formula. Equally one can combine $[A]$ only with $[u]$ and obtain different tableaux.

We can also derive *a small model theorem* by adapting loop checking techniques from [9, 14] and find a model (if any) of size at most $O(|G \cup \{s\}| \times 2^{|G \cup \{s\}|})$.

Proving Correctness of Labeled Transition Systems by Semantic Tableaux

Wolfgang May*

Institut für Informatik, Universität Freiburg, Germany,
may@informatik.uni-freiburg.de

Abstract. The paper presents a method for formally proving correctness of processes specified by transition systems which is based on a tableau calculus for an extended temporal logic. The model-theoretic semantics is given by *labeled* Kripke structures, incorporating information about the actions performed in transitions. Extending first-order CTL for handling action labels, the multi-modal logic MCTL is defined which is well-suited for specifying transition systems and their properties. For MCTL, a tableau semantics and -calulus is presented, allowing formal verification.

1 Introduction

Specifications of processes are often given by transition systems, describing the effects of actions on the state of the process. Proving correctness of such specifications involves an integrated reasoning about static (data) and dynamic (actions) aspects of processes.

Labeled transition systems focus on dynamic acpects, thereby lacking facilities for representing static, data-oriented aspects. In contrast, Kripke structures focus on data-oriented aspects, providing expressive concepts for describing states while no special facilities for expressing action-oriented aspects exist.

As a means to overcome this deficiency, *labeled* Kripke structures provide a suitable model-theoretic semantics for combined modeling of both static and dynamic aspects of processes: Actions are explicitly integrated into Kripke structures by labeling the accessibility relation with the actions executed in the respective transitions.

A suitable multi-modal logic MCTL for labeled Kripke structures is derived from first-order CTL (Computation Tree Logic) by defining more detailed modalities. It is taken into consideration that, when reasoning about processes, different relationships between actions and states have to be investigated: an action is performed *and* some property will hold, or *if* an action is performed, *then* some property will hold.

Dynamic Logic [Har79] is also based on labeled Kripke structures, although it is quite different: Both its modalities query only the action carried out in the next step, corresponding to the focus of DL on *program* verification.

A logic more similar to MCTL, ACTL, has been presented in [DV90]: ACTL is a modal logic which is interpreted over labeled transition systems with propositional interpretations as states, based on *propositional* CTL and propositional

---

* Supported by grant no. GRK 184/1-97 of the Deutsche Forschungsgemeinschaft.

action labels. In contrast, MCTL is designed for first-order logic and first-order actions, also providing more detailed modalities, consisting of a coupling mode and a modality in the classical sense.

In [BMP81] and [Wol85], a tableau method for propositional CTL is presented, closely related to the model checking procedure of [CES86]. The branches of the tableau represent paths in a fictive model. Cycles in the tableau are allowed. After termination, which is guaranteed, eventuality formulas have to be postprocessed. The method cannot be extended to first-order variants because the finite number of possible different states is crucial to their concept.

To overcome these problems, in [MS96] a tableau semantics and calculus for first-order CTL has been developed, making basic changes in the handling of eventualities by introducing the ability to abstract from finitely many states in-between. In the resulting tableau calculus we achieved a 1:1-correspondence of branches of the tableau to Kripke structures.

In the present work, this tableau semantics and calculus is extended to labeled Kripke structures and MCTL by integrating actions into this framework in an intuitive way, allowing a formal verification of transition systems.

The paper is structured as follows: In Section 2 the basic notions of first-order logic, Kripke structures, and temporal logics are reviewed. Section 3 introduces the integration of actions by labeled Kripke structures, and the logic MCTL is defined as an extension of CTL to labeled Kripke structures. Section 4 discusses the representation of labeled Kripke structures by semantical tableaux, identifying the relevant entities and developing a syntax of tableau formulas whereas the calculus is given in Section 5. The conclusion in Section 6 completes the paper.

2 Basic Notions

First Order Logic. Every language of first-order logic includes the symbols ")" and "(", the boolean connectives "¬" and "∨", the quantifier "∃", and an infinite set of variables Var $:= \{x_1, x_2, \ldots\}$. A particular language is given by its *signature* Σ consisting of function symbols and predicate symbols with fixed arities $\text{ord}(f)$ resp. $\text{ord}(p)$. Terms, first-order formulas, and the notions of bound and free variables are defined as usual, free($\mathcal{F}$) denoting the set of variables occurring free in a set $\mathcal{F}$ of formulas.

A *substitution* (over a signature Σ) is a mapping $\sigma : \text{Var} \to \text{Term}_\Sigma$ where $\sigma(x) \neq x$ for only finitely many $x \in \text{Var}$, here denoted by $[\sigma(x)/x]$. Substitutions are extended to terms and formulas as usual.

A *first-order interpretation* $\mathbf{I} = (I, \mathbf{U})$ over a signature Σ consists of a nonempty set $\mathbf{U}$ (*universe*) and a mapping I which maps every function symbol $f \in \Sigma$ to a function $I(f) : \mathbf{U}^{\text{ord}(f)} \to \mathbf{U}$ and every predicate symbol $p \in \Sigma$ to a relation $I(p) \subseteq \mathbf{U}^{\text{ord}(p)}$.

A *variable assignment* is a mapping $\chi : \text{Var} \to \mathbf{U}$. For a variable assignment χ, a variable x, and $d \in \mathbf{U}$, the *modified* variable assignment χ_x^d is identical with χ except that it assigns d to x. Let Ξ denote the set of variable assignments.

Every interpretation induces to an *evaluation* $\mathbf{I} : \text{Term}_\Sigma \times \Xi \to \mathbf{U}$ s.t. $\mathbf{I}(x, \chi) := \chi(x)$ for $x \in \text{Var}$, and $\mathbf{I}(f(t_1, \ldots, t_n), \chi) := (I(f))(\mathbf{I}(t_1, \chi), \ldots, \mathbf{I}(t_n, \chi))$

for $f \in \Sigma$, $\text{ord}(f) = n$ and $t_1, \ldots, t_n \in \text{Term}_\Sigma$. The truth of a formula F in a first-order interpretation $\mathbf{I}$ under a variable assignment χ, $(\mathbf{I}, \chi) \models_{\text{FO}} F$ is defined as usual.

Kripke Structures.

Definition 1. A first-order *Kripke structure* over a signature Σ is a triple $\mathbf{K} = (\mathbf{G}, \mathbf{R}, \mathbf{M})$ where $\mathbf{G}$ is a set of states, $\mathbf{R} \subseteq \mathbf{G} \times \mathbf{G}$ an *accessibility relation*, and for every $g \in \mathbf{G}$, $\mathbf{M}(g) = (M(g), \mathbf{U}(g))$ is a first-order interpretation of Σ with universe $\mathbf{U}(g)$. $\mathbf{G}$ and $\mathbf{R}$ are called the *frame* of $\mathbf{K}$.

A *path* p in a Kripke structure $\mathbf{K} = (\mathbf{G}, \mathbf{R}, \mathbf{M})$ is a sequence $p = (g_0, g_1, g_2, \ldots)$, $g_i \in \mathbf{G}$ with $\mathbf{R}(g_i, g_{i+1})$ holding for all i. It induces a mapping $p : \mathbb{N} \to \mathbf{G}$ with $p(i) = g_i$. Let $p|_i := (g_i, g_{i+1}, \ldots)$.

In this paper, only Kripke structures with constant universe (i.e. $\mathbf{U}(g) = \mathbf{U}(g')$ for all $g, g' \in \mathbf{G}$) are considered. The notion of a *variable assignment* is then defined as in the first-order case.

Definition 2. For a Kripke structure $\mathbf{K}$ over a signature Σ, the state-independent portion $\Sigma^c \subseteq \Sigma$ consists of all function symbols f s.t. $(M(g))(f) = (M(g'))(f)$ for all $g, g' \in \mathbf{G}$ and all predicate symbols p s.t. $(M(g))(p) = (M(g'))(p)$ for all $g, g' \in \mathbf{G}$. This induces a state-independent evaluation $\mathbf{K}(t)$ for $t \in \text{Term}_{\Sigma^c}$.

Temporal Logics.

In [BMP81] and [EH83], the family CTL of temporal logics for branching time is defined in its propositional version for reasoning about Kripke structures in temporal contexts. It uses modal operators $\square$ ("always"), $\lozenge$ ("sometimes"), $\circ$ ("nexttime"), and until, all interpreted as future modalities, and two path-quantifiers A and E. In CTL, two classes of formulas are distinguished: *state formulas* (SF) holding in states, and *path formulas* (PF) holding on paths.

Definition 3. The syntax of CTL-formulas is given as follows:

- Every first-order formula is in SF(CTL).
- $F, G \in$ SF(CTL), $x \in$ Var $\Rightarrow \neg F$, $F \vee G$, $\exists x\, F \in$ SF(CTL),
- $F, G \in$ SF(CTL) $\Rightarrow \circ F$, $(F$ until $G) \in$ PF(CTL),
- $P \in$ PF(CTL) $\Rightarrow \neg P \in$ PF(CTL),
- $P \in$ PF(CTL) $\Rightarrow \mathsf{E}P \in$ SF(CTL),
- Every CTL-state formula is a CTL-formula.

With the abbreviations $F \wedge G := \neg(\neg F \vee \neg G)$, $\forall x\, F := \neg \exists x\, \neg F$, $\lozenge F :=$ true until F, $\square F := \neg \lozenge \neg F$, and $\mathsf{A}P := \neg \mathsf{E} \neg P$, those formulas are also state- resp. path formulas.

The definition shows that in CTL every modality (modal operators and negated modal operators) is immediately preceded by a path-quantifier. CTL* is obtained by weakening this requirement [EH83].

Definition 4. The truth of formulas in a first-order Kripke structure $\mathbf{K} = (\mathbf{G}, \mathbf{R}, \mathbf{M})$ is defined separately for state- and path formulas:

Let $g \in \mathbf{G}$ be a state, $p = (g_0, g_1, \ldots)$ a path in $\mathbf{K}$, A an atomic formula, F and G CTL-state formulas, P a CTL-path formula, and χ a variable assignment:

- $(g, \chi) \models A \quad :\Leftrightarrow (\mathbf{M}(g), \chi) \models_{\text{FO}} A$.
- $(g, \chi) \models \neg F \quad :\Leftrightarrow$ not $(g, \chi) \models F$.

- $(g, \chi) \models F \vee G$ $\quad :\Leftrightarrow$ $(g, \chi) \models F$ or $(g, \chi) \models G$.
- $(g, \chi) \models \exists x \, F$ $\quad :\Leftrightarrow$ there is a $d \in \mathbf{U}$ with $(g, \chi_x^d) \models F$.
- $(p, \chi) \models \circ F$ $\quad :\Leftrightarrow$ $(g_1, \chi) \models F$.
- $(p, \chi) \models F$ until G $:\Leftrightarrow$ there is an $i \geq 0$ such that $(g_i, \chi) \models G$ and for all $j : 0 \leq j < i \ \ (g_i, \chi) \models F$ holds.
- $(p, \chi) \models \neg P$ $\quad :\Leftrightarrow$ not $(p, \chi) \models P$.
- $(g, \chi) \models \mathsf{E}P$ $\quad :\Leftrightarrow$ there is a path $p = (g_0, g_1, \ldots)$ in $\mathbf{K}$ and an i such that $g_i = g$ and $(p|_i, \chi) \models P$.

3 Labeled Kripke Structures

A labeled transition system is a triple $(S, A, \rightarrow)$, where S is a set of states, A is a set of actions, and $\rightarrow$ is a relation over $S \times A \times S$ providing "labeled transitions" $\rightarrow^a$ for each action a. In most cases (cf. [Mil90]), $\rightarrow$ is given by a set of basic rules for elementary actions and a set of rules to derive transitions for composite actions. A multimodal logic for reasoning about labeled transition systems is given in [HM85], using labeled modalities $\diamondsuit$ and $\boxdot$.

Labeled transition systems can even be simply regarded as Kripke structures where the accessibility relation is labeled with actions:

Definition 5. A *labeled Kripke structure* over a signature Σ is a quadruple $\mathbf{K} = (\mathbf{G}, \mathbf{A}, \mathbf{R}, \mathbf{M})$, where $\mathbf{G}$ is a set of elements (states or possible worlds), $\mathbf{A}$ is a set of elements (actions), $\mathbf{R} \subseteq \mathbf{G} \times 2^{\mathbf{A}} \times \mathbf{G}$ is a *labeled accessibility relation* between states, and $\mathbf{M}$ is a function which maps every $g \in \mathbf{G}$ to a first-order interpretation $\mathbf{M}(g)$ over Σ.

A *labeled path* p in a labeled Kripke structure $\mathbf{K} = (\mathbf{G}, \mathbf{A}, \mathbf{R}, \mathbf{M})$ is a sequence $(g_0, a_1, g_1, a_2, g_2, \ldots)$, $g_i \in \mathbf{G}$, $a_i \in 2^{\mathbf{A}}$ such that $(g_i, a_{i+1}, g_{i+1}) \in \mathbf{R}$.

Definition 6. The set $\mathbf{A}$ of actions is built from a signature Σ_A of action symbols with given arities (corresponding to action *schemata*, cf. function and predicate symbols) and the universe $\mathbf{U}$:

- For an n-ary action symbol a and $u_1, \ldots, u_n \in \mathbf{U}$, $a(u_1, \ldots, u_n)$ is an action.

Analogous to first-order logic for querying states, transitions are queried by *action formulas*. There, action symbols take the role of predicate symbols:

Definition 7.

- T is an action formula.
- For an n-ary action symbol $a \in \Sigma_A$ and terms $t_1, \ldots, t_n \in \mathsf{Term}_\Sigma$, $a(t_1, \ldots, t_n)$ is an action formula.
- Action formulas are constructed by the usual logical connectives, quantors and variables: For action formulas E and F, and a variable x, $\neg\mathsf{E}$, $\mathsf{E} \vee \mathsf{F}$, and $\exists x : \mathsf{E}$ are action formulas.

Since action formulas query transitions, the transition labels take the role of interpretations: If an elementary action $a(u_1, \ldots, u_n)$ is carried out in some transition, the "predicate" symbol a evaluates to true on $(u_1, \ldots, u_n)$:

Definition 8. Every transition label $a \in 2^{\mathbf{A}}$ defines an interpretation of action (predicate) symbols. Let $b \in \Sigma_A$.

$$a(b) = \{(u_1, \ldots, u_{\mathrm{ord}(a)}) \in \mathbf{U}^{\mathrm{ord}(b)} \mid b(u_1, \ldots, u_{\mathrm{ord}(b)}) \in a\}$$

Together with an interpretation $\mathbf{I}$ over Σ, every transition label a induces an evaluation of action formulas:

Definition 9.

- $(a, \mathbf{I}, \chi) \models \textsc{t}$.
- For every action symbol $b, t_1, \ldots, t_n \in \text{Term}_\Sigma$ and $a \in 2^{\mathbf{A}}$:

$$(a, \mathbf{I}, \chi) \models b(t_1, \ldots, t_n) \; :\Leftrightarrow \; (\mathbf{I}(t_1, \chi), \ldots, \mathbf{I}(t_n, \chi)) \in a(b) \ .$$

- The evaluation of composite action formulas is defined in the same way as for first-order logic formulas.

3.1 MCTL: Extension of CTL to Labeled Kripke Structures

MCTL is an extension of first-order CTL to labeled Kripke structures, integrating the language of action formulas as defined above into first-order CTL: The modal operators are labeled – in the style of Dynamic Logic [Har79], Hennessy-Milner Logic $\mathcal{HML}$ [HM85], or ACTL [DV90] – with action formulas querying the respective transitions.

An examination of ACTL, which uses the same modalities as CTL, shows that its (labeled) "nexttime"-operator, X, is not its own dual, as it is in CTL: $\neg(\circ_{CTL}F) \Leftrightarrow \circ_{CTL}(\neg F)$, but $\neg(\mathsf{X}_\alpha)F \Leftrightarrow \mathsf{X}_\alpha(\neg F) \vee \neg(\mathsf{X}_\alpha \text{true})$.

This leads to the reflection that there are two relevant coupling modes between actions and facts: One can state that the next transition satisfies α *and* the next state satisfies F, or, that *if* the next transition satisfies α, *then* the next state satisfies F. Thus, in MCTL, the modalities are combinations of a classical modal operator, an action formula, and a coupling mode.

Definition 10. The syntax of MCTL follows the distinction into state- and path formulas known from CTL: MCTL/MCTL* formulas are defined analogously to CTL/CTL* formulas (cf. Section 2) with the additional clause

- For MCTL-state formulas F, G and action formulas $\textsc{e}, \textsc{f}$, $!\circledS F$, $?\circledS F$, $!\boxdot F$, $?\boxdot F$, $!\diamondsuit F$, $?\diamondsuit F$, $!(\textsc{e}, \textsc{f})\mathsf{U}(\textsc{f}, G)$, $?(\textsc{e}, F)\mathsf{U}(\textsc{f}, G)$ are MCTL-path formulas.

Notation: Since for complex action formulas the in-operator-notation becomes impossible, $(\!\textsc{e}\!) := \circledS$, $[\textsc{e}] := \boxdot$, and $\langle \textsc{e} \rangle := \diamondsuit$.

In the following, let $\textsc{e}, \textsc{f}$ be action formulas and F, G MCTL-state formulas. The semantics of the modalities is drawn from the CTL semantics, enriched by the coupling modes:

$\bigcirc$ (**"Nexttime"**):

$!\circledS F$: The transition to the next state satisfies $\textsc{e}$ and in the next state F holds.

$?\circledS F$: If the transition to the next state satisfies $\textsc{e}$, then F holds in the next state.

$\square$ (**"Always"**):

$!\boxdot F$: All subsequent transitions satisfy $\textsc{e}$ and in all subsequent states F holds.

$?\boxdot F$: If some of the subsequent transitions satisfy $\textsc{e}$, then in the states reached by these transitions, F holds.

$\Diamond$ (**"Sometimes"**):

$!\diamondsuit F$: There exists a transition in the future which satisfies $\textsc{e}$ and in the state reached by this transition, F holds.

$?\diamondsuit F$: If there exists a first transition satisfying E, then in the state reached by this transition, F holds.

U ("Until"):

$!(\text{E}, F)\text{U}(\text{F}, G)$: There exists a first transition in the future which satisfies F and in the state reached by it, G holds, and in all transitions resp. states before this transition, E resp. F hold.

$?(\text{E}, F)\text{U}(\text{F}, G)$: There exists a first transition in the future which satisfies F and in the state reached by it, G holds, and for all transitions before this transition which satisfy E, in the state reached by them, F holds.

Formally there is the following definition of semantics:

Definition 11. Let $\mathbf{K} = (\mathbf{G}, \mathbf{A}, \mathbf{R}, \mathbf{M})$ be a labeled Kripke structure with a labeled path $p = (g_0, a_1, g_1, a_2, g_2, \ldots)$, and χ a variable assignment. Then, the semantics of the labeled modalities is

$(p, \chi) \models ! \circledcirc F :\Leftrightarrow (a_1, \mathbf{M}(g_0), \chi) \models \text{E}$ and $(g_1, \chi) \models F$.

$(p, \chi) \models ? \circledcirc F :\Leftrightarrow$ if $(a_1, \mathbf{M}(g_0), \chi) \models \text{E}$, then $(g_1, \chi) \models F$.

$(p, \chi) \models ! \boxdot F :\Leftrightarrow$ for all i: $(a_{i+1}, \mathbf{M}(g_i), \chi) \models \text{E}$ and $(g_{i+1}, \chi) \models F$.

$(p, \chi) \models ? \boxdot F :\Leftrightarrow$ for all i: if $(a_{i+1}, \mathbf{M}(g_i), \chi) \models \text{E}$, then $(g_{i+1}, \chi) \models F$.

$(p, \chi) \models ! \diamondsuit F :\Leftrightarrow$ there is an i s.t. $(a_{i+1}, \mathbf{M}(g_i), \chi) \models \text{E}$ and $(g_{i+1}, \chi) \models F$.

$(p, \chi) \models ? \diamondsuit F :\Leftrightarrow$ if there is a least i s.t. $(a_{i+1}, \mathbf{M}(g_i), \chi) \models \text{E}$,
$\qquad$ then $(g_{i+1}, \chi) \models F$ holds.

$(p, \chi) \models !(\text{E}, F)\text{U}(\text{F}, G) :\Leftrightarrow$
$\qquad$ there is a least i s.t. $(a_{i+1}, \mathbf{M}(g_i), \chi) \models \text{F}$ and $(\mathbf{M}(g_{i+1}), \chi) \models G$,
$\qquad$ and for all $0 \le j < i$, $(a_{j+1}, \mathbf{M}(g_j), \chi) \models \text{E}$ and $(\mathbf{M}(g_{j+1}), \chi) \models F$.

$(p, \chi) \models ?(\text{E}, F)\text{U}(\text{F}, G) :\Leftrightarrow$
$\qquad$ there is a least i s.t. $(a_{i+1}, \mathbf{M}(g_i), \chi) \models \text{F}$ and $(\mathbf{M}(g_{i+1}), \chi) \models G$,
$\qquad$ and for all $0 \le j < i$, if $(a_{j+1}, \mathbf{M}(g_j), \chi) \models \text{E}$ then $(\mathbf{M}(g_{j+1}), \chi) \models F$.

Thus, the coupling mode ! always denotes a conjunction of action and formula whereas coupling mode ? always denotes an implication "if action then formula".

Remark: Roughly, all modalities can also be expressed in CTL, with a state-splitted Kripke structure satisfying $\mathbf{M}(g) \models \text{E} \Leftrightarrow$ all transitions leading to g satisfy E, then e.g. $? \circledcirc F \Leftrightarrow \circ(\text{E} \to F)$, but in contrast to this, MCTL interprets terms occurring in action formulas according to the interpretation of the state where the transition started from, covering the intention of modeling actions even better. Another advantage of modeling transitions separately is that it can also serve as a base for integrating a process calculus, e.g. CCS [Mil90].

Theorem 12. *The modalities of ACTL can be formulated in MCTL:*
$X_\text{E} F \Leftrightarrow !\circledcirc F$, $\quad F_\text{E}U_\text{F}G \Leftrightarrow !(\text{E}, F)\text{U}(\text{F}, G)$, $\quad FF \Leftrightarrow !\diamondsuit F$ *and* $GF \Leftrightarrow !\boxdot F$.

Theorem 13. *The semantics of the modal operators of CTL can be expressed by the labeled modal operators of MCTL with the trivial action predicate* T *and coupling mode* !:

$\circ F \Leftrightarrow !\circledcirc F \qquad , \qquad \diamond F \qquad \Leftrightarrow F \vee !\diamondsuit F \quad ,$

$\Box F \Leftrightarrow F \wedge !\boxdot F \qquad , \qquad F \text{ until } G \Leftrightarrow G \vee (F \wedge !(\text{T}, F)\text{U}(\text{T}, G)) \quad .$

And, similar to CTL,

$\qquad !\diamondsuit F \Leftrightarrow !(\text{T}, \text{true})\text{U}(\text{E}, F) \qquad and \qquad ?\boxdot F \Leftrightarrow \neg(!\diamondsuit(\neg F)) \quad .$

This shows that the versions with coupling mode ! are closely related to classical CTL whereas the coupling mode ? provides an additional aspect of each modality.

Theorem 14. *The duals in MCTL are as follows:*

$$\neg!\circledB F \;\Leftrightarrow\; ?\circledB(\neg F) \qquad and \qquad \neg?\circledB F \;\Leftrightarrow\; !\circledB(\neg F) \;,$$

$$\neg?\boxminus F \;\Leftrightarrow\; !\Diamonddot(\neg F) \qquad and \qquad \neg!\Diamonddot F \;\Leftrightarrow\; ?\boxminus(\neg F) \;,$$

$$\neg?\Diamonddot F \;\Leftrightarrow\; ?\Diamonddot(\neg F) \qquad and \qquad \neg!\boxminus F \;\Leftrightarrow\; !\langle\neg\mathrm{E}\rangle\mathsf{true} \vee \,!\Diamonddot(\neg F)$$

Example 1. $\mathbf{M}(g) \models \mathrm{E}!\; \langle\!\mid a(t_1,\ldots,t_n)\rangle\!\mid$ true states that in g, the execution of action $a(\mathbf{M}(g)(t_1),\ldots,\mathbf{M}(g)(t_n))$ is possible, while $\mathbf{M}(g) \models \mathrm{A}?\; \langle\!\mid a(t_1,\ldots,t_n)\rangle\!\mid$ false states that in g, this action cannot be executed.

$\mathbf{M}(g) \models \mathrm{A}?[a(t_1,\ldots,t_n)]$ false states that from state g, in all future, this action can never be executed.

$\mathbf{M}(g) \models \mathrm{E}!\; \langle\!\mid a(t_1,\ldots,t_n)\rangle\!\mid\, F$ states that by executing $a(\mathbf{M}(g)(t_1),\ldots,\mathbf{M}(g)(t_n))$ (and perhaps some more actions) in g, a state satisfying F is reachable.

$\mathbf{M}(g) \models \mathrm{A}?\langle a(t_1,\ldots,t_n)\rangle\, F$ states that from state g, if $a(\mathbf{M}(g)(t_1),\ldots,\mathbf{M}(g)(t_n))$ is performed in some future transition, a state satisfying F is reached.

Fairness wrt. an action can be expressed in MCTL* by

$$\mathbf{M}(g) \models \mathrm{A}((\Box\Diamond\mathrm{E}!\; \langle\!\mid a(t_1,\ldots,t_n)\rangle\!\mid\, \mathsf{true}) \to (!\langle a(t_1,\ldots,t_n)\rangle\,\mathsf{true}))\;.$$

4 Semantical Tableaux Representation for Labeled Kripke Structures

To achieve a strict distinction between the two graph structures "Kripke structure" and "tableau", the terms "path" and "state" will be used for Kripke structures whereas the terms "branch" and "node" will be used for tableaux. The tableau semantics and -calculus extends the one presented for first-order CTL in [MS96]. Like in traditional tableau proving, for a proof of the validity of a formula F, the inconsistency of the formula $\neg F$ is proven by systematically showing that it is impossible to construct a model for $\neg F$. Thus, the situation from first-order theorem proving to find a model for a given set of formulas occurs multiply: Every state is such a first-order interpretation. For this purpose the well-known first-order tableau calculus is embedded in the temporal tableau calculus which is constructed. From these first-order interpretations, a labeled branching time temporal Kripke structure has to be built.

Therefore it is necessary to describe many individual states as well as the relations between them in the tableau. The latter include the ordering of states on a path together with the connections between different paths.

Thus, three kinds of entities have to be described: Elements of the universe inside states, states, and paths, including transitions. In the chosen semantics these will be explicitly named when their existence is stated by a formula:

- Elements of the universe: as in the first-order tableau calculus, a new constant resp. function symbol is introduced by a δ-rule when an $\exists$-quantor is processed.
- States: states are named when required by an existence formula (e.g. $\Diamonddot F$ or $\circledB F$). In the chosen semantics a newly named state has to be positioned on an existing path, retaining the linear ordering of all states on this path.

- Paths: paths are named when required by an existence formula of the kind
  EP. A newly introduced path is assumed to branch off in the state where its
  existence is claimed.

In general, between two known states there can be many other still unknown
states. These can be named when needed. Thus, a straightforward dissolving of
eventualities at any time is possible.

To allow the naming of states at any position of the model, the descriptions
of paths contain, apart from the (partial) ordering of known states, additional
information about formulas which have to be true in still unknown states on the
segments in-between. These are used when new states are explicitly named.

4.1 Representation

As a conceptional extension of first-order tableaux, every branch of the tableau
corresponds to a complete labeled Kripke structure. Here, apart from the first-
order portion, information about the frame and the transitions has to be coded
in tableau nodes. For distinguishing and naming of states, a tableau calculus
based on the free variable tableau calculus from [Fit90] augmented with *prefixes*
is employed: A state formula F, assumed to be true in a certain state, occurs in
the tableau as *state prefixed formula* $\gamma{:}F$. The paths described in the tableau are
named by *path descriptors*. *Path information formulas* contain the information
about the prefixes situated on this path. Also, *transition prefixed formulas* will
be used, representing information about transitions.

Thus, the signature Σ_T used in the tableau is partitioned into Σ_L (first-order
part), Σ_A (action symbols), and Σ_F (frame part).

Σ_L is obtained by augmenting Σ with a countable infinite set of n-ary skolem
function symbols for every $n \in \mathbb{N}$ and a countable infinite set of variables X_i.

Σ_F consists of a set $\hat{\Gamma}$ of *prefix symbols* and a set $\hat{\Lambda}$ of *path symbols*, each
containing an infinite set of n-ary prefix- resp. path symbols for every $n \in \mathbb{N}$.
The construction of prefixes and path descriptors corresponds to the use of
skolem functions in the first-order tableau calculus. Here the prefix- and path
symbols take the role of the skolem function symbols. With this, the free variables
resulting from invocations of the γ-rule have to be considered. Thus prefixes γ
and path descriptors λ are terms consisting of a prefix symbol $\hat{\gamma}$ resp. a path
symbol $\hat{\lambda}$ of an arity n and an n-tuple of terms as arguments. Additionally, there
is a 0-ary symbol $\hat{\omega}$ that is not a prefix symbol, but is used in a similar way.

Definition 15. Let $\hat{\Gamma}$ be the set of prefix symbols, $\hat{\Lambda}$ the set of path symbols,
Σ^c the state-independent portion of Σ. Then the following sets Σ_L, Γ, and Λ are
simultaneously recursively enumerable:

$\Sigma_L := \Sigma \cup \{f \mid f \text{ an } n\text{-ary skolem function symbol}\} \cup \{f_\gamma \mid f \in \Sigma\backslash\Sigma^c, \gamma \in \Gamma\}$,
with $\operatorname{ord}(f_\gamma) = \operatorname{ord}(f)$ and skolem functions and all f_γ interpreted state-independently, thus

$\Sigma_L^c := \Sigma^c \cup \{f \mid f \text{ an } n\text{-ary skolem function symbol}\} \cup \{f_\gamma \mid f \in \Sigma\backslash\Sigma^c, \gamma \in \Gamma\}$.
The sets of *prefixes* resp. *path descriptors* are given as

$\Gamma := \{\hat{\gamma}(t_1, \ldots, t_n) \mid \hat{\gamma} \in \hat{\Gamma} \text{ is an } n\text{-ary prefix symbol}, t_1, \ldots, t_n \in \operatorname{Term}_{\Sigma_L^c}\}$, and
$\Lambda := \{\hat{\lambda}(t_1, \ldots, t_n) \mid \hat{\lambda} \in \hat{\Lambda} \text{ is an } n\text{-ary path symbol}, t_1, \ldots, t_n \in \operatorname{Term}_{\Sigma_L^c}\}$.

In both sets Γ, $\Lambda \subset \mathsf{Term}_{\Sigma_T}$ of terms it is precisely the leading function symbol which is a prefix- resp. path symbol taken from Σ_F and all argument terms are in $\mathsf{Term}_{\Sigma_L^c}$. Those are interpreted state-independently by $\mathbf{K}$.

An interpretation of Σ_T – describing a labeled Kripke structure – is accordingly partitioned: The interpretation of Σ_L is taken over by the set $\{\mathbf{M}(g) \mid g \in \mathbf{G}\}$ of first-order interpretations, and the symbols of Σ_A are interpreted globally as action predicate symbols. Complementary to this, an "interpretation" of the prefix- and path symbols in Σ_F is defined. The corresponding evaluations map prefixes and path descriptors to the entities described by them:

Definition 16. A P&P-interpretation ("prefixes and paths interpretation") of the sets $\hat{\Gamma}$ and $\hat{\Lambda}$ to a labeled Kripke structure $(\mathbf{G}, \mathbf{A}, \mathbf{R}, \mathbf{M})$ with a constant universe $\mathbf{U}$ and a set $\mathbf{P}(\mathbf{K})$ of paths is a triple $\Omega = (\phi, \pi, \psi)$ where

$\phi : \hat{\Lambda} \to \mathbf{U}^n \to \mathbf{P}(\mathbf{K})$ maps every n-ary $\hat{\lambda} \in \hat{\Lambda}$ to a function $\phi(\hat{\lambda}) : \mathbf{U}^n \to \mathbf{P}(\mathbf{K})$
  resp. $\phi(\hat{\lambda}) : \mathbf{U}^n \to \mathbb{N} \to \mathbf{G}$,

$\pi : \hat{\Lambda} \times (\hat{\Gamma} \cup \{\hat{\infty}\}) \to (\mathbf{U}^n \times \mathbf{U}^m) \to \mathbb{N} \cup \{\infty\}$ is an (in general not total) mapping of pairs of n-ary $\hat{\lambda} \in \hat{\Lambda}$ and m-ary $\hat{\gamma} \in \hat{\Gamma}$ to functions $\pi(\hat{\lambda}, \hat{\gamma}) :$
  $\mathbf{U}^n \times \mathbf{U}^m \to \mathbb{N} \cup \{\infty\}$ with $\pi(\lambda, \gamma) = \infty \Leftrightarrow \gamma = \hat{\infty}$, and

$\psi : \hat{\Gamma} \to \mathbf{U}^n \to \mathbf{G}$ maps every n-ary $\hat{\gamma} \in \hat{\Gamma}$ to a function $\psi(\hat{\gamma}) : \mathbf{U}^n \to \mathbf{G}$.

Ω is organized similarly to a first-order interpretation $\mathbf{I} = (I, \mathbf{U})$ if the corresponding mappings, "universes", and induced evaluations are considered:

$$\Phi = (\phi, \mathbf{P}(\mathbf{K})) \quad , \quad \Pi = (\pi, \mathbb{N} \cup \{\infty\}) \quad , \quad \Psi = (\psi, \mathbf{G}) .$$

Based on ϕ, π, and ψ, the evaluations $\Phi : \Lambda \times \Xi \to \mathbf{P}(\mathbf{K})$ of path descriptors, $\Pi : \Lambda \times (\Gamma \cup \{\hat{\infty}\}) \times \Xi \to \mathbb{N} \cup \{\infty\}$ of pairs of path descriptors and prefixes, and $\Psi : \Gamma \times \Xi \to \mathbf{G}$ of prefixes are defined as follows:
For $\lambda = \hat{\lambda}(t_1, \ldots, t_n) \in \Lambda$ and $\gamma = \hat{\gamma}(s_1, \ldots, s_m) \in \Gamma$ (thus $t_i, s_i \in \mathsf{Term}_{\Sigma_L^c}$),

$$\Phi(\lambda, \chi) := (\phi(\hat{\lambda}))(\mathbf{K}(t_1, \chi), \ldots, \mathbf{K}(t_n, \chi)) \quad ,$$
$$\Pi(\lambda, \gamma, \chi) := (\pi(\hat{\lambda}, \hat{\gamma}))(\mathbf{K}(t_1, \chi), \ldots, \mathbf{K}(t_n, \chi), \mathbf{K}(s_1, \chi), \ldots, \mathbf{K}(s_m, \chi)) \quad , \text{ and}$$
$$\Psi(\gamma, \chi) := (\psi(\hat{\gamma}))(\mathbf{K}(s_1, \chi), \ldots, \mathbf{K}(s_m, \chi)) \quad .$$

Finally, the interpretation of the derived function symbols f_γ is defined state-independently for all $g \in \mathbf{G}$ as

$$(\mathbf{M}(g))(f_\gamma(t_1, \ldots, t_n), \chi) := (\mathbf{M}(\Psi(\gamma, \chi)))(f(t_1, \ldots, t_n), \chi) \quad .$$

Tableau formulas.
On this foundation the syntax used in the tableaux can be worked out: Let $\mathcal{L}$ be the language of state formulas, i.e. SF(MCTL) or SF(MCTL*).
The frame of the Kripke structure is encoded in *path information formulas* of the form $\lambda : [\gamma_0, L_0, \gamma_1, L_1, \ldots, \gamma_n, L_n, \hat{\infty}]$ with $\lambda \in \Lambda$, $\gamma_i \in \Gamma$ and $L_i \in \mathcal{L} \cup \{\circ\}$. Logical formulas occur in the tableau as *state prefixed formulas* of the form $\gamma{:}F$ with $\gamma \in \Gamma$, $F \in \mathcal{L}$, and the same branch of the tableau containing a path information formula $\lambda : [\ldots, \gamma, \ldots]$. Information about transitions is represented by *transition prefixed formulas* $\lambda[\gamma \circ \delta]{:}\mathrm{E}$ with $\lambda \in \Lambda$, $\gamma, \delta \in \Gamma$, E an action formula, and the same branch of the tableau containing a path information formula $\lambda : [\ldots, \gamma, \circ, \delta, \ldots]$.

Following the explicit naming of paths in the calculus, the formulas used internally to the tableau have a more detailed syntax than MCTL/MCTL*-formulas. A syntactic facility to use path descriptors in logical formulas is added: To state the validity of a path formula P on the suffix of a path p (described by a path descriptor λ) beginning in a fixed state g (described by a prefix γ) on that path, the symbol λ can syntactically take the role of a path quantifier. In this role, λ is a *path selector*. This results in the following syntax for node formulas in all tableaux tracing this concept:

Definition 17. • Every atomic formula is a $\mathcal{TL}$-state formula.
- Every action formula is a $\mathcal{TL}$-action formula.
- With F und G $\mathcal{TL}$-state formulas, $\neg F$, $F \wedge G$, $F \vee G$ and $F \to G$ are $\mathcal{TL}$-state formulas.
- With F a $\mathcal{TL}$-state formula and x a variable, $\forall x\, F$ and $\exists x\, F$ are $\mathcal{TL}$-state formulas.
- With F and G $\mathcal{TL}$-state formulas, $\circ F$, $\Box F$, $\Diamond F$, and $(F \text{ until } G)$ are $\mathcal{TL}$-path formulas.
- With F, G $\mathcal{TL}$-state formulas and E, F $\mathcal{TL}$-action formulas, $!\circledS F$, $?\circledS F$, $!\boxdot F$, $?\boxdot F$, $!\diamondsuit F$, $?\diamondsuit F$, $!(\text{E}, F)\mathsf{U}(\text{F}, G)$, $?(\text{E}, F)\mathsf{U}(\text{F}, G)$ are $\mathcal{TL}$-path formulas.
- With P a $\mathcal{TL}$-path formula, $\neg P$ is a $\mathcal{TL}$-path formula.
- With P a $\mathcal{TL}$-path formula, AP and EP are $\mathcal{TL}$-state formulas.
- Every $\mathcal{TL}$-state formula is a $\mathcal{TL}$-pre-node formula.
- With P a $\mathcal{TL}$-path formula and $\lambda \in \Lambda$, λP is a $\mathcal{TL}$-pre-node formula.
- With F a $\mathcal{TL}$-pre-node formula and $\gamma \in \Gamma$ a prefix, $\gamma{:}F$ is a $\mathcal{TL}$-state prefixed formula.
- With E a $\mathcal{TL}$-action formula, $\lambda \in \Lambda$, and $\gamma, \delta \in \Gamma$, $\lambda[\gamma \circ \delta]{:}\text{E}$ is a $\mathcal{TL}$-transition prefixed formula.
- All $\mathcal{TL}$-prefixed formulas are $\mathcal{TL}$-node formulas.
- Every path information formula is a $\mathcal{TL}$-node formula.

Semantics.

Definition 18. For a P&P-interpretation $\Omega = (\phi, \pi, \psi)$, a path information formula $I = \lambda : [\gamma_0, L_0, \gamma_1, L_1, \ldots, \gamma_n, L_n, \hat{\infty}]$ is *consistent with* Ω for a variable assignment χ, if every $\hat{\gamma}$ occurs in I at most once, and for all i

$$\Pi(\lambda, \gamma_0, \chi) = 0 \quad , \quad \Pi(\lambda, \gamma_i, \chi) < \Pi(\lambda, \gamma_{i+1}, \chi) \quad ,$$
$$\text{and} \quad \Psi(\gamma_i, \chi) = \Phi(\lambda, \chi, \Pi(\lambda, \gamma_i, \chi)) \quad .$$

This means that the path $\Phi(\lambda, \chi) = (g_0, a_1, g_1, \ldots)$ of **K** begins in state $g_0 = \Psi(\gamma_0, \chi)$ and passes through the other known states $g_{\Pi(\lambda, \gamma_1, \chi)} = \Psi(\gamma_1, \chi)$, ..., $g_{\Pi(\lambda, \gamma_n, \chi)} = \Psi(\gamma_n, \chi)$ in the specified order.

Definition 19. The relation $\models$ of a labeled Kripke structure $\mathbf{K} = (\mathbf{G}, \mathbf{A}, \mathbf{R}, \mathbf{M})$ with a set $\mathbf{P}(\mathbf{K})$ of labeled paths, a P&P-interpretation Ω, a set $\mathcal{F}$ of formulas and a variable assignment χ to free($\mathcal{F}$) is defined as follows, based on the truth of formulas in labeled Kripke structures, $\models_{\text{MCTL}}$ resp. $\models_{\text{MCTL}^*}$:

- For every state prefixed formula $\gamma : F$, F not containing a path selector:

$$(\mathbf{K}, \Omega, \chi) \models \gamma : F \quad :\Leftrightarrow \quad (\Psi(\gamma, \chi), \chi) \models_{\text{MCTL}} F \quad ,$$

  i.e. in the state corresponding to the prefix γ under variable assignment χ, the (state) formula F holds.

- For every state prefixed formula $\gamma : F$, $F = \lambda P$ containing a (leading) path selector:

$$(\mathbf{K}, \Omega, \chi) \models \gamma : \lambda P \quad :\Leftrightarrow \quad (\Phi(\lambda, \chi)|_{\Pi(\lambda, \gamma, \chi)}, \chi) \models P \quad,$$

i.e. on the suffix of the path $\Phi(\lambda, \chi)$ beginning in the $\Pi(\lambda, \gamma, \chi)$th state (which is $\Psi(\gamma, \chi)$ by consistency), the path formula P holds.

- For every transition prefixed formula $\lambda[\gamma \circ \delta]$:E :

$$(\mathbf{K}, \Omega, \chi) \models \lambda[\gamma \circ \delta] : \mathrm{E} \quad :\Leftrightarrow \quad (\Phi(\lambda, \chi)|_{\Pi(\lambda, \gamma, \chi)}, \chi) \models ! \circledcirc \mathsf{true} \ ,$$

i.e. on the path $\Phi(\lambda, \chi) = (g_0, a_1, g_1, \ldots)$, the action $a_{\Pi(\lambda, \delta, \chi)}$ (which is performed in the transition from $\Psi(\gamma, \chi)$ to $\Psi(\delta, \chi)$) satisfies E.

- For all path information formulas $I = \lambda : [\gamma_0, L_0, \gamma_1, L_1, \ldots, \gamma_n, L_n, \hat{\infty}]$:

$$(\mathbf{K}, \Omega, \chi) \models \lambda : [\gamma_0, L_0, \gamma_1, L_1, \ldots, \gamma_n, L_n, \hat{\infty}]$$

iff I is consistent with Ω for the variable assignment χ, and for all $0 \le i \le n$:

$$L_i = \circ \ \Rightarrow \ \Pi(\lambda, \gamma_{i+1}, \chi) = \Pi(\lambda, \gamma_i, \chi) + 1 \quad,$$
$$L_i \ne \circ \ \Rightarrow \ \text{for all } j \text{ with } \Pi(\lambda, \gamma_i, \chi) < j < \Pi(\lambda, \gamma_{i+1}, \chi) :$$
$$(\Phi(\lambda, \chi, j), \chi) \models L_i \ ,$$

i.e. if $L_i = \circ$, then $\Pi(\lambda, \gamma_i, \chi)$ and $\Pi(\lambda, \gamma_{i+1}, \chi)$ are immediately successive indices, else for all (finitely, but arbitrary many) states g_j situated between $\Phi(\lambda, \chi, \Pi(\lambda, \gamma_i, \chi))$ and $\Phi(\lambda, \chi, \Pi(\lambda, \gamma_{i+1}, \chi))$ on path $\Phi(\lambda, \chi)$, the relation $(g_j, \chi) \models L_i$ holds.

A set $\mathcal{F}$ of path information formulas and prefixed formulas is *valid* in a labeled Kripke structure $\mathbf{K} = (\mathbf{G}, \mathbf{A}, \mathbf{R}, \mathbf{M})$ with a set $\mathbf{P}(\mathbf{K})$ of paths under a variable assignment χ to free($\mathcal{F}$) if there is a P&P-interpretation $\Omega = (\phi, \pi, \psi)$ such that $(\mathbf{K}, \Omega, \chi) \models \mathcal{F}$. Since a branch of a tableau is a set of formulas like this, validity is a relation on labeled Kripke structures and branches.

The construction of Kripke structures and consistent P&P-interpretations to a given set of formulas plays an important role in the proof of correctness.

5 The Tableau Calculus $\mathcal{TL}$

For proving the validity of a formula F, the inconsistency of $\neg F$ is proven: it is shown that there is no labeled Kripke structure $\mathbf{K} = (\mathbf{G}, \mathbf{A}, \mathbf{R}, \mathbf{M})$ with any state $g_0 \in \mathbf{G}$ where F does not hold.
Thus the initialization of the tableau is $\boxed{\overline{0} : \neg F}$.

The tableau calculus is based on the well-known first-order tableau calculus, consisting of α-, β-, γ- and δ-rules and the atomic closure rule [Fit90].

Let F and G be $\mathcal{TL}$-state formulas, A an atomic formula. In the sequel, $F[t/x]$ denotes the formula F with all occurences of x replaced by t.

S-α: $\dfrac{\gamma : F \wedge G}{\begin{array}{l}\gamma : F \\ \gamma : G\end{array}}$	S-γ: $\dfrac{\gamma : \forall x\, F}{\gamma : F[X/x]}$	with X a new variable.
S-β: $\dfrac{\gamma : F \vee G}{\gamma : F \mid \gamma : G}$	S-δ: $\dfrac{\gamma : \exists x\, F}{\gamma : F[f(\mathrm{free}(T))/x]}$	with f a new function symbol and T the current branch.

Analogous for $\neg(F \lor G)$ (α), $\neg(F \land G)$ (β), $\neg\exists x\, F$ (γ), and $\neg\forall x\, F$ (δ).
Similar rules are given to decompose action formulas into their atomic units:

$T-\alpha$:	$\lambda[\alpha \circ \beta] : \text{E} \land \text{F}$	$T-\gamma$:	$\lambda[\alpha \circ \beta] : \forall x\, \text{E}$	with X a new
	$\lambda[\alpha \circ \beta] : \text{E}$		$\lambda[\alpha \circ \beta] : \text{E}[X/x]$	variable.
	$\lambda[\alpha \circ \beta] : \text{F}$			
$T-\beta$:	$\dfrac{\lambda[\alpha \circ \beta] : \text{E} \lor \text{F}}{\lambda[\alpha \circ \beta] : \text{E} \mid \lambda[\alpha \circ \beta] : \text{F}}$	$T-\delta$:	$\dfrac{\lambda[\alpha \circ \beta] : \exists x\, \text{E}}{\lambda[\alpha \circ \beta] : \text{E}[f(\text{free}(T))/x]}$	f and T as for $S-\delta$.

For a substitution σ and a prefix γ, σ_γ is the γ-*localization*, i.e. $\sigma_\gamma(X)$ is obtained from $\sigma(X)$ by replacing every function symbol $f \in \Sigma \backslash \Sigma^c$ by its localized symbol f_γ. So the substitutes in σ_γ contain only function symbols which are interpreted state-independently.

Let σ be a substitution, A, B atomic formulas and E, F atomic action formulas.

Closure by states, $S-\bot$:

$$\frac{\begin{array}{c} \gamma : A \\ \gamma : B \\ \sigma(A) = \neg\sigma(B) \end{array}}{\bot}$$

apply σ_γ to the whole tableau.

Closure by transitions: $T-\bot$:

$$\frac{\begin{array}{c} \lambda[\gamma \circ \delta] : \text{E} \\ \lambda[\gamma \circ \delta] : \text{F} \\ \sigma(\text{E}) = \neg\sigma(\text{F}) \end{array}}{\bot}$$

apply σ_γ to the whole tableau.

For dissolving modalities, the information about the frame of the Kripke structure, which is encoded in the path information formulas, has to be considered. In a single step, a prefixed formula is dissolved "along" a path information formula, inducing the following form of tableau rules:

$$\frac{\begin{array}{c} \text{prefixed formula} \\ \text{path information formula} \end{array}}{\begin{array}{c} \text{prefixed formulas} \\ \text{path information formulas} \end{array}}$$

where the premise takes the latest path information formula on the current branch for the path symbol to be considered. The connection between the prefixed formula being dissolved and the path information formula is established by the prefix and, if it exists, the leading path selector of the prefixed formula. For dissolving prefixed formulas, a path quantifier resp. -selector is broken up together with the subsequent modality.

In the sequel, T denotes the current branch of the tableau, $\hat{\gamma}$ is a new prefix symbol and $\hat{\kappa}$ is a new path symbol. P is a path formula, F is a state formula, and E is an action formula.

- For dissolving a formula of the form $\gamma : \text{E}P$, a path satisfying P is named and the path formula is bound to that path:

$$\frac{\gamma : \text{E}P}{\begin{array}{c} \lambda : [\hat{0}, \dots, \gamma, \dots] \\ \hline \hat{\kappa}(\text{free}(T)) : [\gamma, \text{true}, \hat{\infty}] \\ \gamma : \hat{\kappa}(\text{free}(T))P \end{array}}$$

- Formulas of the form $\gamma : \text{A}P$ are dissolved once for every path information formula on this branch containing the prefix γ.
- Formulas of the form $\gamma : \lambda P$ are dissolved along the path information formula for λ.

In the latter cases, the claim that the state described by the current prefix satisfies some formula is decomposed in some less complex claims:

- Which formulas hold in the current state/transition ?
- Which state should be regarded as the "next relevant state/transition" on the path?
- Which formulas hold in this next relevant state/transition ?
- Which formulas hold in all states/transitions in-between ?

The tableau rules for CTL for formulas which are universally path-quantified or explicitly bound to named paths are as follows.

Transition prefixed formulas are only generated when modalities of type $\bigcirc$ are processed. The other modalities are reduced to these cases.

$\alpha : A!\circledS F$		$\alpha : A!\circledS F$
$\lambda : [\ldots, \alpha, L, \beta, \ldots], \; L \neq \circ$		$\lambda : [\ldots, \alpha, \circ, \beta, \ldots]$
$\lambda : [\ldots, \alpha, \circ, \hat{\gamma}(\mathsf{free}(T)), L, \beta, \ldots]$	if $\beta \neq \hat{\infty}$:	$\beta : F$
$\hat{\gamma}(\mathsf{free}(T)) : L$	$\lambda : [\ldots, \alpha, \circ, \beta, \ldots]$	$\lambda[\alpha \circ \beta] : \mathrm{E}$
$\hat{\gamma}(\mathsf{free}(T)) : F$	$\beta : F$	
$\lambda[\alpha \circ \hat{\gamma}(\mathsf{free}(T))] : \mathrm{E}$	$\lambda[\alpha \circ \beta] : \mathrm{E}$	

$\alpha : A?\circledS F$		$\alpha : A?\circledS F$	
$\lambda : [\ldots, \alpha, L, \beta, \ldots], \; L \neq \circ$		$\lambda : [\ldots, \alpha, \circ, \beta, \ldots]$	
$\lambda:[\ldots, \alpha, \circ, \hat{\gamma}(\mathsf{free}(T)), L, \beta, \ldots]$	if $\beta \neq \hat{\infty}$:	$\lambda[\alpha \circ \beta]:\neg\mathrm{E}$	$\beta : F$
$\hat{\gamma}(\mathsf{free}(T)) : L$	$\lambda : [\ldots, \alpha, \circ, \beta, \ldots]$		$\lambda[\alpha \circ \beta]:\mathrm{E}$
$\lambda[\alpha \circ \hat{\gamma}(\mathsf{free}(T))] : \neg\mathrm{E}$	$\lambda[\alpha \circ \beta] : \neg\mathrm{E}$		
$\lambda:[\ldots, \alpha, \circ, \hat{\gamma}(\mathsf{free}(T)), L, \beta, \ldots]$	if $\beta \neq \hat{\infty}$:		
$\hat{\gamma}(\mathsf{free}(T)) : L$	$\lambda : [\ldots, \alpha, \circ, \beta, \ldots]$		
$\hat{\gamma}(\mathsf{free}(T)) : F$	$\beta : F$		
$\lambda[\alpha \circ \hat{\gamma}(\mathsf{free}(T))] : \mathrm{E}$	$\lambda[\alpha \circ \beta] : \mathrm{E}$		

Due to limited space, for $\mathbb{E}$, $\circledast$, and $(\mathrm{E}, F)\mathsf{U}(\mathrm{F}, G)$, only one coupling mode is presented. Also, the negated modalities and the versions where a path formula is explicitly bound to a path λ are omitted.

$\alpha : A?\circledast F$		
$\lambda : [\ldots, \alpha, L, \beta, \ldots], \; L \neq \circ$		
$\alpha : \lambda!\circledS F$	$\lambda : [\ldots, \alpha, L \cup$	if $\beta \neq \hat{\infty}$:
	$\{A?\circledast F, \lambda! \, \mathsf{C}\neg\mathsf{ED} \; \text{true}\},$	$\lambda : [\ldots, \alpha, L \cup$
	$\hat{\gamma}(\mathsf{free}(T)), L, \beta, \ldots]$	$\{A?\circledast F, \lambda! \, \mathsf{C}\neg\mathsf{ED} \; \text{true}\}, \beta, \ldots]$
	$\alpha : \lambda! \, \mathsf{C}\neg\mathsf{ED} \; \text{true}$	$\alpha : \lambda! \, \mathsf{C}\neg\mathsf{ED} \; \text{true}$
	$\hat{\gamma}(\mathsf{free}(T)) : L$	$\beta : A?\circledast F$
	$\hat{\gamma}(\mathsf{free}(T)) : \lambda!\circledS F$	
		if $\beta = \hat{\infty}$:
		$\lambda : [\ldots, \alpha, L \cup$
		$\{A?\circledast F, \lambda! \, \mathsf{C}\neg\mathsf{ED} \; \text{true}\}, \beta, \ldots]$
		$\alpha : \lambda! \, \mathsf{C}\neg\mathsf{ED} \; \text{true}$

$$\frac{\begin{array}{c}\alpha : \text{A?}\circledast F\\ \lambda : [\ldots, \alpha, \circ, \beta, \ldots]\end{array}}{\alpha : \lambda!\circledast F \quad \bigg|\quad \begin{array}{c}\alpha : \lambda!\,(\lrcorner\neg\text{ED})\ \text{true}\\ \beta : \text{A?}\circledast F\end{array}}$$

$$\frac{\begin{array}{c}\alpha : \text{A!}\boxdot F\\ \lambda : [\ldots, \alpha, L, \beta, \ldots],\ L \neq \circ\end{array}}{\begin{array}{c}\lambda : [\ldots, \alpha, L \cup \{\text{A!}\boxdot F, \lambda!\circledast F\}, \beta, \ldots]\\ \alpha : \text{A!}\circledast F\\ \beta : \text{A!}\boxdot F \quad \text{if } \beta \neq \hat{\infty}\end{array}} \qquad \frac{\begin{array}{c}\alpha : \text{A!}\boxdot F\\ \lambda : [\ldots, \alpha, \circ, \beta, \ldots]\end{array}}{\begin{array}{c}\alpha : \lambda!\circledast F\\ \beta : \text{A!}\boxdot F\end{array}}$$

$$\frac{\begin{array}{c}\alpha : \text{A!}(\text{E}, F)\text{U}(F, G)\\ \lambda : [\ldots, \alpha, L, \beta, \ldots],\ L \neq \circ\end{array}}{\alpha : \lambda!\circledast G \quad\bigg|\quad \begin{array}{c}\lambda : [\ldots, \alpha, L \cup\\ \{\text{A!}(\text{E}, F)\text{U}(F, G),\\ \lambda?\circledast(\neg G), \lambda!\circledast F\}\\ \hat{\gamma}(\text{free}(T)), \circ, L, \beta, \ldots]\\ \alpha : \lambda?\circledast(\neg G)\\ \alpha : \lambda!\circledast F\\ \hat{\gamma}(\text{free}(T)) : L\\ \hat{\gamma}(\text{free}(T)) : \lambda!\circledast G\end{array} \quad\bigg|\quad \begin{array}{c}\text{if } \beta \neq \hat{\infty} :\\ \lambda : [\ldots, \alpha, L \cup\\ \{\text{A!}(\text{E}, F)\text{U}(F, G),\\ \lambda?\circledast(\neg G), \lambda!\circledast F\}, \beta, \ldots]\\ \alpha : \lambda?\circledast(\neg G)\\ \alpha : \lambda!\circledast F\\ \beta : \text{A!}(\text{E}, F)\text{U}(F, G)\end{array}}$$

$$\frac{\begin{array}{c}\alpha : \text{A!}(\text{E}, F)\text{U}(F, G)\\ \lambda : [\ldots, \alpha, \circ, \beta, \ldots]\end{array}}{\alpha : \lambda!\circledast G \quad\bigg|\quad \begin{array}{c}\alpha : \lambda?\circledast(\neg G)\\ \alpha : \lambda!\circledast F\\ \beta : \text{A!}(\text{E}, F)\text{U}(F, G)\end{array}}$$

Correctness. As usual, for proving correctness, a *Substitution Lemma* has to be shown, stating that for a labeled Kripke structure $\mathbf{K}$, a formula F, a term s, a variable assigment χ, a variable X, and some properties $\Theta(F, \chi)$,

$$\Theta([s/X]F, \chi) \;\Leftrightarrow\; \Theta(F, \chi_X^{\mathbf{K}(s,\chi)})$$

holds. The proof of this shows the necessity of the substitutes s of σ_γ in the atomic closure rule being interpreted state-independently (i.e. $s \in \text{Term}_{\Sigma^c}$): Then s has a global interpretation $\mathbf{K}(s, \chi) \in \text{U}(\mathbf{K})$ needed for the modification of χ.

Theorem 20 (Correctness of $\mathcal{TL}$).

(a) *If a tableau $\mathcal{T}$ is satisfiable and $\mathcal{T}'$ is created from $\mathcal{T}$ by an application of any of the rules mentioned above, then $\mathcal{T}'$ is also satisfiable.*

(b) *If there is any closed tableau for $\mathcal{F}$, then $\mathcal{F}$ is unsatisfiable.*

The proof of (a) is done by case-splitting separately for each of the rules. By assumption, there is a labeled Kripke structure $\mathbf{K}$ and a P&P-Interpretation $\Omega = (\phi, \pi, \psi)$ such that for every variable assignment χ there is a branch T_χ in $\mathcal{T}$ with $(\mathbf{K}, \Omega, \chi) \models T_\chi$. In all cases apart from the atomic closure rules, $\mathbf{K}$ and Ω are extended such that they witness the satisfiability of $\mathcal{T}'$. In case of the atomic closure rule the Substitution Lemma guarantees the existence of a branch for every variable assignment to $\text{free}(\mathcal{T}')$. (b) follows directly from (a).

It is well known that first-order CTL is not compact, thus, since MCTL extends CTL, MCTL is also not compact and any calculus for first-order CTL cannot be complete. The calculus is complete modulo inductive properties. For such cases, induction rules for temporal properties and well-founded data-structures have to be included. In this setting, the notion of completeness has to be relativized to that any proof done in a mathematical way can be completely redone formally. Fairness assumptions are treated in the same way as in [MS96].

6 Conclusion

The presented tableau semantics and -calculus integrates data-oriented and action-oriented aspects which occur when reasoning about transition systems, thus allowing a formal verification of processes given as transition systems. Due to the embedding of first-order tableaux, all recent techniques such as universal formulas, free variables, liberalized δ-rule, and equality-handling can fully be made use of. Because of the inherent size of the search space, including many occurrences of inductions, interactive theorem proving seems appropriate. Since transitions are modeled separately, this calculus can also serve as a base for integrating tableau rules for a process calculus, e.g. CCS [Mil90].

Acknowledgements.

The author thanks GEORG LAUSEN and BERTRAM LUDÄSCHER for many fruitful discussions and their help with improving the presentation of this paper.

References

[BMP81] M. Ben-Ari, Z. Manna, and A. Pnueli. The Temporal Logic of Branching Time. In *8th ACM Symp. on Principles of Programming Languages*, 1981.

[CES86] E. M. Clarke, E. A. Emerson, and A. P. Sistla. Automatic Verification of Finite-State Concurrent Systems Using Temporal Logic Specifications. *ACM Transactions on Programming Languages and Systems*, 8(2):244–263, 1986.

[DV90] R. DeNicola and F. Vaandrager. Action versus State Based Logics for Transition Systems. In *Semantics of Systems of Concurrent Processes*, Springer LNCS 469, pp. 407 – 419, 1990.

[EH83] E. A. Emerson and J. Y. Halpern. "Sometimes" and "not never" revisited: On Branching Time versus Linear Time in Temporal Logic. In *10th ACM Symp. on Principles of Programming Languages*, 1983.

[Fit90] M. Fitting. *First Order Logic and Automated Theorem Proving*. Springer, New York, 1990.

[Har79] D. Harel. *First-Order Dynamic Logic*, volume 68 of *LNCS*. Springer, 1979.

[HM85] M. Hennessy and R. Milner. Algebraic Laws for Non-determinism and Concurrency. *Journal of the ACM*, 32:137–161, 1985.

[Mil90] R. Milner. *Operational and Algebraic Semantics of Concurrent Processes*, Ch. 19, pp. 1201–1242 of *Handbook of Theoretical Computer Science*, J. v. Leeuwen, ed., Volume B: Formal Models and Semantics, Elsevier, 1990.

[MS96] W. May and P. H. Schmitt. A Tableau Calculus For First-Order Branching Time Logic. *Intl. Conf. on Formal and Applied Practical Reasoning*, *FAPR'96*, Springer LNCS 1085, pp. 399–413, 1996.

[Wol85] P. Wolper. The Tableau Method for Temporal Logic. *Logique et Analyse*, 28:110–111, 1985.

Tableau Methods for PA-Processes

Richard Mayr

Institut für Informatik, Technische Universität München,
Arcisstr. 21, D-80290 München, Germany;
e-mail: mayrri@informatik.tu-muenchen.de
Phone: +49 89 289-22397
Category (A)

Abstract. PA (Process algebra) is the name that has become common use to denote the algebra with a sequential and parallel operator (without communication), plus recursion. PA-processes are a superset of both Basic Parallel Processes (BPP) [Chr93] and context-free processes (BPA).

We study three problems for PA-processes: The reachability problem, the partial deadlock reachability problem ("Is it possible to reach a state where certain actions are not enabled ?") and the partial livelock reachability problem ("Is it possible to reach a state where certain actions are disabled forever ?"). We present sound and complete tableau systems for these problems and compare them to non-tableau algorithms.

Keywords: tableau systems, temporal logic, process algebras, PA-processes

1 Introduction

The Basic Process Algebra PA is a simple model of infinite state concurrent systems. It has operators for nondeterministic choice, parallel composition and sequential composition. PA-processes and Petri nets are incomparable, meaning that neither model is more expressive than the other one. PA is not a syntactical subset of CCS [Mil89], because CCS does not have an explicit operator for sequential composition. However, as CCS can simulate sequential composition by parallel composition and synchronization, PA is still a weaker model than CCS. PA-processes are a superset of both Basic Parallel Processes (BPP) [Chr93] and context-free processes (BPA).

We study three problems for PA-processes:

1. The reachability problem: "Is a state reachable from the initial state ?"
2. The partial deadlock reachability problem: "Is it possible to reach a state s s.t. no action a in a given set of actions A is enabled ?" (If A is the set of all actions then the state s is a deadlock)
3. The partial livelock reachability problem: "Is it possible to reach a state s s.t. for every state s' reachable from s no action a in a given set of actions A is enabled ?" (If A is the set of all actions then the state s is a deadlock)

We present tableau systems that solve these problems and compare them to non-tableau algorithms, pointing out strengths and weaknesses of tableau techniques. While the reachability problem is NP-complete for PA-processes, the partial deadlock reachability problem and the partial livelock reachability problem only require polynomial time.

In section 2 we define PA-processes. Sections 3, 4 and 5 are about partial deadlock reachability, partial livelock reachability and reachability. The paper closes with a section that summarizes the results, compares tableau systems to iterative algorithms and mentions some related work.

2 Preliminaries

The definition of PA is as follows: Assume a countably infinite set of atomic actions $Act = \{a, b, c, \ldots\}$ and a countably infinite set of process variables $Var = \{X, Y, Z, \ldots\}$. The class of PA expressions is defined by the following abstract syntax

$$E ::= \epsilon \mid X \mid aE \mid E + E \mid E\|E \mid E.E$$

A PA is defined by a family of recursive equations $\{X_i := E_i \mid 1 \leq i \leq n\}$, where the X_i are distinct and the E_i are PA expressions at most containing the variables $\{X_1, \ldots, X_n\}$. We assume that every variable occurrence in the E_i is *guarded*, i.e. appears within the scope of an action prefix, which ensures that PA-processes generate finitely branching transition graphs. This would not be true if unguarded expressions were allowed. For example, the process $X := a + a\|X$ generates an infinitely branching transition graph. For every $a \in Act$ the transition relation $\xrightarrow{a}$ is the least relation satisfying the following inference rules:

$$aE \xrightarrow{a} E \qquad \frac{E \xrightarrow{a} E'}{E + F \xrightarrow{a} E'} \qquad \frac{F \xrightarrow{a} F'}{E + F \xrightarrow{a} F'} \qquad \frac{E \xrightarrow{a} E'}{X \xrightarrow{a} E'}(X := E)$$

$$\frac{E \xrightarrow{a} E'}{E\|F \xrightarrow{a} E'\|F} \qquad \frac{F \xrightarrow{a} F'}{E\|F \xrightarrow{a} E\|F'} \qquad \frac{E \xrightarrow{a} E'}{E.F \xrightarrow{a} E'.F}$$

Alternatively, PA-processes can be represented by a state described by a term of the form

$$G ::= \epsilon \mid X \mid G_1.G_2 \mid G_1\|G_2$$

and a set of rules Δ of the form $X \xrightarrow{a} G$ whose application to states must respect sequential composition. This is described by the following inference rules:

$$X \xrightarrow{a} G \quad \text{if } (X \xrightarrow{a} G) \in \Delta$$

$$\frac{E \xrightarrow{a} E'}{E\|F \xrightarrow{a} E'\|F} \qquad \frac{F \xrightarrow{a} F'}{E\|F \xrightarrow{a} E\|F'} \qquad \frac{E \xrightarrow{a} E'}{E.F \xrightarrow{a} E'.F}$$

We assume w.r. that for every variable X there is at least one rule $X \xrightarrow{a} t$. The property $\exists t'. \ t \xrightarrow{a} t'$ will also be denoted by $t \xrightarrow{a}$. Sequences of actions will be denoted by σ.

Basic Parallel Processes (BPP) are the subset of PA-processes without sequential composition, while context-free processes (BPA) are the subset of PA-processes without parallel composition.

While PA-processes and Petri nets are incomparable, there is a one-to-one correspondence between BPPs and a class of labelled Petri nets, the so called *communication-free nets* [Esp96]. In these nets every transition has exactly one input place with an arc labelled by 1. The translation of a BPP algebra into a communication-free net goes as follows: Introduce a place for each process variable and a transition for each transition rule. For a rule $X \xrightarrow{a} Y_1^{m_1} \| \ldots \| Y_n^{m_n}$ introduce a transition t labelled by a, an arc labelled by 1 leading from place X to t and arcs labelled by m_i leading from t to places Y_i. The other direction is analogous.

3 Partial Deadlock Reachability

Partial deadlock reachability problem:
Instance: A labelled transition system (LTS) with initial state r_0 and a set of atomic actions A.
Question: Is it possible to reach a state r s.t. $\nexists a \in A. \ r \xrightarrow{a}$?

In the special case of $A = Act$ this is the problem if a state of deadlock is reachable.

For general Petri nets the partial deadlock reachability problem is equivalent to the problem of deciding if a state is reachable where certain places are unmarked. This problem has the same complexity as the reachability problem. So the partial deadlock reachability problem which is decidable, but at least EXPSPACE-hard [May84].

The situation is different for Basic Parallel Processes (BPP). While the reachability problem for BPP is NP-complete [Esp95], the partial deadlock reachability problem can be decided in polynomial time. It suffices to prove the property for communication-free nets, as they are equivalent to BPPs.

Now we define A-traps, a generalization of traps for labelled Petri nets. Although they can be defined for general Petri nets, they are only useful for labelled communication-free nets.

Definition 1. Let $N = (S, T, W, L, Act)$ be a labelled communication-free Petri net with places S, transitions T, a weight function W and a labeling function $L : T \rightarrow Act$, where Act is a set of atomic actions. Let $A \subseteq Act$ be a set of actions. An *A-trap* $U \subseteq S$ is a trap of N s.t. $\forall s \in U \ \exists t \in (s\cdot). \ L(t) \in A$.

Lemma 2. *The partial deadlock reachability problem for BPP is decidable in polynomial time.*

Proof Consider the corresponding communication-free net N with initial marking Σ_0. Let M be the maximal A-trap in N. It is possible to reach a state s.t. no action in A is enabled iff M is unmarked by Σ_0.

The maximal A-trap can be computed in polynomial time. Therefore the partial deadlock reachability problem can be decided in polynomial time. □

Later we will show that this result even holds for PA-processes by using a completely different (non-Petri net) technique.

Now we present a sound and complete tableau system that solves the partial deadlock reachability problem for PA-processes. First we define some predicates:

Definition 3. Let $A \subseteq Act$ be a set of actions. Sequences of actions are denoted by σ.

$$S_A(t) :\Longleftrightarrow \exists \sigma, t'. \ t \xrightarrow{\sigma} t'. \ t' \neq \epsilon \wedge \not\exists a \in A. \ t' \xrightarrow{a} \tag{1}$$

$$\mathcal{E}(t) :\Longleftrightarrow \exists t \xrightarrow{\sigma} \epsilon \tag{2}$$

$$D_A(t) :\Longleftrightarrow \exists \sigma, t'. \ t \xrightarrow{\sigma} t'. \ \not\exists a \in A. \ t' \xrightarrow{a} \tag{3}$$

It is clear that $D_A(t) \Longleftrightarrow S_A(t) \vee \mathcal{E}(t)$.

The nodes in the tableau will be sequences of formulae, that are interpreted conjunctively. Such sequences will be denoted by Γ. The branches of the proof tree are interpreted disjunctively. So the tableau will be successful iff there is a successful leaf.

In order to avoid carrying around unnecessary formulae we use the following "cleanup-procedure" after each rule application: If a formula Φ occurs more than once in a sequence Γ, then delete all occurrences but the first. The rules for the construction of the tableau are as follows:

$$\text{D} \quad \frac{D_A(t), \Gamma}{S_A(t), \Gamma \quad \mathcal{E}(t), \Gamma}$$

$$\text{SD} \quad \frac{S_A(t_1.t_2), \Gamma}{S_A(t_1), \Gamma \quad \mathcal{E}(t_1), S_A(t_2), \Gamma}$$

$$\text{PD} \quad \frac{S_A(t_1\|t_2), \Gamma}{S_A(t_1), D_A(t_2), \Gamma \quad D_A(t_1), S_A(t_2), \Gamma}$$

$$\text{ESD} \quad \frac{\mathcal{E}(t_1.t_2), \Gamma}{\mathcal{E}(t_1), \mathcal{E}(t_2), \Gamma}$$

$$\text{EPD} \quad \frac{\mathcal{E}(t_1\|t_2), \Gamma}{\mathcal{E}(t_1), \mathcal{E}(t_2), \Gamma}$$

$$\text{E} \quad \frac{S_A(X), \Gamma}{\Gamma} \quad \text{if } \not\exists a \in A. \ X \xrightarrow{a}$$

$$\text{Step1} \quad \frac{S_A(X), \Gamma}{S_A(t_1), \Gamma \quad \cdots \quad S_A(t_k), \Gamma} \quad \text{where } X \to t_i \text{ and } \exists a \in A. \ X \xrightarrow{a}$$

$$\text{Step2} \quad \frac{\mathcal{E}(X), \Gamma}{\mathcal{E}(t_1), \Gamma \quad \cdots \quad \mathcal{E}(t_k), \Gamma} \quad \text{where } X \xrightarrow{a} t_i$$

$$\text{EE} \quad \frac{\mathcal{E}(\epsilon), \Gamma}{\Gamma}$$

Note that the applicable rule and the result of the rule application is uniquely determined by the sequent.

Proposition 4. *For every valid instance of a rule, the antecedent is true if and only if one of its consequents is true.*

Definition 5. (Termination conditions) A node n in the tableau consisting of a sequence of formulae Γ is a terminal node if either

1. Γ is empty
2. On the path from the root-node to the node n, there is a node n' with a sequence Γ' s.t. $\Gamma' \subseteq \Gamma$. (Every formula in Γ' also occurs in Γ).

Terminals of type 1 are successful, while terminals of type 2 are unsuccessful.

Lemma 6. *The tableau is finite.*

Proof There are only finitely many subterms of the root-term and only finitely many subterms of the terms t on the right hand side of rules $X \xrightarrow{a} t$. So there are only finitely many different sequents in the tableau, because of the "cleanup"-condition. So no branch in the tableau can have infinite length, due to termination condition 2. As the tableau is finitely branching the result follows.　□

Theorem 7. *For every PA-process t and every set of actions $A \subseteq Act$, $D_A(t)$ holds iff there is a successful tableau with root $D_A(t)$.*
Proof

1. If there is a successful tableau, then it has a leaf marked by the empty sequence. This leaf is certainly true. By Proposition 4 the root-formula $D_A(t)$ must be true.
2. If $D_A(t)/S_A(t)/\mathcal{E}(t)$ holds then there is a sequence σ of finite length s.t. $t \xrightarrow{\sigma} t'$ and $\not\exists a \in A.\ t' \xrightarrow{a}$, with $t' \neq \epsilon$ or $t' = \epsilon$, respectively. Choose such a sequence σ of minimal length and let $n := length(\sigma)$. We prove by induction on lexicographically ordered pairs $(n, size(t))$ that if $D_A(t)/S_A(t)/\mathcal{E}(t)$ is true, then there is a successful tableau with root $D_A(t)/S_A(t)/\mathcal{E}(t)$.
   (a) If $n = 0$ then $t = \epsilon$ or $t = (X.t_1) \| t_2$ s.t. $D_A(t_2)$ and $\not\exists a \in A.X \xrightarrow{a}$. ($t_1$ and t_2 may be ϵ) If $t \neq \epsilon$, then by the D-rule we get $S_A(t)$ and by the PD-rule $S_A(X.t_1), D_A(t_2)$. The SD-rule yields $S_A(X), D_A(t_2)$ and by the E-rule we get $D_A(t_2)$. By induction hypothesis there is a correct tableau for $D_A(t_2)$. If $t = \epsilon$, then the D- and EE-rules yield a successful tableau.
   (b) If $n > 0$ then apply the tableau-rules. By Proposition 4 the rules are valid, so least one of the children is true. Also either the terms are smaller or (in case of the Step-rules) the new sequence σ' is shorter. Thus by induction hypothesis there is a successful tableau. This construction can't be stopped by termination condition 2, because this would contradict the minimality of the length of σ.

□

The tableau built by these rules is very redundant and can contain branches of exponential length. However, it is still possible to find a polynomial algorithm for the problem.

Theorem 8. *The partial deadlock reachability problem for PA-processes t is decidable in polynomial time.*

Proof

1. First we decide $\mathcal{E}(X)$ for all variables X. This is done by a marking algorithm. First mark all variables X s.t. there is a rule $X \xrightarrow{a} \epsilon$. If there is a rule $Y \xrightarrow{a} t$ s.t. Y is unmarked so far and all variables in t are marked then mark Y. Do this until no new variable can be marked. This requires polynomial time.
2. Now decide $S_A(X)$ for all variables X. First mark all variables X s.t. $\not\exists a \in A . X \xrightarrow{a}$.
   For all rules $Y \xrightarrow{a} t_1 . t_2$ or $Y \xrightarrow{a} t_1 \| t_2$ s.t. Y is unmarked use the rules D, SD, PD and E and the already acquired knowledge about $\mathcal{E}(Z)$ and $S_A(Z)$ to prove $S_A(Y)$. In addition to the normal failure cases this fails when we encounter unmarked variables. Repeat this until all attempts for all rules fail. This takes polynomial time.
3. As we know the values of $\mathcal{E}(X)$ and $S_A(X)$ for every variable X we can decide $D_A(t)$ in polynomial time with the help of the rules D, SD, PD and E.

□

4 Partial Livelock Reachability

Partial livelock reachability problem:
Instance: A labelled transition system (LTS) with initial state r_0 and a set of atomic actions A.
Question: Is is possible to reach a state r s.t. no state r' that is reachable from r enables any action in A ?

In the special case of $A = Act$ this is the problem if a state of deadlock is reachable.

Now we present a sound and complete tableau system that solves the partial livelock reachability problem for PA-processes. First we define some predicates:

Definition 9. Let $A \subseteq Act$ be a set of actions.

$$\mathcal{E}(t) : \iff \exists t \xrightarrow{\sigma} \epsilon$$
$$En_A(t) : \iff \exists a \in A . t \xrightarrow{a}$$
$$N_A(t) : \iff \not\exists t \xrightarrow{\sigma} t' . En_A(t')$$
$$R_A(t) : \iff \exists t \xrightarrow{\sigma} t' . N_A(t')$$
$$RI_A(t) : \iff \exists t \xrightarrow{\sigma} t' . N_A(t') \wedge \neg\mathcal{E}(t')$$

Intuitively $N_A(t)$ means that t will never be able to do any action from A, and $R_A(t)$ means that a state t' is reachable from t where all actions in A are disabled forever. $RI_A(t)$ means that a state t' is reachable from t that has no terminating computation and where all actions in A are disabled forever. This can also be expressed in temporal logic: Let $A := \{a_1, \ldots, a_n\}$.

$$N_A(t) = t \models \Box(\neg a_1 \land \ldots \land \neg a_n)$$
$$R_A(t) = t \models \Diamond\Box(\neg a_1 \land \ldots \land \neg a_n)$$

The partial livelock reachability problem is to check if some PA-process t satisfies $R_A(t)$.

The branches in the tableau are interpreted conjunctively, so the tableau is successful if all branches are successful. A branch is successful if it ends with a successful leaf. As the rules for the construction of the tableau are nondeterministic, the tableau is not unique. However, $R_A(t)$ holds iff there is at least one successful tableau with root $R_A(t)$.

$$\text{RPD} \quad \frac{R_A(t_1 \| t_2)}{R_A(t_1) \quad R_A(t_2)}$$

$$\text{RSD 1} \quad \frac{R_A(t_1 . t_2)}{R_A(t_2)} \ (\mathcal{E}(t_1))$$

$$\text{RSD 2} \quad \frac{R_A(t_1 . t_2)}{RI_A(t_1)}$$

$$\text{Found} \quad \frac{R_A(X)}{N_A(X)}$$

$$\text{Step R} \quad \frac{R_A(X)}{R_A(t_i)} \ (X \to t_i)$$

$$\text{RIPD 1} \quad \frac{RI_A(t_1 \| t_2)}{RI_A(t_1) \quad R_A(t_2)}$$

$$\text{RIPD 2} \quad \frac{RI_A(t_1 \| t_2)}{R_A(t_1) \quad RI_A(t_2)}$$

$$\text{RISD 1} \quad \frac{RI_A(t_1 . t_2)}{RI_A(t_1)}$$

$$\text{RISD 2} \quad \frac{RI_A(t_1 . t_2)}{RI_A(t_2)} \ (\mathcal{E}(t_1))$$

$$\text{Found RI} \quad \frac{RI_A(X)}{N_A(X)} \ (\neg\mathcal{E}(X))$$

$$\text{Step RI} \quad \frac{RI_A(X)}{RI_A(t_i)} \ (X \to t_i)$$

$$\text{Step N} \quad \frac{N_A(X)}{N_A(t_1) \quad \cdots \quad N_A(t_n)} \ (\neg En_A(X), \ X \to t_i, i = 1, \ldots, n)$$

$$\text{NPD} \quad \frac{N_A(t_1 \| t_2)}{N_A(t_1) \quad N_A(t_2)}$$

$$\text{NSD 1} \quad \frac{N_A(t_1.t_2)}{N_A(t_1)} \ (\neg\mathcal{E}(t_1))$$

$$\text{NSD 2} \quad \frac{N_A(t_1.t_2)}{N_A(t_1) \quad N_A(t_2)}$$

Definition 10. A node in the tableau is a terminal node if the formula is either

1. $N_A(\epsilon)$
2. $N_A(t)$ and a node with the same formula occurred earlier in the branch.
3. $R_A(\epsilon)$
4. $R_A(t)$ and a node with the same formula occurred earlier in the branch.
5. $RI_A(\epsilon)$
6. $RI_A(t)$ and a node with the same formula occurred earlier in the branch.
7. $N_A(X)$ and $En_A(X)$ is true.

Terminals 1,2,3 are successful, while the terminals of types 4,5,6,7 are unsuccessful.

Proposition 11. *If all consequents of an instance of a rule are true and the side conditions are satisfied then the antecedent is true.*

Lemma 12. *If a formula F in the tableau is true and the node is not a terminal node, then there is an instance of a rule with antecedent F and satisfied side conditions s.t. all consequents are true.*

Proof

1. If F is of the form $R_A(t_1.t_2)$, then there are two cases:
   (a) If $\neg\mathcal{E}(t_1)$ then $t_1.t_2$ is equivalent to t_1, and $RI_A(t_1)$. So the rule is RSD 2.
   (b) If $\mathcal{E}(t_1)$ then there are two cases:
      i. $t_1 \rightarrow \epsilon$ and $t_2 \rightarrow t_2'$ with $N_A(t_2')$. So we have $R_A(t_2)$ and the rule is RSD 1.
      ii. $t_1.t_2 \rightarrow t_1'.t_2$ with $N_A(t_1'.t_2)$. If $\mathcal{E}(t_1')$ then $N_A(t_2)$ and thus $R_A(t_2)$ and the rule is RSD 1. If $\neg\mathcal{E}(t_1')$ then $t_1'.t_2$ is equivalent to t_1' and $RI(t_1)$ and the rule is RSD 2.
2. If F is of the form $R_A(X)$ then either we have already reached a state with $N_A(X)$ (rule Found) or we can reach one from one of the successor states t_i that are reachable from X in one step (rule Step R).
3. If F is of the form $RI_A(t_1\|t_2)$ then $t_1 \overset{\sigma_1}{\rightarrow} t_1'$, $t_2 \overset{\sigma_2}{\rightarrow} t_2'$ s.t. $N_A(t_1')$, $N_A(t_2')$ and at least one of t_1', t_2' has no terminating computation. These two cases are covered in rules RIPD 1 and RIPD 2.
4. For the case of $F = RI_A(t_1.t_2)$ the argument is equivalent to the one in case 1a.
5. If $F = RI_A(X)$ then either we have already found the correct X s.t. $N_A(X)$ and X can never terminate ($\neg\mathcal{E}(X)$), or we can find it from at least one of the successor states t_i. The first case is rule Found RI, and the second case is rule Step RI.

6. If $F = N_A(t_1.t_2)$ then there are two cases.
   (a) If $\neg\mathcal{E}(t_1)$ then $t_1.t_2$ is equivalent to t_1 and thus $N_A(t_1)$. The rule is NSD 1.
   (b) If $\mathcal{E}(t_1)$ then both $N_A(t_1)$ and $N_A(t_2)$ must hold. The rule is NSD 2.

All the other rules RPD, Step N and NPD are sound and complete.   □

Lemma 13. *All possible tableaux are finite.*

Proof directly by definition of the rules and termination conditions 2,4 and 6.
□

A node in the tableau is called true iff the formula at this node holds (see Def. 9).

Lemma 14. *If $N_A(t)$ holds then there is a successful tableau with root $N_A(t)$.*

Proof If a node is true then by Lemma 12 and Lemma 13 we can (by choosing the right rules) construct a finite tableau s.t. every node is true. If the root node is $N_A(t)$ then every node in the tableau is of the form $N_A(t')$ for some t'. No branch can terminate with termination condition 7, because every node is true. So every branch must end with a terminal node of type 1 or 2 and is successful.
□

Lemma 15. *If $N_A(t)$ doesn't hold then every tableau with root $N_A(t)$ has an unsuccessful branch.*

Proof If $N_A(t)$ doesn't hold then there is a sequence σ s.t. $t \xrightarrow{\sigma} t'$ and $\exists a \in A. \ t' \xrightarrow{a}$. Choose such a sequence of minimal length. By induction on the length of σ we can prove the existence of a branch ending with a terminal of type 7. In the construction we always choose the rule and its consequent according to σ. The construction cannot terminate by condition 1, because every node in the branch is false and it cannot terminate by condition 2, because σ has minimal length.   □

Lemma 16. *Let t be a PA-term and $A \subseteq Act$. If $R_A(t)$ $(RI_A(t))$ holds then there is a successful tableau with root $R_A(t)$ $(RI_A(t))$.*

Proof There is a sequence σ s.t. $t \xrightarrow{\sigma} t'$ and $N_A(t')$ (and additionally $\neg\mathcal{E}(t')$ if $RI_A(t)$). Choose such a sequence of minimal length. We prove by induction on lexicographically ordered pairs $(length(\sigma), size(t))$ that there is a successful tableau with root $R_A(t)$ $(RI_A(t))$.

If a node is true then by Lemma 12 and Lemma 13 we can (by choosing the right rules) construct a finite tableau s.t. every node is true.

1. If $size(t) = 0$ then $length(\sigma) = 0$ and $t = \epsilon$. In this case $RI_A(t)$ doesn't hold. If $R_A(t)$ holds, then we already have a successful tableau, because $R_A(\epsilon)$ is a successful terminal of type 3.
2. (a) If $size(t) = 1$ then $t = X$, where X is a variable. There are two cases:
   i. If $length(\sigma) = 0$ then apply the rule Found (Found RI) and $N_A(t)$ is true. By Lemma 14 there is a successful tableau.

ii. If $length(\sigma) > 0$ then apply the right instance of the rule Step R (Step RI) s.t. the consequent is true. The size of the new term may be larger, but the length of the rest of the sequence is shorter, so the induction hypothesis applies.

(b) If $size(t) >= 2$ then t is of the form $t_1 \| t_2$ or $t_1.t_2$. Use the right rule from RPD, RSD 1, RSD 2 (RIPD 1, RIPD 2, RISD 1, RISD 2). The new term is smaller and the sequence is certainly not longer. Thus it is possible to apply the induction hypothesis.

This construction cannot be stopped by termination condition 4 or 6, because σ is of minimal length. $\square$

Lemma 17. *Let t be a PA-term and $A \subseteq Act$. If $R_A(t)$ ($RI_A(t)$) doesn't hold then all tableaux with root $R_A(t)$ ($RI_A(t)$) are unsuccessful.*

Proof It suffices to show that there is at least one unsuccessful branch. If a node is false then by Proposition 11 at least one child node must be false. So there is a branch where every node is false. There are two cases:

1. If every node in the branch is of the form $R_A(t')$ or $RI_A(t')$ then the branch must end with an unsuccessful terminal of type 4 or 6.
2. If a node of the form $N_A(t')$ occurs in the branch, then by Lemma 15 the tableau has an unsuccessful branch.

$\square$

Theorem 18. *Let t be a PA-term and $A \subseteq Act$. Then $R_A(t)$ holds iff there is a successful tableau with root $R_A(t)$.*

Proof By Lemma 16 and Lemma 17. $\square$

The complexity of the algorithm obtained from the tableau isn't even in NP, but in PSPACE. On the other hand it is possible to find a polynomial time algorithm that decides $R_A(t)$. This algorithm proceeds in four steps:

1. Decide $\mathcal{E}(X)$ for all variables X.
2. Decide $N_A(X)$ for all variables X using the previously collected information.
3. Decide $R_A(X)$ and $RI_A(X)$ for all X using the previously collected information.
4. Decide $R_A(t)$ using the previously collected information.

The first part uses the algorithm described in the proof of Theorem 8. It requires polynomial time.

For the second part define $P_A(t) := \neg N_A(t)$. Now use a marking algorithm to decide $P_A(X)$ for every X.

Start Mark all variables X s.t. $\exists a \in A.X \xrightarrow{a}$.

Step For every unmarked X and every rule $X \to t$ do if $P'_A(t)$ then mark X. P'_A is defined by

$$P'_A(t_1\|t_2) = P'_A(t_1) \vee P'_A(t_2)$$
$$P'_A(t_1.t_2) = P'_A(t_1) \vee (\mathcal{E}(t_1) \wedge P'_A(t_2))$$
$$P'_A(X) = \text{if } X \text{ is marked then true else false}$$

Repeat Step until no new variable is marked.

$N_A(X)$ is true iff X is not marked. This algorithm also requires polynomial time.

The third part is a marking algorithm that uses two different markings R_A and RI_A. If a variable X is marked by R_A (RI_A) then it means that it is already known that $R_A(X)$ ($RI_A(X)$) is true.

Start If $N_A(X)$ then mark X with R_A. If $N_A(X)$ and $\neg\mathcal{E}(X)$ then mark X with RI_A.

Step For every X and every rule $X \to t$ do: If $R'_A(t)$ then mark X with R_A. If $RI'_A(t)$ then mark X with RI_A. The functions R'_A and RI'_A are defined by

$$R'_A(t_1\|t_2) = R'_A(t_1) \wedge R'_A(t_2)$$
$$R'_A(t_1.t_2) = RI'_A(t_1) \vee (\mathcal{E}(t_1) \wedge R'_A(t_2))$$
$$R'_A(X) = \text{if } X \text{ is marked by } R_A \text{ then true else false}$$

$$RI'_A(t_1\|t_2) = RI'_A(t_1) \wedge R'_A(t_2) \vee R'_A(t_1) \wedge RI'_A(t_2)$$
$$RI'_A(t_1.t_2) = RI'_A(t_1) \vee (\mathcal{E}(t_1) \wedge RI'_A(t_2))$$
$$RI'_A(X) = \text{if } X \text{ is marked by } RI_A \text{ then true else false}$$

Repeat Step until no new variable is marked.

In the last step just use the rules for the functions used in the third step and the previously collected information to decide $R_A(t)$. This also takes polynomial time.

Theorem 19. *Let t be a PA term and $A \subseteq Act$. It is decidable in polynomial time if $R_A(t)$.*

5 The Reachability Problem for PA

We present two algorithms that decide the reachability problem for PA-processes. The first one does a simple bounded search, while the second one is tableau-based and works by decomposition.

First we define the size of an instance of the problem:

Definition 20. For convenience of notation we also write $t_1 \succ t_2$ for $\exists \sigma.\ t_1 \xrightarrow{\sigma} t_2$. An instance of the reachability problem for PA-processes consists of two processes t_1, t_2 and a set of rules Δ of the form $X \xrightarrow{a} t$. The question is if $t_1 \succ t_2$. The size of the instance is $size(t_1) + size(t_2) + \sum_{(X \xrightarrow{a} t) \in \Delta} size(t)$.

Note that this definition is different from the one for BPPs. There terms of the form $Y^n = Y\| \ldots \|Y$ are allowed whose size is $log(n)$, while for PA the size of this term would be n. Even for this weaker definition of the size, the reachability problem is NP-hard [Esp95] (even for BPPs).

Lemma 21. *The reachability problem for PA is in PSPACE.*

Proof Let n be the size of the instance of the problem. It can be shown that if a state t can be reached from the initial state t_0, then it can be reached via a path s.t. the size of every intermediate state t' is bounded by a constant $c \leq O(n^2)$. Details can be found in [May96b]. □

The argument of Lemma 21 is somewhat crude, because it does not take the internal structure of PA-processes into account. By using this structure we can derive a more accurate complexity bound.

Now we present the sound and complete tableau system that solves the reachability problem for PA-processes. The nodes in the tableau will be sequences of formulae, that are interpreted conjunctively. Such sequences will be denoted by Γ. The branches of the proof tree are interpreted disjunctively. So the tableau is successful iff there is a successful leaf. To avoid carrying around unnecessary formulae we use the following "cleanup-procedure" after each rule application: If a formula Φ occurs more than once in a sequence Γ, then delete all occurrences but the first.

$$\text{SP} \quad \frac{t_1.t_2 \succ t_3\|t_4, \Gamma}{t_1 \succ \epsilon, t_2 \succ t_3\|t_4, \Gamma}$$

$$\text{PS} \quad \frac{t_1\|t_2 \succ t_3.t_4, \Gamma}{t_1 \succ t_3.t_4, t_2 \succ \epsilon, \Gamma \quad t_1 \succ \epsilon, t_2 \succ t_3.t_4, \Gamma}$$

$$\text{SS} \quad \frac{t_1.t_2 \succ t_1'.t_2'.\ldots.t_k', \Gamma}{t_1 \succ \epsilon, t_2 \succ t_1'.\ldots.t_k', \Gamma \quad \ldots \quad t_1 \succ t_1'.\ldots.t_i', t_2 = t_{i+1}'.\ldots.t_k', \Gamma \quad \ldots}$$

for $i = 1, \ldots, k-1$ and no t_j' is a sequential composition

$$\text{PP} \quad \frac{t_1\|t_2 \succ t_1'\|t_2'\| \ldots \|t_k', \Gamma}{\ldots t_1 \succ \|M, t_2 \succ \|(\{t_1', \ldots, t_k'\} - M), \Gamma \ldots} \quad \text{for every } M \subseteq \{t_1', \ldots, t_k'\}$$

and no t_i' is a parallel composition

$$\text{Step} \quad \frac{X \succ t, \Gamma}{s_1 \succ t, \Gamma \quad \ldots \quad s_k \succ t, \Gamma} \quad \text{for } X \neq t \text{ and } X \to s_i$$

$$\text{E1} \quad \frac{t \succ t, \Gamma}{\Gamma}$$

$$\text{E2} \quad \frac{t = t, \Gamma}{\Gamma}$$

Note that the applicable rule and the result of the rule application is uniquely determined by the sequent.

Definition 22 A node n consisting of a sequence of formulae Γ is a terminal node if one of the following conditions is satisfied.

1. Γ is empty
2. $\Gamma = \epsilon \succ t, \Gamma'$ and $t \neq \epsilon$.
3. $\Gamma = (t = t'), \Gamma'$ and $t \neq t'$.
4. On the path from the root-node to the node n, there is a node n' with a sequence Γ' s.t. $\Gamma' \subseteq \Gamma$. (Every formula in Γ' occurs in Γ.)

Terminals of type 1 are successful, while terminals of types 2–4 are unsuccessful.

Lemma 23. *The tableau is finite.*

Proof There are only finitely many subterms of the root-term and only finitely many subterms of the terms t on the right hand side of rules $X \xrightarrow{a} t$. So there are only finitely many different sequents in the tableau, because of the "cleanup"-condition. So no branch in the tableau can have infinite length, due to termination condition 4. As the tableau is finitely branching the result follows. □

Now we show the soundness and completeness of the tableau-rules.

Lemma 24. *The antecedent of an instance of a tableau rule is true if and only if at least one of the consequents is true.*

Proof

SP Assume $t_1.t_2 \xrightarrow{\sigma} t_3 \| t_4$. If t_1 is not reduced to ϵ in σ then $t_1.t_2 \xrightarrow{\sigma} t_1'.t_2$ for some t_1'. This cannot be equivalent to $t_3 \| t_4$, because the outermost operator is different. Therefore $\sigma = \sigma_1 \sigma_2$ s.t. $t_1 \xrightarrow{\sigma_1} \epsilon$ and $t_2 \xrightarrow{\sigma_2} t_3 \| t_4$. The other direction is obvious.

PS If $t_1 \| t_2 \xrightarrow{\sigma} t_3.t_4$ then t_1 or t_2 must be reduced to ϵ in σ, because otherwise the outermost operator of the result would be parallel composition instead of sequential composition. So σ must consist of two subsequences σ_1 and σ_2 s.t. $t_1 \xrightarrow{\sigma_1} \epsilon$ and $t_2 \xrightarrow{\sigma_2} t_3.t_4$ or vice versa. The other direction is obvious.

SS If $t_1.t_2 \succ t_1'.t_2'.\ldots.t_k'$ is true, then there are two cases:

1. If t_1 is reduced to ϵ in the rest of the sequence, then $t_2 \succ t_1'.t_2'.\ldots.t_k'$ must be true.
2. If t_1 is not reduced to ϵ in the rest of the sequence, then t_1 must develop into a prefix $t_1'.\ldots.t_i'$ of $t_1'.\ldots.t_k'$. In this case t_2 cannot change in the sequence and must therefore be equal to $t_{i+1}'.\ldots.t_k'$. As $t_2 \neq \epsilon$ it follows that $i < k$.

As no t_j' is a sequential composition, it isn't possible that t_1 develops into a part of a t_j' and t_2 into the rest. The other direction is trivial.

PP If $t_1 \| t_2 \succ t_1' \| t_2' \| \ldots \| t_k'$ then t_1 must develop into the parallel composition of a (possibly empty) subset of $\{t_1', \ldots, t_k'\}$ and t_2 into the parallel composition of the rest. As no t_j' is a parallel composition, it isn't possible that t_1 develops into a part of a t_j' and t_2 into the rest.

The other rules *Step*, E_1 and E_2 are obvious. □

Lemma 25. *A PA-process t_2 is reachable from a PA-process t_1 iff there is a successful tableau with root $t_1 \succ t_2$.*

Proof

1. If the tableau is successful, then there is a successful (empty) leaf that is certainly true. By Lemma 24 the root sequent is also true.
2. If t_2 is reachable from t_1 then there exists a sequence σ of minimal length n s.t. $t_1 \xrightarrow{\sigma} t_2$. We prove by induction on lexicographically ordered pairs $(n, size(t_1) + size(t_2))$ that there is a successful path in the tableau. The base case is trivial. Otherwise apply the tableau rules. By Lemma 24 at least one of the consequents must be true, and by induction hypothesis there must be a successful tableau for it. This construction cannot terminate with termination condition 4, because this would contradict the minimality of the length of σ.

$\square$

The tableau system does not yield an NP-algorithm, because it can contain branches of exponential length. However, this can be avoided by a slight modification. To show containment in NP we modify the tableau system by avoiding some redundant parts.

Theorem 26. *The reachability problem for PA-processes is NP-complete.*

Proof By using the same algorithm as in the proof of Theorem 8 we can decide $\mathcal{E}(X)$ for every variable X in polynomial time. Using this knowledge we can decide $t \succ \epsilon$ using $O(length(t))$ time. Now in the tableau system formulae of the form $t \succ \epsilon$ are decided immediately when they appear, using only polynomial time. Now if $t_1 \succ t_2$ is true, then there is a successful branch of polynomial length. As the formulae on this branch only have polynomial length the branch can be described in polynomial space. We can nondeterministically guess a branch of polynomial size and check in polynomial time if it is a valid successful branch. To do this we use the tableau rules and the polynomial algorithm for deciding $t \succ \epsilon$ on the way.

As the reachability problem is NP-hard even for the weaker model of BPP [Esp95] the result follows. $\square$

6 Conclusion

All three investigated problems for PA-processes have the same complexities as for BPPs. While reachability is NP-complete, partial deadlock reachability and partial livelock reachability can be decided in polynomial time.

Tableau systems often do not yield very efficient algorithms, because the tableaux contain many redundant parts. The tableau in section 3 is an example for this. On the other hand the efficiency of tableau methods can be increased significantly by using small non-tableau algorithms to decide side-conditions during the construction of the tableau. This technique has been used in the modified tableau method for the reachability problem in the proof of Theorem 26.

Tableau techniques are a common tool for the analysis of systems described by Petri nets or process algebras. They have been used for model checking problems for various temporal logics [BEM96, May96c, May96a]. The three problems studied in this paper can be seen as special cases of a general model checking problem for PA-processes. This problem has recently been shown to be decidable with the help of tableau techniques [May96a]. However, the algorithm in [May96a] has an extremely high complexity (non-elementary) and thus it is useful to consider these subproblems here.

Although the tableau-methods presented here are sound and complete, they are not intended to be used as a decision procedure, but rather as a proof method that can be implemented in the framework of a theorem prover with human interaction.

References

[BEM96] J. Bradfield, J. Esparza, and A. Mader. An effective tableau system for the linear time μ-calculus. In F. Meyer auf der Heide and B. Monien, editors, *Proceedings of ICALP'96*, number 1099 in LNCS. Springer Verlag, 1996.

[Chr93] S. Christensen. *Decidability and Decomposition in Process Algebras*. PhD thesis, Edinburgh University, 1993.

[Esp95] Javier Esparza. Petri nets, commutative context-free grammars and basic parallel processes. In Horst Reichel, editor, *Fundamentals of Computation Theory*, number 965 in LNCS. Springer Verlag, 1995.

[Esp96] J. Esparza. More infinite results. In B. Steffen and T. Margaria, editors, *Proceedings of INFINITY'96*, number MIP-9614 in Technical report series of the University of Passau. University of Passau, 1996.

[May84] E. Mayr. An algorithm for the general petri net reachability problem. *SIAM Journal of Computing*, 13:441–460, 1984.

[May96a] Richard Mayr. Model checking pa-processes. Technical Report TUM-I9640, TU-München, December 1996.

[May96b] Richard Mayr. Semantic reachability for simple process algebras. In B. Steffen and T. Margaria, editors, *Proceedings of INFINITY'96*, number MIP-9614 in Technical report series of the University of Passau. University of Passau, 1996.

[May96c] Richard Mayr. A tableau system for model checking petri nets with a fragment of the linear time μ-calculus. Technical Report TUM-I9634, TU-München, October 1996.

[Mil89] R. Milner. *Communication and Concurrency*. Prentice Hall, 1989.

A Tableau Proof System for a Mazurkiewicz Trace Logic with Fixpoints

Peter Niebert[1] and Barbara Sprick[2]

[1]Institut für Informatik,
Universität Hildesheim, Germany
niebert@informatik.uni-hildesheim.de

[2]Fachbereich Informatik, Lehrstuhl 6
Universität Dortmund, Germany
sprick@ls6.informatik.uni-dortmund.de

Abstract. We present a tableau based proof system for νTrTL, a trace based temporal logic with fixpoints. The proof system generalises similar systems for standard interleaving temporal logics with fixpoints. In our case special attention has to be given to the modal rule: First we give a system with an interleaving style modal rule, later we use a technique similar to the sleep set method (known from finite state model checking) to obtain a more efficient proof rule. We briefly highlight the relation of the improved rule with recent advances in tableau systems for classical propositional logic, the *tamed cut* of the system *KE*.

The treatment of fixpoints leads to possibly infinite tableaux, which however can be finitely represented, yielding treelike structures with back loops: we show this using an automata construction. Indirectly we obtain a (known) decidability result.

1 Introduction

Temporal logics are a widely accepted specification and verification formalism in several areas of computer science, in particular in the field of *reactive systems*. The most widely known logic in this area is *linear time temporal logic*, *LTL*: this logic assumes a discrete and totally ordered time structure (natural numbers as time instances) and is based on two modalities, "Next" and "Until". This framework is appropriate for the specification of systems based on global states.

On the other hand a lot of effort went into the development of semantic frameworks for *distributed systems*, typically consisting of several subsystems (*locations*), which work independently and collaborate using a communication mechanism. In such a setting it is more appropriate to use *partially ordered* or *distributed runs* (which coincide with infinite Mazurkiewicz traces [Maz95]) as paradigm.

Recently Thiagarajan has defined *TrPTL* [Thi94] as a seemingly natural *trace based* generalisation of *LTL* to partially ordered runs. The key idea is to interpret temporal operators such as "Until" and in particular "Next" only relatively to the "view" of one location (local time), and it allows to *change the point of view* in order to express properties of several locations. In [Nie95, Huh96] νTrTL, a fixpoint extension of *TrPTL* has been developed. This logic combines the *local Next* modality with the possibility of fixpoint definitions (recursion in formulae), so that e.g. Thiagarajan's *local Until* can be defined as a derived operator. Similar fixpoint extensions are known for standard temporal logics (e.g. νTL [Var88]). The structure of νTrTL is advantageous over *TrPTL* in that it allows a semantically cleaner *changing of the point of view:* νTrTL formulae only look into the future behaviour of a system, while the changing of the point of view in [Thi94] results in an uncontrolled jump into the past. This property of our logic is also essential for the proof system.

There are more reasons to be interested in νTrTL: it is well known, that νTL (the fixpoint extension of *LTL*) is expressively equivalent to Büchi-automata and to the *monadic second order theory of the temporal frames*. It is conjectured (work in progress), that an analogous result also holds for νTrTL, i.e. that this logic is expressively equivalent to the monadic second order theory of distributed runs.

In this work we develop a tableau based proof system for νTrTL. The structural properties of νTrTL allow an elegant formulation of proof rules. Since the treatment of fixpoints in proof systems requires lengthy constructions (see e.g. [Wal95] for the treatment of the propositional μ-calculus), we put the emphasis on the particular aspects of νTrTL. We treat the fixpoints in a system with infinite (depth, not width) tableaux. Using automata constructions it can be shown, that the infinite proofs can be finitely represented (as trees with back loops) and automatically constructed. Thus indirectly we obtain a decision procedure for the satisfiability problem (already directly addressed in [Nie95]).

The most important rule of our proof system is the *modal rule*, i.e. the rule for the treatment of the *Next* modalities. This rule uses a case analyses about the possible next steps occuring at some point in a distributed run and leads to branching in the tableaux similar to the rule for disjunction.

An observation concerning the modal rule interesting in two ways is that the tableau system can be made more efficient by making the cases of the previously mentioned case analysis *mutually exclusive*. On the one hand, we relate this modified modal rule to the *tamed cut* of D'Agostino's and Mondadori's system *KE* [DM94]. On the other hand, the construction we use to make the modal rule more efficient *operationally* behaves similarly to a reduction method used to make finite state model checking more efficient: the sleep set method [GW91].

Sleep sets are one way to exploit the independence of actions in the system in the search for particular states (e.g. a deadlock). Technically, during the depth first search, a set of actions (called sleep set), the exploration of which would be redundant (because they already occurred in another order), is carried around.

The analogy in the tableau system is the exploration of sub tableaux at applications of the modal rule: instead of actions we now have to deal with modalities. However instead of externally keeping track of modalities not to be explored anymore, the modified modal rule introduces formulae like [*a*]**false** (*a* cannot occur) into the sequents. The interplay with the other (standard) rules then *operationally* stops the expansion of redundant tableau parts.

The rest of the paper is structured as follows: In section 2 we give a simplistic example of a distributed system and formally introduce the notion of *distributed runs*, which constitute the semantic models over which νTrTL is interpreted. In section 3 we give the syntax and semantics of νTrTL and illustrate its use in a small example. In section 4 the tableau system is given, the proof of the soundness and completeness is given in section 5. In section 6 we discuss two aspects of automatic proof search: the finite representation of infinite rules and the improved modal rule.

The paper is based on the diploma thesis of the second author [Spr96], which contains a detailed presentation of the present work.

2 Runs of distributed programs

Distributed programs. The location based approach to Mazurkiewicz-Traces considers parallel systems of sequential processes, which communicate via joint actions [Thi94]. Let us first have a look at the Alternating Bit Protocol (ABP) [Mil89] as a small example of a concurrent program: The idea is that two components, a sender S and a receiver R communicate over two (unreliable) channels. The Message Channel transfers messages from S (send) to R (receive) and the Ack(nowledgement) Channel delivers acknowledgements from R (sack) to S (rack), both channels can lose (losem, losea) packets. Each data message sent by S

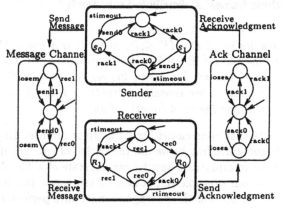

Fig. 1. Alternating Bit Protocol

contains a protocol bit, either 1 or 0 (send1/0). Let us assume the sender sends a 0-message (send0). Before receiving the corresponding acknowledgment (rack0) it can send the 0-message again after a timeout and ignore 1-acknowledgments. After receiving the corresponding ack (rack0) S stops transmitting the current message (send0) and flips the protocol bit to 1 for the next message. The receiver basically works in the same manner: After receiving a message (rec0), R returns an ack to S (sack0). Afterwards R can either resend the ack (sack0) after a timeout or ignore more messages with protocol bit 0 (rec0) before it finally receives a new message with the alternated protocol bit (rec1).

Distributed runs. We will now introduce distributed runs to represent the behaviour of distributed programs. A run is one possible execution of a distributed program, e.g. figure 2 shows one possible run of the ABP example given above.

We will first define the alphabet for distributed runs as the set of actions, which can take place in such a run. Some of these actions are local to only one component (e.g. stimeout, rtimeout in figure 1), while others may belong to more than one component (e.g. send0, rack1). The latter we call "synchronisations" between the components involved in these actions. Actions which take place at different locations such as "send0" and "sack1" are called independent of each other: there is no natural way to observe a causal order between these actions.

Definition 2.1 (alphabet) *Let $K \in \mathbb{N}$ be fixed. Then $Loc = \{1, \ldots, K\}$ denotes a set of locations and $\tilde{\Sigma} = (\Sigma_1, \ldots, \Sigma_K)$ a distributed alphabet, where each Σ_i is a finite, nonempty set of actions of location i. The sets Σ_i may overlap. We define $\Sigma := \bigcup_{1 \leq i \leq K} \Sigma_i$ as the global alphabet of the system. An action a with $a \in \Sigma_i \cap \Sigma_j$ is called a* **synchronisation** *between the locations i and j and $Loc(a) := \{i \in Loc | a \in \Sigma_i\}$ denotes the set of locations which are synchronised by action a.*
Two actions a and b are **independent**, $(a\mathcal{I}b)$, *iff $Loc(a) \cap Loc(b) = \emptyset$.*

It is easy to see, that the independence relation given in 2.1 is irreflexive and symmetric. Thus $(\Sigma, \mathcal{I})$ is a concurrent alphabet in the sense of Mazurkiewicz [Maz95] and we give here a location based approach to Mazurkiewicz traces.

Let us now define distributed runs as a tuple of a frame and interpretations of these frames. A distributed run stands for an execution of a distributed program:

Definition 2.2 (frame) *Let $\tilde{\Sigma}$ be a distributed alphabet. A* **frame** *over $\tilde{\Sigma}$ is a labelled poset $F = (E, \leq, l)$, where E is a countable set of events, $\leq$ a partial order on E and $l : E \to \Sigma$ a labelling-function. Let $E_i = \{e \in E \mid l(e) \in \Sigma_i\}$ be the set of i-events. For each i the restriction $\leq \cap (E_i \times E_i)$ is total, (i.e. the events of one location are causally ordered), and the global order $\leq$ is the least partial order containing the local (total) orders.*

Note that runs can either be finite or infinite. To talk about dynamic behaviour of these static frames we use the notion of configurations. A configurations of a frame gives informations about a certain state of a distributed program. The configuration c_1 in our example run in figure 2 represents the state of the program, in which only the sender and the message channel have performed the action send0, but the other components are still in their initial state. So a configuration contains all the actions of a distributed run, that have occurred so far in the computation.

Definition 2.3 *A* **configuration** *c of a frame F is a finite, with respect to $\leq$ downward closed set of events. $\mathcal{C}_F$ denotes the set of all configurations of F. Two configurations c, c' are i-equivalent ($c \equiv_i c'$), iff $c \cap E_i = c' \cap E_i$. They are A-equivalent ($\equiv_A$) for $A \subseteq Loc$ iff they are i-equivalent for each $i \in A$. We define a* **successor relation** *so that $c \xrightarrow{a} d$ iff $d = c \uplus \{e\}$ with $l(e) = a$.*

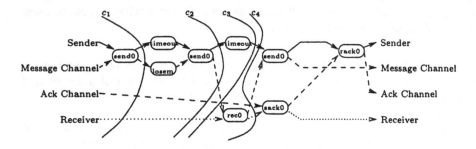

Fig. 2. One distributed run of the alternating bit protocol

Let us have a look at the example again: The configurations c_2 and c_3 in 2 are equivalent with respect to the receiver ($c_2 \equiv_{Receiver} c_3$) but not with respect to the sender ($c_2 \not\equiv_{Sender} c_3$): c_2 evolves into c_3 by performing an action which is local to the sender. From the receivers "point of view" the configuration did not change, the receiver does not "know" whether any of the other components has performed an action or not. Each configuration matches a certain state of the distributed program. At different states of the program, different properties might hold. We call these properties atomic propositions. Each location has its own atomic propositions. E.g.

we could take S_0 as an atomic proposition for "the sender is in S_0" or R_0 as "the receiver is in R_0". Looking at our example, these propositions are satisfied at the configuration c_3 but not at the configuration c_2 (R_0 is still true, but S_0 is not). Note that atomic propositions belonging to one location cannot be changed by actions which take place at other locations: The action which leads from c_2 to c_3 is local to the sender and thus does not have any effect to the proposition R_0, which is local to the receiver. We will now define an interpretation of a frame as a mapping from atomic propositions to the set of configurations:

Definition 2.4 *Let* $\widetilde{P} = (P_1, \ldots, P_K)$ *be a distributed set of local propositions. Here, P_i and P_j are disjoint for $i \neq j$ and P_i denotes the set of propositions affiliated with location i.*
An **interpretation of a frame** *is a mapping* $I : \widetilde{P} \longrightarrow 2^{C_F}$ *such that* $c \equiv_i c'$ *implies that* $c \in I(P)$ *iff* $c' \in I(P)$ *for all* $P \in P_i$, *i.e. the interpretation of propositions of location i depends only on i-events.*

Now we are finally able to define a distributed run as one execution of a distributed program together with an interpretation:

Definition 2.5 *A frame F together with an interpretation I is called a* **distributed run** $M = (F, I)$.

3 A logic for distributed programs

We will now define a logic for the specification of distributed program executions as given in the previous section. The logic, called $\nu TrTL$, is the revised version of a logic first given in [Nie95] and later in [Huh96].

Syntax and Semantics. Let Loc, $\widetilde{\Sigma} = (\Sigma_1, \ldots, \Sigma_K)$ and $\widetilde{P} = (P_1, \ldots, P_K)$ be defined as in section 2.

The propositions from the sets P_i will form a part of the atomic formulae of the logic. Similarly to these *propositional constants*, the meaning of which is given by the interpretation I of a distributed run (F, I), we also need *propositional variables*, written as $X, Y, Z, \ldots \in V$. Just as I gives a meaning to the propositions we furthermore need a *valuation function* $v : V \longrightarrow 2^{C_F}$, i.e. each variable stands for a set of configurations, which is given by v.

The key idea of the logic itself - and of the presentation given here - is that the formulae of the logic look at the configurations, i.e. the global state during a run, from a *local* point of view: Some formulae look at the state of the system from the point of view of a single location (e.g. in the case of local propositions), others may involve a joint look from several locations (e.g. at the very beginning of a run or after a joint action of these locations). This idea is reflected in the syntax by a *family* of sets of formulae Φ_A (looking from the point of view of $A \subseteq Loc$):

Definition 3.1 (Syntax) *The syntax of the logic $\nu TrTL$ consists of sets Φ_A of formulae, where $A \subset Loc$ denotes the type of the formulae in Φ_A, i.e. a set of locations (to which the formulae directly refer). We also write $type(\phi) = A$ for*

$\phi \in \Phi_A$. The set of all formulae is denoted by $\Phi := \bigcup_{A \subseteq Loc} \Phi_A$. The sets Φ_A are defined to be the least sets, such that:

$$\{\text{true}, \text{false}\} \cup \mathcal{V} \subseteq \Phi_\emptyset$$

$$P, \neg P \in \Phi_{\{i;\ P \in \mathcal{P}_i\}}$$

$$\phi \in \Phi_A, \psi \in \Phi_B \Rightarrow \phi \wedge \psi, \phi \vee \psi \in \Phi_{A \cup B}$$

$$i \in Loc(a), A \subseteq Loc(a), \phi \in \Phi_A \Rightarrow \langle a \rangle_i \phi, [a]_i \phi \in \Phi_{\{i\}}$$

$$\phi[X := \phi] \in \Phi_A \Rightarrow \mu X.\phi, \nu X.\phi \in \Phi_A$$

The operators μ and ν bind the variables. By $\phi[X := \psi]$ we denote the formula obtained by substituting all free occurrences of X in ϕ by ψ. A formula that does not contain any free variables is **closed**.

Note that for $A, B \subseteq Loc$ with $A \neq B$ we have $\Phi_A \cap \Phi_B = \emptyset$, i.e. every formula has a *unique type*. We will give some examples of νTrTL formulae later.

Further note that using sets Φ_A of formulae there is a subtle restriction on the the way formulae can be constructed: e.g. within a formula $[a]_i \phi$ the formula ϕ may only directly refer to locations from $Loc(a)$. Hence *changing the point of view* in a formula is only possible via a modality refering to a common action of the *old* and the *new* point of view. We will further comment this issue at the example section.

For convenience we only allow negation of atomic propositions. However the logic is closed under negation, because every operator has its *dual*, and negations can be drawn inside down to the atomic propositions. Let $not(\phi)$ denote the negation of ϕ.

The operator μ defines the least and ν the greatest fixpoint. Since both kinds of fixpoints are often treated equally, we use σ as wild card for both operators.

Definition 3.2 (Semantics of the logic νTrTL) *Let $(F.I)$ be a distributed run and v a valuation function. The semantics of a formula $\phi \in \Phi$ with respect to (F, I) and v is denoted by $[\phi]_v^{(F,I)}$. It is inductively defined by:*

- $[\text{true}]_v^{(F,I)} = C_F$, $[\text{false}]_v^{(F,I)} = \emptyset$ *(all configurations satisfy* **true**, *none* **false***)*
- $[P]_v^{(F,I)} = I(P)$, $[\neg P]_v^{(F,I)} = C_F \setminus [P]_v^{(F,I)}$, $[X]_v^{(F,I)} = v(X)$ *for $P \in \mathcal{P}$, $X \in \mathcal{V}$*
- $[\phi \wedge \psi]_v^{(F,I)} = [\phi]_v^{(F,I)} \cap [\psi]_v^{(F,I)}$, $[\phi \vee \psi]_v^{(F,I)} = [\phi]_v^{(F,I)} \cup [\psi]_v^{(F,I)}$
- $[\langle a \rangle_i \phi]_v^{(F,I)} = \{c \mid \exists r, c'$ *with* $(c \equiv_i c')$ *and* $c' \xrightarrow{a} r$ *and* $r \in [\phi]_v^{(F,I)}\}$
- $[[a]_i \phi]_v^{(F,I)} = \{c \mid \forall c', r$ *with* $(c \equiv_i c')$ *and* $c' \xrightarrow{a} r$ *implies* $r \in [\phi]_v^{(F,I)}\}$
- $[\nu X.\phi]_v^{(F,I)} = \bigcup\{A \mid A \subseteq [\phi]_{v[X:=A]}^{(F,I)}\}$, $[\mu X.\phi]_v^{(F,I)} = \bigcap\{A \mid [\phi]_{v[X:=A]}^{(F,I)} \subseteq A\}$
  where $v[X := A](X) = A$ and for $Y \neq X$ we have $v[X := A](Y) = v(Y)$.

A configuration $c \in C_F$ satisfies a formula ϕ iff $c \in [\phi]_v^{(F,I)}$. A distributed run (F, I) satisfies a formula ϕ iff the initial configuration does, i.e. iff $\emptyset \in [\phi]_v^{(F,I)}$. A distributed program satisfies a formula, iff all of its distributed runs do.

The following observation [Koz83] eases the formulation of our proof-system:

Proposition 1 (guarded formulae). *A formula is ψ **guarded** iff in each subformula of the form $\sigma X.\phi$ of ψ all free occurrences of the fixpoint variable X lie in the scope of a modality $\langle a \rangle_i \phi$ or $[a]_i \phi$. Every formula is equivalent to a guarded formula.*

Sample Properties. In this subsection we want to illustrate the way the logic can express properties of distributed runs. Consider the example run in figure 2.

In section 2 we have already shown at which configurations propositions (as simpliest formulae) hold, so let us now look at a formula with a single modality: $\langle rec0 \rangle_{Receiver} R_0$. Informally we should read this formula as "*From the point of view of the receiver the next action to occur is rec0 and afterwards it will be in state R_0*". Formally this formula holds for instance at the configuration c_1, because that configuration (where the receiver has not even started) is equivalent to c_3 from the receiver's point of view, and in c_3 the action rec_0 is enabled leading to configuration c_4, where (according to the interpretation) R_0 holds. Note that the formula does not say, that rec_0 is immediately enabled in c_1, but will rather be enabled by some preparatory events (*timeout, losem, send0*) of the other locations.

Next we will build up a more complex formula to illustrate the use of fixpoints in formulae. We want to formalise the following property (of all runs of the ABP):

AlwsAckMeansRcvd: "*Whenever the sender participates in a send0-event and eventually (after finitely many stimeout and repeated send0 actions) receives a rack0, then (after the first send0) the message channel will only finitely often lose a message (and receive another send0-request) before transmitting rec0 to the receiver. The receiver then will enter the R_0 state (indicating the reception of a 0-tagged message)*".

We formulate the property using several abbreviations for subformulae (for better readability we use $\phi \rightarrow \psi$ as an abbreviation for $not(\phi) \vee \psi$, S for *Sender* and MC for *MessageChannel*):

$$AlwsAckMeansRcvd = \nu X. \left(AckMeansRcvd \wedge \left(\bigwedge_{a \in \Sigma_s} [a]_S X \right) \right)$$

$$AckMeansRcvd = (\langle send0 \rangle_S \textbf{true} \wedge EvtllyRcvsAck) \rightarrow ChanWillTransmt$$

$$EvtllyRcvsAck = \mu Y.\langle rack0 \rangle_S \textbf{true} \vee \langle stimeout \rangle_S \langle send0 \rangle_S Y \vee \langle rack1 \rangle_S Y$$

$$ChanWillTransmt = \mu Z.\langle rec0 \rangle_{MC} R_0 \vee \langle losem \rangle_{MC} \langle send0 \rangle_{MC} Z.$$

Note how recursion is used in the above formulae: to formulate "always" we use a greatest fixpoint, which corresponds to an infinite unfolding of the formula. In the other cases we deal with *finite recursion*, i.e. we use a least fixpoint to allow an arbitrary but finite expansion of the formula: e.g. we may only cycle finitely often through the *losem − send0*-cycle in ChanWillTransmt until finally we have to commit to the action $rec0$.

In the formulation of the property we walk along a *causal chain* (i.e. a maximal totally ordered set) of events, which is typical for this logic. Also note that the logic does not directly allow us to say things involving changes of the point of view except via common actions (in the example via *send0* from Sender to Message Channel, and via *rec0* from Message Channel to Receiver). For instance we cannot directly say "when the sender receives a *rack0* the receiver is in state R_0".

This does not mean a general restriction of the expressiveness of the logic for the specification of properties of *complete runs*, only it may not be possible to write down "global" invariants (for example) in a compositional way.

4 A tableaux system

Let us now define a tableau proof system for the validity of formulae of νTrTL. Let Γ be a finitary set of formulae. We will call $\Gamma \vdash$ a *sequent*. Given a distributed run (F, I) and a configuration $c \in \mathcal{C}_F$, the configuration c satisfies a sequent $\Gamma \vdash$ iff a formula $\gamma \in \Gamma$ exists such that $c \notin [\gamma]_v^{(F,I)}$.

For these sequents we want to define a tableau proof system, which accepts a sequent iff the sequent is valid and rejects it otherwise. Here validity is defined as follows:

Definition 4.1 *A sequent $\Gamma \vdash$ is valid iff $\forall (F, I), v : \bigcap_{\gamma \in \Gamma} [\gamma]_v^{(F,I)} = \emptyset$*

Definition 4.2 (rules) *In the following let $\Gamma_{\langle\rangle}$ ($\Gamma_{[\,]}$) $\subseteq \Gamma$ denote the set of formulae in Γ of the form $\langle a\rangle_i \phi$ ($[a]_i \phi$) $\in \Gamma$, and let Γ_P be the set of propositions in Γ. Let T be the following set of tableau-rules, which are divided into three groups: axioms, logical rules and a modal rule.*

Axioms and logical rules

$$\frac{\Gamma, P, \neg P \vdash}{} \quad (1) \qquad\qquad \frac{\Gamma, \text{false} \vdash}{} \quad (2) \qquad\qquad \frac{\Gamma, \text{true} \vdash}{\Gamma \vdash} \quad (3)$$

$$\frac{\Gamma, \phi \wedge \psi \vdash}{\Gamma, \phi, \psi \vdash} \quad (4) \qquad\qquad \frac{\Gamma, \phi \vee \psi \vdash}{\Gamma, \phi \vdash \quad \Gamma, \psi \vdash} \quad (5) \qquad\qquad \frac{\sigma X.\alpha, \Gamma \vdash}{\alpha[X := \sigma X.\alpha], \Gamma \vdash} \quad (6)$$

Modalities

$$\frac{\Gamma_P, \Gamma_{\langle\rangle}, \Gamma_{[\,]} \vdash}{\Gamma_{a_1} \vdash, \; \Gamma_{a_2} \vdash, \ldots \Gamma_{a_n} \vdash} \quad (7)$$

In the modal rule $\Gamma_{\langle\rangle}$ must be non-empty and Σ' and Γ_a are defined as follows:

- $a_i \in \Sigma', \Sigma' := \{a \in \Sigma | \exists \phi \in \Gamma_{[\,]}, \Gamma_{\langle\rangle} \text{ s.t. } type(\phi) \in Loc(a)\}$
- $[a]_i\phi \in \Gamma_{[\,]} \Rightarrow \phi \in \Gamma_a, \langle a\rangle_i \phi \in \Gamma_{\langle\rangle} \Rightarrow \phi \in \Gamma_a$
- $[b]_i\phi \in \Gamma_{[\,]}, i \in Loc(a) \Rightarrow \text{true} \in \Gamma_a, \langle b\rangle_i\phi \in \Gamma_{\langle\rangle}, i \in Loc(a) \Rightarrow \text{false} \in \Gamma_a$
- $[b]_i\phi \in \Gamma_{[\,]}, i \notin Loc(a) \Rightarrow [b]_i\phi \in \Gamma_a, \langle b\rangle_i\phi \in \Gamma_{\langle\rangle}, i \notin Loc(a) \Rightarrow \langle b\rangle_i\phi \in \Gamma_a$
- $P \in \Gamma_P, P \in \mathcal{P}_i, i \notin Loc(a) \Rightarrow P \in \Gamma_a$

The axioms and logical rules are standard. The goal of such a rule is valid, iff all the subgoals are valid. In rule 6 we unwind a fixpoint; only the use of this rule can lead to infinite tableau-paths.

Let us now have a more detailed look at the rule 7. While the rules 1-6 are local to one configuration (a configuration c satisfies the goal of a rule, if it satisfies at least one of the subgoals), the modal rule refers to a step from one configuration c to the next configuration c' by adding one event. This explains, why the modal rules may only be applied after none of the other (local) rules is applicable any more. In the case of $\Gamma_{\langle\rangle} = \emptyset$ every configuration in which no more actions can occur can satisfy Γ (depending on the interpretation) Thus we have to make sure, that $\Gamma_{\langle\rangle} \neq \emptyset$. Note that the fixpoints are guarded. Thus, always after applying a finite number of rules 1-5, a modal rule has to be applied. Just as the or-rule 5 investigates the two possibilities to satisfy a disjunction, the modal rule does a case-analysis depending on which actions could be performed next in an arbitrary configuration of an arbitrary run. According to this choice and the semantics of the logic each formula in the goal is linked to a corresponding formula in each subgoal.

In a modal step a with $i \in Loc(a)$, propositions $P \in \mathcal{P}_i$ will be deleted because their interpretation in c' can become different from the one in c. Propositions $P \notin \mathcal{P}_i$ are not concerned by such a step and will thus not be deleted.

Definition 4.3 (tableaux) *Given a set of formulae Γ. A tableau $\mathcal{T}$ for Γ is a labelled tree $< K, L >$, where K is a tree and L is the labelling function, such that*
- the root of K is labelled with $\Gamma \vdash$,
- if $L(n)$ is a tableau-axiom, then n is a leaf of K
- if $L(n)$ is not an axiom, then the children of n in K are created and labelled according to the tableau-rules. $L(n)$ is the goal and the labels of the children the subgoals of a tableau-rule.

An acceptance condition is needed such that a tableau is accepted iff the sequent is valid. In the case of finite tableaux, we can accept, if all leaves are labelled with axioms. Hence the proof-system is already sound and complete for logic νTrTL without fixpoints. Because of the induction-rule 6 which unwinds fixpoints, we can also create infinite paths. These paths require a different acceptance condition. For this we have to face two problems: One is the satisfaction of fixpoint-formulae, the second is a particular fairness problem.

To solve the first problem we need the following definition which allows us to observe a particular formula over a path:

Definition 4.4 (trace, μ-trace) *Let $\mathcal{T}$ be a tableau for $\Gamma \vdash$ and $\sigma = v_1, v_2, \ldots$ be an infinite path in the tableau, i.e. v_{i+1} is a child of v_i. A trace on the path σ is each sequence of formulae $(\alpha_1, \alpha_2, \ldots, \alpha_n)$ such that $\alpha_j \in L(v_k)$ and α_{j+1} is either*

1. α_j, if the formula α_j is not reduced by the rule applied in node v_k or
2. if α_j was reduced, then α_{j+1} is one of the formulae produced out of α_j.

A least fixpoint formula $\mu X.\phi = \alpha_i$ is regenerated from α_i to $\alpha_j, i < j$ on a trace, if $\mu X.\phi = \alpha_i$ derives $\mu X.\phi = \alpha_j$ in such a way, that $\mu X.\phi$ is a subformula of each $\alpha_k, i < k \leq j$. We call a trace on which a least fixpoint formula is regenerated infinitely often, a μ-trace.

Considering the second problem, we define a particular notion of fairness for paths. Paths in which a formula $\langle a \rangle_i \psi$ is never evaluated are called *unfair:*

Definition 4.5 (fair tableau paths) *A tableau-path $\sigma = (v_0, v_1, v_2, \ldots.)$ is called fair, iff for each i and for each $\langle a \rangle_i \phi \in v_i$ there exists a node $v_j \in \sigma$ with $j \geq i$, such that the modal rule is applied to v_j and there exists a b-child of v_j with $b \in \Sigma_i$. A path, which is not fair, is called* unfair.

Definition 4.6 (acceptance-condition) *Given a tableau $\mathcal{T}$ for a sequent $\Gamma \vdash$. The tableau will accept, iff every leaf is labelled with an axiom, and every infinite tableau-path is either unfair or contains an infinite μ-trace.*

The requirement for finite paths is easy to see, for infinite paths we will give a detailed explanation a bit further on in this paper. The given rules together with the acceptance condition form a sound and complete proof-system for the logic νTrTL. We will prove this in the next section.

Theorem 2 (soundness and completeness). *Let Γ be a set of formulae and let $\mathcal{T}$ be a tableau for the sequent $\Gamma \vdash$. The tableau $\mathcal{T}$ accepts $\Gamma \vdash$ iff $\Gamma \vdash$ is valid.*

5 Soundness and Completeness

We will now give the outline of the soundness and completeness proofs of the proof-system given above (for more details see [Spr96]). In the "standard" way, the soundness and completeness proofs are done by induction over tableau paths: a leaf is valid iff it is an axiom, and an inner goal is valid, iff all its subgoals are valid.

In the case of the proof system without the rules for fixpoints (i.e. a proof system for a reduced logic without the fixpoint operators) we could have applied the induction method for the soundness and completeness proof. But due to the fixpoint rules we have to deal with possibly infinite paths (which might be accepted as well) in the tableau. Thus this induction method cannot be used here.

Instead we will give a global proof based on the correspondence of tableau paths and distributed runs. As a formal framework we use an established tool for fixpoint logics, the Streett/Emerson theorem [ES89]: it gives a different characterisation for the logical satisfaction of fixpoints in the propositional μ-calculus, and it can be adapted to several fixpoint-frameworks. The idea is first to weaken the semantics of μ and ν to arbitrary (not least or greatest) fixpoints, represented in derivation graphs (vaguely corresponding to the interleaved representation of distributed runs) and to then regain the proper semantics by a separate criterion.

We will now use this characterisation to prove soundness and completeness.

Definition 5.1 (derivation graph) *A derivation graph $DG((F, I), d_0, \Gamma)$ for a distributed run (F, I) with the configuration d_0 and a set of formulae Γ is a tuple $(V, \rightsquigarrow)$, with $V \subseteq \Phi \times \Phi$, $\rightsquigarrow \subseteq V \times V$, where $(V, \rightsquigarrow)$ is minimal, such that the following holds (we implicitly assume $(c, \phi), (c', \phi') \in V$ when we write $(c, \phi) \rightsquigarrow (c', \phi')$):*

- $(d_0, \gamma) \in V$ $\forall \gamma \in \Gamma$
- *For $(c, P) \in V, P \in \mathcal{P}$ and $c \notin I(P)$ we get $(c, P) \rightsquigarrow (c, \mathbf{false})$*
- *For $(c, \neg P) \in V, P \in \mathcal{P}$ and $c \in I(P)$ we get $(c, \neg P) \rightsquigarrow (c, \mathbf{false})$*
- *For $(c, \phi \vee \psi) \in V$ we either get $(c, \phi \vee \psi) \rightsquigarrow (c, \phi)$ or $(c, \phi \vee \psi) \rightsquigarrow (c, \psi)$*
- *For $(c, \phi \wedge \psi) \in V$ we get $(c, \phi \wedge \psi) \rightsquigarrow (c, \phi)$ and $(c, \phi \wedge \psi) \rightsquigarrow (c, \psi)$*
- *For $(c, [a]_i \phi) \in V, c \xrightarrow{a} c'$ we get $(c, [a]_i \phi) \rightsquigarrow (c', \phi)$*
- *For $(c, [b]_i \phi) \in V, c \xrightarrow{a} c', a \in \Sigma_i$ we get $(c, [a]_i \phi) \rightsquigarrow (c', \mathbf{true})$*
- *For $(c, [b]_i \phi) \in V, c \xrightarrow{a} c', a \notin \Sigma_i$ we get $(c, [a]_i \phi) \rightsquigarrow (c', [a]_i \phi)$*
- *For $(c, \langle a \rangle_i \phi) \in V, c \xrightarrow{a} c'$ we get $(c, \langle a \rangle_i \phi) \rightsquigarrow (c', \phi)$*
- *For $(c, \langle b \rangle_i \phi) \in V, c \xrightarrow{a} c', a \in \Sigma_i$ we get $(c, \langle b \rangle_i \phi) \rightsquigarrow (c', \mathbf{false})$*
- *For $(c, \langle b \rangle_i \phi) \in V, c \xrightarrow{a} c', a \notin \Sigma_i$ we get $(c, \langle b \rangle_i \phi) \rightsquigarrow (c', \langle b \rangle_i \phi)$*
- *For $(c, \sigma X.\alpha(X)) \in V$ we get $(c, \sigma X.\alpha(X)) \rightsquigarrow (c, \alpha(\sigma X.\alpha(X)))$*

Definition 5.2 (correct) *A derivation graph is called* correct, *iff*

1. *it does not contain a node $(c, \mathbf{false}) \in DG((F, I), d_0, \Gamma)$,*
2. *for each node $(c, \langle a \rangle_i \phi) \in DG((F, I), d_0, \Gamma)$ we have $(c, \langle a \rangle_i \phi) \rightsquigarrow^* (d, \langle a \rangle_i \phi) \rightsquigarrow (d', \phi')$ and $(d', \phi') \in DG((F, I), d_0, \Gamma)$ with $\langle a \rangle_i \phi \neq \phi'$,*
3. *it does not contain any path with an infinite regeneration of a least fixpoint.*

This correctness definition of derivation graphs needs some further explanations. Item 1 makes sure that all leaves of the derivation graph are satisfied. Since the conditions on derivation graphs are necessarily local (single step), but the semantics of $\langle\ \rangle_i$ refers to global (multi step) jumps, we need the additional condition 2 for resolving $\langle\ \rangle_i$ formulae. Condition 3 is the heart of the original Streett/Emerson theorem ensuring the proper semantics of least fixpoints.

Thus we recast the Streett/Emerson theorem to our framework as follows:

Theorem 3 (Streett/Emerson [ES89]). *A configuration d_0 of a distributed run (F, I) satisfies a formula ϕ iff there exists a correct derivation graph for $((F, I), d_0, \phi)$*

The proof of theorem 2 will be divided into two parts. We will show both directions separately. First we will show that if a tableau $\Gamma \vdash$ accepts a sequent, the sequent is valid. Then, in a second part, we will show that a sequent rejected by the tableau is not valid. Note that for every sequent there exists either a rejecting or an accepting tableau. For finite tableaux the soundness and completeness can easily be shown by induction over the length of the tree. Showing the soundness and completeness for infinite tableaux requires some more work.

Lemma 4. *For $\Gamma \vdash$ not valid there exist rejected paths in any tableau for $\Gamma \vdash$.*

Proof (sketch): Let Γ_0 be a set of formulae and let (F, I) be a distributed run. The configuration d_0 of (F, I) satisfies all formulae in Γ_0.
Now we choose a fair execution of the distributed run. e.g. using round robin. With theorem 3 we know, that there exists a correct derivation graph for $(F, I), d_0, \Gamma_0$.
We then have to show that the tableau $\mathcal{T}$ built for Γ_0 either contains a leaf which is not labelled with an axiom or it has at least one infinite *fair* path without a $\mu - trace$. With the help of the derivation graph we will inductively identify either this leaf or the infinite path $\pi = v_0, v_1, v_2, \ldots$ in the tableau. During this process we assume that each v_i corresponds to a configuration $c_i \in \mathcal{C}_F$ which is given as interleaving in the sequence of actions taken in modal rules along the path from the root to v_i. Further more we assume that for each $\gamma \in \Gamma(v_i)$, (c_i, γ) belongs to $DG((F, I), d_0, \Gamma_0)$. This invariant is easily seen to carry over for rule applications with only one subgoal v_{i+1} (the rules 1, 4, 6). For rules with several subgoals we choose the successor v_{i+1} according to the derivation graph. Let $\Gamma(v_i) = \Gamma_i, \phi \vee \psi$ and let the or-rule be applied next. Depending on $\Gamma(v_{i+1}) = \Gamma_i, \phi$ or $\Gamma(v_{i+1}) = \Gamma_i, \psi$ we either choose the left or the right successor of v_i.
Similarly, we proceed for the modal rule. If this process ends at a leaf v of $\mathcal{T}$, it cannot be an axiom (with $P, \neg P \in \Gamma(v)$) because our invariant would obviously lead to a violation of condition 1 of definition 5.2. On the other hand the path π would be fair by construction, and from the invariant and because of condition 3 of 5.2 the path would not contain an infinite μ-trace. Hence, the tableau will not accept $\Gamma \vdash$.

Lemma 5. *Let Γ_0 be a finite set of formulae and let $\mathcal{T}$ be a tableau starting from Γ_0. If $\mathcal{T}$ contains a rejected path then $\Gamma_0 \vdash$ is not valid and there exists a distributed run (F, I) such that the initial configuration d_0 of (F, I) satisfies $\Gamma_0 \vdash$.*

Proof (sketch): The proof of this lemma requires a bit more effort. Let Γ_0 be fixed and let $\pi = v_1, v_2, \ldots$ be a rejected path in a tableau $\mathcal{T}$ for Γ_0. The idea is to construct a distributed run (F, I) with a correct derivation graph $DG((F, I), d_0, \Gamma_0)$. First, we have to construct a distributed run (F, I) such that its initial configuration satisfies $\Gamma_0 \vdash$.

- From π we first construct the frame F by considering the sequence of actions taken in applications of the modal rule in π as one interleaving of the distributed run. This interleaving also corresponds to a sequence of configurations $\theta = c_1, c_2, c_3, \ldots$.

- To construct the interpretation of configurations occurring in θ it is sufficient to set all propositions occurring just before the next application of a modal rule to **true**. Unfortunately, not all configurations $c \in C_F$ are met along the path. But for each $i \in Loc$ there exists at least one i-equivalent configuration $c' \in \theta$ from which the interpretation of propositions $P \notin \mathcal{P}_i$ can be adopted consistently.

The main part now is the construction of the derivation graph.

- For configurations $c \in \theta$ and formulae ϕ, such that (c, ϕ) already belongs to the derivation graph, the definition of successor nodes is basically the reverse of the identification of the successor goal in the soundness proof. For other configurations $c' \notin \theta$ a similar method as for the interpretation construction is used: If the type of ϕ is A, it is possible to show that there exists an A-equivalent configuration $c \in \theta$.

- A problem not mentioned so far occurs with the definition of successors of nodes $(c, \phi_1 \vee \phi_2)$: While there can only exist one successor $(c, \phi_l), l \in \{1, 2\}$ in the derivation graph, it can happen that the or-rule 5 is applied differently on several occurrences of $\phi_1 \vee \phi_2$ on the section of the tableau path relevant to (c, ϕ_l). However, it is easy to construct from any rejected path in the tableau another rejected path in the same tableau, where disjunctions $\phi_1 \vee \phi_2$ are always resolved in the same way between two applications of the modal rule. Hence, in the above construction we can assume to have such a path.

- Finally we observe that the constructed derivation graph is correct: condition 1 in definition 5.2 is satisfied as the derivation graph and the interpretation are constructed consistently. Condition 2 is satisfied because for a finite path we get a finite derivation graph without $\langle \rangle_i$ at the leaves, whereas for an infinite path the condition follows from the fairness of π. Condition 3 is inherited from the absence of μ-traces in the path.

6 Automated proof search

The taming of the modal rule. Although the logic is defined on traces, the proof system works in an interleaving fashion (in particular the modal rule). Thus the question arises whether we can apply partial order reductions as known from model checking ([Pel93, GW91]) to the proof-search. Obviously each tableau path represents one interleaving of a distributed run. Thus equivalent interleavings should either both accept or both reject the path. Consider two actions $a, b \in \Sigma, aIb$, and a configuration c. As a and b are independent, the order in which a and b are executed should not make any difference, since both ways lead to the same configuration c' in a distributed run. Thus, to investigate, whether c' satisfies an arbitraty formula ϕ, it is not necessary to follow both paths leading to the same configuration. In model checking the so called sleep-set algorithm ([GW91]) is developed to keep track of this problem. There, a sleep-set belonging to a node v in the state-graph of a system specifies all directions, which do not have to be developed from that node. An action a in the sleep-set of a node v means, that the a-path outgoing from v need not be investigated. Instead of actually adding sleep-sets to tableau-nodes, we can naturally (i.e. logically) incorporate an equivalent of the sleep-set method into the modal rule. Consider the modal rule as given in rule 7. In this rule we do a case analysis between all actions in Σ: a_1 or a_2 or ... or a_n. By using the modal rule like this we produce some redundant paths in the tableau, as each interleaving of a trace is represented in the tableau. But now we can change this case analysis to an exclusive one by defining a total order on actions $a_1 < a_2 < ... < a_n$. The case analysis will now be: a_1 or not a_1 but a_2 or ... or not $\Sigma \setminus \{a_n\}$ but a_n. This order needs to be fixed for one application of the modal rule. Due to a particular fairness problem which we will discuss below we have to change the order in a fair way (e.g. Round Robin) to make sure, that we do not produce unfair paths.

We can logically code this in the modal rule by the following modification (where Γ_{a_i}, Σ' are defined as before): Let $A_i = \{a \in \Sigma' | (a, a_i) \in \mathcal{I}, a < a_i\}$. And let $[a]\textbf{false}$ abbreviate $\bigvee_{i \in Loc(a)} [a]_i\textbf{false}$ and $[A]\textbf{false}$ abbreviate $\bigwedge_{a \in A} [a]\textbf{false}$. Then the reduced modal rule 8 is defined as:

$$\frac{\Gamma_P, \Gamma_{\langle \rangle}, \Gamma_{[]} \vdash}{\Gamma_{a_1} \vdash \quad [A_2]\textbf{false}, \Gamma_{a_2} \vdash \quad ... \quad [A_n]\textbf{false}, \Gamma_{a_n} \vdash} \tag{8}$$

This modification turns out to be the logical encoding of sleep-sets into the modal rule. Consider a modal step with action a_j. Any action $a_i < a_j, (a_i, a_j) \in \mathcal{I}$, which was not enabled before the execution of a_j is not enabled after the modal step either. Thus we add action a_i to the set A_j which can be viewed as the "sleep-set".

The method we use here is not only the logical encoding of the sleep-sets which are a partial order reduction method known from model-checking. The idea, to make the branches mutually exclusive, is similar to the one shown in [DM94]. There, a tableau-system (called **KE**) with an analytic *cut*-rule (**PB**) is given, where the *cut*-rule does not contradict the subformula principle. The idea of that paper is to separate the branching from the logical rules, such that the rule for disjunction is linear and the branching takes place only in the cut-rule(**PB**):

$$\frac{}{A \mid \neg A}(PB) \qquad \frac{A \vee B, \neg A}{B}(E \vee 1) \qquad \frac{A \vee B, \neg B}{A}(E \vee 2)$$

Our reduced modal rule could also be seen as the combination of a built-in analytic cut-rule **PB** (only there the case analysis is done on subformulae and not on actions, as we do it here) and a modified modal rule.

Let $\langle b \rangle \textbf{true}$ be an abbreviation for $\bigwedge_{i \in Loc(b)} \langle b \rangle_i \textbf{true}$ and $[b]\textbf{false}$ an abbreviation for $\bigvee_{i \in Loc(b)} [b]_i \textbf{false}$. Then in our system the analogue of the cut-rule **PB** would be for $b \in \Sigma'$: "*do b or (do) not do b*" (rule 9). We can then immediately apply the modified modal rule 10 to the left subgoal of rule 9 (modified in the sense, that we have separated the case-analysis from the modal rule such that we now only have got one subgoal):

$$\frac{\Gamma \vdash}{\langle b\rangle\mathbf{true}, \Gamma \vdash \quad [b]\mathbf{false}, \Gamma \vdash} \quad (9) \qquad\qquad \frac{\langle b\rangle\mathbf{true}, \Gamma_P, \Gamma_{\langle\,\rangle}, \Gamma_{[\,]} \vdash}{\Gamma_b \vdash} \quad (10)$$

Here Γ_b is constructed as defined in 7. To the right subgoal we would apply rule 9 again for the "next" action in Σ'. The following figure will illustrate how our *new* modal rule 8 can be seen as a combination of the rules 9 and 10.

$$\Gamma \vdash$$

$$\frac{}{\langle a_1\rangle\mathbf{true}, \Gamma \vdash \qquad\qquad\qquad\qquad\qquad [a_1]\mathbf{false}, \Gamma \vdash}$$

$$\frac{}{\Gamma_{a_1} \vdash} \qquad \frac{}{[a_1]\mathbf{false}, \langle a_2\rangle\mathbf{true}, \Gamma \vdash \qquad [a_1]\mathbf{false}, [a_2]\mathbf{false}, \Gamma \vdash}$$

$$\frac{}{[A_2]\mathbf{false}, \Gamma_{a_2} \vdash} \qquad\qquad \vdots$$

$$\vdots \qquad\qquad\qquad\qquad \frac{}{[A_n]\mathbf{false}, \Gamma_{a_n} \vdash}$$

In rule 9 we only create a subgoal for an action $a \in \Sigma'$.

A particular fairness problem occurs on infinite paths: we must find a fair method for finding a representative for an infinite trace. Consider e.g. an infinite trace with infinitely many a's but only one b. Let a and b be independent. With $a < b$ the above rule would put a asleep forever on every b-path such that there is no representation of the trace $[a^\omega b]$ We can take care of this problem by redefining the total ordering of actions in each modal step with a fair method (e.g. Round Robin).

It seems that this reduction will often lead to significantly smaller proofs (with the usual worst case exceptions). However, we have not practically checked this.

Handling infinite tableaux. Our tableau system often requires infinite proofs, so one might ask for practical use of the system. With the help of tree automata (see [Tho90] for an introduction) we can give a decision method for the existence of accepting tableaux. More precisely we can construct tree automata running on tableaux by regarding them as trees with sequents as node labels[1] in the obvious way. The construction is sketched in the appendix, for further details see [Spr96].

Theorem 6. *For each sequent $\Gamma \vdash$ we can canonically construct a Rabin-tree-automaton $\mathcal{A}_T$ recognising exactly the accepting tableaux for $\Gamma \vdash$.*

These automata give us a finitary representation of proofs: instead of constructing tableaux with infinite paths we now represent the proof as a tree with backloops.

Definition 6.1 (regular tableaux) *An infinite tree $\mathcal{T}$ is said to be regular if there are only finitely distinct non-isomorphic subtrees $\mathcal{T}'$ in $\mathcal{T}$. An infinite tableau is called regular, iff it is a regular tree.*

Obviously regular trees and tableaux can be finitely represented. Since the emptiness problem for tree automata is decidable (see [Tho90]) and moreover in the nonempty case the decision procedure returns a regular tree, we can deduce the following:

Corollary 7. *The set of sequents $\Gamma \vdash$ with an accepting tableau is decidable. If there exists any accepting tableau for $\Gamma \vdash$, then there also exists a regular one.*

---

[1] Note that for an initial sequent $\Gamma_0 \vdash$ only a finite and easy to determine set of sequents $\Gamma \vdash$ can occur in any tableau, the Fischer-Ladner closure.

305

Acknowledgemets

We thank Michaela Huhn, R. Ramanujam and P.S. Thiagarajan for fruitful discussions. Rajeev Goré made very valuable comments to the presentation of the paper and has pointed us to the connection with tamed cut. Part of this work was financially supported by the Human Capital and Mobility Cooperation Network "EXPRESS" (Expressivity of Languages for Concurrency).

References

[Dam92] Mads Dam. Fixpoints of Büchi automata. In *International Conference on the Foundations of Software Technology and Theoretical Computer Science (FST&TCS)*, Lecture Notes in Computer Science, pages 39–50, 1992.

[DM94] Marcello D'Agostino and Marco Mondadori. The taming of the cut. Classical refutations with analytic cut. *Logic and Computation*, 4(3):285–319, 1994.

[ES89] E. Allen Emerson and Robert S. Streett. An automata theoretic decision procedure for the propositional mu-calculus. *Information and Computation*, 81:249–264, 1989.

[GW91] Patrice Godefroid and Pierre Wolper. A partial approach to model checking. In *IEEE Symposium on Logic in Computer Science*, volume 6, pages 406–415, 1991.

[Huh96] Michaela Huhn. Action refinement and property inheritance in systems of sequential agents. In U. Montanari and V. Sassone, editors, *International Conference on Concurrency Theory (CONCUR)*, volume 1119 of *Lecture Notes in Computer Science*, pages 639–654. Springer-Verlag, 1996.

[Koz83] Dexter Kozen. Results on the propositional μ-calculus. *Theoretical Computer Science*, 27:333–354, 1983.

[Maz95] Antoni Mazurkiewicz. Introduction to trace theory. In Volker Diekert and Grzegorz Rozenberg, editors, *The Book of Traces*, chapter 1, pages 1–42. World Scientific, 1995.

[Mil89] Robin Milner. *Communications and Concurrency*. Prentice-Hall, 1989.

[MP92] Manna and Pnueli. *The Temporal Logic of Reactive and Concurrent Systems*. Springer-Verlag, 1992.

[Nie95] Peter Niebert. A ν-calculus with local views for systems of sequential agents. In *MFCS*, volume 969 of *Lecture Notes in Computer Science*, 1995.

[Pel93] Doron Peled. All from one, one for all: on model checking using representatives. In *International Conference on Computer Aided Verification (CAV)*, Lecture Notes in Computer Science, 1993.

[Spr96] B. Sprick. Ein Beweissystem für die modale Tracelogik ν-TrTl und seine Optimierung durch Halbordnungsreduktionen. Master's thesis, Universität Hildesheim, 1996.

[Thi94] P.S. Thiagarajan. A trace based extension of Linear Time Temporal Logic. In *IEEE Symposium on Logic in Computer Science (LICS)*, volume 9, 1994.

[Tho90] Wolfgang Thomas. Automata on infinite objects. In J. van Leeuwen, editor, *Handbook of Theoretical Computer Science*, volume B, chapter 4, pages 133–191. Elsevier Science Publishers B.V., 1990.

[Var88] Moshe Y. Vardi. A temporal fixpoint calculus. In *ACM Symposium on Principles of Programming Languages*, pages 250–259, 1988.

[Wal95] Igor Walukiewicz. A complete deductive system for the μ-calculus. BRICS Report Series RS-95-6, Danish Center for Basic Research in Computer Science (BRICS), 1995.

A Construction of automata recognising tableaux

Here we give a brief description of a Rabin tree automaton which for input $\Gamma \vdash$ recognises exactly the accepting tableaux for $\Gamma \vdash$. The required automaton is a product of a nondeterministic tree automaton $\mathcal{A}_{tree}$ recognising tableaux and a deterministic Rabin-word-automaton $\mathcal{A}_L$ recognising accepted tableau-paths.

Tree-automaton recognising tableaux: Its states are sequents $\Gamma \vdash \in Seq(\Gamma_0)$ and its transition relation is defined by the tableau-rules. $\mathcal{A}_{tree}$ nondeterministically "guesses" a tableau and checks that the input tableau matches the guess. Any tree meeting the transition relation is accepted by the automaton.

Word-automaton recognising accepting tableau paths: This automaton is constructed as the union of three sub-automata which all run on tableau-paths: $\mathcal{A}_{fin}$ accepts finite paths ending with an axiom, $\mathcal{A}_{unfair}$ recognises unfair paths, and $\mathcal{A}_\mu$ accepts exactly all paths with a $\mu - trace$. As the construction of $\mathcal{A}_{fin}$ is trivial, we only describe the construction of $\mathcal{A}_{unfair}$ and $\mathcal{A}_\mu$ and show how to construct their union and the product with the tree automaton.

Automaton recognising unfair paths: For $\mathcal{A}_{unfair}$ we need two states q_{in_i} and q_i for each location $i \in Loc$. In q_{in_i} we check, if there exists a formula $\langle \, \rangle_i$ in the next input sequent Γ. If it does the automaton goes over to q_i where it stays until the next action at location i is performed and then it goes over to $q_{in_{i+1}}$. Otherwise it directly goes over to $q_{in_{i+1}}$. This automaton accepts, if it meets q_i infinitely often while it meets q_{in_i} only finitely many times for any i as there is no more action at location i though $\langle \, \rangle_i$ is still in the input sequent.

Automaton recognising μ-traces: $\mathcal{A}_\mu$ checks for the existence of a μ-trace. Given here as a nondeterministic Büchi-automaton $\mathcal{A}'_\mu$ it can clearly be determinised with Safra's determinization construction. $\mathcal{A}'_\mu$ "guesses" one formula from each sequent along the input sequence so that the sequence of chosen formulae forms a trace and ensures the trace being a μ-trace.

Safras's construction can be used for determization or, as shown in [Wal95], a deterministic (Rabin) automaton can be constructed directly. Both ways lead to a deterministic Rabin-automaton for the acceptance of μ-traces.

Union of $\mathcal{A}_{fin}$, $\mathcal{A}_{unfair}$, and $\mathcal{A}_\mu$: The states of this union-automaton $(\mathcal{A}_L)$ are the products of the sub-automata with the tuple of all three initial states as new initial state. A state (q_1, q_2, q_3) goes over to (q'_1, q'_2, q'_3) iff each state q_i in the sub automaton goes over to q'_i with the same input. The Rabin-acceptance condition of $\mathcal{A}_L$ can be defined as a product of all three sub-acceptance conditions.

Product-automaton recognising accepting tableaux: The idea is to let $\mathcal{A}_L$ run simultaneously on all paths of the tree. Thus we first convert the deterministic word-automaton $\mathcal{A}_L$ into a tree-automaton $\mathcal{A}_{tree(L)}$. $\mathcal{A}_{tree(L)}$ is then running on trees with each path in the tree being a sequence of sequents and accepts, iff all paths of the tree are accepted by $\mathcal{A}_L$. Finally we compose the product of the tree automaton $\mathcal{A}_{tree(L)}$ and the tree automaton $\mathcal{A}_{tree}$, to obtain the intersection of both tree languages. The states of this (final) product automaton are again sequents $\Gamma \vdash \in Seq(\Gamma_0)$. The acceptance condition is the same as for $\mathcal{A}_{tree(L)}$ and the intersection of the tranistion relations of both automata gives the new transition relation.

This tree automaton finally gives us a finitary representation of a proof and thus a basis for the construction of a finitary Hilbert-style proof system as in [Wal95].

ileanTAP: An Intuitionistic Theorem Prover

Jens Otten*

Fachgebiet Intellektik, Fachbereich Informatik
Technische Hochschule Darmstadt
Alexanderstr. 10, 64283 Darmstadt, Germany
`jeotten@informatik.th-darmstadt.de`

Abstract. We present a Prolog program that implements a sound and complete theorem prover for first-order intuitionistic logic. It is based on free-variable semantic tableaux extended by an additional string unification to ensure the particular restrictions in intuitionistic logic. Due to the modular treatment of the different logical connectives the implementation can easily be adapted to deal with other non-classical logics.

1 Introduction

Intuitionistic logic, due to its constructive nature, has an essential significance for the derivation of verifiably correct software. Unfortunately it is much more difficult to prove a theorem in intuitionistic logic than finding a classical proof for it. Whereas there are many classical provers there exists only very few (published) implementations of theorem provers for first-order intuitionistic logic (e.g. [8, 9]).

The following implementation was inspired by the classical prover leanTAP [1, 2]. leanTAP is based on free-variable tableaux [4], works for formulae in non-clausal form and reach its considerable performance by a very compact representation and an optimized Skolemization. To extend leanTAP to deal with intuitionistic logic one possibility is to add (or modify) the clauses implementing the corresponding tableau rules appropriately. This would of course not lead to a very efficient implementation, since the non-permutabilities between certain intuitionistic rules cause a large search space. A lot of additional strategies have to be added in this case (as done in [8]).

The following approach solves this problem in a more sophisticated way. In classical provers usually *term unification* and Skolemization is used to express the non-permutabilities between the quantifier rules (due to the "eigenvariable condition" in the sequent calculus). To handle the non-permutabilities between certain intuitionistic rules in a similar way we use a specialed *string unification* and extend the Skolemization accordingly. The basis of this approach was invented by Wallen who developed a matrix characterization for some non-classical logics [10].

In the following implementation we first use a leanTAP like technique for *path checking* to prove the *classical* validity of a given formula. Afterwards we try to unify the so-called *prefixes* of those atoms closing the branches of the tableau proof found in the first step. If this additional string unification succeeds the formula is *intuitionistically* valid. We present some performance results and show how to modify the code to deal with other non-classical logics.

The source code of ileanTAP can be obtained free of charge from the author.[2]

---

* The author is supported by the Adolf Messer Stiftung
[2] Or via web http://aida.intellektik.informatik.th-darmstadt.de/~jeotten/ileantap/

2 The Program

We assume the reader to be familiar with free-variable tableaux [4] and the leanTAP code (see [1, 2]) as well as with some details of Wallen's approach (see [10] or [7, 5]). The following Prolog implementation is of course not as lean and compact as the original code of leanTAP. The logical connectives and quantifiers of intuitionistic logic need a separate treatment and we can not make use of any negation normal form.[3] We divide the description of the implementation into the parts "path checking" and "T-string unification". Whenever possible we will use the notation of leanTAP.[4] For the syntax of formulae we use the logical connectives "~" (negation), "," (conjunction), ";" (disjunction), "=>" (implication), "<=>" (equivalence), the quantifiers all X:F (universal) and ex X:F (existential), and Prolog terms for atomic formulae. For example to express the formula $\forall x \exists y (\neg q \wedge p(y) \Rightarrow p(x))$ we use the Prolog term all X:ex Y:(~q,p(Y)=>p(X)).

2.1 Path Checking

The technique of path checking is similar to the one used in leanTAP. In order to get a compact code and to allow an easy adaptation to other logics (see conclusion) we decided to use two predicates instead of one: prove and fml.

The predicate fml is used to specify the particular characteristics of each logical connective or quantifier:

fml(F,Pol,Pre1,FreeV,S,F1,F2,FUnE,FUnE1,FrV,PrV,Lim1,Lim2,V,PrN,Cp,Cp1)

succeeds if F is a first-order formula but not atomic. The parameter Pol is its polarity (either 0 or 1), Pre1 its prefix, FreeV a list of its "free" quantifier- and prefix-variables and S its unique position in the formula tree. The parameters FrV and PrV are lists of free quantifier-variables and prefix-variables of the current branch, respectively.

The parameters F1, F2, FUnE and FUnE1 are of the form (Formula,Polarity) and are bound to Prolog terms as follows: F1 is the first (or only) subformula of F (possibly later bound to a copy of the subformula), F2 is the second subformula of F if F is a β-formula (otherwise []), FUnE is the second subformula of F if F is an α-formula (otherwise []), FUnE1 is bound to the formula F (and its polarity) itself if F is a γ-formula. If F is a γ-formula Lim1 is bound to FrV and V to the variable (strictly speaking to a copy of it later on) which is quantified in F. If the position of F belongs to a prefix-variable Lim2 is bound to PrV. If necessary PrN will be bound to a new prefix-character of F. The parameter Cp contains a term which has to be copied later on and bound to the parameter Cp1. The following 13 clauses define the corresponding characteristics of the intuitionistic connectives and quantifiers:

```
fml((A,B),   1, _,_,_,(A,1),[],(B,1),[],          _,_,  [],[],[],[],  [],[]).
fml((A,B),   0, _,_,_,(A,0),(B,0),[],[],           _,_,  [],[],[],[],  [],[]).
fml((A;B),   1, _,_,_,(A,1),(B,1),[],[],           _,_,  [],[],[],[],  [],[]).
fml((A;B),   0, _,_,_,(A,0),[],(B,0),[],           _,_,  [],[],[],[],  [],[]).
fml((A<=>B),P1,_,_,_,(((A=>B),(B=>A)),P1),[],[],[],_,_,[],[],[],[],[],[]).
fml((A=>B),  1,_,_,_, (C,0),(D,1),[],((A=>B),1),_,PrV,[],PrV,[],_,A:B,C:D).
fml((A=>B),  0,_,FV,S,(B,0),[],(A,1),[],           _,_,  [],[],[],S~FV,[],[]).
```

---

[3] In intuitionistic logic we have, e.g., $\neg\neg A \not\equiv A$ and $A \Rightarrow B \not\equiv \neg A \vee B$.

[4] But notice that we *prove* a formula and do not *refute* its negation.

```
fml((~A),    1,_,_,_, (C,0),[],[],((~A),1),      _,PrV,[],PrV,[],_,   A,C ).
fml((~A),    0,_,FV,S,(A,1),[],[],[],            _,_,  [],[],[],S~FV,[],[]).
fml(all X:A,1,_,_,_, (C,1),[],[],(all X:A,1),FrV,PrV,FrV,PrV,Y,_,X:A,Y:C).
fml(all X:A,0,Pr,FV,S,(C,0),[],[],[],_,_,_,[],[],[],S~FV,(X,A),(S~[]~Pr,C)).
fml(ex X:A, 1,Pr,FV,S,(C,1),[],[],[],_,_,_,[],[],[],[],  (X,A),(S~FV~Pr,C)).
fml(ex X:A, 0,_,_,_, (C,0),[],[],(ex X:A,0), FrV,_,  FrV,[],Y,[],X:A,Y:C).
```

We use a similar technique for Skolemization as lean*TAP*. That is we replace
the quantified variable by the Skolem-term S~FV~Pr in the current formula F
where S is the position of F (in the formula tree), FV its free quantifier- and
prefix-variables and Pr its prefix (which we need later on). The prefix-constants
have a similar format, namely S~FV.[5]

The predicate actually performing the proof search is

```
prove([(F,Pol),Pre,FreeV,S],UnExp,Lits,FrV,PrV,VarLim,[PU,AC])
```

It succeeds if there is a (classical) closed tableau for F. The parameters UnExp
and Lits represent lists of formulae not yet expanded and the atomic formulae on
the current branch, respectively. The parameter VarLim is a positive integer used
to initiate backtracking (in order to obtain completeness within Prolog's depth-
first search). In case of success, PU is bound to the prefixes of the atomic formulae
which have closed the tableau, i.e. it contains pairs of prefixes Pre1=Pre2. The
parameter AC is bound to a list containing the free variables (which might be
bound to Prolog terms) of the proven formula and the corresponding prefixes,
i.e. it contains pairs of [Variable,Prefix]. The other parameters have been
explained before.

```
prove([(F,Pol),Pre,FreeV,S],UnExp,Lits,FrV,PrV,VarLim,[PU,AC]) :-
    fml(F,Pol,Pre1,FreeV,S,F1,F2,FUnE,FUnE1,FrV,PrV,Lim1,Lim2,V,PrN,Cp,Cp1),
    !, \+length(Lim1,VarLim), \+length(Lim2,VarLim),
    copy_term((Cp,FreeV),(Cp1,FreeV)),   append(Pre,[PrN],Pre1),
    (FUnE =[] -> UnEx2=UnExp  ; UnEx2=[[FUnE,Pre1,FreeV,r(S)]|UnExp]),
    (FUnE1=[] -> UnExp1=UnEx2 ; append(UnEx2,[[FUnE1,Pre,FreeV,S]],UnExp1)),
    (var(V)    -> FV2=[V|FreeV], FrV1=[V|FrV], AC2=[[V,Pre1]|AC1] ;
                  FV2=FreeV,     FrV1=FrV,     AC2=AC1),
    (var(PrN) -> FreeV1=[PrN|FV2], PrV1=[PrN|PrV] ; FreeV1=FV2, PrV1=PrV),
    prove([F1,Pre1,FreeV1,l(S)],UnExp1,Lits,FrV1,PrV1,VarLim,[PU1,AC1]),
    (F2=[] -> PU=PU1, AC=AC2 ;
    prove([F2,Pre1,FreeV1,r(S)],UnExp1,Lits,FrV1,PrV1,VarLim,[PU2,AC3]),
    append(PU1,PU2,PU), append(AC2,AC3,AC)).
```

It depends on the actual formula F which steps are performed to expand F.
After checking whether the depth-bound VarLim is reached, we make a copy of
the specified Prolog term.[6] This is used to make a copy of those formulae which
have to be kept among the unexpanded formulae or to insert a Skolem-term into
a formula (in each case the free variables FreeV are not renamed). The current
prefix is extended and the list of formulae not yet expanded is extended accordingly. We also have to add the quantifier- or prefix-variable to the corresponding
lists, before expanding the subformulae.

---

[5] This "liberalized" Skolemization is in fact correct and complete. Just consider this
Skolemization as an technique to check if the *reduction ordering* is acyclic and consider
appropriate copies of the corresponding subformulae.

[6] Lim1/Lim2 are used to restrict the first-order/intuitionistic *multiplicity*.

```
prove([(Lit,Pol),Pre|_],_,[[(L,P),Pr|_]|Lits],_,_,_,[PU,AC]) :-
    ( Lit=L, Pol is 1-P, (Pol=1 -> PU=[Pre=Pr] ; PU=[Pr=Pre]), AC=[]) ;
    prove([(Lit,Pol),Pre|_],[],Lits,_,_,_,[PU,AC]).
prove(Lit,[Next|UnExp],Lits,FrV,PrV,VarLim,PU_AC) :-
    prove(Next,UnExp,[Lit|Lits],FrV,PrV,VarLim,PU_AC).
```

The last two clauses remain almost unchanged from lean*TAP*. The first clause closes the current branch if the two atomic formulae Lit and L (with different polarity) can be unified.[7] The list PU of prefixes to be unified is extended accordingly. The last clause adds Lit to the list of atomic formulae on the current branch and selects another formula from those not yet expanded.

2.2 T-String Unification

After finding a closed tableau we have to unify the prefixes of those atomic formulae which have closed the tableau. These prefixes are stored in the list PU. A *prefix* of an atomic formula is a string and essentially describes the position of this formula in the formula tree (see [10]). Furthermore an *additional condition* on all (universal) variables has to be checked.[8] These variables are stored in the list AC.

The predicate t_string_unify(PU,AC) succeeds if all prefixes in PU can be unified and the additional condition for all variables in AC holds. In this case an intuitionistic proof can be obtained from the (classical) proof found in the first step. Otherwise we have to look for an other (classical) proof.

The two prefixes S and T are unified using the predicate tunify(S,[],T).[9] A description of this predicate together with all the theoretical details are explicitly explained in [6].

```
t_string_unify([],AC)        :- addco(AC,[],final).
t_string_unify([S=T|G],AC)   :- flatten([S,_],S1,[]), flatten(T,T1,[]),
                                tunify(S1,[],T1), addco(AC,[],t),
                                t_string_unify(G,AC).
tunify([],[],[]).
tunify([],[],[X|T])          :- tunify([X|T],[],[]).
tunify([X1|S],[],[X2|T])     :- (var(X1) -> (var(X2), X1==X2);
                                (\+var(X2), X1=X2)), !, tunify(S,[],T).
tunify([C|S],[],[V|T])       :- \+var(C), !, var(V), tunify([V|T],[],[C|S]).
tunify([V|S],Z,[])           :- V=Z, tunify(S,[],[]).
tunify([V|S],[],[C1|T])      :- \+var(C1), V=[], tunify(S,[],[C1|T]).
tunify([V|S],Z,[C1,C2|T])    :- \+var(C1), \+var(C2), append(Z,[C1],V),
                                tunify(S,[],[C2|T]).
tunify([V,X|S],[],[V1|T])    :- var(V1), tunify([V1|T],[V],[X|S]).
tunify([V,X|S],[Z1|Z],[V1|T]) :- var(V1), append([Z1|Z],[Vnew],V),
                                tunify([V1|T],[Vnew],[X|S]).
tunify([V|S],Z,[X|T])        :- (S=[]; T\=[]; \+var(X)) ->
                                append(Z,[X],Z1), tunify([V|S],Z1,T).
```

---

[7] Note that for this purpose we have to use *sound* unification (i.e. with occurs check).

[8] Let σ_Q and σ_J be the term-/prefix-substitution. For all free variables u and all Skolemized variables v occurring in $\sigma_Q(u)$ the condition $|\sigma_J(\mathrm{prefix}(u))| \geq |\sigma_J(\mathrm{prefix}(v))|$ must hold.

[9] The variables representing the prefixes may be instantiated with nested lists. Therefore the predicate flatten is necessary which can be implemented as follows:
```
flatten(A,[A|B],B) :- (var(A); A=_^_), !.    flatten([],A,A).
flatten([A|B],C,D) :- flatten(A,C,E), flatten(B,E,D).
```

The predicate `addco` is used to check the interaction condition between the term- and prefix-substitution mentioned above.

```
addco(X,_,_).              :- (var(X); X=[[]]), !.
addco([[X,Pre]|L],[],Ki)   :- !, addco(X,Pre,Ki), addco(L,[],Ki).
addco(_^_^Pre1,Pre,Ki)     :- !, ( Ki=final -> flatten(Pre1,S,[]),
                                    flatten(Pre,T,[]), append(S,_,T) ;
                                    \+ \+ t_string_unify([Pre1=Pre],[]) ).
addco(T,Pre,Ki)            :- T=..[_,T1|T2],!, addco(T1,Pre,Ki), addco(T2,Pre,Ki).
addco(_,_,_).
```

The following goal succeeds if F can be proven *intuitionistically* without using more than `VarLim` (quantifier- and prefix-) variables on each branch:

```
prove([(F,0),[],[],1],[],[],[],[],VarLim,[PU,AC]), t_string_unify(PU,AC).
```

3 Performance

Although the presented implementation is comparably short its performance seems to be quite good. We will show some experimental results comparing ileanTAP with the tableau prover ft [8]. We also provide the timing of lean*TAP* to point out the correlation between lean*TAP* and ileanTAP. All three provers perform an iterative deepening. The following problems are taken from [8] and measured on a Sun SPARC10 (times are given in seconds; "–" means that no proof was found within 150 seconds).

No.	ft	ilean TAP	lean TAP	No.	ft	ilean TAP	lean TAP	No.	ft	ilean TAP	lean TAP
1.1	0.03	0.07	0.02	3.3	0.01	0.05	0.02	6.6	< 0.01	0.02	< 0.01
1.2	0.56	0.12	0.05	3.4	< 0.01	0.17	0.02	6.7	< 0.01	0.02	< 0.01
1.3	–	0.35	0.08	3.5	2.66	–	0.65	6.8	0.01	0.02	0.02
1.4	0.01	0.05	< 0.01	4.1	10.53	–	2.40	6.9	0.01	0.03	0.03
1.5	1.81	0.13	0.02	4.2	11.78	–	4.30	6.10	< 0.01	0.10	0.03
1.6	18.67	0.20	0.07	5.1	< 0.01	0.08	< 0.01	6.11	< 0.01	0.03	< 0.01
1.7	0.16	0.03	0.02	5.2	0.03	3.17	0.30	6.12	0.04	–	0.02
1.8	–	0.13	0.08	5.3	0.76	121.98	8.05	6.13	0.07	4.26	0.07
2.1	1.73	–	–	6.1	< 0.01	< 0.01	< 0.01	6.14	0.01	0.10	0.03
2.2	3.61	–	–	6.2	0.01	0.48	0.60	6.15	0.04	–	0.15
2.3	6.08	–	–	6.3	0.01	0.10	< 0.01	7.1	< 0.01	0.02	< 0.01
3.1	< 0.01	0.02	< 0.01	6.4	< 0.01	0.08	0.02	7.2	0.05	0.48	< 0.01
3.2	0.07	1.72	< 0.01	6.5	< 0.01	0.15	< 0.01	7.3	3.98	32.85	0.03

Table 1. Performance of ileanTAP with eclipse Prolog Version 3.5.2

ileanTAP yields very good results on the problems of group 1 ("alternations of quantifiers"), 3 ("Pelletier's problems 39–43") and 6 ("simple"). It has a similar performance on the problems of group 5 ("unify") and 7 ("problematic"), but behaves very poorly on problems of group 2 ("append") and 4 ("existence"). The problems of group 2 are even too hard for the classical prover lean*TAP*.

Altogether, a remarkable result keeping in mind that ft consists of about 200 kbytes C-source code (whereas ileanTAP has a size of about 4 kbytes).[10]

---

[10] The prover in [9] is a bit slower than ft on the simpler problems. It cannot solve problem no. 5.3 but prove all other problems in less than one second.

4 Conclusion

We presented a Prolog implementation for a first-order intuitionistic theorem prover. We encode the additional non-permutabilities arising in intuitionistic logic in a sophisticated way namely by an additional string unification which yields a quite good performance. Due to the compact code the program can easily be modified for special purposes or applications.

Since the particular characteristics of the logical connectives and quantifiers are specified in a separate way, it is easy to adapt the prover to other non-classical logics. Replacing the last 8 clauses of the predicate fml by the following clauses

```
fml((A=>B), 1,_,_,_, (A,0),(B,1),[],[],          _,_,  [],[],[],[], [],[]).
fml((A=>B), 0,_,_,_, (B,0),[],(A,1),[],           _,_,  [],[],[],[], [],[]).
fml((~A),  1,_,_,_, (A,0),[],    [],[],           _,_,  [],[],[],[], [],[]).
fml((~A),  0,_,_,_, (A,1),[],    [],[],           _,_,  [],[],[],[], [],[]).
fml(all X:A,1,_,_,_, (C,1),[],[],(all X:A,1),FrV,_, FrV,[],Y,[],X:A,Y:C).
fml(all X:A,0,Pr,FV,S,(C,0),[],[],[],_,_, [],[],[],[],(X,A),(S^FV^Pr,C)).
fml(ex X:A, 1,Pr,FV,S,(C,1),[],[],[],_,_, [],[],[],[],(X,A),(S^FV^Pr,C)).
fml(ex X:A, 0,_,_,_, (C,0),[],[],(ex X:A,0), FrV,_, FrV,[],Y,[],X:A,Y:C).
```

and adding the following 4 clauses

```
fml([](A),  1,_,_,_, (C,1),[],[],([](A),1),_,PrV,[], PrV,[],_,     A,C ).
fml([](A),  0,_,FV,S,(A,0),[],[],[],        _,_, [], [], [],S^FV, [],[]).
fml(<>(A),  1,_,FV,S,(A,1),[],[],[],        _,_, [], [], [],S^FV, [],[]).
fml(<>(A),  0,_,_,_, (C,0),[],[],(<>(A),0),_,PrV,[], PrV,[],_,     A,C ).
```

immediately yields a program implementing a theorem prover for the first-order modal logic S4 (where "[]" and "<>" represent the corresponding modal operators). Modifying the algorithm for T-string unification accordingly leads also to provers for the modal logics D, D4, S5, and T (see [6, 7] for details).

Of course, there is still room for further research. The current implementation is not a decision procedure for the propositional intuitionistic logic (which is decidable), since we need *multiplicities* already in this fragment. For example it would be interesting to integrate some techniques from [3] to get a decision procedure for the propositional fragment of intuitionistic logic.

References

1. B. BECKERT AND J. POSEGGA. leanT*A*P: Lean Tableau-Based Theorem Proving. *Proc. CADE-12*, LNAI 814, pp. 793–797, Springer Verlag, 1994.
2. B. BECKERT AND J. POSEGGA. leanT*A*P: Lean Tableau-based Deduction. *Journal of Automated Reasoning*, 15(3):339–358, 1995.
3. R. DYCKHOFF. Contraction-free Sequent Calculi for Intuitionistic Logic. *Journal of Symbolic Logic*, 57(3):795–807, 1992.
4. M. C. FITTING. *First-Order Logic and Automated Theorem Proving*. Springer Verlag, 1990.
5. J. OTTEN, C. KREITZ. A Connection-Based Proof Method for Intuitionistic Logic. *Proc. 4th TABLEAUX Workshop*, LNAI 918, pp. 122–137, 1995.
6. J. OTTEN, C. KREITZ. T-String-Unification: Unifying Prefixes in Non-Classical Proof Methods. *Proc. 5th TABLEAUX Workshop*, LNAI 1071, pp. 244–260, 1996.
7. J. OTTEN, C. KREITZ. A Uniform Proof Procedure for Classical and Non-Classical Logics. *KI-96: Advances in Artificial Intelligence*, LNAI, Springer Verlag, 1996.
8. D. SAHLIN, T. FRANZEN, S. HARIDI. An Intuitionistic Predicate Logic Theorem Prover. *Journal of Logic and Computation*, 2(5):619–656, 1992.
9. T. TAMMET. A Resolution Theorem Prover for Intuitionistic Logic. *Proc. CADE-13*, LNAI, Springer Verlag, 1996.
10. L. WALLEN. *Automated deduction in nonclassical logic*. MIT Press, 1990.

Simplifying and Generalizing Formulae in Tableaux. Pruning the Search Space and Building Models

Nicolas Peltier

Laboratory LEIBNIZ-IMAG
46, Avenue Félix Viallet 38031 Grenoble Cedex FRANCE
Nicolas.Peltier@imag.fr
Phone: (33) 04-76-57-46-59

Abstract. A powerful extension of the tableau method is described. It consists in a new simplification rule allowing to prune the search space and a new way of extracting a model from a given (possibly infinite) branch. These features are combined with a former method for simultaneous search for refutations and models. The possibilities of the new method w.r.t. the original one are clearly stated. In particular it is shown that the method is able to build model for each formula having a model expressible by equational constraints.

1 Introduction

The construction and the use of *models* or *counter-examples* are crucial techniques widely used in all aspects of human reasoning. Incorporating such abilities into automated theorem provers is therefore a very natural idea, which has been considered since the beginning [13, 21]. Nevertheless, it is not until the nineties that feasible methods have been proposed to *automatically* build models for first order formulae. The first works in this direction were those of Wos and Winker [24], using an existing theorem prover as an assistant in the interactive search for a model. Since 1990, other methods have been proposed. Most of them are based on enumeration and backtracking (see for example [22, 25]). Their general principle is to perform an exhaustive search on the set of interpretations on a *finite* domain. Other methods, based on the use of the resolution rule, have also been designed. They rely on techniques restricted to some particular decidable classes of formulae: Monadic and Ackermann classes [23] and positively decomposed formulae [10]. Manthey and Bry [17] designed SATCHMO, based on splitting of non-Horn clauses and backtracking, and working bottom-up.[1]

Using tableaux-based methods for building models of first-order formulae is a very natural idea. Indeed, the building of a tableau for a given formula A corresponds exactly to the enumeration of all the possible Herbrand models of A: each open branch of the tableau can be associated in a very natural way to a Herbrand model of the initial formula. It is worth noting that the system

---

[1] A parallel and optimized version of SATCHMO is described in [12].

SATCHMO or MGTP can be seen as a particular refinement of a tableaux-based method (see for example [3, 1]), called the *positive unit hyper-resolution tableaux*.

The main problem of this approach is that, of course, the open branches may be infinite, which prevents the method from exhibiting the model in all but some *very particular cases*. Existing tableaux-based approaches (or refinement of them) will therefore not terminate and will not be able to detect the satisfiability of the formula. In order to overcome this limitation, one has to find a way to *detect* such infinite branches and to *build* the corresponding model (and express it *finitely*).

In this work we define an extension of tableaux based on two principles. The first one is to use the information inferred by the method (i.e. the set of formulae contained in a given branch of the tableau) in order to simplify formulae occurring in this branch. The second one consists in using *generalization* in order to extract a model from an open branch of the tableau.

The features of our approach can be combined with any tableau-based method. In this work we incorporate them to the tableaux-based model building method RAMCET defined by Caferra and Zabel [8] and we show that our approach allows to enhance significantly the power of the original method.

The general principle of the model building method originally defined in [7, 8] is very simple: to set conditions that allow the application of the rules of the classical refutations procedure (resolution, tableaux...) *or that prevent their application*. These conditions are coded by *constraints* associated to formulae in order to restrict the domain of their variables. A nice feature of this approach is its generality. It can be used to extend any existing calculus, in a very natural way. However, it has been essentially developed for resolution: this leads to the method RAMC (Refutation And Model Construction), that has been further extended and widely studied in [2, 4, 6, 5]. This approach has also been applied to the tableaux calculus [8]. The corresponding procedure was called RAMCET (for Refutation And Model Construction with Extended Tableaux).

Several works have incorporated *unification* to semantic tableaux [20]. Nevertheless this approach is *not* sufficient for searching for models in a systematic way. In [8], it is proposed to extend tableaux with *equational constraints* (they also take in account *dis-unification*). This extension corresponds to including in the object level reasoning usually done in an ad-hoc manner in the meta-level. The principle of the method is quite different from the classical approach for tableaux-based model building. Instead of extracting a model from a *finite* open branch (as a by-product of the method), the principle is to try to build a model *during the search*. The main advantage of this feature is that it can prune the search space of the refutation procedure and may build models for formulae for which the classical approach does not terminate. The properties of the original method (i.e. soundness and refutational completeness) are preserved.

In this work, we extend the method RAMCET in two directions. First we define a *new simplification rule* allowing to prune the search space. Second, we propose a *new method for extracting a model from an open (possibly infinite) branch*. The new method is based on generalization techniques and inductive reasoning.

It is proven to be *strictly more powerful* than the simpler one proposed in the original work [8].

The models built by the method are expressed by equational formulae interpreted on the finite terms algebra (i.e. where equality is interpreted as *syntactic equality* on the set of ground terms). These models are called "eq-models" (or eq-interpretations). The class of formulae having an eq-model is called *Eqmodel*.

We prove that our method is a decision procedure for the class *Eqmodel*, i.e. *our method build models for each formulae having an eq-model*. This result neither holds for the original method in [8], nor for the resolution-based calculus RAMC, which shows evidence of the interest of our work. As a corollary this implies that our new method is a decision procedure for a wide range of decidable classes of first-order logic and strictly subsumes all existing methods for building Herbrand models, based on *atomic representations*, that are a particular case of eq-interpretations.

Finally we point out some possible improvement of this work. In particular we mention that this approach can be very easily extended to other model representation formalism (for example terms with integer exponents, terms grammars...).

Due to space restriction, proofs are not included, they can be found in [19].

2 Preliminaries

We assume the reader is familiar with the standard notions of first-order logic and automated deduction. For the sake of clarity, we review some notions necessary for the understanding of our work.

Let Ω be a set of *predicate symbols* and Σ a set of *functional symbols*. a is an arity function mapping each element of $\Sigma \cup \Omega$ to a natural integer (a function symbol b such that $a(b) = 0$ is called a *constant*). Let V be a set of variables (we assume that Σ, Ω, V share no element). We denote by $\tau(\Sigma, V)$ the set of terms build as usual on the signature Σ and the set of variables V. $\tau(\Sigma, \emptyset)$ is simply denoted by $\tau(\Sigma)$. We note $\overline{x}, \overline{y}, \ldots$ tuples of variables. The notion of *substitution* is defined as usual. The result of applying a substitution σ to an expression (term, formula, etc.) E is denoted by $E\sigma$. For any term t we denote by $Var(t)$ the set of variables of t.

Equational formulae play an important role in our method. We recall below some basic definitions concerning equational formulae (definition and solvability) [9]. A *equational formula* is a first-order formula (built on the set of symbols $\wedge, \vee, \neg, \exists, \forall$) containing only the predicate "=". A substitution σ *validates* an equational formula P iff $P\sigma$ is valid in the finite tree algebra (in the usual sense i.e. evaluated to "true"). The set of substitutions σ that validates P is called the set of *solutions* of P and is noted $Sol(P)$. An equational formula X such that $Sol(X) = \emptyset$ (resp. $Sol(\neg X) = \emptyset$) is said to be *unsatisfiable* (resp. *valid*). The validity problem of equational formulae is known to be decidable and algorithms have been given in order to find the solution of a given equational formula (see [9].

The set of *constrained formula* (of *c-formula* for short) is inductively defined as the least set containing $A \vee A'$, $A \wedge A'$, $\neg A$, $\forall \overline{x} : \mathcal{X}.A$, $\exists \overline{x} : \mathcal{X}.A$, $P(t_1, \ldots, t_n)$, where A, A' are c-formulae, t_i ($1 \leq i \leq n$) are terms, $P \in \Omega$, $a(P) = n$ and $x \in \mathcal{V}$, $\mathcal{X}$ is an equational formula (called a *constraint*). A formula of the form $P(\overline{t})$ or $\neg P(\overline{t})$ is called a *literal*. If A is a literal A^c denotes its complement (i.e., the literal with the same predicate symbol and arguments than A but with different sign).

For any c-formula (or equational formula) A we naturally denote by $Var(A)$ the set of its free variables. A formula (resp. term) A is said to be *ground* iff $Var(A) = \emptyset$.

Throughout this paper formulae will be denoted by A, B, F and sets of formulae by $\mathcal{A}, \mathcal{B}, \mathcal{F}$.

Unless otherwise stated, the interpretations we will consider are *partial* (i.e. they do *not* necessarily associate a truth value to each formulae but only to some of them)[2].

We recall below the definition of *partial Herbrand interpretation* [7]. A partial Herbrand interpretation is a set of ground literals, such that if $P \in \mathcal{I}$ then $\neg P \notin \mathcal{I}$. If moreover for all P, either $P \in \mathcal{I}$ or $\neg P \in \mathcal{I}$, then $\mathcal{I}$ is *total*.

Definition 1. Let $\mathcal{I}$ be a total interpretation. $\mathcal{I}$ *validates* a ground formula A (noted $\mathcal{I} \models A$) iff one of the following conditions holds.

- $A = \top$.
- $A = P(\overline{t})$ and $P(\overline{t}) \in \mathcal{I}$.
- $A = A_1 \vee A_2$ and $\mathcal{I} \models A_1$ or $\mathcal{I} \models A_2$.
- $A = A_1 \wedge A_2$ and $\mathcal{I} \models A_1$ and $\mathcal{I} \models A_2$.
- $A = \neg A'$ and $\mathcal{I} \not\models A'$
- $A = \forall x : \mathcal{X}.A'$ and for all ground substitutions $\sigma \in \mathcal{S}ol(\mathcal{X})$, $\mathcal{I} \models A'\sigma$.
- $A = \exists x : \mathcal{X}.A'$ and there exists a ground substitution $\sigma \in \mathcal{S}ol(\mathcal{X})$ such that $\mathcal{I} \models A'\sigma$.

A partial interpretation $\mathcal{I}$ is an *extension* of another partial interpretation $\mathcal{J}$ iff $\mathcal{J} \subseteq \mathcal{I}$. A partial interpretation $\mathcal{I}$ *validates* a c-formula A iff any total extension (i.e. all extensions that are total interpretations) of $\mathcal{I}$ validate A.

Two formulae A and B are *equivalent* (noted $A \equiv B$) iff for all total interpretation $\mathcal{I}$, $\mathcal{I} \models (A \Leftrightarrow B)$.

The following theorem states basic properties of c-formulae (that are straightforward consequences of definition 1).

Theorem 2. We have

$(\mathcal{P}_1)$ $\forall x : \mathcal{X} \vee \mathcal{Y}.A \equiv \forall x : \mathcal{X}.A \wedge \forall x.\mathcal{Y}.A$;
$(\mathcal{P}_2)$ $\forall x : x = t \wedge \mathcal{X}.A \equiv \forall x : \mathcal{X}.A\{x \rightarrow t\}$;
$(\mathcal{P}_3)$ $\forall x : \bot.A \equiv \top$.

---

[2] They can be seen as 3-valued interpretations, the third value meaning "undefined".

Representing Herbrand interpretations

The goal of our method is, given a conjecture C (a c-formula) to look *simultaneously* for a refutation of C or a *Herbrand model* $\mathcal{M}$ of C. The model $\mathcal{M}$ is in general infinite (if Σ contains at least one functional symbol of arity greater than 0).

The first problem we have to solve is *to find a suitable formalism for representing such interpretations.* Indeed, the Herbrand models are in general infinite, thus *cannot* be simply represented by sets of atoms. In this work, Herbrand interpretations are represented by using *equational constraints.* More precisely:

A partial Herbrand interpretation $\mathcal{I}$ is said to be an *eq-interpretation*, iff for each predicate P of arity n there exist two equational formulae noted $\mathcal{I}_P^+$ and $\mathcal{I}_P^-$ and n variables $x_1, \ldots, x_n$, such that

- $P(x_1, \ldots, x_n)\sigma$ is true in $\mathcal{I}$ iff $\sigma \in \mathcal{S}ol(\mathcal{I}_P^+)$;
- $\neg P(x_1, \ldots, x_n)\sigma$ is true in $\mathcal{I}$ iff $\sigma \in \mathcal{S}ol(\mathcal{I}_P^-)$.

3 Facts

Before describing our method, we need to introduce a few definitions and state some interesting properties of c-formulae. We first introduce the notion of *fact.* A *fact* is a formula of the form $\forall \overline{x}.A$, where A is a literal. ($\overline{x}$ maybe an empty string, and A may contain variables that are not in $\overline{x}$). A fact is called *ground* iff $\forall \overline{x}.A$ is ground.

If $F = \forall \overline{x}.A$ is a fact, we note F^c the fact $\forall \overline{x}.A^c$ (the complementary of F). Similarly, if $\mathcal{F}$ is a set of facts, $\mathcal{F}^c$ is the set $\{F^c/F \in \mathcal{F}\}$. Two facts A and B are said to be *unifiable* iff A and B are of the form $A = \forall \overline{x} : \mathcal{X}.P(\overline{t})$ and $B = \forall \overline{y} : \mathcal{Y}.P(\overline{s})$ (resp. $A = \forall \overline{x} : \mathcal{X}.\neg P(\overline{t})$ and $B = \forall \overline{y} : \mathcal{Y}.\neg P(\overline{s})$) and $\overline{t} = \overline{s} \wedge \mathcal{X} \wedge \mathcal{Y}$ has a solution. We say that $A \succ B$ iff A is of the form $\forall \overline{x} : \mathcal{X}.P(\overline{s})$ and B is of the form $\forall \overline{y} : \mathcal{Y}.P(\overline{s})$ (resp. $A = \forall \overline{x} : \mathcal{X}.\neg P(\overline{t})$ and $B = \forall \overline{y} : \mathcal{Y}.\neg P(\overline{s})$) and $\forall \overline{y}.\exists \overline{x}.\neg \mathcal{Y} \vee (\mathcal{X} \wedge \overline{t} = \overline{s})$ is valid.

Sets of facts will be used to record the information about a model of the initial formula in an open branch. They also allow to *simplify a given formula* as shown in the next section.

3.1 Simplifying formulae in a context

Let A be a formula, $\mathcal{B}$ be a finite set of facts. The formulae $\mathcal{F}^+(A, \mathcal{B})$ and $\mathcal{F}^-(A, \mathcal{B})$ are defined as follows (by induction on the set of formulae).

$$\mathcal{F}^+(P(\overline{t}), \mathcal{B}) = \bigvee_{\forall \overline{x}: \mathcal{X}.P(\overline{s}) \in \mathcal{B}} \exists \overline{x}.\overline{s} = \overline{t} \wedge \mathcal{X}$$

$$\mathcal{F}^-(P(\overline{t}), \mathcal{B}) = \bigvee_{\forall \overline{x}: \mathcal{X}.\neg P(\overline{s}) \in \mathcal{B}} \exists \overline{s}.\overline{x} = \overline{t} \wedge \mathcal{X}$$

$$\mathcal{F}^+(\neg A, \mathcal{B}) = \mathcal{F}^-(A, \mathcal{B})$$

$$\mathcal{F}^-(\neg A, \mathcal{B}) = \mathcal{F}^+(A, \mathcal{B})$$
$$\mathcal{F}^+(A \vee A', \mathcal{B}) = \mathcal{F}^+(A, \mathcal{B}) \vee \mathcal{F}^+(A', \mathcal{B})$$
$$\mathcal{F}^-(A \vee A', \mathcal{B}) = \mathcal{F}^-(A, \mathcal{B}) \wedge \mathcal{F}^-(A', \mathcal{B})$$
$$\mathcal{F}^+(A \wedge A', \mathcal{B}) = \mathcal{F}^+(A, \mathcal{B}) \wedge \mathcal{F}^+(A', \mathcal{B})$$
$$\mathcal{F}^-(A \wedge A', \mathcal{B}) = \mathcal{F}^-(A, \mathcal{B}) \vee \mathcal{F}^-(A', \mathcal{B})$$
$$\mathcal{F}^+(\exists x.A, \mathcal{B}) = \exists x.\mathcal{F}^+(A, \mathcal{B})$$
$$\mathcal{F}^-(\exists x.A, \mathcal{B}) = \forall x.\mathcal{F}^-(A, \mathcal{B})$$
$$\mathcal{F}^+(\forall x.A, \mathcal{B}) = \forall x.\mathcal{F}^+(A, \mathcal{B})$$
$$\mathcal{F}^-(\forall x.A, \mathcal{B}) = \exists x.\mathcal{F}^-(A, \mathcal{B})$$

Intuitively, $\mathcal{F}^+(A, \mathcal{B})$ and $\mathcal{F}^-(A, \mathcal{B})$ state conditions *forcing* the formula A to be *true* (respectively *false*) in any interpretation in which all the facts in $\mathcal{B}$ are true.

Theorem 3. *If $\sigma \in Sol(\mathcal{F}^+(A, \mathcal{B}))$ then $\mathcal{B} \models A\sigma$. If $\sigma \in Sol(\mathcal{F}^-(A, \mathcal{B}))$ then $\mathcal{B} \models \neg A\sigma$.*

The formula $\mathcal{F}^+(A, \mathcal{B})$ can be used to *simplify* the initial formula A.

Let A be a formula and let $\mathcal{B}$ be a finite set of facts. Let A be a formula of the form $\forall \overline{x}.C$. We call the formula below the *simplification of A in $\mathcal{B}$* and denote it by $A^+(\mathcal{B})$: $\forall \overline{x} : \neg \mathcal{F}^+(C, \mathcal{B}) \wedge \neg \mathcal{F}^-(C, \mathcal{B}).C \wedge \forall \overline{x} : \mathcal{F}^+(C, \mathcal{B}).\top \wedge \forall \overline{x} : \mathcal{F}^-(C, \mathcal{B}).\bot$.

Theorem 4. *We have $\mathcal{B} \cup \{A\} \equiv \mathcal{B} \cup \{A^+(\mathcal{B})\}$.*

Example 1. Let $A = \forall y.Z(y) \wedge (\forall x, y.P(x, y))$ and $\mathcal{B} = \{Z(u), \forall v.P(v, v)\}$. Then $A^+(\mathcal{B}) = \forall y : y \neq u.Z(y) \wedge \forall y : y = u.\top \wedge (\forall x, y : x \neq y.P(x, y) \wedge \forall x, y : x = y.\top)$. This formula can be simplified into $\forall y : y \neq u.Z(y) \wedge \forall x, y : x \neq y.P(x, y)$.

3.2 Sets of facts as partial interpretations

Obviously, if a given set of facts $\mathcal{F}$ is ground (i.e. contains no free variables) and *satisfiable* (i.e. contains no contradiction) it can be seen as the representation of a partial Herbrand interpretation: a ground literal L is true in the interpretation if there exists a fact $A \in \mathcal{F}$ such that L and A are unifiable.

Example 2. Let $\mathcal{B} = \{\forall x.P(x), \forall z.R(z, z).\forall x, y : x \neq y.\neg R(x, y)\}$. $\mathcal{B}$ is the representation of the Herbrand interpretation defined as follows: $P(t)$ is true for all term t and $R(x, y)$ is true iff $x = y$.

The formulae $\mathcal{F}^+(A, \mathcal{B})$ and $\mathcal{F}^-(A, \mathcal{B})$ will express in this particular case sufficient (but in general non necessary) conditions on the free variables such that A is true (resp. false) in the interpretation denoted by $\mathcal{B}$. In other words, $\mathcal{F}^+(A, \mathcal{B})$ (resp. $\mathcal{F}^-(A, \mathcal{B})$) means "$A$ is **true** (resp. **false**) in $\mathcal{B}$".

For any satisfiable set of ground facts $\mathcal{B}$, we note $\mathcal{I}(\mathcal{B})$ the corresponding partial interpretation.

4 An extended tableaux method: RAMCET-2

In this section we describe our method. The rules are similar (though the presentation differs on several aspects) to those given in [8]. For the sake of simplicity, we assume that the initial formula is in *skolemized negation normal form*[3] (this simplifies the writing of the rules).

A *position* is a sequence of integers of $\mathbb{N} \setminus \{0\}$. "." denotes the concatenation of sequences and Λ denotes the empty sequence. If p and q are two positions, we said that p is a prefix of q (note $p \leq q$) iff there exists a position p' such that $q = p.p'$. p is a *proper prefix* of q (noted $p < q$) iff $p \leq q$ and $p \neq q$. If p is a position in a term t, $t_{|p}$ denotes the term at position p in t. A *labeled c-formula* is a pair (p, A), where p is a position (the *label*) and A a c-formula. Labels are used to code the position of the c-formula A in the tableau.

Let S be a set of labeled formula. Let p be a position. We note $SF(p, S)$ the set $\{A/(p, A) \in S\}$.

A *tableau* $\mathcal{T}$ is a pair $(\mathcal{S}, \mathcal{P})$ where $\mathcal{S}$ is a set of labeled c-formulae, and $\mathcal{P}$ is an equational c-formula (noted $Eqpb(\mathcal{T})$). $\mathcal{P}$ express conditions on the free variables of $\mathcal{T}$.

Let $\mathcal{S}$ be a tableau. We denote by $\mathcal{L}abel(\mathcal{S})$ the set of positions $\mathcal{L}abel(\mathcal{S}) = \{p/\exists A.(p, A) \in \mathcal{S}\}$. A label p is said to be a *leaf* of a tableau $\mathcal{S}$, iff there does not exist a position $q \in \mathcal{L}abel(\mathcal{S})$ such that $p < q$. A *branch* of a tableau $\mathcal{S}$ is a leaf position in $\mathcal{L}abel(\mathcal{S})$. A branch p in a tableau S is said to be *closed* if $\bot \in SF(p, S)$. A tableau is said to be *closed* iff all its branches are closed.

We denote by $\mathcal{M}od(\mathcal{T}, p)$ the set of facts that occur at a position $q \leq p$ in $\mathcal{T}$. Now we have all we need to define our method on a formal basis.

4.1 Rules of the method

The rules R_{and}, R_{or} are simply the standard rules of semantic tableaux for dealing with conjunctions and disjunctions adapted to the particular formalism of our method.

$$R_{and}: \frac{(\mathcal{S} \cup \{(p, A \wedge B)\}, \mathcal{P})}{(\mathcal{S} \cup \{(p, A), (p, B)\}, \mathcal{P})}$$

$$R_{or} : \frac{(\mathcal{S} \cup \{(p, A \vee B)\}, \mathcal{P})}{(\mathcal{S} \cup \{(p.1, A), (p.2, B)\}, \mathcal{P})} \text{If } p \text{ is a leaf of the tableau.}$$

The following rule is necessary for dealing with formulae that do not occur at a leaf position in the tableau.

$$\frac{(\mathcal{S} \cup \{(p, A)\}, \mathcal{P})}{(\mathcal{S} \cup \{(p.i, A)/i \in \mathbb{N}, p.i \in \mathcal{L}abel(\mathcal{S})\}, \mathcal{P})} \text{If } p \text{ is not a leaf.}$$

---

[3] i.e. each existentially quantified variable is replaced by a new functional symbol and negations are shifted innermost in the formula.

The rule R_{in} introduces new free variables into the tableau. It corresponds to the usual γ-rule. One can also introduce terms instead of variables in the tableau.

$$R_{in}: \frac{(S \cup \{(p, \forall x : \mathcal{X}.A)\}, \mathcal{P})}{(S \cup \{(p, \forall x : \mathcal{X} \wedge x \neq y.A), (p, A\{x \to y\})\}, \mathcal{P} \wedge \exists x.y = x \wedge \mathcal{X})}$$

If y is a term (it can be for example a new variable) and $\mathcal{P} \wedge \exists x.y = x \wedge \mathcal{X} \not\equiv \bot$.

$$R_{unif}: \frac{(S \cup \{(p, \forall \overline{x}_1 : \mathcal{X}_1.\neg P(\overline{t}_1)), (q, \forall \overline{x}_2 : \mathcal{X}_2.P(\overline{t}_2))\}, \mathcal{P})}{(S \cup \{(p, \forall \overline{x}_1 : \mathcal{X}_1.\neg P(\overline{t}_1)), (q, \bot)\}, \mathcal{P} \wedge \exists \overline{x}_1, \overline{x}_2.\mathcal{X}_1 \wedge \mathcal{X}_2 \wedge \overline{t}_1 = \overline{t}_2)}$$

Where $p \leq q$, $\mathcal{P} \wedge \exists \overline{x}_1, \overline{x}_2.\mathcal{X}_1 \wedge \mathcal{X}_2 \wedge \overline{t}_1 = \overline{t}_2 \not\equiv \bot$

Simplification rules The following rules are used to simplify the tableau, i.e. to delete the closed branches.

$$\frac{(S \cup \{(p.\bot), (q, A)\}, \mathcal{P})}{(S \cup \{p.\bot\}, \mathcal{P})} \text{If } p \leq q.$$

$$\frac{(S \cup \{(p.A)\}, \mathcal{P})}{(S \cup \{p.\bot\}, \mathcal{P})} \text{If for all } q > p, \; SF(q, S) = \{\bot\}.$$

The following rule allows to remove c-formulae whose constraints are not compatible with the current global constraints $\mathcal{P}$.

$$\frac{(S \cup \{(p.\forall \overline{x} : \mathcal{X}.A)\}, \mathcal{P})}{(S, \mathcal{P})} \text{If } \mathcal{X} \wedge \mathcal{P} \text{ is unsatisfiable.}$$

$$\frac{(S \cup \{(p.\forall \overline{x} : \mathcal{X}.\bot)\}, \mathcal{P})}{(S \cup \{(p.\bot)\}, \mathcal{P} \wedge \exists \overline{x}.\mathcal{X})} \text{If } \mathcal{P} \wedge \exists \overline{x}.\mathcal{X} \text{ has a solution.}$$

4.2 General dissubsumption

Redundancy checking is an essential part of any efficient refutation procedure. For example for resolution methods, subsumption is used to delete clauses that are instances of existing ones. In [7] a powerful extension of the subsumption rule, called the *dissubsumption rule* is introduced. Its principle is to add to a given formulae conditions preventing it from being an instance of another clauses. For example, assume that the set of clauses contains the c-clause $P(a)$. Then the clause $P(x) \vee Q(x)$ can be replaced by the clause with constraints $[\![P(x) \vee Q(x) \; : x \neq a]\!]$ (meaning that $P(x) \vee Q(x)$ is true if $x \neq a$). The constraint $x \neq a$ corresponds to the conditions preventing $P(x) \vee Q(x)$ from being an instance of $P(a)$.

We propose here to extend this rule to general constrained formulae (not only to clauses). For doing that, we take advantage of the partial model previously generated in a given branch in order to simplify formulae in this branch.

The principle is quite simple. Let A be a formula occurring in a given branch p. Let $\mathcal{M} = \mathcal{M}od(\mathcal{T}, p)$ be the set of facts corresponding to this branch. Then we know by Theorem 4 that

$$\mathcal{M} \cup \{A\} \equiv \mathcal{M} \cup \{A^+(\mathcal{M})\}$$

Hence we can remove the formula A from the branch and replace it by the formula $A^+(\mathcal{M})$. This idea can be formalized by the following rule.

$$\frac{(\mathcal{S} \cup \{(p, F[A]_q)\}, \mathcal{P})}{(\mathcal{S} \cup \{(p, F[A^+(\mathcal{M})]_q)\}, \mathcal{P})}\text{Where } \mathcal{M} = \mathcal{M}od(\mathcal{T}, p).$$

Example 3. Let $\mathcal{T}$ be a tableau. Assume that $\mathcal{T}$ contains in a branch b a formula $A = \forall x. P(f(x)) \vee \forall y. R(x, y)$ and the set of facts $\{\forall u. P(u), R(a, a)\}$. We have $\mathcal{F}^-(A, \mathcal{B}) = \bot$ and $\mathcal{F}^+(A, \mathcal{B}) = (\exists u. u = f(x)) \vee \forall y. (x = a \wedge y = a) \equiv \exists u. u = f(x)$. Then the formula A is deleted from the branch and replaced by

$$A' = \forall x : \forall u. f(x) \neq u. P(x) \vee \forall y. y \neq a \vee x \neq a. R(x, y) \wedge \forall x : \bot. \bot \wedge \forall x : \exists u. u = f(x).\mathsf{T}.$$

A' is equivalent to $\forall x : \forall u. f(x) \neq u. P(x) \vee \forall y. y \neq a \vee x \neq a. R(x, y)$ Since $\forall u. x \neq u$ has no solution, A' can be removed from the branch.

4.3 Properties of RAMCET-2

Definition 5. Let $(S, \mathcal{P})$ be a tableau and $\mathcal{I}$ an interpretation. $\mathcal{I}$ *validates* $(S, \mathcal{P})$ iff there exists a branch p in S and a solution σ in $\mathcal{S}ol(\mathcal{P})$ such that $\mathcal{I} \models SF(p, S)\sigma$.

Lemma 6. *Let $\mathcal{I}$ be an interpretation. Let $(S, \mathcal{X})$ be a tableau and let $(S', \mathcal{X}')$ be a tableau obtained from $(S, \mathcal{X})$ by applying the rules in Section 4.1. Then $\mathcal{I} \models (S, \mathcal{X}) \Rightarrow \mathcal{I} \models (S', \mathcal{X}')$.*

Theorem 7. *Let $\mathcal{T}$ be a tableau built from $((\Lambda. A), \mathsf{T})$. If $\mathcal{T}$ is closed then A is unsatisfiable.*

Theorem 8. *The tableau method using the rules introduced in Section 4.1 is refutationally complete.*

5 Model Building

We present in this section a new method for extracting models of some non closed branches. This method is far more general than the one used in [8]. It is based on generalization techniques: in order to extract a model from a given branch b, we try to *generalize* the partial model generated so far in the branch b, that is to say to take advantage of the set of facts belonging to b in order to find the interpretation $\mathcal{M}$ corresponding to the branch b. Then it will only remain to check whether $\mathcal{M}$ is a model of the initial formula.

In order to describe it on a formal basis, we need to introduce the concept of *representation set*. Intuitively speaking representation sets are sets of positive ground facts allowing to represent partial eq-interpretations. An eq-interpretation can be represented using a given representation set B by giving the value of each fact in B in the interpretation.

A representation set is a set of positive ground facts. Let B be a representation set. An interpretation $\mathcal{I}$ is said to be B-representable iff there exists a partial mapping α from B into $\{\top, \bot\}$ such that

$$\mathcal{I} = \{\neg A / A \in B, \alpha(A) = \bot\} \cup \{A / A \in B, \alpha(A) = \top\}$$

In this case $\mathcal{I}$ is noted $\mathcal{I}(B, \alpha)$. B-representable interpretations are similar to *settings* [16].

Representation sets will be used to guide the building of a model. The next subsections describe how to generate representation sets and how to use them for finding the model corresponding to a given branch.

5.1   Building representation sets

Let p be a positive integer. For any formula A we denote by A_p the set of expressions (terms of formulae) occurring at a position q in A such that $|q| \leq p$. We denote by $\mathcal{R}_p$ be the following system of rewriting rules (operating on sets of facts).

$\{\forall \overline{x} : C.A\} \rightarrow \{\forall \overline{x} : C \wedge t = s.A, \forall \overline{x}.C \wedge t \neq s.A\}$
If $(t, s) \in A_p^2$ and $C \wedge t = s \not\equiv \bot$ and $C \wedge t \neq s \not\equiv \bot$.
$\{\forall \overline{x} : C.A\} \rightarrow \{\forall \overline{x}, x_1, \ldots, x_n : C \wedge x = f(x_1, \ldots, x_n).A / f \in \Sigma, n = a(f)\}$
If $x \in \mathcal{V}$, $x \in A_p$ and $x_1, \ldots, x_n$ are new variables.

Theorem 9. *The nondeterministic application of the rules in $\mathcal{R}_p$ terminates.*

We will consequently note $\mathcal{S}_p$ a set of fact obtained by applying the system $\mathcal{R}_p$ on the set of facts $\{\forall \overline{x}.P(\overline{x}) / P \in \Omega\}$ (this set is not necessarily unique, but all the possible sets share some properties that we will use in the following).

The representation set $\mathcal{S}_p$ has an interesting property stated by the following theorem.

Theorem 10. *Let $\mathcal{I}$ be an eq-interpretation. Then $\mathcal{I}$ is $\mathcal{S}_p$-representable for some p. Moreover if an eq-interpretation is $\mathcal{S}_p$-representable then it is $\mathcal{S}_q$-representable for all $q \geq p$.*

5.2   Generalizing sets of facts

Representation sets will be used as a basis for extracting a model of a given open branch.

Let A be a fact and B be a set of facts. We note $A \bowtie B$ iff there exists $B \in B$ such that A and B are unifiable. Let $\mathcal{A}$ and B be two sets of facts. $\mathcal{A}$ is said to be B-*generalizable* iff for all $A \in \mathcal{A}$ either $A \not\bowtie B$ or $A \not\bowtie B^c$ (i.e. iff A is not simultaneously unifiable with an element of $\mathcal{B}$ or with the complementary of an element of $\mathcal{B}$).

Then, the $\mathcal{B}$-*generalization* of a set of facts $\mathcal{A}$ (noted $gen(\mathcal{B}, \mathcal{A})$) is the set of facts $\mathcal{I}(\mathcal{B}, \alpha)$ where α is defined as follows.

- $\alpha(F) = \top$ iff there exists a fact $F' \in \mathcal{A}$ such that F and F' are unifiable.
- $\alpha(F) = \bot$ iff there exists a fact $F' \in \mathcal{A}^c$ such that F and F' are unifiable.

Example 4. Let $\mathcal{B} = \{\forall x.P(x), \forall y.R(y)\}$. Let $\mathcal{A} = \{P(a), \neg R(b)\}$. $\mathcal{A}$ is $\mathcal{B}$-generalizable and we have $gen(\mathcal{B}, \mathcal{A}) = \{\forall x.P(x), \forall y.\neg R(y)\}$. Indeed, $\forall x.P(x)$ and $P(a)$ are unifiable and $\forall y.R(y)$ and $R(b)$ are unifiable.

$\mathcal{A}' = \{P(a), \neg P(b)\}$ is not $\mathcal{B}$-generalizable since $\forall x.P(x)$ is unifiable with both $P(a)$ and $P(b)$.

Let p be a branch of a tableau $\mathcal{T}$.

$gen(p, \mathcal{T}, \sigma, \mathcal{B})$ is a set $gen(SF(p, \mathcal{T})\sigma, \mathcal{B})$ where σ is a substitution of $Eqpb(\mathcal{T})$[4].

5.3   A more powerful method for model building

We first define a procedure *Extract* that extracts a model of a given branch. Informally speaking *Extract* takes as input a tableau $\mathcal{T}$ and an integer n and aims at choosing a branch in the tableau $\mathcal{T}$ and at finding (by using the function *gen*) a $\mathcal{S}_n$-representable interpretation $\mathcal{M}$ corresponding to this branch.

Procedure *Extract*:
%   Extracting a **Model** of a given tableau %
INPUT:
        A tableau $\mathcal{T}$
        A formula A
        An integer n
OUTPUT:
        A model of A
        or $\emptyset$
begin
  Compute a solution σ of $Eqpb(\mathcal{T})$
  choose a position p in $\mathcal{T}$
  $S := \mathcal{S}_n$
  if $\mathcal{M}od(\mathcal{T}, p)$ is S-generalizable
  then $\mathcal{M} := gen(p, \mathcal{T}, \sigma, S)$
  else return$(\emptyset)$
  if $\mathcal{F}^+(A, \mathcal{M}) \equiv \top$
  return$(\mathcal{M})$
  else return$(\emptyset)$
end

---

[4] In [9] an algorithm is described for finding a solution of a given satisfiable equational formula.

The intuitive principle of our method can be summarized as follows. We use an integer n (initially fixed to 0), and try to find a S_n-representable model of A. For doing that we build a tableau for F and use the procedure *Extract* for extracting the interpretation. If no model is found then we have to increase the value of n, until finding a model, or a refutation, of A. It is worth noting that when building the tableau, we only need to apply the rules of Section 4.1 on S_n-generalizable branches. Indeed, if a branch b is not S_n-generalizable then we know that it cannot correspond to a S_n-generalizable model of A. Hence adding new information in b would be irrelevant. This strategy preserves refutational completeness, since any fair application of the method will increase the value of n at a level sufficient for closing the branch.

Procedure RAMCET2:

% **Refutation And Model Construction with Eextended Tableau** – version 2 %

INPUT:

A formula A

OUTPUT:

UNSATISFIABLE or SATISFIABLE or (SATISFIABLE & A peq–model).

begin
    Let $\mathcal{T} = (\{(\Lambda, A)\}, \top)$
    Let $n := 0$
    while $(\mathcal{T} \not\equiv \perp \wedge \mathit{Extract}(\mathcal{T}, n) = \emptyset)$
    begin
        if there exists a rule ρ applicable on a S_n-generalizable branch;
        then modify $\mathcal{T}$ according to ρ.
        else $n := n + 1$
    end
    if $\mathcal{T} \equiv \perp$
    then return(UNSATISFIABLE)
    else return(SATISFIABLE & $\mathit{Extract}(\mathcal{T}, n)$)
end.

In practice the computation of the model should of course be done incrementally (for example, it is not necessary to recompute the set S_n at each time).

6   The class *Eqmodel*

We now state the main technical result of our paper, showing the impact of our approach.

The class *Eqmodel* is the set of formulae having an eq-interpretation.

Theorem 11. *Let* $A \in$ *Eqmodel. Any fair application of* RAMCET-2 *terminates on* A *and returns an eq-model of* A.

As a corollary our method is a decision procedure for all classes of formulae in *Eqmodel*. This implies that our method is strictly more powerful than the method by Fermüller and Leitsch [10] or the model generator SATCHMO that are based on the use of atomic representations.

In particular our method is a decision procedure for a wide range of decidable classes: OCC1N, PVD, KPOD [11, 10, 15] and more generally any class decidable by hyperresolution. Indeed it is proved in [5] that each satisfiable formula in these classes has an eq-model. More generally, let $\mathcal{C}$ be a class of formulae. If any satisfiable formula in $\mathcal{C}$ has an eq-model then RAMCET is a decision procedure for $\mathcal{C}$. This is the case for example for the class RRSD introduced in [14]. It is worth mentioning that RRSD cannot be decided by hyperresolution nor by A-ordered-resolution [14].

7 Example

We give below one example illustrating our method. Because of space limitation, we only give the main step of the method, i.e. the generation of the model from the branch of the tableau.

Example 5. Let A be the following formula (on the signature $\Sigma = \{a, f\}$).
$\forall x, y. P(x, y) \Leftrightarrow P(f(x), f(y)) \wedge \neg P(a, f(x)) \wedge P(a, a)$
Let $p = 0$. The reader can easily check that $\mathcal{S}_p$ is the set of facts $\mathcal{B} = \{\forall x, y : x \neq y. P(x, y), \forall x. P(x, x)\}$.
Then the procedure *Extract* returns the interpretation $\mathcal{M}$

$$\{\forall x, y : x \neq y : \neg P(x, y), \forall x. P(x, x)\}.$$

Indeed, $\forall x, y : x \neq y : P(x, y)$ is unifiable with $P(a, f(x))$ whose complementary belongs to $\mathcal{B}$ and $\forall x. P(x, x)$ is unifiable with $P(a, a)$.
$\mathcal{M}$ is obviously a model of A.
Though very simple, this example is interesting because

- A does not have any finite model;
- existing approaches cannot deal with this formula. Indeed they will generate the infinite branch $\{P(a, a), P(f(a), f(a)), P(f(f(a)), f(f(a))), \ldots\}$ but will *not* be able to infer the inductive consequence $P(x, x)$.

Therefore, no other (published) method is able to build a model for A.

We have presented a powerful extension of a tableaux-based model building method. This extension consists in a new method for extracting a model of a given branch, based on generalization techniques. This improvement allows to enlarge the class of constructible models. We prove that our method is a decision procedure for the class of formulae having an eq-model.

We would like to emphasize the *modularity* of our approach. Our procedure for extracting an eq-model is *independent* from the procedure used for generating the sets of facts. Hence one can very easily combine our approach with any existing method for enumerating Herbrand models (such that existing tableaux theorem provers, SATCHMO, HYPER-TABLEAUX...).

Moreover the techniques presented can be extended to more powerful model representation formalisms. Indeed, the formalism of equational constraints used in this paper is clearly not expressive enough for some applications. It does not allow for example to express the model defined by the very simple formula $P(a) \wedge (P(x) \Leftrightarrow \neg P(f(x)))$. In order to overcome this problem a solution is to *extend the language* used in the constraints by using for example tree automata or terms with integer exponents [18]. One of the more interesting features of our method is that it can be very easily extended to other formalisms. We have just to change the definition of representation set, for example by adding new rules in the system $\mathcal{R}_p$ in order to take into account the new features of the extended language.

The following are the main lines of future work.

- Extend the model building method proposed here to first-order formula with equality. The main problem is that the formalism of eq-interpretation cannot be used in this case (since the formula may have no models on the finite tree algebra). Hence we have to find new representations of interpretation with "good" decidability properties (see [2] for more details).
- Improve our current implementation and study the practical performances of our method.
- Try to identify more precisely the classes decidable by RAMCET-2. It seems that there is no syntactical caracterization of the class *Eqmodel*, however one could find a syntactical caracterization of some subclass of it.

Acknowledgments

We would like to thank Ricardo Caferra and Stéphane Demri for a careful reading of an earlier version of this paper, and the anonymous referees for their pertinent and precise comments.

References

1. P. Baumgartner, U. Furbach, and I. Niemela. Hyper-tableaux. In *Logics in AI, JELIA'96*. Springer, 1996.
2. C. Bourely, R. Caferra, and N. Peltier. A method for building models automatically. Experiments with an extension of Otter. In *Proceedings of CADE-12*, pages 72–86. Springer, 1994. LNAI 814.
3. F. Bry and A. Yahya. Minimal model generation with positive unit hyper-resolution tableaux. In *Proceeding of Tableaux'96*, LNAI 1071, pages 143–159. Springer, 1996.
4. R. Caferra and N. Peltier. Model building and interactive theory discovery. In *Proceeding of Tableaux'95*, LNAI 918, pages 154—168. Springer, 1995.
5. R. Caferra and N. Peltier. Decision procedures using model building techniques. In *Proceeding of Computer Science Logic, CSL'95*, pages 130–144. Springer, LNCS 1092, 1996.
6. R. Caferra and N. Peltier. A significant extension of logic programming by adapting model buildings rules. In *Proc. of Extensions of Logic Programming 96*, pages 51–65. Springer, LNAI 1050, March 1996.

7. R. Caferra and N. Zabel. Extending resolution for model construction. In *Logics in AI, JELIA '90*, pages 153–169. Springer, LNAI 478, 1990.

8. R. Caferra and N. Zabel. Building models by using tableaux extended by equational problems. *Journal of Logic and Computation*, 3:3–25, 1993.

9. H. Comon and P. Lescanne. Equational problems and disunification. *Journal of Symbolic Computation*, 7:371–475, 1989.

10. C. Fermüller and A. Leitsch. Hyperresolution and automated model building. *Journal of Logic and Computation*, 6(2):173–203, 1996.

11. C. Fermüller, A. Leitsch, T. Tammet, and N. Zamov. *Resolution Methods for the Decision Problem*. LNAI 679. Springer, 1993.

12. M. Fujita and Hasegawa. A model generation theorem prover in kl1 using a ramified stack algorithm. In *Proceedings of 8th International Conference Symp. Logic Programming*, pages 1070—1080, 1991.

13. H. Gelernter, J. Hansen, and D. Loveland. Empirical explorations of the geometry theorem-proving machine. In J. Siekmann and G. Wrightson, editors, *Automation of Reasoning, vol. 1*, pages 140–150. Springer, 1983. Originally published in 1960.

14. S. Klingenbeck. *Counter Examples in Semantic Tableaux*. PhD thesis, University of Karlsruhe, 1996.

15. A. Leitsch. Deciding clause classes by semantic clash resolution. *Fundamenta Informaticae*, 18:163–182, 1993.

16. D. W. Loveland. *Automated Theorem Proving: A Logical Basis*, volume 6 of *Fundamental Studies in Computer Science*. North Holland, 1978.

17. R. Manthey and F. Bry. SATCHMO: A theorem prover implemented in Prolog. In *Proc. of CADE-9*, pages 415–434. Springer, LNCS 310, 1988.

18. N. Peltier. Increasing the capabilities of model building by constraint solving with terms with integer exponents. Research Report IMAG, No 966-I. To appear in the Journal of Symbolic Computation, November 1996.

19. N. Peltier. Simplifying and generalizing formulae in tableaux. pruning the search space and building models (long version). http://leibniz.imag.fr/ATINF/Nicolas.Peltier, 1996.

20. S. Reeves. Semantic tableaux as framework for automated theorem-proving. In *AISB Conference*, pages 125–139, 1987.

21. J. R. Slagle. Automatic theorem proving with renamable and semantic resolution. *Journal of the ACM*, 14(4):687–697, October 1967.

22. J. Slaney. Finder (finite domain enumerator): Notes and guides. Technical report, Australian National University Automated Reasoning Project, Canberra, 1992.

23. T. Tammet. Using resolution for deciding solvable classes and building finite models. In *Baltic Computer Science*, pages 33–64. Springer, LNCS 502, 1991.

24. S. Winker. Generation and verification of finite models and counter-examples using an automated theorem prover answering two open questions. *Journal of the ACM*, 29(2):273–284, April 1982.

25. J. Zhang. Search for models of equational theories. In *Proceedings of ICYCS-93*, pages 60–63, 1993.

A Framework for Using Knowledge in Tableau Proofs

Benjamin Shults

The University of Texas at Austin, Austin TX 78712, USA

Abstract. The problem of automatically reasoning using a knowledge base containing axioms, definitions and theorems from a first-order theory is recurrent in automated reasoning research. Here we present a sound and complete method for reasoning over an arbitrary first-order theory using the tableau calculus. A natural, well-motivated and simple restriction (implemented in IPR) to the method provides a powerful framework for the automation of the selection of theorems from a knowledge base for use in theorem proving [22]. The restrictions are related to semantic resolution restrictions and the set-of-support restriction in resolution, and to hyper-tableaux and the weak connection condition in tableaux. We also present additional tableau rules used by the IPR prover for handling some equality which is not complete but is sufficient for handling the problems in its intended domain of problem solving.

1 Introduction

The rules presented in this paper allow an automatic theorem proving program to prove theorems in some theories in which rewriting and Horn techniques are not particularly well-suited. An implementation, IPR, of a restriction to the rules given in this paper has proved the following exercise in Topology [1]

$$(\forall S)(\text{Hausdorff}(S) \supset \text{closed-in}(\text{top-to-class}(\text{the-diagonal-of}(S)), S \times_\tau S))$$

given a knowledge base whose size is on the order of 20 sequents where the proof requires the use of 12 of the sequents from the knowledge base. The following seven formulas produce all twelve of the sequents needed in the proof. They actually produce more than twelve, but only twelve are used in the proof.

Definition of diagonal:
$$(\forall X)(\forall S)(X \in \text{top-to-class}(\text{the-diagonal-of}(S)) \leftrightarrow$$
$$(\exists A)(A \in \text{top-to-class}(S) \wedge X = \langle A, A \rangle))$$

Axiom of ordered pairs:
$$(\forall A)(\forall B)(\forall C)(\forall D)(\langle A, B \rangle = \langle C, D \rangle \supset (B = D \wedge A = C))$$

Definition of disjoint:
$$(\forall A)(\forall B)(\text{disjoint}(A, B) \leftrightarrow \neg(\exists Y)(Y \in A \wedge Y \in B))$$

Definition of product:
$$(\forall X)(\forall S)(\forall T)(X \in S \times T \leftrightarrow (\exists A)(\exists B)(A \in S \wedge B \in T \wedge X = \langle A, B \rangle))$$

Part of the definition of product of topological spaces:
$$(\forall X)(\forall S)(\forall T)(X \in \text{top-to-class}(S \times_\tau T) \supset$$
$$(\exists A)(\exists B)(A \in \text{top-to-class}(S) \wedge B \in \text{top-to-class}(T) \wedge X = \langle A, B \rangle)))$$

Part of the definition of Hausdorff:
$$(\forall X)(\text{Hausdorff}(X) \supset (\forall A)(\forall B)((A \in \text{top-to-class}(X) \wedge B \in \text{top-to-class}(X) \wedge$$
$$A \neq B) \supset (\exists G1)(\exists G2)(\text{open-in}(G1, X) \wedge \text{open-in}(G2, X) \wedge A \in G1 \wedge$$
$$B \in G2 \wedge \text{disjoint}(G1, G2))))$$

Part of the definition of closed:
$$(\forall X)(\forall A)((\forall y)((y \in \text{top-to-class}(X) \wedge y \notin A) \supset$$
$$(\exists G)(y \in G \wedge \text{open-in}(G, X) \wedge \text{disjoint}(G, A))) \supset \text{closed-in}(A, X))$$

Notice that $\times_\tau$ is the product topology on the product of topological spaces, whereas $\times$ is the simple cartesian product of sets. Also the-diagonal-of(S) represents a subspace (rather than a subset) of $S \times_\tau S$.

The proof was found, pared down to its shortest form and printed in English by an implementation of the framework presented here in about 30 seconds. Other examples of traditionally difficult problems solved by IPR are given elsewhere [21] including one in which over 100 related theorems were present in the knowledge base. The reader will notice that this problem requires some equality reasoning. In Section 4 we briefly describe the equality mechanisms used by IPR.

Axioms, theorems and definitions are stored in the knowledge base in the form of sequents rather than clauses, formulas or rewrite rules. The sequents are broken down toward conjunctive normal form, but existential quantifiers are not removed.

The notation we use is that which is used by Smullyan and Fitting [24, 10]. We use the free-variable semantic tableau method for theorem proving. The ε-rule for applying knowledge from the knowledge base is consistent with any of the liberalized δ-rules [12, 8, 2] and with mixed rigid and universal variables [7].

Section 2 describes the rules which are used in forming the sequents in the knowledge base and includes soundness and completeness results. Section 3 describes the ε-rule which allows sequents from the knowledge base to be used in a tableau proof. Soundness and completeness of the combination of the rules in Sections 2 and 3 is proved. In Section 4, we describe the set-of-support restriction and other strategies which make IPR successful.

2 Knowledge-Base Construction

In this section, we define the special sequent calculus rules that are used in the process of forming the knowledge base. The rules are the same as the rules for transforming a formula into conjunctive normal form except that we do not skolemize.

The rules given in this section are used to take a formula and put it into the knowledge base. Section 3 describes how sequents from the knowledge base are used in the proof of a new formula.

First we mention the motivation for storing information in this form. The form is closely related to clausal form but it has certain advantages and disadvantages. We also present an illustrative example of a set of sequents being formed from an input definition. Finally, we prove certain soundness and completeness results which are needed later.

2.1 Motivation

The framework presented in this paper was developed to support an automated theorem proving program which is intended to have certain properties. These desired properties influenced the development of the framework. The two important properties we wanted the program to have were the ablity to prove theorems in mathematics using knowledge and that it be easy to use by someone who is not an expert in ATP techniques. Therefore, we want it to be very easy for the user to understand the contents of the knowledge base and to understand how knowledge is applied. This was the initial motivation for avoiding skolemization. Since then, we have found some other advantages and disadvantages to using this format.

Efficiency. The first advantage we mention for our framework for storing knowledge is efficiency. This efficiency comes about in two ways. First, since formulas are stored at a slightly higher level, matching can occur earlier in the theorem proving process. Second, and for the same reason, the knowledge base is smaller than if formulas were taken all the way to clausal form.

Completeness. The main *disadvantage* to this framework is that when we implement the set-of-support strategy, completeness is lost. This is explained more in Section 4.

Readability. The advantage which most motivated the adoption of the form which knowledge takes in this framework is readability. Part of the problem non-experts have in reading formulas in clausal form is easily relieved by writing the clauses in "Kowalski" or sequent form. However, this does not take care of the more serious logical problem caused by skolem functions.

Example 1. Consider the example of the definition of a continuous function which we want to put into the knowledge base.

$$(\forall f)(\forall S)(\forall T)(\text{continuous-from-to}(f, S, T) \leftrightarrow$$
$$(\text{a-function-from-to}(f, \text{coerce-to-class}(S), \text{coerce-to-class}(T)) \wedge$$
$$(\forall G)(\text{open-in}(G, T) \supset \text{open-in}(\text{the-set-inverse-image}(f, G), S))))$$

Here is this formula in Kowalski form:

continuous-from-to$(f, S, T) \supset$
a-function-from-to$(f, \text{coerce-to-class}(S), \text{coerce-to-class}(T))$

continuous-from-to$(f, S, T) \wedge$ open-in$(G, T) \supset$
open-in(the-set-inverse-image$(f, G), S)$

a-function-from-to(f, coerce-to-class(S), coerce-to-class(T)) $\supset$
    open-in($G(f, S, T), T$)$\vee$
    continuous-from-to(f, S, T)

a-function-from-to(f, coerce-to-class(S), coerce-to-class(T))$\wedge$
      open-in(the-set-inverse-image($f, G(f, S, T)$), S) $\supset$
    continuous-from-to(f, S, T)

The first two clauses are perfectly readable and sensible.[1] However, the second two are difficult to read. What's worse, after they are read, they don't make sense on their own. And finally, they do not follow from the input formula, that is to say, there are models of the original formula which are not models of the last two clauses.

These facts make a formula in clausal form much more difficult to understand and relate to the formulas which the user input. In Section 2.2 we present this example again using the rules we introduce below.

2.2 Rules in Sequent Form

Here we present the rules for breaking formulas into sequents to be put into the knowledge base. The rules are the same as the rules for converting into conjunctive normal form with the two exceptions that we do not skolemize and we use sequent notation.

Because we use sequents, the presentation of the rules in this format may seem strange. We stick as closely as possible to Smullyan's unified notation. In Smullyan's notation (see Chapter XI, Section 1 of *First-Order Logic* [24]) formulas in the antecedent of a sequent correspond to positive formulas on a branch of a tableau and formulas in the consequent of a sequent correspond to negative formulas on a branch of a tableau. Therefore, an existentially quantified formula in the consequent of a sequent is a γ-formula.

Recall that if Γ and Δ are sets for formulas, then a sequent, $\Gamma \rightarrow \Delta$, is thought of as asserting that the disjunction of the formulas in Δ logically follows from the conjunction of the formulas in Γ.

Suppose X is an axiom, definition or theorem of a theory expressed as a closed sentence. In order to include X in the knowledge base, we build a sequent tableau rooted at the sequent $\rightarrow X$, that is, the sequent which asserts the truth of X.

Smullyan's rules for propositional connectives (α- and β-formulas) are used in this context. For example, if we want the formula $A \supset (C \wedge B)$ to be in the knowledge base, we begin with the sequent $\rightarrow A \supset (C \wedge B)$ and apply the ordinary α- and β-rules from the sequent calculus so that the sequents $A \rightarrow C$ and $A \rightarrow B$ end up in the knowledge base. Notice that this is just the clausal or Kowalski form of the original formula.

---

[1] The reader who is unfamiliar with these predicates and functions can simply imagine that they are undefined. No knowledge of mathematical topology is assumed.

However, the quantifier removal rules must be reversed. (When Fitting [10] describes clausifying, he takes the opposite approach so that the original formula is negated, propositional rules are reversed and the quantifier removal rules are left the same. Our approach is more appropriate to the semantics of sequents with which Fitting was not concerned in this context. Furthermore, in the present context, we do not negate the formulas being clausified since they are our assumptions.) Following the semantics of sequents, if the formula we want to have in the knowledge base is of the form $(\forall x)P(x)$, then we begin by asserting $\rightarrow (\forall x)P(x)$. But, while this is called a δ-formula in Smullyan's unified notation, we clearly want to apply a γ-rule and introduce a new free variable to a formula of this form.

Summarizing, in the context of building a sequent tableau during the construction of the knowledge base, in addition to the α- and β-rules (propositional rules), we apply the free variable γ-rule once to each δ-formula and eliminate the δ-formula from the new sequent. In order to avoid confusion, we call this the δ^{-1}-rule.

Definition 1 δ^{-1}-rule.

$$\frac{\Gamma \rightarrow \Delta, A(x')}{\Gamma \rightarrow \Delta, (\forall x)A(x)} \qquad \frac{A(x'), \Gamma \rightarrow \Delta}{(\exists x)A(x), \Gamma \rightarrow \Delta}$$

where x' is a new variable.

Furthermore, if the top-level quantifier of a γ-formula can be "pushed inward" to obtain an equivalent formula, then that may be done. We call this the γ^{-1}-rule although it has other names such as "anti prenexing." When all possible α-, β-, γ^{-1}- and δ^{-1}-rules have been applied, we will call the resulting sequent tableau a *finished KB-tableau*. The sequents on the leaves of the finished KB-tableau are added to the knowledge base. In the remainder of the paper, we say that a sequent s is *in a knowledge base* if and only if s is a leaf of the finished KB-tableau of some first-order formula.

The following example of an application of the method illustrates its effectiveness at simulating some aspect of the way humans think of theorems.

Example 2. Suppose we want the definition of a continuous function to be in the knowledge base. We begin with the desired definitional formula and put it in the conclusions of a sequent. In other words, we form the sequent which asserts the truth of the formula.

$$\rightarrow \left\{ \begin{array}{l} (\forall f)(\forall S)(\forall T)(\text{continuous-from-to}(f, S, T) \leftrightarrow \\ (\text{a-function-from-to}(f, \text{coerce-to-class}(S), \text{coerce-to-class}(T)) \wedge \\ (\forall G)(\text{open-in}(G, T) \supset \text{open-in}(\text{the-set-inverse-image}(f, G), S)))) \end{array} \right\} \quad (1)$$

After the application of all α-, β- and δ^{-1}-rules, the following three sequents are obtained:

$$\text{continuous-from-to}(f, S, T) \rightarrow$$
$$\text{a-function-from-to}(f, \text{coerce-to-class}(S), \text{coerce-to-class}(T)) \quad (2)$$

$$\left\{ \begin{array}{c} \text{continuous-from-to}(f, S, T), \\ \text{open-in}(G, T) \end{array} \right\} \rightarrow \text{open-in(the-set-inverse-image}(f, G), S) \quad (3)$$

$$\left\{ \begin{array}{c} \text{a-function-from-to}(f, \text{coerce-to-class}(S), \text{coerce-to-class}(T)), \\ (\forall G)(\text{open-in}(G, T) \supset \text{open-in(the-set-inverse-image}(f, G), S)) \end{array} \right\} \rightarrow \quad (4)$$
$$\text{continuous-from-to}(f, S, T)$$

Therefore, the finished KB-tableau for the original formula in sequent 1 has three leaves (the sequents 2, 3 and 4) which are added to the knowledge base. Notice that sequents 2 and 3 are exactly the first two clauses from example 1 while sequent 4 encodes the information from the last two clauses in example 1 in a more comprehensible, readable and logical way.

We prove that this procedure for creating a knowledge base is both sound and complete. By soundness, we mean that the original formula logically implies all of the sequents which are leaves of its finished KB-tableau. By completeness, we mean that the conjunction of the leaves of a finished KB-tableau logically imply the formula at the root.

Theorem 2 Soundness of the Knowledge Base. *If a sequent s is a leaf of a finished KB-tableau rooted at* $\rightarrow \phi$ *then every model of* ϕ *satisfies s.*

The proof is routine by induction on the length of the branch between s and the root.

Theorem 3 Completeness of the Knowledge Base. *Suppose* ϕ *is a closed formula and S is the set of leaves of the finished KB-tableau for* ϕ. *Then every model of S satisfies* ϕ.

This proof is routine. By induction we prove the stronger result that the conjunction of the leaves of the tree entail the root at every stage of the construction of the tree.

3 The ε-rule

Here we describe a general and complete rule for applying sequents from a knowledge base in the context of a free-variable semantic tableau proof [24, 10]. We illustrate the use of the rule by proving a theorem of topology. We then show that the rule is sound and we present a systematic procedure using this rule and describe the proof of the completeness of that procedure. In Section 4, we describe a set-of-support strategy and other strategies applied by the IPR implementation in guiding the use of the ε-rule.

In the rules of the tableau calculus, the symbol naming the rule (e.g. α) is also used to represent the formula to which the rule is to be applied. Since the ε-rule is applied to a *set* of formulas, we let ε stand for any (possibly empty) set of formulas on a branch of a tableau. Each element of ε must be either a γ-formula or a literal. An application of the ε-rule is an application of a sequent in the knowledge base to the formulas in ε.

Let ε be a subset of a branch of a tableau, and $s = U \to V$ a sequent in the knowledge base. In order to apply the ε-rule, ε and s must satisfy the following conditions:

- s must have its free variables renamed if necessary so that they are disjoint from the free variables in the tableau,
- $\varepsilon = H \cup G$ where H is the set of positive formulas in ε and G is the set of negative formulas in ε and
- there are subsets U' and V' of U and V respectively and a most general idempotent substitution, σ, such that
  - $H\sigma = U'\sigma$ and
  - $\{\psi\sigma | \neg\psi \in G\} = V'\sigma$.

For the statement of the ε-rule, we let $M = V\sigma \setminus V'\sigma$ and $N = U\sigma \setminus U'\sigma$. Loosely speaking, M contains exactly the formulas in V (the conclusions in s) which were not successfully unified with any formula in G (the negative formulas in ε) and similarly for N with respect to the hypotheses and positive formulas. We will also let σ' denote the subset of σ containing only those substitutions of variables in ε.

Definition 4 ε-rule. Where $M = \{M_1, \ldots, M_m\}, N = \{N_1, \ldots, N_n\}$ and σ' have been determined as described above we apply the following tableau rule (with provisions below) to every branch containing ε:

$$\frac{\varepsilon}{M_1 \mid \cdots \mid M_m \mid \neg N_1 \mid \cdots \mid \neg N_n}$$

- none of the new formulas already occurs on the branch,
- the tree substitution rule must be applied with σ' and
- if $M = N = \emptyset$ then the branch is closed by the application of the rule.

The provisions are stated assuming that depth-first unification is being used. We will discuss below how they need to be restated if breadth-first unification is being used. The first provision is not needed for soundness but it—like subsumption in a resolution prover—does not take away any first-order completeness and is therefore generally useful. This is what is called a *regularity* condition [15]. The ε-rule would be unsound without the second provision.

If breadth-first unification is being used, then the provisions change. It goes against the idea of breadth-first unification to apply σ' across the entire tree until it is known to be the substitution which finishes the proof. However, it is not enough simply to apply the substitution σ' to the formulas being introduced by the ε-rule because some formula in $M \cup N$ may not contain any occurrence of some variable replaced in σ'. Here is a solution which works and maintains the goals of breadth-first unification: the substitution σ' need not be applied across the tableau but must be associated with the formulas added by the rule and their descendents. In subsequent applications of the ε-rule, the substitution formed must be composed with the substitutions associated with each of the formulas in ε and the composition of those substitutions must be associated with the formulas created by the ε-rule. Furthermore, if a contradiction is sought between two formulas on a branch, the substitution which unifies them must be composed with the substitutions associated with the two formulas.

The first provision given in the depth-first ε-rule may destroy completeness if breadth-first unification is used since the two identical formulas may have different substitutions associated with them. Regarding the third provision, it should be understood that instead of closing a branch we label the appropriate node with σ'.

Example 3. Here we illustrate the use of the ε-rule in a fairly simple example. The challenge is part of the 101st labeled theorem from John Kelley's *General Topology* [13]. This is Theorem 19 on page 147.

If $\prod_A^\tau X$ is locally compact then for every a, X_a is locally compact.

We will prove this in the knowledge base containing the following sequents:

$$\rightarrow \text{continuous-from-onto}(\pi_a, \prod_A^\tau X, X_a) \tag{1}$$

$$\rightarrow \text{open-from-onto}(\pi_a, \prod_A^\tau X, X_a) \tag{2}$$

$$\left\{ \begin{array}{c} \text{open-from-onto}(f, A, B), \\ \text{continuous-from-onto}(f, A, B), \\ \text{locally-compact}(A) \end{array} \right\} \rightarrow \text{locally-compact}(B) \tag{3}$$

Each of these sequents comes from a statement earlier in Kelley's book. Here, $\prod_A^\tau X$ represents the product topology where X is a bijection from the index set A to a set of topologies. We use $\prod^\tau$ rather than $\prod$ to distinguish the topology from the underlying set.

In Figure 1 we use parentheses to distinguish a variable, such as X, from a skolem constant, such as (X). We construct the tableau in Figure 1 using the ordinary free-variable tableau rules until we reach node 5.

After node 5 has been added, no more ordinary tableau rules may be applied so we apply an ε-rule. The set ε will contain the only two literals in the tableau (4 and 5). We will apply the sequent 3 with the substitution $\sigma = \{\prod_{(A)}^\tau (X)/A, (X)_{(a)}/B\}$. Applying the ε-rule, we split the tableau into two branches as shown. The left branch is closed by applying the ε-rule with ε containing only formula 6. We use the sequent 2 and the substitution $\sigma = \{(A)/A, (X)/X, (a)/a, \pi_{(a)}/f\}$. The right branch is closed by applying ε-rule with ε containing formula 7. We use sequent 1 and the same σ as was used on the left branch. Both of these applications of the ε-rule close the respective branches and the substitutions used are identical. Therefore the tableau is closed by the substitution $\sigma' = \{\pi_{(a)}/f\}$.

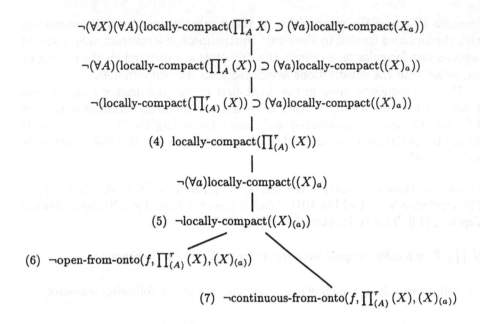

Fig. 1. The tableau for example 3.

This proof is found by the IPR implementation even in the presence of a knowledge base containing over 100 sequents related to the concepts of products, product topologies, (locally) compact spaces, etc. See Section 4 for a description of the strategy used by IPR.

Definition 5 *S*-**tableau.** If S is the set of sequents from the knowledge base used in applications of the ε-rule in the construction of a tableau, then the tableau is called an *S-tableau.*

If we let S denote the set of sequents used in this proof, (i.e. sequents 1, 2 and 3) then our soundness result will show that since the S-tableau is closed, the theorem proved is true in every model of the formulas input to create the sequents in S.

Definition 6 *z.* If s is a leaf of a finished KB-tableau rooted at the sequent $\to \phi$ where ϕ is a closed formula, then we let the function $z : s \mapsto \phi$ identify the formula at the root. Furthermore, if S is a set of sequents in the knowledge base, then we let $z(S)$, as is usually done in mathematics, denote the set $\{\phi \mid \text{for some } s \in S, \phi = z(s)\}$.

Using this function z, we can now state the meaning of soundness as follows: if an S-tableau with root labeled $\neg\phi$ is closed, then ϕ is true in every model of $z(S)$. The following theorem is useful in the proof of soundness of the ε-rule.

Lemma 7. *Suppose S is a set of sequents in the knowledge base. Suppose T is an S-tableau whose root is labeled by a closed formula ϕ and there is an interpretation, $\mathcal{M}$, such that $\models_{\mathcal{M}} S$ and $\models_{\mathcal{M}} \phi$. Then $\models_{\mathcal{M}} T$.*

The standard induction on the sequence of tableaux constructed between ϕ and T demonstrates the truth of this lemma.

Theorem 8 Soundness. *If T is a closed S-tableau whose root is labeled by the closed formula $\neg\phi$, then every model of $z(S)$ satisfies ϕ.*

Proof. Since T is closed, T is not satisfiable. Therefore, $\neg\phi$ cannot be satisfied by any model of S by Lemma 7. Thus every model of S satisfies ϕ. Since every model of $z(S)$ satisfies S by Theorem 2, every model of $z(S)$ satisfies ϕ.

Now we present a simple systematic tableau procedure using the ε-rule which is complete for any first-order theory. This systematic procedure makes no effort to be efficient. See Section 4 for a description of a restriction to the rule which destroys completeness but gives good success. More efficient complete procedures will be mentioned in Section 5.

Definition 9 Most General ε-Rule Application. An application of the ε-rule in which $\varepsilon = \emptyset$ and $s = U \rightarrow V$ is any sequent in the knowledge base with variables renamed so as not to conflict with variables in the tableau is called a *most general application* of the ε-rule. When we make such an application, we say that we apply the most general ε-rule.

When we apply the most general rule, $M = V, N = U$ and $\sigma' = \epsilon$. Under these circumstances, the application of the ε-rule is equivalent to adding the sequent s to the branch.

Suppose we are given a set Φ of first-order sentences and that they have been used to create a knowledge base, S. That is, S is the set of leaves of the KB-finished tableaux for the formulas in Φ. Suppose that each sequent in the knowledge base, S, is labeled with a counter initialized to 0. To prove that every model of Φ satisfies ϕ we build a systematic tableau following these steps. At step 0, we initialize the tableau with the formula $\neg\phi$ and we set $q = 1$. Every formula in the tableau has two labels on it: {used|unused} and a *copies* label which holds an integer. Each new formula starts out unused and with its copies set to 0. At step $n+1$, if there is an unused α-, δ- or β-formula, then we apply the appropriate rule and label the formula used. Otherwise, if there is a γ-formula whose copies label is less than q, then we apply the γ-rule and increment the copies label of the γ-formula. Otherwise, if there is a sequent in the knowledge base S whose counter is less than q, then we apply the most general ε-rule for that sequent to every branch of the tableau on which it is permissible by the provisions. Then we increment the counter of that sequent. If none of these rules can be applied, then we increment q and continue. An S-tableau constructed by these rules is *finished* if it is infinite or if q is incremented twice without any other rule being applied.

The completeness proof follows the well-known proofs [10] and the changes are straightforward. We include here the statement of the version of Hintikka's Lemma needed in the proof.

Definition 10 Hintikka set for a knowledge base. Let L be a first-order language and S be a knowledge base (in the language.) That is, S is the set of leaves of the finished KB-tableaux for some satisfiable set of first-order sentences. A set H of formulas of L is a *Hintikka set for the knowledge base S* (with respect to L) provided H satisfies the following conditions:

H_0 No conjugate pair occurs in H

H_1 $\perp, \neg\top \notin H$

H_2 If $\alpha \in H$ then $\alpha_1, \alpha_2 \in H$

H_3 If $\beta \in H$ then $\beta_1 \in H$ or $\beta_2 \in H$

H_4 If $\gamma \in H$ then $\gamma(t) \in H$ for every closed term t of L

H_5 If $\delta \in H$ then $\delta(t) \in H$ for some closed term t of L

H_6 For every sequent $s \in S$ if $\mathbf{x}$ is the vector of free variables in s and

$$s = \{U_i(\mathbf{x})|1 \leq i \leq u\} \to \{V_i(\mathbf{x})|1 \leq i \leq v\}$$

then for every vector $\mathbf{t}$ of closed terms in L, some element of $\{\neg U_i(\mathbf{t})|1 \leq i \leq u\} \cup \{V_i(\mathbf{t})|1 \leq i \leq v\}$ is in H.

Lemma 11 Hintikka's Lemma. *Every Hintikka set for the knowledge base S (with respect to a language with a non-empty set of closed terms) is satisfiable by a model of S.*

The ordinary proof of Hintikka's Lemma goes through [10]. The model constructed is easily seen to be a model for S.

Theorem 12 Completeness. *Given the set Φ of first-order formulas, suppose S is the set of leaves of the finished KB-tableaux for the formulas of Φ. If every model of Φ satisfies ϕ, then there is a closed S-tableau for $\neg\phi$.*

Proof. Since every model of S satisfies Φ by the completeness of the knowledge base (Theorem 3), every model of S satisfies ϕ. If the finished S-tableau constructed by the systematic procedure given above for $\neg\phi$ has an open branch B, then any ground instance of that branch is a Hintikka set for S. Thus some model of S satisfies B by Hintikka's Lemma. Hence, this model satisfies $\neg\phi$. But this contradicts the fact that every model of S satisfies ϕ. Thus there can be no open branch on the finished S-tableau for $\neg\phi$. So the finished S-tableau for $\neg\phi$ must be closed.

4 IPR

IPR is a theorem-proving system which uses the ε-rule. The IPR system prefers applications of the ε-rule which add as few formulas to the tableau as possible [22]. In this section, we describe how IPR restricts the use of the ε-rule and we give the additional rules used by IPR for handling equality because equality is used in the example in Section 1.

Restricting the ε-rule. The IPR system is not complete over a first-order theory. Instead, it is selective in its choice of theorems to apply and how to apply them. It does this by taking advantage of the representation of the knowledge in the knowledge base and common-sense strategies in applying the ε-rule.

Recall that in the ε-rule the number of branches added to the tableau depends directly on the number of formulas in the sequent from the knowledge base which were not unified with formulas on the branch. An obvious way to try to improve the performance of a fetching algorithm would be to have it prefer to apply the ε-rule in such a way that the number of formulas added to the tableau is small. In the limiting case, when there are no formulas added to the tableau, the branch is closed by the application of the ε-rule. IPR gives this preference the highest weight.

A second natural way to improve performance is to have the fetcher prefer to apply the ε-rule when the number of formulas which *do* match is large. This makes the fetcher prefer to apply theorems which tend to have more in common with the theorem being proved.

In addition to the above, IPR prefers to apply the ε-rule when the substitution σ' has small cardinality and is not complex. This is another way of helping the prover select theorem applications which have more to do with the problem at hand.

Since IPR is intended to work in domains in which rewriting techniques are not strong, it prefers not to apply the same theorem in the same way many times.

Given the framework we have set up for storing and applying lemmas, it is very easy to implement these strategies. Each of these strategies can be reworded in terms of the information available to the theorem application (ε-) rule. The most important strategies are implemented simply by preferring to apply theorems for which $n + m$ (the number of formulas added to the tableau) is small. This is the criterion which is given the most weight by IPR. This "$n + m$" strategy, together with the framework for storing knowledge, seem to be the main contributors to the success of IPR.

The IPR program applies another important restriction to the ε-rule which is not explicitly mentioned above.

Set-of-support Restriction With the exception of the application of theorems involving equality, when a theorem is applied, $\varepsilon \neq \emptyset$.

That is to say that some formula from the theorem being applied must be unified with some formula in the tableau. (The handling of equality in theorem fetching is not described in detail here.) This restriction destroys completeness because of the fact that theorems are not stored in the form of clauses. Examples which exploit this incompleteness are generally unnatural syntactic tricks.

Example 4. Suppose the axiom $(\exists x)(P(x) \wedge Q(x))$ is entered into the knowledge base. Without using an empty ε, we can prove neither $(\exists x)(Q(x) \wedge P(x))$ nor $(\exists x)Q(x)$ from this knowledge base even though they are logical consequences.

However, by giving up a bit of completeness and resigning degenerate examples as these, a method may be found which is very successful on more natural examples.

These are the strategies used by IPR in fetching lemmas. It is remarkable that the framework allows for such simple strategies to have impressive results [21]. The kind of reasoning which IPR handles seems to be representative of a large area of reasoning which has been difficult for other methods.

Equality. IPR combines two fast, incomplete methods for handling equality. The first, which we call Brown's Rule, is a limited substitution method [9]. The second is an incomplete E-unification method.

Inference Rule 13 (Brown's Rule) If a branch contains a formula of the form $s = t$, where one side of the equality (say s) is a skolem term which does not occur in the term on the other side (t), then add to the branch a copy of every formula on the branch containing s and replace all occurrences of s in the new formulas with t.

It turns out that this single, efficient inference rule is sufficient for the proof of the example given in Section 1 as well as all of the problems given in [21].

IPR supplements this restricted substitution of equals with some E-unification. IPR's method of E-unification is complete for ground equality. This is accomplished by the ground congruence closure technique [17, 18, 16].

Some simple equality reasoning is also applied when fetching theorems [23].

5 Closely Related Work

Most of the closely-related work in automating the use of a knowledge base relies on the use of rewriting or Horn clauses. IPR does not control rewriting very well and therefore does not perform well on the difficult challenges on which these other methods show their strength. However, IPR does perform well on some problems which do not require (or allow) rewriting techniques.

The work of David Barker-Plummer on his Gazing algorithm [3, 4] is very closely related to the work here. The formulas entered into Gazer's knowledge base are translated into "rewrite normal form" which is similar to the form taken by sequents in IPR's knowledge base although, in Gazer, the formulas are assumed to be nearly Horn. Gazer uses abstraction planning to plan the entire proof. This gives Gazer an advantage over IPR on problems which require reasoning which has the nature of rewriting but Gazer's planner can only find linear proofs.

The framework for storing knowledge presented here is also related to Letz' work [14]. Letz' developed the use of full clausal form into tableaux proofs. Haehnle [11] published some further restrictions to Letz' framework. Haehnle's weak connection condition corresponds closely to the set-of-support strategy presented in Section 4. Baumgartner's hyper-tableaux method [6] is a restriction to the clausal tableaux and is closely related to this author's ε-rule. The framework and strategies presented here were all developed independently in 1993 and 1994 [19, 20]. Connections between these methods were recognized embarrassingly late partly due to the fact that this author was not using clausal form. Reports on the present research have not been widely disseminated previously because the author wanted to have a capable implementation, which could prove some interesting examples, prior to making a significant announcement.

Frank Brown [9] used what was essentially an incomplete sequent-based prover with free variables to prove theorems in set theory. Brown's system used theorems primarily as rewrite rules.

Baumgartner has worked to allow provers to use certain types of first-order theories by means of theory reasoning [5]. A theory reasoning system needs a background reasoner for the theories of interest. They are particularly useful for Horn theories.

Many semi-automatic and interactive theorem provers and proof checkers are able to reason in first- and higher-order theories with the help of information and direction from the user. The current research has focused on completely automatic proof discovery.

Acknowledgements

I am grateful to Larry Hines and Reiner Hähnle for their comments and suggestions on drafts of this paper.

References

1. R. Abraham, J. E. Marsden, and T. Ratiu. *Manifolds, Tensor Analysis, and Applications*. Springer-Verlag, second edition, 1988.
2. Matthias Baaz and Christian G. Fermüller. Non-elementary speedups between different versions of tableaux. In Peter Baumgartner, Reiner Hähnle, and Joachim Posegga, editors, *Proc. 4th Workshop on Deduction with Tableaux and Related Methods, St. Goar, Germany*, LNCS 918, pages 217–230. Springer-Verlag, 1995.
3. Dave Barker-Plummer. Gazing: An approach to the problem of definition and lemma use. *Journal of Automated Reasoning*, 8(3):311–344, 1992.
4. Dave Barker-Plummer and Alex Rothenberg. The Gazer theorem prover. In D. Kapur, editor, *Proceedings of the Eleventh International Conference on Automated Deduction*, volume 607 of *Lecture Notes in Computer Science*, pages 726–730. Springer-Verlag, 1992.
5. P. Baumgartner. A model elimination calculus with built-in theories. In H.-J. Ohlbach, editor, *Proceedings of the 16th German AI-Conference (GWAI-92)*, pages 30–42. Springer-Verlag, 1992. LNAI 671.
6. P. Baumgartner, U. Furbach, and I. Niemelä. Hyper Tableaux. In *JELIA 96*. European Workshop on Logic in AI, Springer, LNCS, 1996.
7. Bernhard Beckert. A completion-based method for mixed universal and rigid E-unification. In Alan Bundy, editor, *Proc. 12th Conference on Automated Deduction CADE, Nancy/France*, LNAI 814, pages 678–692. Springer-Verlag, 1994.
8. Bernhard Beckert, Reiner Hähnle, and Peter H. Schmitt. The *even more* liberalized δ-rule in free variable semantic tableaux. In Georg Gottlob, Alexander Leitsch, and Daniele Mundici, editors, *Proceedings of the third Kurt Gödel Colloquium KGC'93, Brno, Czech Republic*, volume 713 of *Lecture Notes in Computer Science*, pages 108–119. Springer-Verlag, August 1993.

9. Frank M. Brown. Towards the automation of set theory and its logic. *Artificial Intelligence*, 10(3):281–316, 1978.

10. Melvin C. Fitting. *First-Order Logic and Automated Theorem Proving*. Springer-Verlag, New York, second edition, 1996.

11. Reiner Hähnle and Stefan Klingenbeck. A-ordered tableaux. *Journal of Logic and Computation*, 1996, to appear. Available as Technical Report 26/95 from University of Karlsruhe, Department of Computer Science, URL: ftp://ftp.ira.uka.de/pub/uni-karlsruhe/papers/techreports/1995/1995-26.ps.gz.

12. Reiner Hähnle and Peter H. Schmitt. The liberalized δ-rule in free variable semantic tableaux. *Journal of Automated Reasoning*, 13(2):211–222, October 1994.

13. John Kelley. *General Topology*. The University Series in Higher Mathematics. D. Van Nostrand Company, 1955.

14. R. Letz, K. Mayr, and C. Goller. Controlled integration of the cut rule into connection tableau calculi. *Journal of Automated Reasoning*, 13(3):297–338, December 1994.

15. Reinhold Letz, Johann Schumann, Stephan Bayerl, and Wolfgang Bibel. SETHEO: A high-perfomance theorem prover. *Journal of Automated Reasoning*, 8(2):183–212, 1992.

16. David McAllester. Grammar rewriting. In D. Kapur, editor, *Proceedings of the Eleventh International Conference on Automated Deduction*, volume 607 of *Lecture Notes in Artificial Intelligence*, pages 124–138. Springer-Verlag, June 1992.

17. G. Nelson and D. Oppen. Fast decision procedures based on congruence closure. *Journal of the ACM*, 27(2):356–364, 1980.

18. Robert E. Shostak. An algorithm for reasoning about equality. *Comm. of the ACM*, 21(2):583–585, 1978.

19. Benjamin Shults. The creation and use of a knowledge base of mathematical theorems and definitions. Technical Report ATP-127, The University of Texas at Austin, 1995.

20. Benjamin Shults. A framework for the creation and use of a knowledge base of mathematical theorems and definitions. Technical Report ATP-127a, The University of Texas at Austin, 1995.

21. Benjamin Shults. Challenge problems in first-order theories. *Association of Automated Reasoning Newsletter*, (34):5–8, October 1996.

22. Benjamin Shults. Intelligent use of a knowledge base in automated theorem proving. Technical Report TR AI96-252, Univ. of Texas at Austin, Computer Science Dept., Artificial Intelligence Lab, 1996.

23. Benjamin Shults. Ipr source code and examples. URL: ftp://ftp.cs.utexas.edu/pub/bshults/provers/IPR-latest.tar.gz, 1996.

24. Raymond M. Smullyan. *First-Order Logic*. Dover Publications, New York, second corrected edition, 1995. First published 1968 by Springer-Verlag.

A Sequent Calculus for Reasoning in Four-Valued Description Logics

Umberto Straccia

Istituto di Elaborazione della Informazione
Via S. Maria, 46-56126 Pisa ITALY
straccia@iei.pi.cnr.it

Abstract. *Description Logics* (DLs, for short) provide a logical reconstruction of the so-called frame-based knowledge representation languages. Originally, four-valued DLs have been proposed in order to develop expressively powerful DLs with tractable subsumption algorithms. Recently, four-valued DLs have been proposed as a model for (multimedia) document retrieval. In this context, the main reasoning task is instance checking. Unfortunately, the known subsumption algorithms for four-valued DLs, based on "structural" subsumption, do not work with respect to the semantics proposed in the DL-based approach to document retrieval. Moreover, they are unsuitable for solving the instance checking problem, as this latter problem is more general than the subsumption problem. We present an alternative decision procedure for four-valued DLs with the aim to solve these problems. The decision procedure is a sequent calculus for instance checking. Since in general the four-valued subsumption problem can be reduced to the four-valued instance checking problem, we obtain a decision procedure for the subsumption problem. Some related complexity results will be presented.

1 Introduction

In the last decade a substantial amount of work has been carried out in the context of *Description Logics* (DLs, for short), which provide a logical reconstruction of the so-called frame-based knowledge representation languages, with the aim of providing a simple well-established denotational semantics to capture the meaning of the most popular features of structured representation of knowledge [20]. *Concepts, roles* and *individuals* are the basic building blocks of these logics. Concepts are expressions which collect the properties, described by means of roles, of a set of individuals. From a logical point of view, concepts can be seen as unary predicates, whereas roles are interpreted as binary predicates. Typically concepts are structured into a hierarchy, induced by the subsumption relation and interpreted as set containment: a concept C subsumes a concept D, iff the set of individuals denoted by the concept D is a subset of the set of individuals denoted by C, *i.e.* in symbols $D \preceq C$. It is a common opinion that subsumption checking is an important reasoning task in DLs. This has motivated a large body of research on the problem of subsumption checking in different DLs [3, 6, 13, 25, 27].

A DL *knowledge base* is a set of *assertions*. An assertion states either that an individual a is an instance of a concept C, written $C(a)$, or that two individuals a and b are related by means of a role R, written $R(a, b)$. The basic inference task with knowledge bases is *instance checking* and amounts to verify whether the individual a is an instance of the concept C with respect to the knowledge base Σ, *i.e.* in symbols $\Sigma \models C(a)$.

Originally, four-valued DLs have been proposed in order to develop expressively powerful DLs with tractable subsumption [21, 22, 23]. It has been shown that the four-valued subsumption relation captures an important subset of two-valued subsumption relationships.

More recently, DLs and their four-valued versions have been proposed in the area of (multimedia) *Document Retrieval* (DR, for short) [12, 17, 18, 19, 26]. In this context, a document base represents a collection of documents described by means of a set of assertions Σ. A query is specified by means of a concept C describing the set of documents to be retrieved. The retrieval of a document identified by the individual d is determined by checking whether $\Sigma \models C(d)$. Four-valued DLs have been proposed as a DR model fulfilling two basic *desiderata* not satisfied by classical DLs: (*i*) since retrieval is defined as the task of retrieving the documents that are *relevant* to the user's information need, the required DL should enforce a notion of entailment in which premises need be *relevant* to conclusions, and to a stronger extent than classical "material" logical implication does; (*ii*) since the *content representation* of a collection of documents cannot be considered as a consistent set of assertions an adequate DR model must be capable of tolerating inconsistencies without loosing its deductive capabilities. Four-valued DLs, which are mainly inspired on [2, 15], satisfy the above points since they (*i*) inherit from Relevance Logic [1] a flavour of "relevance" between premises and conclusion and (*ii*) are paraconsistent and allow *liberal reasoning* [28]. Liberal reasoning has been shown to be suitable for document retrieval purposes [19].

Effectively deciding whether $\Sigma \models C(a)$ requires a calculus. The only known decision procedures in the context of four-valued DLs are subsumption algorithms. These algorithms perform, in an efficient way, "structural" subsumption [3] and are directly inspired on Levesque's algorithm for propositional tautological entailment [15][1]. Unfortunately, they do not work with respect to the semantics proposed in the DL-based DR model. Moreover, they are unsuitable for solving the instance checking problem, as the instance checking problem is a more general problem than the subsumption problem. In fact, in most DLs the subsumption problem can easily reduced to the instance checking problem, as $D \preceq C$ iff $\{D(a)\} \models C(a)$ (where a is an individual) holds[2] and there are cases in which the instance checking problem belongs to a higher complexity

---

[1] Let α and β be two propositional formulae in conjunctive normal form. Then α tautological entails β iff for each conjunct β_j of β there is a conjunct α_i of α such that $\alpha_i \subseteq \beta_j$, where α_i and β_j are in clausal form. Tautological entailment between two formulae α and β in conjunctive normal form can be verified in time $O(|\alpha||\beta|)$.

[2] See [24] for those cases in which the reduction is not immediate.

class than the corresponding subsumption problem [7].

This paper presents an alternative decision procedure for four-valued DLs with the aim to solve these problems. The decision procedure is a *sequent calculus* for instance checking in four-valued DLs. Since generally the subsumption problem in it can be easily reduced to the instance checking problem, we obtain a decision procedure for subsumption checking for four-valued DLs.

The rest of this paper is organised as follows. In the next section we will briefly recall syntax and two-valued semantics of $\mathcal{ALC}$, a representative DL[3]. In Section 3 we will give to it four-valued semantics, whereas Section 4 highlights some properties of the logic. In Section 5 we will present a calculus for instance checking for four-valued DLs and Section 6 presents some related complexity results. Section 7 concludes.

2 A quick look to Description Logics

The building blocks of the language are primitive concepts (denoted by the letter A), primitive roles (denoted by the letter R) and individuals (denoted by the letter a and b). Complex concepts (C and D) are built out of primitive symbols via the language constructors by means of the syntax rule:

$$C, D \rightarrow A|C \sqcap D|C \sqcup D|\neg C|\forall R.C|\exists R.C$$

For example, the complex concept $\mathtt{Order} \sqcap \forall \mathtt{Sender.CarVendor}$ is obtained combining the primitive concepts $\mathtt{Order}$ and $\mathtt{CarVendor}$ and the primitive role $\mathtt{Sender}$ by the conjunction ($\sqcap$) and the universal quantification ($\forall$) constructors. An *interpretation* $\mathcal{I}$ consists of a domain $\Delta^{\mathcal{I}}$ (a non-empty set) and of a function mapping a primitive concept into a subset of $\Delta^{\mathcal{I}}$, mapping a primitive role into a subset of $\Delta^{\mathcal{I}} \times \Delta^{\mathcal{I}}$ and mapping different individuals into different elements of its domain $\Delta^{\mathcal{I}}$. The interpretation of complex concepts is obtained by appropriately combining the interpretation of their components, *i.e.* $(C \sqcap D)^{\mathcal{I}} = C^{\mathcal{I}} \cap D^{\mathcal{I}}$, $(C \sqcup D)^{\mathcal{I}} = C^{\mathcal{I}} \cup D^{\mathcal{I}}$, $(\neg C)^{\mathcal{I}} = \Delta^{\mathcal{I}} \setminus C^{\mathcal{I}}$, $(\forall R.C)^{\mathcal{I}} = \{d \in \Delta^{\mathcal{I}} : \text{for all } d', \text{ if } (d, d') \in R^{\mathcal{I}} \text{ then } d' \in C^{\mathcal{I}}\}$, and $(\exists R.C)^{\mathcal{I}} = \{d \in \Delta^{\mathcal{I}} : \text{for some } d', (d, d') \in R^{\mathcal{I}} \text{ and } d' \in C^{\mathcal{I}}\}$.

The subsumption relation between two concepts is defined in terms of set containment: a concept C *subsumes* a concept D, written $D \preceq_2 C$, iff for all interpretations $\mathcal{I}$, $D^{\mathcal{I}} \subseteq C^{\mathcal{I}}$.

An assertion is of type $C(a)$ or of type $R(a, b)$, where C, R and a, b are a concept, a role and two individuals, respectively. A *primitive assertion* (resp. *negated primitive assertion*) is an assertion of type $A(a)$ (resp. $\neg A(a)$). An interpretation $\mathcal{I}$ satisfies an assertion $C(a)$ just in case $a^{\mathcal{I}} \in C^{\mathcal{I}}$, whereas $\mathcal{I}$ satisfies an assertion $R(a, b)$ whenever $(a^{\mathcal{I}}, b^{\mathcal{I}}) \in R^{\mathcal{I}}$. Given a set of assertions Σ, a concept C and an individual a, then Σ *entails* $C(a)$, written $\Sigma \models_2 C(a)$, iff every interpretation satisfying all the assertions in Σ also satisfies $C(a)$. It is easily

---

[3] Although we restrict our attention to $\mathcal{ALC}$, our framework can be applied to other DLs as well.

verified that in $\mathcal{ALC}$, $D \preceq_2 C$ iff $\{D(a)\} \models_2 C(a)$ (where a is an individual) holds. Hence, subsumption can be reduced to instance checking in $\mathcal{ALC}$. This property holds for most DLs presented in the literature and in particular it will hold for all DLs we will consider in this paper.

3 Four-valued Description Logics

The four-valued semantics for $\mathcal{ALC}$ is based on a four-valued semantics similar as in [23], but we will also present a different choice for the semantics of the ($\forall$) constructor.

As usual, the four truth values are the elements of $2^{\{t,f\}}$, the powerset of $\{t,f\}$, i.e. $\{t,f\}$, $\{\}$, $\{t\}$ and $\{f\}$. These values are known as *contradiction*, *unknown*, *true* and *false*, respectively. An *interpretation* $\mathcal{I} = (\Delta^{\mathcal{I}}, \cdot^{\mathcal{I}})$ consists of a non empty set $\Delta^{\mathcal{I}}$ (the *domain* of $\mathcal{I}$) and a function $\cdot^{\mathcal{I}}$ (the *interpretation function* of $\mathcal{I}$) such that (i) $\cdot^{\mathcal{I}}$ maps every concept into a function from $\Delta^{\mathcal{I}}$ to $2^{\{t,f\}}$; (ii) $\cdot^{\mathcal{I}}$ maps every role into a function from $\Delta^{\mathcal{I}} \times \Delta^{\mathcal{I}}$ to $2^{\{t,f\}}$; (iii) $\cdot^{\mathcal{I}}$ maps every individual into an element of $\Delta^{\mathcal{I}}$ and (iv) $a^{\mathcal{I}} \neq b^{\mathcal{I}}$, if $a \neq b$. Given an interpretation $\mathcal{I}$, the *positive extension* of a concept C, written $[\![C^{\mathcal{I}}]\!]^{+}$, is defined as the set $\{d \in \Delta^{\mathcal{I}} : t \in C^{\mathcal{I}}(d)\}$, whereas the *negative extension* of a concept C, written $[\![C^{\mathcal{I}}]\!]^{-}$, is defined as the set $\{d \in \Delta^{\mathcal{I}} : f \in C^{\mathcal{I}}(d)\}$. The positive and negative extension of a role is defined similarly. It is easily verified that a two-valued interpretation is an interpretation $\mathcal{I}$ such that for every concept C, $[\![C^{\mathcal{I}}]\!]^{-} = \Delta^{\mathcal{I}} \setminus [\![C^{\mathcal{I}}]\!]^{+}$ and for all roles R, $[\![R^{\mathcal{I}}]\!]^{-} = \Delta^{\mathcal{I}} \setminus [\![R^{\mathcal{I}}]\!]^{+}$. Unlike two-valued semantics, the positive extension and the negative extension need not to be complement of each other.

The extensions of concepts have to meet certain restrictions, designed so that the formal semantics respects the informal meaning of concepts and roles. They are: $[\![(C \sqcap D)^{\mathcal{I}}]\!]^{+} = [\![C^{\mathcal{I}}]\!]^{+} \cap [\![D^{\mathcal{I}}]\!]^{+}$, $[\![(C \sqcap D)^{\mathcal{I}}]\!]^{-} = [\![C^{\mathcal{I}}]\!]^{-} \cup [\![D^{\mathcal{I}}]\!]^{-}$, $[\![(C \sqcup D)^{\mathcal{I}}]\!]^{+}$ $= [\![C^{\mathcal{I}}]\!]^{+} \cup [\![D^{\mathcal{I}}]\!]^{+}$, $[\![(C \sqcup D)^{\mathcal{I}}]\!]^{-} = [\![C^{\mathcal{I}}]\!]^{-} \cap [\![D^{\mathcal{I}}]\!]^{-}$, $[\![(\neg C)^{\mathcal{I}}]\!]^{+} = [\![C^{\mathcal{I}}]\!]^{-}$, $[\![(\neg C)^{\mathcal{I}}]\!]^{-}$ $= [\![C^{\mathcal{I}}]\!]^{+}$, $[\![(\exists R.C)^{\mathcal{I}}]\!]^{+} = [\![(\neg \forall R.\neg C)^{\mathcal{I}}]\!]^{+}$, $[\![(\exists R.C)^{\mathcal{I}}]\!]^{-} = [\![(\neg \forall R.\neg C)^{\mathcal{I}}]\!]^{-}$. It is worth noting that the semantics for the ($\exists$) constructor is given in terms of the ($\forall$) constructor. For it, we present two different semantics:

Type A: for each $d, d' \in \Delta^{\mathcal{I}}$

$$t \in (\forall R.C)^{\mathcal{I}}(d) \text{ iff } \forall\, e \in \Delta^{\mathcal{I}}, t \in R^{\mathcal{I}}(d,e) \text{ implies } t \in C^{\mathcal{I}}(e)$$
$$f \in (\forall R.C)^{\mathcal{I}}(d) \text{ iff } \exists\, e \in \Delta^{\mathcal{I}}, t \in R^{\mathcal{I}}(d,e) \text{ and } f \in C^{\mathcal{I}}(e)$$

Type B: for each $d, d' \in \Delta^{\mathcal{I}}$

$$t \in (\forall R.C)^{\mathcal{I}}(d) \text{ iff } \forall\, e \in \Delta^{\mathcal{I}}, f \in R^{\mathcal{I}}(d,e) \text{ or } t \in C^{\mathcal{I}}(e)$$
$$f \in (\forall R.C)^{\mathcal{I}}(d) \text{ iff } \exists\, e \in \Delta^{\mathcal{I}}, t \in R^{\mathcal{I}}(d,e) \text{ and } f \in C^{\mathcal{I}}(e)$$

Notice that type B semantics is used in [23], whereas type A semantics is used in [19].

In the following, the default will be the semantics of type A. A concept C is *equivalent* to a concept D, written $C \equiv_4^A D$, iff $[C^{\mathcal{I}}]^+ = [D^{\mathcal{I}}]^+$, for every interpretation $\mathcal{I}$. A concept C *subsumes* a concept D, written $D \preceq_4^A C$, iff $[D^{\mathcal{I}}]^+ \subseteq [C^{\mathcal{I}}]^+$, for every interpretation $\mathcal{I}$. A concept C is *coherent* iff there is an interpretation $\mathcal{I}$ such that $[C^{\mathcal{I}}]^+ \neq \emptyset$.

With respect to assertions, we have the following definitions. An interpretation $\mathcal{I}$ *satisfies an assertion* $C(a)$, iff $t \in C^{\mathcal{I}}(a^{\mathcal{I}})$, whereas $\mathcal{I}$ *satisfies an assertion* $R(a, b)$ iff $t \in R^{\mathcal{I}}(a^{\mathcal{I}}, b^{\mathcal{I}})$. Given two $\mathcal{ALC}$ assertions α and β, α is *equivalent* to β, written $\alpha \equiv_4^A \beta$, iff for every interpretation $\mathcal{I}$, $\mathcal{I}$ satisfies α iff $\mathcal{I}$ satisfies β. An interpretation $\mathcal{I}$ *satisfies* (is a *model* of) a knowledge base Σ iff $\mathcal{I}$ satisfies all assertions in Σ. A knowledge base Σ *entails* an assertion $C(a)$, written $\Sigma \models_4^A C(a)$, iff all models of Σ satisfy $C(a)$. Finally, all the above definitions are given for the case of two-valued semantics and type B semantics too. In these cases we will use $\cdot_2$ and $\cdot_4^B$, respectively, in place of $\cdot_4^A$. For instance, we will write $\Sigma \models_4^B C(a)$, if Σ entails $C(a)$ with respect to type B semantics and we will write $C \equiv_2 D$ iff C and D are equivalent with respect to two-valued semantics.

4 Some properties of the semantics

We will not go into a detailed discussion since a detailed one can be found in [19].

Just note that in our four-valued logic, every concept is coherent, whereas this is not true with respect to two-valued semantics, *e.g.* $A \sqcap \neg A$ is not a "two-valued" coherent concept. Similarly, every knowledge base is satisfiable.

As in two-valued $\mathcal{ALC}$, it is easily verified that $D \preceq_4^A C$ iff $\{D(a)\} \models_4^A C(a)$, and $D \preceq_4^B C$ iff $\{D(a)\} \models_4^B C(a)$, where a is an individual. Therefore, in four-valued $\mathcal{ALC}$ the subsumption problem can be reduced to the instance checking problem. This property holds for all four-valued DLs we will analyse in this paper.

Reasoning in our logic is *sound* with respect to two-valued semantics. In fact, it is easily verified that for all knowledge bases Σ, for all assertions α and for all concepts C and D, $D \preceq_4^A C$ implies $D \preceq_2 C$ and $\Sigma \models_4^A \alpha$ implies $\Sigma \models_2 \alpha$. From $A \sqcap \neg A \preceq_2 B$ and $A \sqcap \neg A \npreceq_4^A B$, it follows that $\preceq_4^A \subset \preceq_2$ and $\models_4^A \subset \models_2$ hold.

In [19, 23] it has already shown that $\preceq_4^B$ and $\models_4^A$ capture an *interesting* subset of $\preceq_2$ and $\models_2$, respectively. In what follows we will show which inferences, licensed in two-valued semantics, are left out by our semantics. It is quite obvious that the so-called paradoxes of logical implication, *i.e.* $\{C(a), \neg C(a)\} \models_2 D(b)$ and $\emptyset \models_2 (C \sqcup \neg C)(a)$ do not hold in our type A and type B semantics. Generally, modus ponens is not a valid inference rule in our logic, *i.e.* $\{(C \sqcap (\neg C \sqcup D))(a)\} \nvDash_4^A D(a)$. But, the semantics of type A allows a restricted form of modus ponens, called *modus ponens on roles*: for all concepts C and D, for any role R, and for all individuals a, b, $\{(\forall R.C)(a), R(a, b)\} \models_4^A C(b)$ and $(\exists R.C) \sqcap (\forall R.D) \preceq_4^A \exists R.(C \sqcap D)$. This kind of inference is not allowed in type B semantics. Hence, $\models_4^B \neq \models_4^A$ and $\preceq_4^B \neq \preceq_4^A$ hold. In the next section we will see that type B semantics is weaker than type A semantics, by showing that $\models_4^B \subseteq \models_4^A$. Thus, $\preceq_4^B \subset \preceq_4^A$ and $\models_4^B \subset \models_4^A$ hold. Unfortunately, the above additional inference has a cost in terms of computational complexity. In fact, as we will see in Section 6, checking whether $\Sigma \models_4^A C(a)$ is harder than checking whether $\Sigma \models_4^B C(a)$.

Reasoning by cases does not work within our type A and type B semantics. Consider the following knowledge base $\Sigma' = \{P(p), HS(p, c1), HS(p, c2), F(c1, c2),$ $F(c2, c3), I(c1), \neg I(c3)\}$, where P, HS, F and I stand for **Person, HasStudent, Friend** and **Ill**, respectively. Consider now the assertion $\alpha = (\exists HS.(I \sqcap \exists F.\neg I))(p)$. It can be verified that $\Sigma' \not\models_4^A \alpha$ and $\Sigma' \models_2 \alpha$ hold. In fact, $\Sigma' \models_2 \alpha$ since in each two-valued interpretation $\mathcal{I}$ satisfying Σ', either $I(c2)$ is true or $I(c2)$ is false. On the other hand, $\Sigma' \not\models_4^A \alpha$ holds since it could be the case $I^{\mathcal{I}}(c2^{\mathcal{I}}) = \emptyset$. That is, we are uncertain about c2's illness: we have no "relevant" information about c2's illness. In [19] it has been shown that $\Sigma' \not\models_4^A \alpha$ is considered suitable for document retrieval purposes.

It is worth noting that there is no "top" concept within type A and type B semantics, *i.e.* there is no concept C such that $\emptyset \models_4^A C(a)$, for all individuals a. This plays an important role in the example above. In fact, suppose that we would admit a "top" concept $\top$ with semantics $[\![\top^{\mathcal{I}}]\!]^+ = \Delta^{\mathcal{I}}$ and $[\![\top^{\mathcal{I}}]\!]^- = \emptyset$. If we replace in Σ' and α above, I and $\neg$I with $\exists R.\top$ and $\forall R.A$, respectively, then it can be verified that $\Sigma'' \models_4^A \alpha'$ and $\Sigma'' \not\models_4^B \alpha'$ hold, where Σ'' and α' are the result of the substitutions. This is due to the fact that $\Sigma'' \models_4^A \alpha'$ relies on the relations $I \sqcup \neg I \equiv_2 \top$ and $\exists R.\top \sqcup \forall R.A \equiv_4^A \top$, whereas $\exists R.\top \sqcup \forall R.A \not\equiv_4^B \top$. As we will see in Section 6, admitting a top concept changes the computational complexity of instance checking.

5 A sequent calculus for entailment

In the following, we will assume type A semantics as the default semantics.

Effectively deciding whether, given a knowledge base Σ and an assertion α, $\Sigma \models_4^A \alpha$, requires a calculus. The one we have developed is a *sequent calculus* [9, 11].

The main idea behind our calculus for entailment is that in order to prove $\Sigma \models_4^A \alpha$ we try to prove the "sequent" $\Sigma \rightarrow \alpha$. Without loss of generality we can restrict our attention to assertions in *negation normal form* (NNF), *i.e.* the negation constructor can be only in front of primitive symbols[4]. Furthermore, we add to the alphabets of concepts, roles and individuals another alphabet of symbols, called *variables* (denoted by x and y). The alphabet of *objects* is the union of the alphabets of variables and individuals (objects are denoted by v and w). An interpretation $\mathcal{I}$ is extended to variables by mapping them into an element of its domain $\Delta^{\mathcal{I}}$.

A *sequent* is an expression of the form $\Gamma \rightarrow \Delta$, where $\Gamma = \{\gamma_1, \ldots, \gamma_n\}$ and $\Delta = \{\delta_1, \ldots, \delta_m\}$ are finite sets of assertions, with $n + m \geq 1$. Γ is called the

---

[4] It is easy to see that every assertion can be transformed in linear time and space into an equivalent assertion in NNF, respectively. Note that in $\mathcal{ALC}$ roles are already in NNF.

antecedent and Δ is called the *consequent*. A sequent $\{\gamma_1, \ldots, \gamma_n\} \to \{\delta_1, \ldots, \delta_m\}$ is *satisfiable* iff there is an interpretation $\mathcal{I}$ such that if $\mathcal{I}$ satisfies *all* assertions in $\{\gamma_1, \ldots, \gamma_n\}$ then $\mathcal{I}$ satisfies *some* assertion in $\{\delta_1, \ldots, \delta_m\}$. A sequent $\Gamma \to \Delta$ is *valid* iff all interpretations satisfy $\Gamma \to \Delta$. A sequent $\Gamma \to \Delta$ is *falsifiable* iff it is not valid. Just note that $\Sigma \models_4^A \alpha$ iff the sequent $\Sigma \to \alpha$ is valid. For ease of notation we will often omit braces and operations of set-theoretic union, thus writing *e.g.* γ, Δ, Γ in place of $\{\gamma\} \cup \Delta \cup \Gamma$.

An *axiom (of type AX)* is a sequent of the form $\alpha, \Gamma \to \Delta, \alpha$. From the definition it is immediate to see that all axioms are valid sequents. A sequent calculus is based on a number of *rules of inference* operating on sequents. Rules fall naturally into two categories: those operating on assertions occurring in the antecedent, and those operating on assertions occurring in the consequent. Every rule consists of one or two "upper" sequents called *premises* and of a "lower" sequent called *conclusion*. The rules of the calculus for entailment with respect to type A semantics are defined as follows:

$$(\sqcap \to) \quad \frac{C(v), D(v), \Gamma \to \Delta}{(C \sqcap D)(v), \Gamma \to \Delta} \qquad (\to \sqcap) \quad \frac{\Gamma \to \Delta, C(v) \quad \Gamma \to \Delta, D(v)}{\Gamma \to \Delta, (C \sqcap D)(v)}$$

$$(\sqcup \to) \quad \frac{C(v), \Gamma \to \Delta \quad D(v), \Gamma \to \Delta}{(C \sqcup D)(v), \Gamma \to \Delta} \qquad (\to \sqcup) \quad \frac{\Gamma \to \Delta, C(v), D(v)}{\Gamma \to \Delta, (C \sqcup D)(v)}$$

$$(\forall \to) \quad \frac{(\forall R.C)(v), R(v, w), C(w), \Gamma \to \Delta}{(\forall R.C)(v), R(v, w), \Gamma \to \Delta} \qquad (\to \forall) \quad \frac{R(v, x), \Gamma \to \Delta, D(x)}{\Gamma \to \Delta, (\forall R.D)(v)}$$

$$(\exists \to) \quad \frac{R(v, x), C(x), \Gamma \to \Delta}{(\exists R.C)(v), \Gamma \to \Delta} \qquad (\to \exists) \quad \frac{R(v, w), \Gamma \to \Delta, (\exists R.C)(v), C(w)}{R(v, w), \Gamma \to \Delta, (\exists R.C)(v)}$$

where x is a new variable (called also *eigenvariable*) and v, w are objects. Of course, in order to prevent infinite applications, the $(\forall \to)$ rule can be applied only if $C(w) \notin \Gamma$. Similarly for the $(\to \exists)$ rule.

A deduction can easily be represented as a tree (growing upwards): a *deduction tree* is a tree whose nodes are each labelled with a sequent and in which a sequent labelling a node may be obtained through the application of a rule of inference to the sequents labelling their children nodes. The sequent labelling the root of a deduction tree is called *conclusion* of the deduction tree[5]. A *proof tree* is a deduction tree whose leaves are labelled with an axiom. A sequent $\Gamma \to \Delta$ is *provable*, written $\Gamma \vdash_4^A \Delta$, iff there is a proof tree of which it is the conclusion. We will write $\Gamma \vdash_4^B \Delta$ and $\Gamma \vdash_2 \Delta$, whenever the calculus refers to axioms and rules for type B semantics and two-valued semantics, respectively. A proof of a sequent $\Gamma \to \Delta$ proceeds backward, by constructing a proof tree with root $\Gamma \to \Delta$ and applying the rules until each branch reaches an axiom. For example, the proof of the valid sequent $(C \sqcup D)(a), E(a) \to (E \sqcap (C \sqcup D))(a)$ starts with a root node labelled $(C \sqcup D)(a), E(a) \to (E \sqcap (C \sqcup D))(a)$, applies the $(\sqcup \to)$ rule to it, generates two new nodes labelled with the sequents

---

[5] Trees are represented reversed, *i.e.* the root is at the bottom.

$C(a), E(a) \to (E \sqcap (C \sqcup D))(a)$ and $D(a), E(a) \to (E \sqcap (C \sqcup D))(a)$ and proceeds backward until each branch reaches an axiom. The deduction tree is shown below.

$$\frac{\dfrac{C(a), E(a) \to C(a), D(a)}{C(a), E(a) \to E(a) \quad C(a), E(a) \to (C \sqcup D)(a)}}{\dfrac{C(a), E(a) \to (E \sqcap (C \sqcup D))(a)}{}} \qquad \frac{\dfrac{D(a), E(a) \to C(a), D(a)}{D(a), E(a) \to E(a) \quad D(a), E(a) \to (C \sqcup D)(a)}}{D(a), E(a) \to (E \sqcap (C \sqcup D))(a)}$$

$$(C \sqcup D)(a), E(a) \to (E \sqcap (C \sqcup D))(a)$$

5.1 Provability, Soundness and Completeness

As first, it can be verified that any deduction tree is finite. Therefore, the deduction of a sequent terminates after a finite number of rule applications. Soundness of the calculus, *i.e.* if the sequent $\Gamma \to \Delta$ is provable then $\Gamma \to \Delta$ is valid, can easily be proven by observing that every axiom is valid and that for each of the rules, the conclusion of a rule is valid iff all premises of the rule are valid.

Completeness is proved by means of Hintikka sets (sets of signed assertions) [8, 9]. Let T and NT be two new symbols not appearing in any considered alphabet. *Signed assertions of type-a, type-b, type-c and type-d* and their *components* are defined as follows (C, D are concepts, R is a role and v, w are objects):

Type-a assertion α^a	Components α_1^a	α_2^a	Type-b assertion α^b	Components α_1^b	α_2^b
$T(C \sqcap D)(v)$	$TC(v)$	$TD(v)$	$T(C \sqcup D)(v)$	$TC(v)$	$TD(v)$
$NT(C \sqcup D)(v)$	$NTC(v)$	$NTD(v)$	$NT(C \sqcap D)(v)$	$NTC(v)$	$NTD(v)$
Type-c assertion α^c	Components α_1^c	α_2^c	Type-d assertion α^d	Components α_1^d	α_2^d
$T(\forall R.C)(v)$	$TR(v, w)$	$TC(w)$	$T(\exists R.C)(v)$	$TR(v, w)$	$TC(w)$
$NT(\exists R.C)(v)$	$TR(v, w)$	$NTC(w)$	$NT(\forall R.C)(v)$	$TR(v, w)$	$NTC(w)$

Let α be an assertion. Then $T\alpha$ and $NT\alpha$ are called *conjugated signed assertions*. We define then satisfaction of signed assertions as follows. Let $\mathcal{I}$ be an interpretation and α an assertion. Then $\mathcal{I}$ *satisfies* $T\alpha$ iff $\mathcal{I}$ satisfies α, whereas $\mathcal{I}$ *satisfies* $NT\alpha$ iff $\mathcal{I}$ does not satisfy α.

Let H be a set of objects. A *Hintikka set S with respect to H* is a set of signed assertions such that the following conditions hold for all signed assertions α^a, α^b, α^c, α^d of type a, b, c, d, respectively: (*i*) no conjugated signed primitive assertions or conjugated signed negated primitive assertions are in S; (*ii*) if a type-a assertion α^a is in S, then both α_1^a and α_2^a are in S; (*iii*) if a type-b assertion α^b is in S, then either α_1^b is in S or α_2^b is in S; (*iv*) if a type-c assertion α^c is in S, then for every object $w \in H$, if α_1^c is in S then α_2^c is in S; (*v*) if a type-d assertion α^d is in S, then there is at least one object $w \in H$ such that both α_1^d and α_2^d are in S. It can easily be shown that

Lemma 1. *Every Hintikka set S w.r.t. a set of objects H is satisfiable in an interpretation with domain H.* ⊣

Theorem 2. *A sequent $\Gamma \to \Delta$ is valid w.r.t. type A semantics iff $\Gamma \vdash_4^A \Delta$, where $\vdash_4^A$ is determined by axioms of type AX and the rules $(\sqcap \to)$, $(\to \sqcap)$, $(\sqcup \to)$, $(\to \sqcup)$, $(\forall \to)$, $(\to \forall)$, $(\exists \to)$ and $(\to \exists)$.* ⊣

Proof. If there is a proof tree for $\Gamma \to \Delta$ then, from the correctness of the rules we have that $\Gamma \to \Delta$ is valid. Otherwise, pick up a deduction tree not being a proof tree, and which cannot be expanded any more. Therefore, there is a path from the conclusion to a non axiom leaf of the tree. Let LHS be the union of all assertions occurring in the left hand side of each sequent along that path and RHS be the union of all assertions occurring in the right hand side of any such sequent. Let $S = \{T\alpha : \alpha \in LHS\} \cup \{NT\alpha : \alpha \in RHS\}$. Let H be the set of objects appearing in S. It can be proven that S is a Hintikka set. From Lemma 1 it follows that S is satisfiable with respect to H. Since $\Gamma \subseteq LHS$ and $\Delta \subseteq RHS$, it follows that $\Gamma \to \Delta$ is falsifiable. □

With respect to type B semantics, the rules for the $(\forall)$ constructor no longer preserve validity between conclusion and premise. We can easily modify these rules in order to obtain completeness with respect to type B semantics. In fact, let $(\forall_s)$ be the rule (x is a new variable)

$$(\forall_s) \quad \frac{(\forall R.C)(v), C(x), \Gamma \to \Delta, (\forall R.D)(v), D(x)}{(\forall R.C)(v), \Gamma \to \Delta, (\forall R.D)(v)}$$

It is easy to see that these rules reflects the type B semantics of the $(\forall)$ constructor in the sense that a conclusion is valid iff its premise is valid. If we consider the axioms and rules for type A semantics where the rules $(\forall \to)$ and $(\to \forall)$ are replaced with the rule $(\forall_s)$, then, similarly as for Theorem 2, it can be shown that

Theorem 3. *A sequent $\Gamma \to \Delta$ is valid w.r.t. type B semantics iff $\Gamma \vdash_4^B \Delta$, where $\vdash_4^B$ is determined by axioms of type AX and the rules $(\sqcap \to)$, $(\to \sqcap)$, $(\sqcup \to)$, $(\to \sqcup)$, $(\forall_s)$, $(\exists \to)$ and $(\to \exists)$.* ⊣

Since the $(\forall_s)$ rule is a special case of the $(\to \forall)$ rule, it follows that $\Gamma \vdash_4^B \Delta$ implies $\Gamma \vdash_4^A \Delta$. Therefore, $\models_4^B \subseteq \models_4^A$ (strictness of the inclusion follows from Section 4).

Completeness with respect to standard two-valued $\mathcal{ALC}$ is obtained by simply adding the following rules to our calculus:

$$(\neg \to) \quad \frac{\Gamma \to \Delta, A(v)}{\neg A(v), \Gamma \to \Delta} \qquad (\to \neg) \quad \frac{\Gamma \to \Delta, \neg A(v)}{A(v), \Gamma \to \Delta}$$

where A is a primitive concept. It is easily verified that for both rules the conclusion is valid iff its premise is valid. Hence, validity is preserved.

Theorem 4. *A sequent $\Gamma \to \Delta$ is valid w.r.t. two-valued semantics iff $\Gamma \mathrel{\vdash_2} \Delta$, where $\mathrel{\vdash_2}$ is determined by axioms of type AX and the rules $(\sqcap \to)$, $(\to \sqcap)$, $(\sqcup \to)$, $(\to \sqcup)$, $(\forall \to)$, $(\to \forall)$, $(\exists \to)$, $(\to \exists)$, $(\neg \to)$ and $(\to \neg)$.* ⊣

It should be noted that another complete calculus with respect to two-valued semantics is obtained by considering axioms (*of type AX'*) of the form $\neg A(v), A(v), \Gamma \to \Delta$ and only the left hand side rules. The following theorem can easily be shown.

Theorem 5. *Let $\mathrel{\vdash'_2}$ be the derivation relation determined by axioms of type AX' and the rules $(\sqcap \to)$, $(\sqcup \to)$, $(\forall \to)$ and $(\exists \to)$. Then*

1. *if C and D are two concepts, then $C \preceq_2 D$ iff $C \sqcap \neg D$ is not coherent;*
2. *a concept C is coherent iff the knowledge base $\{C(a)\}$ is satisfiable, where a is an individual;*
3. *a knowledge base Σ is satisfiable iff the sequent $\Sigma \to \emptyset$ is not valid;*
4. *a sequent $\Gamma \to \emptyset$ is valid iff $\Gamma \mathrel{\vdash'_2} \emptyset$.* ⊣

Therefore, the above theorem establishes that the well known (refutation based) constraint propagation method [25] devised for reasoning in two-valued DLs is a special case within our framework, *i.e.* the decision procedure based on $\mathrel{\vdash'_2}$ is *exactly the same* decision procedure as the standard constraint propagation method for two-valued knowledge base satisfiability checking.

6 Computational complexity

In the following, $\mathcal{AL}$ is the DL with syntax rule[6]

$$C, D \to \top \,|\, \bot \,|\, A \,|\, \neg A \,|\, C \sqcap D \,|\, \forall R.C \,|\, \exists R$$

$\mathcal{ALE}$ is $\mathcal{AL}$ plus qualified existential quantification $\exists R.C$. $\mathcal{ALE}_1^-$ is the DL with syntax rule

$$C, D \to A \,|\, \neg A \,|\, C \sqcap D \,|\, \forall R.C \,|\, \exists R.C$$

and $\mathcal{ALE}_2^-$ is $\mathcal{ALE}_1^-$ plus unqualified existential quantification $\exists R$.

It is well known that the (in)coherence problem is PSPACE-complete for two-valued $\mathcal{ALC}$ [25]. By adopting the same reduction technique as for two-valued $\mathcal{ALC}$, the following theorem can be shown.

Theorem 6. *The instance checking problem and the subsumption checking problem w.r.t. type A semantics are PSPACE-complete for $\mathcal{ALC}$.* ⊣

---

[6] The 2-valued extension of $\exists R$ is $\{d \in \Delta^{\mathcal{I}} : \exists d' \in \Delta^{\mathcal{I}}$ such that $(d, d') \in R^{\mathcal{I}}\}$. We use $\bot$ as a macro for $\neg \top$.

Proof. We recall that the PSPACE-hardness of the (two-valued) coherence problem has been shown by means of a reduction of the validity problem for quantified boolean formulae [25], which is PSPACE-complete [10]. As in [25], let $P.M$ be a quantified boolean formula, where P (the prefix) is a sequence of quantification over boolean variables and M (the matrix) is a set of clauses (a clause is a set of boolean variables). For instance, $\forall x \exists y.(x \vee \neg y) \wedge \neg y$ is a valid boolean formula, where $P = \forall x \exists y$ and $M = (x \vee \neg y) \wedge \neg y$[7]. Consider the reduction of $P.M$ into the $\mathcal{ALC}$ concept $[P.M] = [P]^0 \sqcap [C_1]^0 \sqcap \ldots \sqcap [C_n]^0$, as in [25]. It has been shown that $P.M$ is a valid boolean formula iff $[P.M]$ is coherent. Consider $[P]^0 \sqcap \neg D$, where $D = D_1 \sqcup \ldots \sqcup D_1$ and D_i is the NNF of $\neg [C_i]^0$. It follows that $[P.M]$ is incoherent iff $[P] \sqcap \neg D$ is incoherent iff $\{[P](a)\} \models_2 D(a)$, where a is an individual. Therefore, by completeness it follows that $\{[P](a)\} \models_2 D(a)$ iff $[P](a) \hspace{0.2em}\sim\hspace{-0.9em}\mid_2 D(a)$. It can be verified that any deduction $[P](a) \hspace{0.2em}\sim\hspace{-0.9em}\mid_2 D(a)$ does not rely on the rules $(\neg \to)$ and $(\to \neg)$. Hence, $[P](a) \hspace{0.2em}\sim\hspace{-0.9em}\mid_2 D(a)$ iff $[P](a) \hspace{0.2em}\sim\hspace{-0.9em}\mid_4^A D(a)$. Therefore, $[P.M]$ is incoherent iff $\{[P](a)\} \hspace{0.2em}\sim\hspace{-0.9em}\mid_4^A D(a)$. As a consequence, instance checking is a PSPACE-hard problem. Analogously, $[P.M]$ is incoherent iff $[P] \preceq_4^A D$. Hence, subsumption checking is a PSPACE-hard problem too.

Furthermore, checking whether $\Sigma \hspace{0.2em}\sim\hspace{-0.9em}\mid_4^A C(a)$ can be done in polynomial space. This is obtained by adopting a *trace rule* in place of the $(\exists \to)$ rule which is identical to the trace rule for the $(\exists)$ constructor in the case of the two-valued constraint propagation technique [25]. Therefore, instance checking is a PSPACE-complete problem with respect to type A semantics. Since subsumption can be reduced to instance checking, it follows that the subsumption problem is PSPACE-complete too. □

This result shows that modus ponens on roles is effectively sufficient to get PSPACE-hardness. Moreover, from the proof it follows that the theorem holds for $\mathcal{ALC}$ without the negation constructor $(\neg)$ too.

By using the same reduction as described in [7], where PSPACE-completeness of the instance checking problem with respect to two-valued $\mathcal{ALE}$ is shown, it can be verified that Theorem 6 holds for the language $\mathcal{ALE}_2^-$ too.

Theorem 7. *The instance checking problem w.r.t. type A semantics is PSPACE-complete for $\mathcal{ALE}_2^-$.* ⊣

Similarly, by proceeding as in [5], where NP-completeness of the two-valued subsumption problem is shown for $\mathcal{ALE}$ and $\mathcal{FLE}^-$, it can be proven that

Theorem 8. *Subsumption checking w.r.t. type A semantics is a NP-complete problem for $\mathcal{ALE}_2^-$.* ⊣

Hence, with respect to type A semantics instance checking is strictly more difficult than subsumption for $\mathcal{ALE}_2^-$.

By using type B semantics, modus ponens on roles is not a valid inference rule. As a consequence we can prove that checking the falsifiability of sequents

---

[7] For readability, we write $(x \vee \neg y) \wedge \neg y$ in place of $\{\{x, \neg y\}, \{\neg y\}\}$.

can be done in non-deterministic polynomial time. Since propositional logic is a sub language of $\mathcal{ALC}$ and since it is well known that propositional tautological entailment is a coNP-complete problem, it follows that:

Theorem 9. *The instance checking problem and the subsumption problem w.r.t. type B semantics are coNP-complete for $\mathcal{ALC}$.* ⊣

As noted in Section 4, type B semantics has a weaker entailment relation than type A semantics. On the other hand, from a computational point of view the computational complexity switches from coNP to PSPACE.

Since in certain circumstances reasoning by cases does not hold, the instance checking problem can be even in P. Suppose that Σ is formed out by $\mathcal{AL}$ assertions, and consider an assertion $C(a)$ which is an $\mathcal{ALE}_1^-$ assertion.

As shown in [7], the instance checking problem with respect to two-valued semantics is coNP-hard in this case. Whereas, it can be shown that

Theorem 10. *Let Σ be formed out by $\mathcal{AL}$ assertions and let C be an $\mathcal{ALE}_1^-$ concept. Then checking whether $\Sigma \models_4^A C(a)$ and checking whether $\Sigma \models_4^B C(a)$ can be done in polynomial time.* ⊣

Proof Sketch. The proof is based on the fact that the $(\to \exists)$ rule can be replaced by the rule

$$(\to_T \exists) \quad \frac{R(v,w), \Gamma \to \Delta, C(w)}{R(v,w), \Gamma \to \Delta, (\exists R.C)(v)}$$

and the $(\forall_s)$ rule (in the case of type B semantics) can be replaced by the rule

$$(\forall_s T) \quad \frac{C(x), \Gamma \to \Delta, D(x)}{(\forall R.C)(v), \Gamma \to \Delta, (\forall R.D)(v)}$$

and observing that in all rules Δ is always empty. Therefore, the number and the dimension of the leaves of a deduction tree, and the number of deduction trees to be generated are bounded by the dimension of the root node. □

It is worth noting that, from [7] it follows that whenever unqualified existential concepts of type $\exists R$ are allowed to occur in C, the theorem does not hold for type A semantics (again, we rely on the relation $\exists R \sqcup \exists R.A \equiv_4^A \top$).

Theorem 11. *Checking whether $\Sigma \models_4^A C(a)$, where Σ is formed out by $\mathcal{AL}$ assertions and C is an $\mathcal{ALE}_2^-$ concept, is a coNP-hard problem.* ⊣

Theorem 10 is certainly a positive result, as the $\exists R.C$ construct is very useful in the query language. This result, and in general a sequent calculus approach, suggests a detailed investigation about the computational complexity of the instance checking problem in those cases where the knowledge base language and the query language are different.

The idea of distinguishing between the knowledge base language and the query language is certainly not new in the DL area [4, 14] and has given positive results. We claim that the use of a sequent calculus framework could be better, due to its intrinsic distinction between knowledge base part and query part, than the up to now refutation based methods. As a consequence, decision procedures based on $\vdash_2$ seems to be more suitable than the (refutation based) decision procedure based on $\vdash_2'$.

A practical consideration: From an inspection on proof trees with respect to type A and type B semantics the following theorem can easily be shown[8].

Theorem 12. *Let $\Gamma \rightarrow \Delta$ be a sequent. If there is no common role symbol, concept symbol or object symbol between Γ and Δ, then $\Gamma \rightarrow \Delta$ is not provable w.r.t. type A and type B semantics.* ⊣

Hence, suppose that at step n of a proof we are considering sequent $\Gamma \rightarrow \Delta$. We can split Γ into two partitions Γ_1 and Γ_2, where Γ_2 is the set of assertions of Γ with no concept, role and object symbols appearing in Δ and $\Gamma_1 = \Gamma \setminus \Gamma_2$. Now, Γ_2 will be never involved in an axiom at step $n + i$ ($i \geq 0$). That is, Γ_2 is a subset of the *irrelevant* facts in Γ with respect to Δ. In fact, the following holds: $\Gamma \vdash_4^A \Delta$ iff $\Gamma_1 \cup \Gamma_2 \vdash_4^A \Delta$ iff $\Gamma_1 \vdash_4^A \Delta$. Similarly for $\vdash_4^B$. Since for huge knowledge bases, as they are in the DL-based DR model, certainly $|\Gamma_2| \gg |\Gamma_1|$ holds, a significant improvement of the deduction process can be obtained.

7 Conclusions

Four-valued DLs were introduced mainly for computational reasons, *i.e.* for designing DLs with tractable subsumption algorithms. More recently, four-valued DLs were presented in the context of multimedia document retrieval in order both to deal with inconsistent knowledge bases and to model an entailment relation with a flavour of "relevance" between premises and conclusion. Unfortunately, the known subsumption algorithms for four-valued DLs, based on "structural" subsumption, do not work with respect to type A semantics. Moreover, they are unsuitable for solving the instance checking problem, as instance checking is more difficult than subsumption.

We have presented an alternative calculus for reasoning in four-valued DLs with the aim to solve the above problems. It is based on a sequent calculus solving the instance checking problem and, thus, the subsumption problem. Moreover, we have seen that the standard constraint propagation methods for two-valued knowledge base satisfiability are a special case within our framework.

We have presented several complexity results with respect to type A and type B semantics. These show that the price we must pay for a stronger four-valued

---

[8] This theorem can be seen as a consequence of a four-valued version of Craig's interpolation theorem [8, 9].

semantics, *i.e.* of type A, than type B semantics is a worsening of the computational complexity. Moreover, the complexity results show that weakening classical semantics does generally not imply an improvement of the computational complexity. We have also presented a positive result where reasoning is tractable for both type A and type B semantics. It suggests us to explore those cases in which the knowledge base language differs from the query language. A sequent calculus seems to be a good candidate in such cases, as it distinguishes between assertions of the knowledge base and the query, independently of the chosen semantics (two-valued, four-valued).

Acknowledgements

This work has been carried out in the context of the project FERMI 8134 - "Formalization and Experimentation in the Retrieval of Multimedia Information", funded by the European Community under the ESPRIT Basic Research scheme.

References

1. Alan R. Anderson and Nuel D. Belnap. *Entailment - the logic of relevance and necessity.* Princeton University Press, Princeton, NJ, 1975.
2. Nuel D. Belnap. A useful four-valued logic. In Gunnar Epstein and J. Michael Dunn, editors, *Modern uses of multiple-valued logic*, pages 5–37. Reidel, Dordrecht, NL, 1977.
3. Alexander Borgida. Structural subsumption: What is it and why is it important? In *AAAI Fall Symposium:Issues in Description Logics*, pages 14–18, 1992.
4. Martin Buchheit, A. Manfred Jeusfeld, Werner Nutt, and Martin Staudt. Subsumption between queries to object-oriented databases. *Information Systems*, 19(1):33–54, 1994. Special issue on Extending Database Technology, EDBT'94.
5. Francesco M. Donini, Bernhard Hollunder, Maurizio Lenzerini, A. Marchetti Spaccamela, Daniele Nardi, and Werner Nutt. The complexity of existential quantification in concept languages. *Artificial Intelligence*, 2-3:309–327, 1992.
6. Francesco M. Donini, Maurizio Lenzerini, Daniele Nardi, and Werner Nutt. Tractable concept languages. In *Proceedings of IJCAI-91, 12th International Joint Conference on Artificial Intelligence*, pages 458–463, Sidney, Australia, 1991.
7. Francesco M. Donini, Maurizio Lenzerini, Daniele Nardi, and Andrea Schaerf. From subsumption to instance checking. Technical Report 15.92, Università degli studi di Roma "La Sapienza". Dipartimento di informatica e sistemistica, Rome, Italy, 1992.
8. Melvin Fitting. *First-Order Logic and Automated Theorem Proving.* Springer-Verlag, 1990.
9. Jean H. Gallier. *Logic for Computer Science: Foundations of Automatic Theorem Proving.* Harper & Row Publishers, New York, 1986.
10. Michael R. Garey and David S. Johnson. *Computers and intractability. A guide to the theory of NP-completeness.* Freeman and Company, New York, NY, 1979.
11. G. Gentzen. Untersuchungen über das logische Schliessen. *Mathematische Zeitschrift*, 39:176–210,405–431, 1935.

12. C. A. Gobel, C. Haul, and S. Bechhofer. Describing and classifying multimedia using the description logic GRAIL. In *Proceedings of the SPIE Conference on Storage and Retrieval for Still Images and Video Databases IV (SPIE-95)*, pages 132–143, San Jose, CA, February 1995.

13. Bernhard Hollunder, Werner Nutt, and Manfred Schmidt-Schauß. Subsumption algorithms for concept description languages. In *Proc. of ECAI-90, 9th European Conference on Artificial Intelligence*, pages 348–353, Stockholm, Sweden, 1990.

14. Maurizio Lenzerini and Andrea Schaerf. Concept languages as query languages. In *Proc. of the 9th Nat. Conf. on Artificial Intelligence (AAAI-91)*, pages 471–476, 1991.

15. Hector J. Levesque. A logic of implicit and explicit belief. In *Proc. of the 4th Nat. Conf. on Artificial Intelligence (AAAI-84)*, pages 198–202, Austin, TX, 1984.

16. Hector J. Levesque and Ronald J. Brachman. Expressiveness and tractability in knowledge representation and reasoning. *Computational Intelligence*, 3:78–93, 1987.

17. Carlo Meghini, Fabrizio Sebastiani, Umberto Straccia, and Costantino Thanos. A model of information retrieval based on a terminological logic. In *Proceedings of SIGIR-93, 16th International Conference on Research and Development in Information Retrieval*, pages 298–307, Pittsburgh, PA, 1993.

18. Carlo Meghini and Umberto Straccia. Extending a description logic to cope with the completeness of multimedia documents. In *Proc. of the 12th European Conf. on Artificial Intelligence (ECAI-96): Workshop on Knowledge Representation for Interactive Multimedia Systems*, pages 42–50, Budapest, Hungary, 1996.

19. Carlo Meghini and Umberto Straccia. A relevance terminological logic for information retrieval. In *Proceedings of SIGIR-96, 19th International Conference on Research and Development in Information Retrieval*, pages 197–205, Zurich, Switzerland, 1996.

20. Bernhard Nebel. *Reasoning and revision in hybrid representation systems.* Springer, Heidelberg, FRG, 1990.

21. Peter F. Patel-Schneider. A four-valued semantics for frame-based description languages. In *Proceedings of AAAI-86, 5th Conference of the American Association for Artificial Intelligence*, pages 344–348, Philadelphia, PA, 1986.

22. Peter F. Patel-Schneider. A hybrid, decidable, logic-based knowledge representation system. *Computational Intelligence*, 3:64–77, 1987.

23. Peter F. Patel-Schneider. A four-valued semantics for terminological logics. *Artificial Intelligence*, 38:319–351, 1989.

24. Andrea Schaerf. Reasoning with individuals in concept languages. *Data and Knowledge Engineering*, 13(2):141–176, 1994.

25. Manfred Schmidt-Schauß and Gert Smolka. Attributive concept descriptions with complements. *Artificial Intelligence*, 48:1–26, 1991.

26. Fabrizio Sebastiani. A probabilistic terminological logic for modelling information retrieval. In *Proceedings of SIGIR-94, 17th ACM International Conference on Research and Development in Information Retrieval*, pages 122–130, Dublin, IRL, 1994. Published by Springer Verlag, Heidelberg, FRG.

27. Fabrizio Sebastiani and Umberto Straccia. A computationally tractable terminological logic. In *Proceedings of SCAI-91, 3rd Scandinavian Conference on Artificial Intelligence*, pages 307–315, Roskilde, Denmark, 1991.

28. Gerd Wagner. Ex contradictione nihil sequitur. In *Proc. of the 12th Int. Joint Conf. on Artificial Intelligence (IJCAI-91)*, pages 538–543, Sydney, Australia, 1991.

Tableaux for Functional Dependencies and Independencies

Duminda Wijesekera , M. Ganesh, Jaideep Srivastava and Anil Nerode[t]
Dept. of Computer Science, Univ. of Minnesota Minneapolis, MN 55455.
Dept. of Mathematics, Cornell University, Ithaca, NY 14853[t]
e-mail: {wijesek|ganesh|srivasta}@cs.umn.edu, anil@math.cornell.edu

Abstract. In this paper we show the application of prefixed-signed tableaux to functional dependencies and independencies. Although functional dependencies and independencies are first-order formulae, a custom-made tableau proof procedure is more efficient than a general purpose tableau for first-order logic.

1 Introduction

Success of database technology has brought about the accumulation of large amounts of data which hold hidden facts and knowledge about objects that have been modelled. Naturally, there has been a quest to extract knowledge out of such data to increase profitability. Although the general problem of knowledge acquisition has been an active area of research in Artificial Intelligence (AI), what is novel in the current context is the necessity to use it together with available database technologies such as a general query processing framework. This area of research, commonly referred to as *data mining* [1], has in recent times been very actively pursued in various application domains such as association rules [3, 10, 2], frequent episodes in sequence [14], and integrity constraints [16, 15].

It has been rightly pointed out that data mining has a vast collection of techniques and algorithms, but is in need of an unifying conceptual framework [12, 11]. Such a framework needs to address three key conceptual aspects of the problem. The first is the inductive nature of knowledge acquisition, which is characteristic of passing judgments about the general from particulars. The second is the probability and the risks involved in interpreting such judgments in similar circumstances, because the consumer of such knowledge must be made aware of scope and limitations of its validity. The third is the deductive (logical) aspect involved in knowledge acquisition, because some facts are consequences of others and consequently, (modulo some philosophical standpoints) knowledge of an antecedent entails knowledge of its consequents. There are proposals that satisfy such conceptual requirements which draw upon techniques from AI, statistics and database systems and are based on probabilistic logic [12].

One application of data mining that has received considerable attention over the years, and therefore stands out as a candidate testbed for general frameworks in the area, is mining for integrity constraints. Data dependencies are one such class of integrity constraints. They specify semantic dependencies between data attributes in relations, that are not usually given by the data tables. Data dependencies of various kinds were defined and investigated as a means of specifying and enforcing known relationships between entities in a database. Database relations in which given types of dependencies hold among entities result in specific normal forms [5], thereby making cleaner and more modular data tables. Such modularity is necessary to maintain proper semantics during insert, delete and update operations [19]. In this paper, our attention is focussed on mining for a specific type of data dependency, commonly referred to as functional dependency (FD). Functional dependencies specify that values of some data attributes depend on values of other data attributes. A set of attributes Y is said to be functionally dependent on a set of attributes X (denoted $X \to Y$) if any two rows that have the same values for attributes in X, also have the same values for attributes in Y.

The deductive aspect of the problem of mining for FDs has been solved by a sound and complete axiomatization, now referred to as *Armstrong's axioms* [4]. However, in applications where the designer is unaware of existing functional dependencies, they need to be mined out of raw data. In such situations, a naive *mining engine* runs through data tables, looking for existing functional dependencies. Such a basic mining engine can be significantly enhanced by using two optimizations. The first is that, if at any stage in the mining process, if there is a mechanism to derive consequences of already mined FDs, then looking for these can be omitted. The second is that, at any stage if the mining engine keeps tracks of already invalidated FDs, then it need not mine for their logical consequences. Also mining for invalid FDs, i.e. functional independencies (FI), takes only two rows with appropriate values. Hence, having a fast deductive engine to derive both FIs and FDs greatly enhances the speed of a mining engine. This paper proposes to fulfill that need in one way. Although the use of such negative knowledge about FDs, i.e. functional independencies, has been proposed [6, 13] to aid in the mining problem, but no accurate and complete axiomatization have been given. A semantic proof of completeness of an equivalent axiomatization is given in [17]. In a related paper [20] we give a syntactic proof of a complete axiomatization. In this paper we explore tableaux for that system, called the *FI-system*, thereby providing a fast and more intuitive method to derive consequences of already mined functional dependencies and independencies.

In developing tableaux for functional dependencies and independencies, two facts need to be considered. The first is that both dependencies and independencies are first order sentences with equality, because B is said to be functionally dependent on A if for all rows t_1, t_2 in a data table $t_1[A] = t_2[A]$ implies that $t_1[B] = t_2[B]$. Furthermore, B is said to be functionally independent of A if B is not functionally dependent on A. Due to its high complexity, and generally non-terminating nature of tableaux for classical predicate logic, we decided to develop a custom made tableau system to be used in a general purpose data-mining engine, that is currently being designed.

As it will become clear later, as opposed to tableau systems for classical predicate logic. our system is able to process what amounts to *existential witnesses* at the initial root node of the tableau. Furthermore, we only have two levels of nesting in our formulae. These characteristics make our implementation task easier, simpler, and with more room for optimizations.

In database schema design, the designer of the schema has to declare the functional dependencies that exist among the attributes in a relation. In practice, as an aid to such designers, it has been found useful to produce a relation that satisfies *only* those functional dependencies that have been specified. In response to this need, a tool referred to as *Design-By-Example* has been developed [7]. We believe that this tool also can benefit from using the counter-examples produced by the tableaux, or produce a logic-based alternative to *Design-By-Example*. As opposed to other proof methods, a failed proofs in a tableau produce a counter example.

2 Syntax and Semantics

In this section we present the syntax and semantics of dependency theory. In practice, functional dependencies and independencies relate sets of attributes from a collection of relational symbols that are taken from a relation schema. Each relational symbol also ranges over relations with a finite number of rows. In our construction, we do not impose any such restrictions, although we assume that all data attributes relate to one data table, known as a universal relation. [18].

2.1 Syntax

Our syntax consists of the following components:

1. U is the set of all attributes.

2. Subsets of attributes (i.e., subsets of U) are denoted by upper case letters (possibly subscripted).

3. Attribute values are denoted by lower case letters (possibly subscripted) from corresponding attribute sets.

4. Sentences are of the form $(X \to Y)$ and $(X \not\to Y)$, where X and Y are sets of attributes as given in item 2. Connectives $\to$ and $\not\to$ denote, respectively, dependencies and independencies between sets of attributes.

2.2 Semantics

A model in our syntax is a data table that has all elements of U as attributes. For the purposes of this work, we assume that the database consists of a universal relation (i.e. all data tables in a database as one data table). Rows i and j are respectively denoted by t_i and t_j. The set of values of attributes corresponding to the attribute set A in row t_i is denoted by $t_i[A]$.

Definition 1 (Satisfaction) *Let $\mathcal{T}$ be a model and A, B be sets of attributes. Then:*

1. *We say a data table (model) $\mathcal{T}$ satisfies the functional dependency $(A \to B)$ (Notation: $\mathcal{T} \models (A \to B)$) [8], iff for all rows i and j of $\mathcal{T}$ if $t_i[A] = t_j[A]$, then $t_i[B] = t_j[B]$.*

2. *We say a data table (model) $\mathcal{T}$ satisfies the functional independency, $(A \nrightarrow B)$ (Notation: $\mathcal{T} \models (A \nrightarrow B)$) if $\mathcal{T} \not\models (A \to B)$, i.e there are two rows i and j of $\mathcal{T}$ with $t_i[A] = t_j[A]$ and $t_i[B] \neq t_j[B]$.*

3 Tableau Rules

In this section, we develop tableaux for functional dependencies and independencies. Our tableaux are prefixed-signed tableaux similar to, but not exactly, those in the literature; e.g. in [9]. The differences are in the·rules that we use. As will be seen shortly, the language of our tableaux is richer than the language used in dependency theory.

In our development, we consider a tableau proof as a failed attempt to systematically construct a counter-example. In our case, the syntax consists of functional dependencies and independencies.

As stated, for a data table $\mathcal{T}$ to satisfy a functional independency $(A \nrightarrow B)$, where A and B are attribute sets, there must be two rows t_1, t_2 such that both t_1 and t_2 have the same attribute values for A, but different attribute values for B. In order to indicate such pairs of rows we use prefixes in our tableaux.

3.1 Tableaux

In order to develop tableaux, we need further syntax. Towards fulfilling this goal, we define signed formulae and prefixed-signed formulae as follows.

Definition 2 (Signed Formulae and Prefix Notation) *Suppose U is the set of all attributes. Then:*

- **Signed Formulae:** $T(A \nrightarrow B), F(A \nrightarrow B), T(A \to B), F(A \to B), T(A), F(A)$ *are called signed formulae for any sets of attributes $A, B \subseteq U$.*

- **Structure of Prefixes:** *Prefixes are integers denoted by small-case greek letters and Σ denotes the set of prefixes.*

- **Prefixed-Signed Formulae:** *If ρ is a prefix, and ψ a signed formula we say that $\rho(\psi)$ is a prefixed-signed formula.*

- **Sets of Prefixed-Signed Formulae:** *Suppose S is a set of signed formulae, and ρ is a prefix. Then $\rho(S) = \{\rho(s) : s \in S\}$ is said to be a prefixed set of signed formulae, prefixed by ρ.*

We call the collection of all formulae, signed formulae, and sets of prefixed-signed formulae the extended language. For a given set U of all attributes and a given set of prefixes Σ, the extended language is named $L(U, \Sigma)$.

Definition 3 (Semantic Mappings) *Suppose $\mathcal{T}$ is a universal relation where all attributes in $\mathcal{T}$ are from U, Σ is a set of prefixes, and $L(U, \Sigma)$ is the extended language. Then we define a semantic mapping $\mathcal{F}$ to be a partial function from Σ to pairs of rows of $\mathcal{T}$. The pair $\mathcal{F}(\rho)$ is denoted by $(\mathcal{F}(\rho)(1), \mathcal{F}(\rho)(2))$.*

Now we use semantic mappings to interpret our extended language $L(U, \Sigma)$.

Definition 4 (Interpreting the Extended Language $L(U, \Sigma)$) *Suppose $\mathcal{T}$ is a universal relation where all attributes in $\mathcal{T}$ are from U, Σ is a set of prefixes, $L(U, \Sigma)$ is the extended language and $\mathcal{F}$ is a semantic mapping from Σ to $\mathcal{T} \times \mathcal{T}$. Then extend $\models$ to $L(U, \Sigma)$ as follows:*

1. *$\mathcal{T} \models T(A)$ if for any two rows t_1 and t_2 in T, $t_1[A] = t_2[A]$.*

2. *$\mathcal{T} \models F(A)$ if $\mathcal{T} \not\models T(A)$*

3. *$\mathcal{T} \models \rho(T(A))$ if $\mathcal{F}(\rho)(1)[A] = \mathcal{F}(\rho)(2)[A]$*

4. *$\mathcal{T} \models \rho(F(A))$ if $\mathcal{T} \not\models \rho(T(A))$.*

5. *$\mathcal{T} \models \rho(S)$, where S is a set of signed formulae if $\mathcal{T} \models \rho(s)$ for each $s \in S$.*

6. *$\mathcal{T} \models T(A \rightarrow B)$ if $\mathcal{T} \models (A \rightarrow B)$, $\mathcal{T} \models F(A \rightarrow B)$ if $\mathcal{T} \not\models (A \rightarrow B)$, $\mathcal{T} \models T(A \nrightarrow B)$ if $\mathcal{T} \models (A \nrightarrow B)$, $\mathcal{T} \models F(A \nrightarrow B)$ if $\mathcal{T} \not\models (A \nrightarrow B)$.*

With the above interpretation of the extended language we can now define tableaux as a tree in the following.

Definition 5 (Tableau) *A tableau is a finite branching tree where:*

1. *The root node contains $T(\emptyset)$, and possibly some other prefixed-signed formulae.*

2. *The nodes contain sets of signed formulae. e.g. $T(A), F(B), T(A \nrightarrow B), F(A \rightarrow B)$, or a set of prefixed-signed formulae. e.g. $\rho(T(A), F(B))$.*

The tableau is constructed by recursively applying either denotation, branch update or expansion rules as follows.

1. **Denotation Rule:**
   (I) and (II): *At a node n, if $F(A \rightarrow B)$ or $T(A \nrightarrow B)$ are on the path p from the root to n, then extend p by adding two nodes, the second a child of the first and the first a child of n, that have $\rho(T(A)), \rho(F(B))$, respectively on them where ρ has not appeared on p as a prefix.*

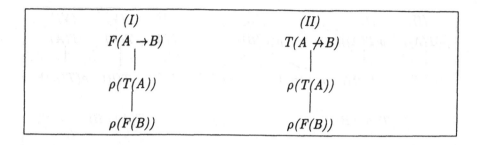

2. Branch Update Rule:

(I) and (II): At leaf n, where p is the path from n to the root, if $T(A \to B)$ or $F(A \not\to B)$, and $\rho(T(A))$ appear on nodes in p, then extend p by adding a child node to n that has $\rho(T(B))$ on it.

(III) and (IV): At leaf n, where p is the path from n to the root, if $T(A \to B)$ or $F(A \not\to B)$, and $T(A)$ appear on nodes in p, then extend p by adding a child node to n that has $T(B)$ on it.

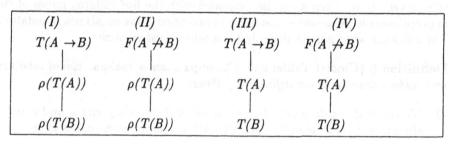

3. Expansion Rule:

(I) At leaf n, where p is the path from n to the root, if $\rho(T(A))$ appears on a node in p, and $C \subseteq A$, then extend p by adding a child node to n that has $\rho(T(C))$ on it.

(II) At a node n, where p is the path from n to the root, if $\rho(T(A))$ and $\rho(T(B))$ appear on nodes in p, then extend p by adding a a child node to n that has $\rho(T(A \cup B))$ on it.

(III) At leaf n, where p is the path from n to the root, if $\rho(F(B))$ appears on a node in p, then extend p by adding m children nodes of n, where the ith child node has $\rho(F(B_i))$ for each $i \leq m$, where $B_1, \ldots B_m$ are the singleton subsets of B.

(IV) and (V) These expansion rules are counterparts of **(I)** and **(II)** respectively without the prefixes.

(VI) At leaf n, where p is the path from n to the root, if $T(A)$ appears on a node in p and ρ appears as a prefix of some signed formula on p, then extend p by adding a child node to n that has $\rho(T(A))$ on it.

$$
\begin{array}{ccccccc}
(I) & (II) & (III) & (IV) & (V) & (VI) \\
\rho(T(A)) & \rho(T(A)) & \rho(F(B)) & T(A) & T(A) & T(A) \\
| & | & & | & | & | \\
\rho(T(C)) & \rho(T(B)) & \rho(F(B_1)) \cdots \rho(F(B_n)) & T(C) & T(B) & \rho(T(A)) \\
& | & & & | & \\
& \rho(T(A \cup B)) & & & T(A \cup B) &
\end{array}
$$

In these rules $C \subseteq A$, and $B_1 \ldots B_n$ are the singleton subsets of B.

Following standard usage, tableaux rules are to be interpreted as follows. For example, suppose we want to apply a branch update rule. Assume that n is a leaf in the tableau and $T(A \to B)$, $\rho(T(A))$ appear on nodes on the path from n to the root. Then n can be extended by adding a node with $\rho(T(B))$. To generalize, suppose n is a leaf in a tableau, and nodes in the path from the root to n contain the formula/set-of-formulae that are the non-leaf nodes of the rules given above. Then n can be expanded with the leaf-children nodes of the appropriate rule. Also notice that tableau extension is (as usual) not mandatory. The rules say what *may* be done, but not what *must* be done.

Definition 6 (Closed Tableaux) *Closed-path on a tableau, closed-tableaux, and tableau deductions are defined as follows:*

1. *We say that a path p in a tree is closed if there is a prefix ρ and a set of attributes A where $\rho(T(A))$ and $\rho(F(A))$ appear on some nodes in p.*

2. *We say that a tableau is closed if all paths in it are closed.*

3. *We say that a closed-tableau with a root node $\{T(\phi_1), \ldots T(\phi_n), F(\phi_0)\}$ is a proof of ϕ_0 from $\phi_1, \ldots \phi_n$. If there is a proof of ϕ_0 from $\phi_1, \ldots \phi_n$ then we say that $\phi_1, \ldots \phi_n$ proves ϕ_0 which is denoted by $\{\phi_1, \ldots \phi_n\} \vdash \phi_0$. Here each ϕ_i for $0 \leq i \leq n$ is either a functional independency or a functional dependency.* ·

Closed-paths in tableaux correspond to semantic impossibilities. We formally prove so in the next theorem.

Lemma 1 Non-Satisfiability of Closed Paths. *Suppose p is a closed path and let Σ be the prefixes on p. Then $\mathcal{T} \not\models \{s : s \in p\}$ for any model (table) $\mathcal{T}$, and any semantic mapping $\mathcal{F}$ from Σ to pairs of rows of $\mathcal{T}$.*

Proof: By Definition 6, there is a set of attributes, A, where $\rho(T(A)), \rho(F(A)) \in p$. If $\mathcal{T} \models \rho(T(A)), \rho(F(A))$, then for some rows t_i, t_j of $\mathcal{T}$, we have that $t_i[A] = t_j[A]$ and $t_i[A] \neq t_j[A]$, where $\mathcal{F}(\rho)(1) = t_i$ and $\mathcal{F}(\rho)(2) = t_j$; contradiction. ∎

4 Correctness of Tableau Extension Rules

In this section, we show that our tableau rules are correct with respect to the intended interpretation of $L(U, \Sigma)$. We do so by showing that an application of an extension rule to a satisfiable open path results in a satisfiable open path.

Lemma 2 Correctness of Tableau Rules. *Suppose p is a path in a tableau from the root to a leaf n and all formulae of the extended language $L(U, \Sigma)$ on p are satisfiable by a model $\mathcal{T}$, where the semantic mapping $\mathcal{F}$ maps the prefixes on p to pairs of rows in $\mathcal{T}$. Then there is an extension p' of p, and an extension $\mathcal{F}'$ of $\mathcal{F}$ that satisfies all formulae on p'.*

Proof: By case analysis on the proof rule used.

Denotation Rule: Suppose $F(A \to B)$, or $T(A \not\to B)$ is in p and there is a table $\mathcal{T}$ that satisfies $F(A \to B)$. Then there are rows t_i, t_j of $\mathcal{T}$ with $t_i[A] = t_j[A]$ and $t_i[B] \neq t_j[B]$. By extending $\mathcal{F}$ such that $\mathcal{F}(\rho) = (t_i, t_j)$, for any prefix ρ that did not appear on p, we get that all formulae on p' are satisfiable.

Branch Update Rule: For Rule (I), suppose $T(A \to B)$ and $\rho(T(A))$ are on p, and they are satisfiable in $\mathcal{T}$. Then there are rows t_i, t_j satisfying $\mathcal{F}(\rho) = (t_i, t_j)$ and $t_i[A] = t_j[A]$. Then, because $\mathcal{T} \models (A \to B)$ and $t_i[A] = t_j[A]$, we get that $\mathcal{T} \models t_i[B] = t_j[B]$.

For Rule (II), if $\mathcal{T} \models F(A \not\to B)$, then $\mathcal{T} \models (A \to B)$, and the result follows by the previous case.

For Rule (III) and Rule (IV), suppose $T(A \to B)$ or $F(A \not\to B)$, and $T(A)$ are on p, and they are satisfiable in $\mathcal{T}$. Then for any two rows t_i, t_j we have $t_i[A] = t_j[A]$. Then, because $\mathcal{T} \models (A \to B)$ and $t_i[A] = t_j[A]$, we get that $t_i[B] = t_j[B]$, for any t_i, t_j.

Expansion Rule: For Rule (I), suppose $\rho(T(A))$ is on p and $C \subseteq A$, and $\mathcal{T} \models \rho(T(A))$. Then by definition, $\mathcal{F}(\rho)(1)[A] = \mathcal{F}(\rho)(2)[A]$. For any $C \subseteq A$, we get $\mathcal{F}(\rho)(1)[C] = \mathcal{F}(\rho)(2)[C]$, satisfying $\mathcal{T} \models \rho(T(C))$. Similarly for Rule (IV).

For Rule (II), suppose $\rho(T(A)), \rho(T(B))$ is on p and $\mathcal{T} \models \rho(T(A)), \rho(T(B))$. Then by definition, $\mathcal{F}(\rho)(1)[A] = \mathcal{F}(\rho)(2)[A]$, and $\mathcal{F}(\rho)(1)[B] = \mathcal{F}(\rho)(2)[B]$, and hence $\mathcal{F}(\rho)(1)[A \cup B] = \mathcal{F}(\rho)(2)[A \cup B]$, and consequently, $\mathcal{T} \models \rho(T(A \cup B))$. Similarly for Rule (V).

For Rule (III), suppose $\rho(F(B))$ is on p, and $\{B_i : i \leq n\}$ are the singleton subsets of B. Since $\mathcal{T} \models \rho(F(B))$ we have $\mathcal{F}(\rho)(1)[B] \neq \mathcal{F}(\rho)(2)[B]$. Then there is a singleton subset B_i of B satisfying $\mathcal{F}(\rho)(1)[B_i] \neq \mathcal{F}(\rho)(2)[B_i]$. Let p' be the extension of p that includes $\rho(F(B_i))$. Then $\mathcal{T} \models \rho(F(B_i))$, and hence all formulae on p' are satisfiable by $\mathcal{T}$.

For Rule (VI), suppose $T(A)$ is on path p and $\mathcal{T} \models T(A)$. Therefore, for any two rows t_i, t_j we have $t_i[A] = t_j[A]$. By extending $\mathcal{F}$ such that $\mathcal{F}(\rho) = (t_i, t_j)$, for any prefix ρ that did not appear on p, we get that extended path p' with $\rho(T(A))$ is satisfiable in $\mathcal{T}$. ∎

Next, to prove the correctness of tableau rules, we show that stringing a series of *semantically correct* extensions result in a *semantically correct* tableaux.

Theorem 3 Correctness of Tableau Deductions. *Suppose* $\{\phi_1, \ldots \phi_n\} \vdash \phi_0$, *then for every model (table)* $\mathcal{T}$, *if* $\mathcal{T} \models \phi_i$ *for each* $1 \leq i \leq n$, *then* $\mathcal{T} \models \phi_0$.

Proof: Suppose that there is a tableau proof t of ϕ_0 from $\{\phi_1, \ldots, \phi_n\}$, and that the claim of the theorem is false. Consequently, there is a table $\mathcal{T}$ such that $\mathcal{T} \models \phi_i$ for each $1 \leq i \leq n$; but $\mathcal{T} \not\models \phi_0$. Hence, $\mathcal{T} \models T(\phi_i)$ for each $1 \leq i \leq n$ and $\mathcal{T} \models F(\phi_0)$. Then by Lemma 2, there is a path, say p, of t that is satisfied by $\mathcal{T}$. Nevertheless, because $\{\phi_1, \ldots \phi_n\} \vdash \phi_0$ and p is closed, this contradicts Lemma 1. ∎

5 Completeness of Tableaux

A tableau proof is an attempt to find a model for a collection of formulae. In order for its failure to be considered a proof, because tableaux are non-deterministic, we provide a systematic way of developing tableaux, called *systematic tableaux*. In a sense, it constitutes a procedure to find a model, if one exists at all, and when it fails, its failure is considered a proof. If systematic tableaux achieve this end, the procedure is called a *complete-systematic tableau procedure*. Our systematic tableau construction is as follows:

Definition 7 (Attended and Unattended formulae) *We say that a prefixed-signed formula* ψ *on a path* p *requires attention of type 0, through 7, if they are of the following form, and satisfy the given condition. In the systematic tableau procedure, we perform the appropriate action, as follows.*

Form	Attention Type	Condition	Action
$\rho(T(A))$	0	$T(A \to B)$ or $F(A \not\to B)$ on p, but $\rho(T(B))$ not on p	add a child node with $\rho(T(B))$
$\rho(T(A))$	1	$\rho(T(A))$ on p but $\rho(T(A'))$ not on p for some subset A'	add a child node with $\rho(T(A'))$
$\rho(T(A))$	2	$\rho(T(B))$ on p, but $\rho(T(A \cup B))$ not on p	add a child node with $\rho(T(A \cup B))$
$T(A)$	3	$T(A \to B)$ or $F(A \not\to B)$ on p, but $T(B)$ not on p	add a child node with $T(B)$
$T(A)$	4	$T(A)$ on p but $T(A')$ not on p for some subset A'	add a child node with $T(A')$
$T(A)$	5	$T(B)$ on p, but $T(A \cup B)$ not on p	add a child node with $T(A \cup B)$
$T(A)$	6	ρ appears on p but $\rho(T(A))$ does not.	add child node with $\rho(T(A))$ on it.
$\rho(F(A))$	7	$\rho(F(A))$ on p, but $\rho(F(A'))$ not on p for any singleton subset A' of A	add children nodes with $\rho(F(A'))$ for each singleton subset A' of A

The systematic tableau is constructed in stages as follows:

1. **At Stage 0:** For all formulae of the type $F(A \to B)$ and $T(A \not\to B)$ apply Denotation Rules I and II respectively, with a new prefix for each instance. Then those formulae are declared to have *received attention.*

2. **At Stage 8k + i > 0, where** $0 \le i \le 7$: For every open path p from the root, look for the node that is closest to the root that has a formula that *requires attention of type i*, and if found, take the appropriate action by appending the leaf with a child node/ children nodes, as given in the table above and declare that formula to have *received attention* of the appropriate type.

In order to show that our tableau system is complete, we show that for a given set of formulae, if its complete-systematic tableau has an open branch, then there is a model for the given set of formulae. This is proved by showing that any open branch of a complete-systematic tableau can lead to the construction of a downward-saturated set (as defined shortly), which can be used to produce a model.

5.1 Downward-Saturated Sets

In this section we define downward-saturated sets, a collection of formulae used to build syntactic models. Thereafter, we show that prefixed-signed formulae on an open path in a complete-systematic tableau constitute a downward-saturated set, and is a candidate to be used in constructing a model.

Definition 8 (Downward-Saturated Set) *We say that a set S of prefixed-signed formulae is a downward-saturated set if it satisfies the following properties.*

1. *It is not the case that $\rho(T(A)) \in S$ and $\rho(F(A)) \in S$ for any prefix ρ and any set of attributes A.*

2. *If $F(A \to B) \in S$ or $T(A \not\to B) \in S$ then there is an integer prefix ρ such that $\rho(T(A), F(B)) \in S$.*

3. *If $T(A \to B) \in S$ or $F(A \not\to B) \in S$, then for every prefix ρ, if $\rho(T(A)) \in S$ then $\rho(T(B)) \in S$.*

4. *If $\rho(T(A)) \in S$, then $\rho(T(A')) \in S$ for every subset A' of A.*

5. *If $\rho(T(A), T(B)) \in S$, then $\rho(T(A \cup B)) \in S$.*

6. *If $T(A \to B) \in S$ or $F(A \not\to B) \in S$, and $T(A) \in S$, then $T(B) \in S$.*

7. *If $T(A) \in S$, then $T(A') \in S$ for every subset A' of A.*

8. *If $T(A), T(B) \in S$, then $T(A \cup B) \in S$.*

9. *If $T(A) \in S$, then $\rho(T(A)) \in S$ for all $\rho \in \Sigma$.*

10. *If $\rho(F(B)) \in S$, then $\rho(F(B_i)) \in S$ for some singleton subset B_i of B.*

Now we show that every open branch of a systematic tableau is a downward-saturated set.

Theorem 4 Downward-Saturated Sets from Tableaux. *Formulae on nodes of an open branch of a systematic tableau constitute a downward-saturated set, provided that every formula that require attention of any type has received attention.*

Proof: Suppose P is an open branch of a systematic tableau t. Let S be the set of prefixed-signed formulae on p. Then we show that S is a downward-saturated set as follows.

1. $\rho(T(A), F(A)) \notin S$, because P is an open branch.

2. Suppose $F(A \to B) \in S$ or $T(A \not\to B) \in S$. Then $F(A \to B)$ or $T(A \not\to B)$ must be at the root of t. Then, by stage 0 of the construction of the systematic tableau, there is a prefix ρ such that $\rho(T(A)), \rho(F(B))$ is a node, say n, on t. Since the construction at stage 0 produced only single branching extensions, n must be on p.

3. Suppose $T(A \to B) \in S$ or $F(A \not\to B) \in S$ and $\rho(T(A)) \in S$. Then there is a stage $8k + 0 > 0$ in the construction where $\rho(T(A))$ receives attention of type 0, and consequently, $\rho(T(B)) \in S$.

4. Suppose $\rho(T(A)) \in S$, and $A' \subset A$. Then at some stage $8k + 1 > 0$, $\rho(T(A))$ receives type 1 attention and $\rho(T(A')) \in S$.

5. Suppose $\rho(T(A)), \rho(T(B)) \in S$. Then at some stage $8k + 2 > 0$ $\rho(T(A))$ receives type 2 attention with respect to $\rho(T(B))$ and consequently $\rho(T(A \cup B)) \in S$.

6. Suppose $T(A \to B) \in S$ or $F(A \not\to B) \in S$ and $T(A) \in S$. Then there is a stage $8k + 3 > 0$ in the construction where $T(A)$ receives attention of type 3, and consequently, $T(B) \in S$.

7. Suppose $\rho(T(A)) \in S$, and $A' \subset A$. Then at some stage $8k + 4 > 0$, $\rho(T(A))$ receives type 4 attention and $\rho(T(A')) \in S$.

8. Suppose $T(A), T(B) \in S$. Then at some stage $8k + 5 > 0$ $T(A)$ receives type 5 attention with respect to $T(B)$ and consequently $T(A \cup B) \in S$.

9. Suppose $T(A) \in S$. Then at some stage $8k + 6 > 0$, $T(A)$ receives type 6 attention with respect to the index ρ and consequently, $\rho(T(A)) \in S$.

10. Suppose $\rho(F(B)) \in S$. Then at some stage $8k + 7 > 0$, $\rho(F(B)) \in S$ receives type 7 attention, and then there are nodes added as children to the leaf with $\rho(F(B_i))$ on the ith child, for each singleton subset B_i of B . Since p will have one such child node, there is some i with $\rho(F(B_i)) \in S$. ∎

In order to produce a model from a downward-saturated set we need to produce a complete-downward-saturated set.

Definition 9 (Complete-Downward-Saturated Set) *We say that a downward-saturated set S is a complete-downward-saturated set if for every prefix $\rho \in \Sigma$ and every singleton attribute set A of U, either $\rho(T(A)) \in S$ or $\rho(F(A)) \in S$.*

Theorem 5 Constructing Complete-Downward-Saturated Sets. *Suppose S is a downward-saturated set. Then there is a complete-downward-saturated set S' satisfying $S' \supseteq S$.*

Proof: Suppose S is a downward-saturated set. Then define S' as: $S' = \{\rho(F(A)) : \rho(T(A)), \rho(F(A)) \notin S\} \cup S$. Then $S' \supseteq S$ is a complete-downward-saturated set. It can be seen that S' is a downward-saturated set, because S was. ∎

5.2 Constructing Syntactic Models

In this section we show how to construct a model from a complete-downward-saturated set.

Theorem 6 Models from Complete-Downward-Saturated Sets. *Suppose S is a complete-downward-saturated set. Then there is a model (table) T and a semantic mapping $\mathcal{F}$ from Σ to $T \times T$ such that for every extended formula ψ in $L(U, \Sigma)$, $\psi \in S$ if and only if $T \models \psi$.*

Proof: First we construct the table T and then show that it satisfies $\psi \in S$ if and only if $T \models \psi$. To do so assume that each attribute $A_i \in U$ can take a countable number of values, say $\{a_{i,j} : j \geq 1\}$.
Construction:

1. Let $\Sigma' = \{\rho_i : i \geq 0\}$ be a listing of prefixes $\rho \in \Sigma$ used in S. For each $\rho_i \in \Sigma'$ create two rows ρ_i, ρ'_i in T.

2. Define $U_0 = \{A : T(A) \in S\}$. Suppose $U_0 = \{A_j : 1 \leq j \leq n\}$.

   For each attribute $A_j \in U_0$, and each row $t \in T$, let $t[A_j] = a_{j,0}$, where t is some ρ_i or ρ'_i.

3. For each attribute $A_j \in U \setminus U_0$:

   (a) If $\rho_i(T(A_j)) \in S$ let $\rho_i[A_j] = \rho'_i[A_j] = a_{j,2i}$.

   (b) If $\rho_i(F(A_j)) \in S$ let $\rho_i[A_j] = a_{j,2i}$ and $\rho'_i[A_j] = a_{j,2i+1}$.

4. Define the semantic mapping $\mathcal{F} : \Sigma' \to T \times T$ as $\mathcal{F}(\rho_i) = (\rho_i, \rho'_i)$, where ρ'_i is as given in (1).

Proof that Our Construction Works:

1. To show that if $\rho_i(T(A)) \in S$ then $\mathcal{T} \models \rho_i(T(A))$: If $\rho_i(T(A)) \in S$, then for each singleton attribute A_j of A, $\rho_i(T(A_j)) \in S$. Hence $\mathcal{F}(\rho_i)(1)[A_j] = \mathcal{F}(\rho_i)(2)[A_j] = a_{j,2i}$ for all singleton subsets A_i of A, and consequently, $\mathcal{F}(\rho_i)(1)[A] = \mathcal{F}(\rho_i)(2)[A]$. Hence $\mathcal{T} \models \rho_i(T(A))$.

2. To show that if $\rho_i(F(A)) \in S$ then $\mathcal{T} \models \rho(F(A))$: By property (10) of downward-saturated sets, $\rho_i(F(A_j)) \in S$ for some singleton subset A_j of A. Hence, by construction, $\mathcal{F}(\rho_i)(1)[A_j] = a_{j,2i}$ and $\mathcal{F}(\rho_i)(2)[A_j] = a_{j,2i+1}$. Hence $\mathcal{T} \models \rho_i(F(A))$.

3. To show that if $T(A) \in S$ then $\mathcal{T} \models T(A)$: If $T(A) \in S$, then for each singleton attribute A_j of A, $T(A_j) \in S$. Consequently, $\mathcal{F}(\rho_i)(1)[A_j] = \mathcal{F}(\rho)(2)[A_j] = a_{j,0}$ for all singleton subsets A_i of A, and all $\rho_i \in \Sigma'$. Consequently, $\mathcal{F}(\rho_i)(k_1)[A] = \mathcal{F}(\rho'_i)(k_2)[A]$ for any $k_1, k_2 \in \{1, 2\}$. Hence $\mathcal{T} \models T(A)$.

4. To show that if $T(A \to B) \in S$, or $F(A \not\to B) \in S$, then $\mathcal{T} \models (A \to B)$, $\mathcal{T} \not\models (A \not\to B)$ respectively: Suppose $T(A \to B) \in S$ or $F(A \not\to B) \in S$. We need to show that $\mathcal{T} \models (A \to B)$. In order to do so assume that there are rows t_i, t_k of $\mathcal{T}$ such that $t_i[A] = t_k[A]$. Then there are integer prefixes ρ_1, ρ_2 such that rows t_i, t_k are either of $\rho_1, \rho'_1, \rho_2, \rho'_2$.

   Case 1: Suppose $t_i = \rho_1$ and $t_k = \rho'_1$. Then by construction, $\rho_1(T(A)) \in S$, as if not, because S is a complete-downward-saturated set, $\rho_1(F(A)) \in S$, and consequently, $\rho_1(F(A_j)) \in S$ for some singleton subset A_j of A. Then by (2), $t_i[A_j] \neq t_k[A_j]$, contradicting $t_i[A] = t_k[A]$. Hence $\rho_1(T(A)) \in S$ and consequently by property (3) of downward-saturated sets, $\rho_1(T(B)) \in S$, and consequently by proofs in (1), $t_i[B] = t_k[B]$.

   Case 2: Suppose $t_i = \rho_1$ and $t_2 = \rho_2$ or ρ'_2. Then by construction, $t_i[A] = t_k[A]$ if and only if $A_j \in U_0$ for each singleton subset of A. Hence, by property (8) of downward saturated sets, $A \in U_0$, and therefore $T(A) \in S$. Consequently, because S is a downward-saturated set, $T(B) \in S$, and therefore $B_j \in U_0$ for each singleton subset of B_j of B. Therefore, $t_i[B_j] = t_k[B_j]$ for each singleton subset B_j of B. Hence $t_i[B] = t_k[B]$, satisfying $\mathcal{T} \models (A \to B)$.

5. Suppose $F(A \to B)$ or $T(A \not\to B) \in S$. Then by property (2) of downward-saturated sets, there is a prefix ρ such that $\rho(T(A), F(B)) \in S$. Then by proofs in (1) and (2), $\mathcal{F}(\rho)(1)[A] = \mathcal{F}(\rho)(2)[A]$ and $\mathcal{F}(\rho)(1)[B] \neq \mathcal{F}(\rho)(2)[B]$. Hence $\mathcal{T} \models F(A \to B), T(A \not\to B), (A \not\to B)$. ∎

5.3 Completeness Theorem

In this section, we prove the completeness of our tableaux system by showing that an open branch of a complete-systematic tableau can be expanded to a complete-downward-saturated set, and thereby constructing a model.

Theorem 7 Completeness Theorem. *Our tableau proof system is complete for functional dependencies and independencies: i.e. For all models $\mathcal{T}$, if $\mathcal{T} \models \psi_0$ whenever $\mathcal{T} \models \{\psi_i : 1 \leq i \leq n\}$, then $\{\psi_i : 1 \leq i \leq n\} \vdash \psi_0$, where $\{\psi_i : 0 \leq i \leq n\}$ is a collection of dependencies and independencies.*

Proof: Suppose $\{\psi_i : 0 \leq i \leq n\}$ is a collection of dependencies and independencies, and $\{\psi_i : 1 \leq i \leq n\} \nvdash \psi_0$. Then the systematic tableau with $\{T(\psi_i) : 1 \leq i \leq n\} \cup \{F(\psi_0)\}$ has an open branch, say p, on which all formulas that need attention have received attention of the appropriate type. Then, by Theorem 4, formulae on p form a downward-saturated set, say S. Then $\{T(\psi_i) : 1 \leq i \leq n\} \cup \{F(\psi_0)\} \subseteq S$. Then by Theorem 5, there is a complete-downward-saturated set $S' \supseteq S$. Consequently, by Theorem 6, there is a model, say $\mathcal{T}$, that satisfy $\mathcal{T} \models \psi$ if $T(\psi) \in S'$ and $\mathcal{T} \nvDash \psi$ if $F(\psi) \in S'$.

Accordingly, $\mathcal{T} \models \{\psi_i : 1 \leq i \leq n\}$ but $\mathcal{T} \nvDash \psi_0$, leading to a contradiction. Consequently, for all models $\mathcal{T}$, if $\mathcal{T} \models \psi_0$ whenever $\mathcal{T} \models \{\psi_i : 1 \leq i \leq n\}$, then $\{\psi_i : 1 \leq i \leq n\} \vdash \psi_0$. ∎

6 Conclusions

In this paper, we have presented a tableau proof procedure custom made for functional dependencies and independencies. We have done so by interpreting them over a universal relation. In our tableaux, we handle existential and universal witnesses for rows in the universal relation in a way that is both necessary to interpret our connectives and as efficient as possible for implementational purposes.

This project has been carried out in a larger context of implementing a generic, logic based data-mining engine. It is hoped that implementationally efficient logical methods such as tableaux can be used in the context of an emerging application area such as data-mining. Our general research agenda is directed towards this goal.

References

1. R. Agrawal, T. Imielinski, and A. Swami. Database Mining: A Performance Perspective. *IEEE Transactions on Knowledge and Data Engg.*, 5(6):914–925, December 1993.
2. Rakesh Agrawal and Ramakrishnan Srikant. Fast Algorithms for Mining Association Rules. In *Proc. of 20th VLDB Conference*, pages 487–499, Santiago, Chile, 1994.
3. Rakesh Agrawal and Ramakrishnan Srikant. Mining Generalized Association Rules. In *Proc. of the 21st VLDB Conference*, Zurich, Switzerland, 1995.
4. W. W. Armstrong. Dependency Structure of Database Relationships. In *Proc. IFIP 74*, pages 580–583. North Holland, Amsterdam, 1974.
5. C. Beeri, P.A. Bernstein, and N. Goodman. A Sophisticate's Introduction to Database Normalization Theory. *Proc. of the Fourth Int'l. Conf. on Very Large Data Bases*, pages 113–124, 1978.

6. Siegfired Bell. Discovery and Maintainence of Functional Dependencies. In *Proc. of the First KDD Conference*, pages 27–32, Montréal, Canada, 1995.

7. Dina Bitton, Heikki Mannila, and Kari-Jouko Räihä. Design-By-Example: A design Tool for Relational Databases. Technical Report TR 85-692, Dept of Computer Science, Cornell University, 1985.

8. Ramez A. Elmasri and Shamkant B. Navathe. *Fundamentals of Database Systems*. Benjamin Cummings, Redwood City, CA, 2nd edition, 1994.

9. Melvin Fitting. *Proof Methods for Modal and Intuitionistic Logic*. Riedel, Amsterdam, 1983.

10. Maurice Houtsma and Arun Swami. Set Oriented Mining for Association Rules. In *Proc. of 11th Int'l Conf. on Data Engg. (ICDE)*, Taipei, Taiwan, 1995.

11. Tomasz Imielinski. From File Mining to Database Mining. In *ACM-SIGMOD DMKD Workshop*, Montréal, Canada, June 1996.

12. Mansfield Jaeger, Heikki Mannila, and Emil Weydert. Data Mining as Selective Theory Extraction in Probabilistic Logic. In *ACM-SIGMOD DMKD Workshop*, Montréal, Canada, June 1996.

13. J. M. Janas. Covers for Functional Independencies. In *Conference on Database Theory*, volume 338. Springer-Verlag, Lecture Notes in Computer Science, 1988.

14. Heikki Mannila, H Toivonen, and A. I. Verkamo. Discovering Frequent Eposides in Sequences. In *Proc. of the First KDD Conference*, pages 210–215, Montréal, Canada, 1995.

15. Sujal Parikh, M Ganesh, and Jaideep Srivastava. Parallel Data Mining for Functional Dependencies. Technical Report TR 96-033, Dept of Computer Science, University of Minnesota, 1996.

16. Bernhard Pfahringer and Stefan Kramer. Compression-Based Evaluation of Partial Determinations. In *Proc. of the First KDD Conference*, pages 234–239, Montréal, Canada, 1995.

17. B. Thalheim. Logical Relational Database Design Tool Using Different Classes of Dependencies. *Journal of New Generation Computing Systems*, 1(3):211–228, 1988.

18. Jeffrey D. Ullman. *Database and Knowledge-Base Systems*. Computer Science press, Rockville, Maryland, 1988.

19. Moshe Vardi. Fundamentals of Dependency Theory. pages 171–224. Computer Science Press, 1988.

20. Duminda Wijesekera, M. Ganesh, and Jaideep Srivastava. A Complete Axiomatization for Functional Independencies. Technical Report TR 96-066, Dept. of Computer Science, Univ. of Minnesota, Minneapolis, MN 55455, 1996.

Author Index

Lecture Notes in Artificial Intelligence (LNAI)

Vol. 1071: P. Miglioli, U. Moscato, D. Mundici, M. Ornaghi (Eds.), Theorem Proving with Analytic Tableaux and Related Methods. Proceedings, 1996. X, 330 pages. 1996.

Vol. 1076: N. Shadbolt, K. O'Hara, G. Schreiber (Eds.), Advances in Knowledge Acquisition. Proceedings, 1996. XII, 371 pages. 1996.

Vol. 1079: Z. W. Raś, M. Michalewicz (Eds.), Foundations of Intelligent Systems. Proceedings, 1996. XI, 664 pages. 1996.

Vol. 1081: G. McCalla (Ed.), Advances in Artificial Intelligence. Proceedings, 1996. XII, 459 pages. 1996.

Vol. 1083: K. Sparck Jones, J.R. Galliers, Evaluating Natural Language Processing Systems. XV, 228 pages. 1996.

Vol. 1085: D.M. Gabbay, H.J. Ohlbach (Eds.), Practical Reasoning. Proceedings, 1996. XV, 721 pages. 1996.

Vol. 1087: C. Zhang, D. Lukose (Eds.), Distributed Artificial Intelligence. Proceedings, 1995. VIII, 232 pages. 1996.

Vol. 1093: L. Dorst, M. van Lambalgen, F. Voorbraak (Eds.), Reasoning with Uncertainty in Robotics. Proceedings, 1995. VIII, 387 pages. 1996.

Vol. 1095: W. McCune, R. Padmanabhan, Automated Deduction in Equational Logic and Cubic Curves. X, 231 pages. 1996.

Vol. 1104: M.A. McRobbie, J.K. Slaney (Eds.), Automated Deduction – Cade-13. Proceedings, 1996. XV, 764 pages. 1996.

Vol. 1111: J. J. Alferes, L. Moniz Pereira, Reasoning with Logic Programming. XXI, 326 pages. 1996.

Vol. 1114: N. Foo, R. Goebel (Eds.), PRICAI'96: Topics in Artificial Intelligence. Proceedings, 1996. XXI, 658 pages. 1996.

Vol. 1115: P.W. Eklund, G. Ellis, G. Mann (Eds.), Conceptual Structures: Knowledge Representation as Interlingua. Proceedings, 1996. XIII, 321 pages. 1996.

Vol. 1126: J.J. Alferes, L. Moniz Pereira, E. Orlowska (Eds.), Logics in Artificial Intelligence. Proceedings, 1996. IX, 417 pages. 1996.

Vol. 1137: G. Görz, S. Hölldobler (Eds.), KI-96: Advances in Artificial Intelligence. Proceedings, 1996. XI, 387 pages. 1996.

Vol. 1147: L. Miclet, C. de la Higuera (Eds.), Grammatical Inference: Learning Syntax from Sentences. Proceedings, 1996. VIII, 327 pages. 1996.

Vol. 1152: T. Furuhashi, Y. Uchikawa (Eds.), Fuzzy Logic, Neural Networks, and Evolutionary Computation. Proceedings, 1995. VIII, 243 pages. 1996.

Vol. 1159: D.L. Borges, C.A.A. Kaestner (Eds.), Advances in Artificial Intelligence. Proceedings, 1996. XI, 243 pages. 1996.

Vol. 1160: S. Arikawa, A.K. Sharma (Eds.), Algorithmic Learning Theory. Proceedings, 1996. XVII, 337 pages. 1996.

Vol. 1168: I. Smith, B. Faltings (Eds.), Advances in Case-Based Reasoning. Proceedings, 1996. IX, 531 pages. 1996.

Vol. 1171: A. Franz, Automatic Ambiguity Resolution in Natural Language Processing. XIX, 155 pages. 1996.

Vol. 1177: J.P. Müller, The Design of Intelligent Agents. XV, 227 pages. 1996.

Vol. 1187: K. Schlechta, Nonmonotonic Logics. IX, 243 pages. 1997.

Vol. 1188: T.P. Martin, A.L. Ralescu (Eds.), Fuzzy Logic in Artificial Intelligence. Proceedings, 1995. VIII, 272 pages. 1997.

Vol. 1193: J.P. Müller, M.J. Wooldridge, N.R. Jennings (Eds.), Intelligent Agents III. XV, 401 pages. 1997.

Vol. 1195: R. Trappl, P. Petta (Eds.), Creating Personalities for Synthetic Actors. VII, 251 pages. 1997.

Vol. 1198: H. S. Nwana, N. Azarmi (Eds.), Software Agents and Soft Computing: Towards Enhancing Machine Intelligents. XIV, 298 pages. 1997.

Vol. 1202: P. Kandzia, M. Klusch (Eds.), Cooperative Information Agents. Proceedings, 1997. IX, 287 pages. 1997.

Vol. 1208: S. Ben-David (Ed.), Computational Learning Theory. Proceedings, 1997. VIII, 331 pages. 1997.

Vol. 1209: L. Cavedon, A. Rao, W. Wobcke (Eds.), Intelligent Agent Systems. Proceedings, 1996. IX, 188 pages. 1997.

Vol. 1211: E. Keravnou, C. Garbay, R. Baud, J. Wyatt (Eds.), Artificial Intelligence in Medicine. Proceedings, 1997. XIII, 526 pages. 1997.

Vol. 1216: J. Dix, L. Moniz Pereira, T.C. Przymusinski (Eds.), Non-Monotonic Extensions of Logic Programming. Proceedings, 1996. XI, 224 pages. 1997.

Vol. 1221: G. Weiß (Ed.), Distributed Artificial Intelligence Meets Machine Learning. Proceedings, 1996. X, 294 pages. 1997.

Vol. 1224: M. van Someren, G. Widmer (Eds.), Machine Learning: ECML-97. Proceedings, 1997. XI, 361 pages. 1997.

Vol. 1227: D. Galmiche (Ed.), Automated Reasoning with Analytic Tableaux and Related Methods. Proceedings, 1997. XI, 373 pages. 1997.

Vol. 1228: S.-H. Nienhuys-Cheng, R. de Wolf, Foundations of Inductive Logic Programming. XVII, 404 pages. 1997.

Lecture Notes in Computer Science